Natural and Gauge Natural Formalism for Classical Field Theorie

A Geometric Perspective including Spinors and Gauge Theories

by

Lorenzo Fatibene
University of Torino,
Department of Mathematics, Italy

and

Mauro Francaviglia
University of Torino,
Department of Mathematics, Italy

SPRINGER-SCIENCE+BUSINESS MEDIA, B.V.

A C.I.P. Catalogue record for this book is available from the Library of Congress.

ISBN 978-90-481-6461-5 ISBN 978-94-017-2384-8 (eBook)
DOI 10.1007/978-94-017-2384-8

Printed on acid-free paper

Natural and Gauge Natural Formalism for Classical Field Theories

Mauro dedicates this book
to the memory of
his beloved wife Anna.

Contents

Foreword	xi
Introduction	xv
Acknowledgments	xxi
Part I The Geometric Setting	
Introduction	3
1. FIBER BUNDLES	9
1 Definition of Fiber Bundles	10
2 Fibered Coordinates	22
3 Particular Classes of Fiber Bundles	23
4 Fiber Bundles with Structure Group	34
5 Canonical Constructions of Bundles	39
6 Vector Fields on a Bundle	46
Exercises	50
2. JET BUNDLES	53
1 The Jet Bundles	53
2 Prolongation of Morphisms and Sections	57
3 Contact Forms	60
4 The Infinite Jet Bundle	63
5 Generalized Vector Fields	67
6 Lie Derivative of Sections	70
7 Examples	72

8 Differential Equations 74
Exercises 77

3. PRINCIPAL BUNDLES AND CONNECTIONS 79
1 Principal Bundles and Their Right Action 80
2 Invariant Vector Fields 81
3 Local Expression of Principal Automorphisms 82
4 Infinitesimal Generators of Automorphisms 85
5 Bundle Connections 88
6 Induced Connections on Associated Bundles 98
7 Examples 99
8 Structures Induced by a Connection 105
Exercises 112

4. NATURAL BUNDLES 113
1 Natural Bundles 113
2 Natural Morphisms 118
3 Lie Derivatives of Sections 120
4 Examples 122
Exercises 124

5. GAUGE NATURAL BUNDLES 125
1 Gauge Natural Bundles 125
2 Gauge Natural Prolongations of a Principal Bundle 128
3 The Infinitesimal Generators of Automorphisms (IGA) 135
4 Lie Derivatives of Sections 138
5 Examples 139
Exercises 142

Part II The Variational Structure of Field Theories
Introduction 145

6. THE LAGRANGIAN FORMALISM 151
1 The Lagrangian and the Action Functional 152
2 The Variational Morphisms 155
3 Euler-Lagrange Equations 169
4 Poincaré-Cartan Form 173
5 Symmetries and Nöther's Theorem 180
6 Examples 187
7 Conserved Quantities in Field Theory 204
8 BRST Transformations 213
Exercises 217

7. NATURAL THEORIES 219
1 Definition of Natural Theories 220
2 Bianchi Identities and Superpotentials 224
3 Conserved Quantities 226
4 Geodesics 233
5 General Relativity 234
6 On Maxwell and Yang-Mills Naturality 247
7 Gravity and Bosonic Matter 251
8 Metric-Affine Formalism for Gravity 265
Exercises 268

8. GAUGE NATURAL THEORIES 269
1 Motivations 270
2 Gauge Natural Field Theories 272
3 Generalized Bianchi Identities and Superpotentials 273
4 Conserved Quantities 275
5 Yang-Mills Theories 277
6 Gauge Natural Formulation of Hilbert Lagrangian 280
7 Vielbein Formulation of Gravitational Theories 281
8 Kosmann Lift 288
Exercises 291

Part III Spinor Fields

Introduction 295

9. SPIN STRUCTURES AND SPIN FRAMES 299
   1 Clifford Algebras and Spin Groups 299
   2 Spin Structures 304
   3 Example: Spin Structures on Spheres 307
   4 Spin Frames 314
   5 The Bundle of Spin Frames 317
   6 Other Formulations of General Relativity 319
   7 Kosmann Lift 321
   Exercises 323

10. SPINOR THEORIES 325
   1 Dirac Spinor Field 325
   2 Neutrinos 329
   3 (Reduced) Electroweak Model 332
   4 Anticommuting Majorana Spinor Fields 339
   5 Wess-Zumino Model 343
   6 Rarita-Schwinger Model 345
   Exercises 347

Final Word 349

References 353

Bibliography 356

Analitic Index 359

Foreword

This book is devoted to present a unifying geometric framework for Mechanics and Classical Field Theory, as it has been developed by us in collaboration with the group of Mathematical Physics in Torino, as well as other groups spread all over the World.

The explicit aim is to show that there is a single unifying geometrical language for all Physics of fundamental interactions, a unifying language which allows one to treat on equal footing all structural issues of Lagrangian field theories, from field equations to symmetries and conservation laws. This language is the language of *gauge natural bundles*, a piece of Mathematics which has been developed fairly recently (see [KMS93]). We shall show that exactly this language allows to discuss at once all known classical Field Theories, including General Relativity, Gauge Theories, their mutual interactions, all kinds of Bosonic and Fermionic matter, both in standard Physics and in their supersymmetric versions.

Moreover, what makes this language a *necessary* tool for understanding the structure of classical field theory is the fact that at least for theories which are not purely covariant or pure gauge theories it provides extra terms in conservation laws which corresponds to exact symmetries of Nature which are totally hidden when using other incomplete mathematical instruments. These extra terms do not change already existing currents but they reveal the existence of further important conservation laws, i.e. all those related to genuine gauge-natural interactions. This is particularly evident when working with spinors since spinors are the prototype of objects which are not purely covariant nor purely gauge.

Fiber bundles will be extensively used as a framework for Variational Calculus which both Field Theory and Mechanics rely on. Despite *(natural) fiber bundles* have been first introduced as a language for Differential Geometry, they *spontaneously* arise in Field Theory and Mechanics as well. Furthermore, natural bundles encode the properties of the base manifolds on which they lie on, thence providing *"just"* a useful and elegant language for Differential Geometry. On the contrary, the configuration bundle of a physical field theory generally encodes more information than the one related to the spacetime base manifold, since the configuration bundle is often a non-natural bundle (e.g. it may be, and it often is, a *gauge natural bundle*). In this way, fiber bundles may be considered more important and essential to physical applications rather than to Differential Geometry itself.

We shall hereafter present a self-contained introduction to bundle theory and global variational calculus for Lagrangian systems, as well as some extensive applications to Field Theory and Mechanics. The reader is assumed to know the modern elementary formalism of manifolds. We also assume the reader to be familiar with elementary differentiable structures which can be expressed in terms of tensor fields and connections on manifolds (e.g. Riemannian or pseudo-Riemannian structures together with their metric connections, the Riemann tensor and its contractions, the Ricci tensor and the Ricci scalar). Also the language of differential forms on manifolds and their theory of integration on compact submanifolds will be taken for granted.

As for every piece of work in Mathematical Physics, we also assume the reader to have an idea (even a faint one) of physical phenomenology (in Mechanics as well as in Physics of Fundamental Interactions and General Relativity) which, even if it not strictly necessary, could certainly help the reader to better understand the examples we shall present.

On the contrary, a knowledge of bundle formalism is not required here at all, since it will be extensively introduced in Part I, together with the modern perspective about Differential Geometry based on natural bundles, which are introduced and discussed. Non-natural structures will be also introduced (principal bundles, principal connections, curvature of a principal connection, forms valued in a Lie algebra, etc.). They provide the necessary framework for a geometric formulation of gauge theories. Finally, gauge natural bundles are discussed, as mathematical structures which encompass the mathematics required to deal at once with covariant objects and gauge covariant objects in mutual interaction.

This introduction to fiber bundles is more detailed with respect to what is usually found in physically minded literature, though it is not exhaustive at the mathematical level. In fact, topics have been selected on a more pedagogical level, having in mind the applications contained in the second and third part of the book. Because of this we have left out many other important topics which are interesting for applications to physics (e.g. solitonic solutions, methods from differential topology and so on) for which we refer to the specialistic literature, which is already fully satisfactory. Although these further topics are important to some specific problems we believe they should be left out in a first introduction to the new subject, which is in fact the problem of a unifying structural language. On the other hand we made an effort to expand the topics presented also with examples and details, often providing both the intrinsic and the local coordinate expressions, in order to better relate our work to the current literature. It is our opinion that, despite the length of the presentation, this feature will be appreciated by graduate students in particular and in general by all researchers who will approach bundle theory for the first time or want to revisit it in a more complete way. Experts should simply skip the material they are familiar with, once they familiarized with the notation introduced (and which will be used in the second and third part).

In Part II the geometric framework for the calculus of variations on fiber bundles is first introduced. Structures devoted to the description of dynamics (Poincaré-Cartan forms, first variation formula, etc.) are also defined. There we shall keep a generic viewpoint, so that all the structures are described in their most general formulation, without *any* physical restriction or requirement. Nöther's theorem about conserved quantities in field theories is also presented. As important examples we then analyze in detail Mechanics of holonomic (possibly time-dependent) systems, test particles in Special Relativity and gauge (Yang-Mills) theories, which are the core of the modern understanding of the physics of fundamental interactions. We also present a brief account about BRST transformations. Then natural and gauge natural theories are defined on gauge natural bundles and their conservation laws are investigated in detail. Natural theories suitably generalize General Relativity while gauge natural theories generalize gauge theories and encompass all physically relevant examples of fields. Particularly worth noticing is the theorem which guarantees the existence of superpotentials in any gauge natural theory. We shall here present General Relativity in its purely metric second order formulation, the first order formulation with a reference background, the metric-affine *à la* Palatini and the frame formulations. Yang-Mills fields coupled to gravity are also considered in detail. A complete discussion is devoted to the different formulations of Yang-Mills-like theories and their physical and mathematical equivalence. We shall also consider a number of exact vacuum solutions of General Relativity (Schwarzschild, Kerr and BTZ solutions) as well as a number of possible couplings to different types of Bosonic matter (scalar Klein-Gordon, vector Klein-Gordon and Proca fields).

Part III is an introduction to spinor fields, which as we already said before, are a genuine example of fundamental physical fields which do not admit in general a purely natural or purely gauge formulation, but can be fully understood as dynamical fields on non-trivial spacetime manifolds only on gauge natural grounds. We shall introduce the geometric structures used to define spinor fields on curved and topologically non-trivial spaces. The standard theory of spinors on curved spaces is first presented; it describes the evolution of spinor fields in a fixed *external* gravitational field which is completely unaffected by spinors themselves. Then we provide the necessary structures which will allow us to build up a theory for the genuine interaction between spinor fields and gravity, without any restriction on dimension or signature, neither too stringent topological requirement on the spacetime manifold M. As explicit *pedagogical* examples, the spin structures on spheres are also presented. As an application, we shall present a gauge natural formulation of Dirac spinors and Weyl neutrinos. As an example dealing with a fairly complicated system we shall present the Weinberg-Salam electroweak model and the Wess-Zumino model which deals with anticommuting spinors.

As a result all field theories of fundamental physics (including the interaction between General Relativity, gauge theories and all kinds of matter described by tensor densities or spinors) are described altogether and at once in the gauge natural framework. We stress once more that only a fully gauge natural treatment allows to discover that the set of dynamical symmetries is larger than the one obtainable working with standard natural or purely gauge methods.

This project has grown by collecting and organizing the contents of Lectures in Mathematical Physics we gave in Torino in the years 1992-2000 for undergraduate and graduate students. It includes as well many research results of our group, so that it now probably contains too much material for a single course. We tried however to maintain the original didactical approach so that it will be simple to extract from the book the material for a course at an undergraduate level. The contents are of interest also for mathematically oriented researchers who will probably be able to skip some of the introductory parts finding an exposition of the principal features of modern Theoretical Physics in the familiar language of Differential Geometry. On the other hand we believe that physically oriented researchers may find here a number of motivations to better understand the structures introduced by mathematicians when speaking about Physics. A particular effort is in fact devoted to show the relations between the mathematical structures introduced so far and the very basic physical assumptions about *what* and *how* we know of Nature.

Introduction

The modern perspective on Theoretical Physics is largely based on the classical concept of field theory. Even quantum field theory is based on quantization prescriptions which originate in a classical field theory (though of course symbols and quantities are readily re-interpreted from a quantum perspective). A better founded understanding of the structure of the classical theory is necessary, though probably not sufficient, to justify, understand and motivate what is later done at a quantum level.

Two aspects of classical field theories have proved to be essential to the modern perspective on the structure of the physical world: symmetries and conserved quantities (or currents).

All fundamental theories (from General Relativity to Yang-Mills theories) have a huge symmetry group. This group results in both an underdetermination and overdetermination of field equations which is in turn deeply related to what can and what cannot be physically observed (or equivalently, what is a genuine property of the physical world and what is just a convention of the observer).

Conserved quantities are the integral counterparts of symmetries because of Nöther theorem. Many important physical quantities are directly related to conserved quantities: energy, angular momentum, electric charges, and so on. The Nöther theorem provides in fact an explicit (and algorithmic) correspondence between (Lagrangian) symmetries and conserved quantities.

Now the various prescriptions of (Lagrangian) symmetry used in applications are extremely sensitive to both the kinematical and dynamical details of the theory under investigation. In particle physics (possibly on a curved background) one usually has a fixed metric background on spacetime which has to be preserved by symmetries and deformations. On the contrary, in General Relativity one is free to change spacetime coordinates at will.

In Yang-Mills theories, depending on global topological situations, one is sometimes able to decouple the *coordinate transformations* from the so-called *pure gauge transformations* which are *vertical* (i.e. they do not move the spacetime point, simply acting on the fields).

When spinor fields are taken into account, all these problems appear at the same time. In this case the role of spacetime transformations is mixed with vertical transformations through certain precise combinations which leave the Lagrangian covariant, none of the original transformations being a symmetry by itself. The situation with spinor fields, which is more or less clear on a Minkowski (hence topologically trivial) spacetime, becomes extremely cumbersome in more general cases. In the current literature strict topological conditions (e.g. parallelizability) are often imposed in order to carry out the global dynamics in a physically satisfactory way. In this book we aim at showing (in Part III) that the conditions needed for a physical meaningful theory of spinor fields are fairly less restrictive, *provided the symmetry group is defined to act in the appropriate way* and the corresponding conserved quantities are correctly interpreted. This last interpretation is necessarily based on proving that some traditional notions (historically developed on Minkowski background) have to be abandoned in favor of a generalized set of notions which directly lead us to the gauge natural formulation of field theories. We refer for this to Part II.

Other kinds of symmetries (e.g. supersymmetries and BRST transformations) are defined on the space of configurations of a field theory and used to get conservation laws. These cases are significantly different from a bundle perspective though they obey a generalization of Nöther theorem. When this sort of transformations (called *higher order transformations*) are included in the framework, then the correspondence between symmetries and conservations laws expressed by Nöther theorem becomes surjective, as Emmy Nöther noticed herself.

Let us briefly sketch by a worked example the different possible attitudes to find conservation laws in a given theory. We use more or less standard notation (for which we refer the reader to the relevant Chapters of this book). Let us consider the Lagrangian

$$L = [R - \tfrac{1}{4}\mathbf{F}_{\mu\nu} \cdot \mathbf{F}^{\mu\nu} + \tfrac{i}{2}(\bar{\psi}\gamma^\mu \nabla_\mu \psi - \nabla_\mu \bar{\psi}\gamma^\mu \psi) - m\bar{\psi}\psi]\sqrt{g}\, \mathrm{d}s \qquad (1)$$

This Lagrangian informally describes a gravitational sector encoded by the scalar curvature R of a metric $g_{\mu\nu}$ (or equivalently of a moving frame e^μ_a), a Yang-Mills sector described by the (matrix) curvature $\mathbf{F}_{\mu\nu}$ of a (matrix) gauge potential $\mathbf{A}_\mu$ (with values in the Lie algebra of a (matrix) semisimple gauge group G), and a Fermi sector described by the spinor field ψ with respect to the representation induced by the Dirac matrices γ^μ.

For the gravitational sector we easily get the identity:

$$\nabla_\epsilon \left[g^{\alpha\beta}\, \pounds_\xi \Gamma^\epsilon_{\alpha\beta} - g^{\alpha\epsilon}\, \pounds_\xi \Gamma^\lambda_{\alpha\lambda} - R\, \xi^\epsilon\right] = -\Big(R_{\alpha\beta} - \tfrac{1}{2} R\, g_{\alpha\beta}\Big) \pounds_\xi g^{\alpha\beta} \qquad (2)$$

We refer to Section 7.5 for the details and notation. The r.h.s. provides vacuum Einstein equations while the l.h.s. is a conservation law. If the spacetime vector ξ^β is a Killing vector, the current is conserved (even off-shell, i.e. along each field configuration). For any spacetime vector field the current is conserved on-shell (i.e. along all solutions of field equations).

In Minkowski spacetime (as well as on any *a priori* fixed curved background $g_{\mu\nu}(x)$) the gravitational sector is frozen by the choice of the background. In the Yang-Mills sector (see Section 8.5) one has gauge symmetries

$$\mathbf{A}'_\mu = \gamma^{-1} \cdot [\mathbf{A}_\mu \cdot \gamma + \mathrm{d}_\mu \gamma] \tag{3}$$

as well as all isometries of the fixed background

$$\mathbf{A}'_\mu = \bar{J}^\nu_\mu \mathbf{A}_\nu \tag{4}$$

where $\bar{J}^\nu_\mu$ is the inverse Jacobian of a spacetime transformation. In general, when non trivial topologies are allowed (e.g. when dealing with solitonic solutions) and/or the gravitational sector is switched on again, this splitting between isometries and pure gauge transformations is only local even if the Lagrangian is globally covariant with respect to (generalized) gauge transformations

$$\mathbf{A}'_\mu = \bar{J}^\nu_\mu \gamma^{-1} \cdot [\mathbf{A}_\nu \cdot \gamma + \mathrm{d}_\nu \gamma] \tag{5}$$

In the case (3) one has the conserved current

$$\mathcal{E}^\lambda = \sqrt{g}\, \mathbf{F}^{\mu\lambda} \nabla_\mu \xi(x) \tag{6}$$

which is commonly found in the literature. Here $\xi(x)$ is any (matrix) point dependent (local) quantity with values in the Lie algebra of the gauge group. In the case (4) one obtains

$$\mathcal{E}^\lambda = -\sqrt{g}\, (\mathbf{F}^{\mu\lambda} \cdot \mathbf{F}_{\mu\nu} - \tfrac{1}{4} \delta^\lambda_\nu \mathbf{F}_{\mu\sigma} \mathbf{F}^{\mu\sigma}) \xi^\nu \equiv -\hat{H}^{\lambda}{}_{\nu} \xi^\nu \tag{7}$$

where ξ^ν are the components of a Killing vector, while in the case (5) one obtains the more general identity

$$\nabla_\nu \Big(\sqrt{g}\, (\mathbf{F}^{\mu\nu} \cdot \pounds_\Xi \mathbf{A}_\mu + \tfrac{1}{4} \xi^\nu \mathbf{F}_{\alpha\beta} \cdot \mathbf{F}^{\alpha\beta}) \Big) = -\mathbb{E}^\mu(L_{YM}) \pounds_\Xi \mathbf{A}_\mu + \tfrac{1}{2} \hat{H}_{\mu\nu}(L_{YM}) \pounds_\xi g^{\mu\nu} \tag{8}$$

Here $\hat{H}_{\mu\nu}$ is the standard energy-momentum tensor density of Yang-Mills field, while $\mathbb{E}^\mu(L_{YM})$ are Yang-Mills field equations. One can get rid of the last term on the r.h.s. in two different ways: with ξ^μ being a Killing vector of $g^{\mu\nu}$ (i.e. $\pounds_\xi g^{\mu\nu} = 0$) or by evaluation on a solution when the gravitational sector is re-established. In any event the quantity

$$\mathcal{E} = \sqrt{g}\, \left(\mathbf{F}^{\mu\nu} \pounds_\Xi \mathbf{A}_\mu + \tfrac{1}{4} \xi^\nu \mathbf{F}_{\alpha\beta} \mathbf{F}^{\alpha\beta} \right) \mathrm{d}s_\nu \tag{9}$$

is conserved, being evaluated along a solution and/or because ξ^ν is a Killing vector. If one restricts (artificially) to a vertical deformation (i.e. to a deformation along the fibers) the current (6) is obtained. If one considers purely horizontal (local) transformations (which globally exists only in some special cases, e.g. on Minkowski space) the current (7) is recovered. In either way, the quantity (8) is conserved, it encompasses both cases (6) and (7) and it applies globally even in non-trivial topological situations. When we consider the gravitational and the gauge sector together then the current (9), together with the one from (2), is conserved for any ξ^ν, even if it is not a Killing vector. This important result seems to be known only to a restricted audience.

Furthermore, in general the Nöther current (7) cannot be globalized (in all cases which have no global gauge fixing). In these cases, the current (9) is still well-defined and it is the only global way of defining the energy-momentum and the angular momentum in a manifestly covariant way.

On the spinor sector the situation is even more cumbersome. In Minkowski space (which is both maximally symmetric and parallelizable) one obtains the currents

$$\mathcal{E} = \tfrac{i}{2}\xi^\nu(\bar{v}\gamma^\lambda\nabla_\nu v - \nabla_\nu\bar{v}\gamma^\lambda v) - \xi^\lambda\mathcal{L}_{\mathrm{D}} \tag{10}$$

with ξ^λ being a Killing vector (see e.g. [BD64]). However, there is a much bigger class of conserved currents (see Section 10.1), namely

$$\begin{aligned}&\nabla_\lambda\left[\tfrac{i}{2}\left((\bar{v}\gamma^a\pounds_\Xi v) - (\pounds_\Xi\bar{v}\gamma^a v) + \tfrac{1}{2}(\bar{v}\gamma^{abc}v)e_{b\mu}\pounds_\Xi e^\mu_c\right)e^\lambda_a - \xi^\lambda\mathcal{L}_{\mathrm{D}}\right] = \\ &= \hat{H}_{\mu\nu}e^{a\nu}\pounds_\Xi e^\mu_a - \mathbb{E}\,\pounds_\Xi v - \pounds_\Xi\bar{v}\,\bar{\mathbb{E}}\end{aligned} \tag{11}$$

where e^μ_c is a *(moving) frame* for the metric $g_{\mu\nu}$, $\hat{H}_{\mu\nu}$ is the so-called *energy-momentum* tensor density of the spinor field and $\mathbb{E}$ are Dirac equations (see Part III for a geometrical interpretation of the objects involved). Furthermore in the literature about spinors on curved spacetimes one usually sets $\xi^{ab}_V = \nabla^{[a}\xi^{b]}$ for the vertical part of the deformation (which is a global – though not natural – prescription) to get a conserved current for each spacetime vector field, even if it is not a Killing vector. Notice that in all cases the currents considered obey a conservation law and they are built via some sort of generalized Nöther theorem starting from different classes of symmetries. Again the r.h.s. vanishes on solution and/or for Killing vectors ($\pounds_\Xi e^\mu_a = 0$).

In each context mentioned above (as well as in Mechanics where many aspects drastically simplify) slightly different formulations of the Nöther theorem are usually introduced to adapt to the particular kinematical and dynamical situation. We shall prove that all these cases can be obtained by the "same" Nöther theorem proved once for all in the framework of gauge natural field theories. All the cases mentioned above are particular cases of this general geometrical framework.

In this book we shall introduce a unifying geometrical language able to keep under control the differences arising in various cases, which in each case reproduces the framework developed *ad hoc* for the particular situation.

Furthermore, we shall prove that in general, in any gauge natural theory the Nöther currents are not only closed along solutions because of conservations, but also exact (along solutions) regardless of the cohomology of spacetime.

We believe that finding the common features of the various partial *ad hoc* frameworks introduced in different situations to deal with conservation laws is important from a foundational point of view. In fact the common features (which in this case will be shown to be also sufficient for a physically satisfactory interpretation) will be certainly more relevant and fundamental than the features unessentially introduced to deal with a particular physical situation. After having taken confidence with these new prescriptions all applications become just a matter of calculation and do not belong to the subjects developed hereafter.

Acknowledgements

We are deeply indebted to Marco Ferraris for invaluable discussions. We are also grateful to Enrico Bibbona, Marco Godina, Silvio Mercadante, Gianluca Pacchiella and Marco Raiteri for comments. We wish also to thank our students who first experienced the lectures of ours which have been collected and expanded in this book.

PART I

THE GEOMETRIC SETTING

C'era un gran rumore negli universi. Generazioni di stelle nascevano e morivano sotto lo sguardo di telescopi assuefatti, fortune elettromagnetiche venivano dissipate in un attimo, sorgevano imperi d'elio e svanivano civiltà molecolari, gang di gas sovraeccitati seminavano il panico, le galassie fuggivano rombando dal loro luogo d'origine, i buchi neri tracannavano energia e da bolle frattali nascevano universi dissidenti, ognuno con una legislazione fisica autonoma.

S. Benni, Elianto

There was a huge noise around the universes. Generations of stars were born and were dying under the look of addicted telescopes, electromagnetic fortunes were being dissipated in a moment, helium empires were rising and molecular civilizations were fading, gangs of over-excited gas were spreading panic, the roaring galaxies were fleeing away from their place of origin, the black holes were knocking energy back and from fractal bubbles dissident universes were born, each of them with its own physical legislation.

Introduction

We shall hereafter introduce fiber bundles together with the structures which are needed to deal with Mechanics and Field Theories. Historically speaking, fiber bundles arose to exhibit linear connections on manifolds as *global geometrical objects*. Since then, bundles have been extensively used in Differential Geometry and the Calculus of Variations on manifolds, together with their applications to Physics, as well as in Algebraic Geometry and Global Analysis. Therefore Bundle Theory is important in its own, as well as it is a general language which is useful in many branches of pure Mathematics and Mathematical Physics.

As far as physical applications are concerned, namely what is of interest for us in this book, the knowledge of *field* as a physical entity (which has its own energy, momentum and angular momentum) is due to the work of Faraday and Maxwell about electromagnetism. Since then field theory has become one of the cornerstones which our understanding of Nature is built on.

A *field* is, by a naive definition, an entity which assigns to each point of spacetime the value of some physical quantity. Let us denote by F the set of all the values that a field can assume at a spacetime point; then *configurations* on an open set U of spacetime M are (locally) described by maps $\varphi : U \longrightarrow F$. If we assume F to have a topology along with a differential structure, we can require φ to be a continuous and differentiable map.

Under this mild and general hypothesis we can already have a geometrical perspective on field theories. In fact, giving a map $\varphi : U \longrightarrow F$ is equivalent to give its *graph* $\{(x, \varphi(x)) \in U \times F\}$ which is in turn equivalent to the map $\hat{\varphi} : U \longrightarrow U \times F$ given by $\hat{\varphi} : x \mapsto (x, \varphi(x))$. The product $U \times F$, together with its natural projection $\mathrm{pr}_1 : U \times F \longrightarrow U$, is called a *trivial bundle over* U. Then we can regard $U \times F$ as the local model of a *bundle* having F as a fiber over spacetime M, i.e. a manifold having local topology $U \times F$, which is obtained by glueing together over M trivial pieces, in much the same way a manifold M is obtained by glueing together local pieces which are identical to a model (say $\mathbb{R}^m$). As M can have a global topology which is fairly different from the topology of its *model* $\mathbb{R}^m$, a bundle will in general have a global topology which differs sensibly from the trivial one $M \times F$. In this way $\hat{\varphi}$ is the local expression for what will be called a *section* of the bundle. Then we can say that the field configurations *are* the (local or global) sections of the bundle which is then appropriately called the *configuration bundle*.

At a first glance this way of introducing bundles in field theory may seem to be unsubstantial: one could in fact argue that only trivial bundles be really important for Physics. If this were the case, the bundle framework would have no real physical basis and bundles would not be worth introducing in Field Theories.

However, the bundle formulation of field theory is not at all motivated by just seeking a full mathematical generality; on the contrary it is just an *empirical* consequence of physical situations that concretely happen in Nature. One among the simplest of these situations may be that of a particle constrained to move on a sphere, here denoted by S^2; the physical state of such a dynamical system is described by providing both the position of the particle *and* its momentum, which is a tangent vector to the sphere. In other words, the state of this system is described by a point of the so-called *tangent bundle* TS^2 of the sphere, which can be shown to be non-trivial, i.e. it has a global topology which differs from the (trivial) product topology of $S^2 \times \mathbb{R}^2$. When one seeks for solutions of the relevant equations of motion some local coordinates have to be chosen on the sphere, e.g. stereographic coordinates covering the whole sphere but a point (let us say the *north pole*). On such a coordinate neighbourhood (which is contractible to a point being a diffeomorphic copy of $\mathbb{R}^2$) there exists a trivialization of the corresponding portion of the tangent bundle of the sphere, so that the relevant equations of motion can be *locally* written in $\mathbb{R}^2 \times \mathbb{R}^2$. At the global level, however, together with the equations, one should give some boundary conditions which will ensure regularity in the north pole. As is well known, different inequivalent choices are possible; these boundary conditions may be considered as what is left in the local theory out of the non-triviality of the configuration bundle TS^2.

Moreover, much before modern gauge theories or even more complicated new field theories, the theory of General Relativity is the ultimate proof of the need of a bundle framework to describe physical situations. Among other things, in fact, General Relativity assumes that spacetime is not the "simple" Minkowski space introduced for Special Relativity, which has the topology of $\mathbb{R}^4$. In general it is a Lorentzian four-dimensional manifold possibly endowed with a complicated global topology. On such a manifold, the choice of a trivial bundle $M \times F$ as the configuration bundle for a field theory is mathematically unjustified as well as physically wrong in general. In fact, as long as spacetime is a contractible manifold, as Minkowski space is, all bundles on it are forced to be trivial; however, if spacetime is allowed to be topologically non-trivial, then trivial bundles on it are just a small subclass of all possible bundles among which the configuration bundle can be chosen. Again, given the *base* M and the *fiber* F, the non-unique choice of the topology of the configuration bundle corresponds to different global requirements.

A simple purely geometrical example can be considered to sustain this claim. Let us consider $M = S^1$ and $F = (-1, 1)$, an interval of the real line $\mathbb{R}$; then there exist (at least) countably many "inequivalent" bundles other than the trivial one $\text{Mö}_0 = S^1 \times F$, i.e. the cylinder, as shown in *Fig.1*.

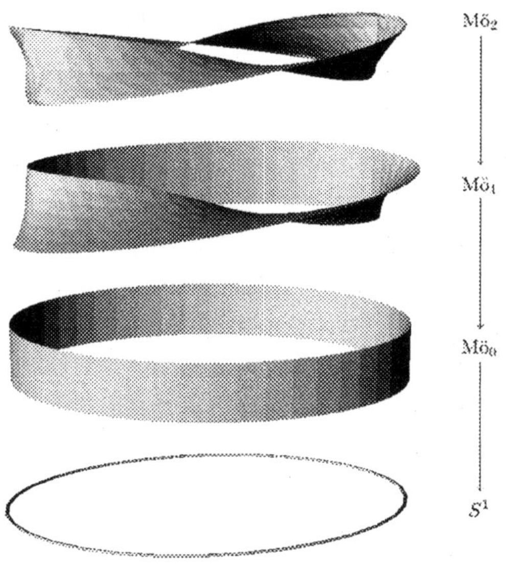

Fig. 1 – Bundles on S^1.

Furthermore the word "inequivalent" can be endowed with different meanings. The bundles shown in the figure are all inequivalent as embedded bundles (i.e. there is no diffeomorphism of the ambient space transforming one into the other) but the even ones (as well as the odd ones) are all equivalent among each other as abstract (i.e. not embedded) bundles (since they have the same transition functions).

The bundles Mö_n (n being any positive integer) can be obtained from the trivial bundle Mö_0 by cutting it along a fiber, twisting n-times and then glueing again together. The bundle Mö_1 is called the *Moebius band* (or *strip*). All bundles Mö_n are canonically fibered on S^1, but just Mö_0 is trivial. Differences among such bundles are global properties, which for example imply that the even ones $\mathrm{M\ddot{o}}_{(2k)}$ allow never-vanishing sections (i.e. field configurations) while the odd ones $\mathrm{M\ddot{o}}_{(2k+1)}$ do not. Even if the Möbius strip seems to be just a playful tool in pure Geometry, nevertheless its understanding is crucial for Physics, especially in connection with the problem of *orientability*.

We shall thence introduce the geometrical notation which will be used hereafter as a framework for Field Theories and Mechanics. They provide the same notation used for Differential Geometry and General Relativity, as well. Of course our presentation has not to be intended as a complete exposition of the mathematical theory of fiber bundles, but just as a pragmatic treatment which allows us to introduce, as quickly as possible, all the concepts we shall need below. The reader will find some reference at the beginning of each Chapter.

Chapters

Chapter 1

FIBER BUNDLES

Abstract *Fiber bundles*, their *morphisms* and *sections* are defined. *Transition functions* are introduced and used to describe the global structure of a bundle and to build it out of local data. Particular classes of fiber bundles (*affine bundles*, *vector bundles* and *principal bundles*) are introduced. Bundles are then provided with the further structure of *fiber bundles with structure group*, which will be used below as prototypes of configuration bundles in Field Theory. The *Hopf bundle*, the *tangent bundle* and the *frame bundle* are considered all along in detail as examples.

References

We assume the reader is already familiar with the basic notions of Differential Geometry (manifolds, mappings, Lie groups, vector fields, differential forms, exterior calculus, etc.). For a quick reference see e.g. [F88], [GS87], [W80]. For a more detailed treatment see e.g. [CB68], [CBWM82].

We shall also use at a basic level the language of categories (see [ML71], [HS79]), sheaf theory (see [T75], [P99]) and some (differential and algebraic) topology (see [GP74], [S66], [MS63]).

As a technical reference for the bundle material we refer to the classical books of [KN63], [S51].

For a general discussion on the use of bundle theory in theoretical physics we refer to [BM94], [B81], [N90], [MM92].

1. Definition of Fiber Bundles

A fiber bundle over a manifold[1] M is, loosely speaking, a space B the local topology of which is the topology of Cartesian products $U_\alpha \times F$, where $\{U_\alpha\}_{\alpha \in I}$ is an open covering of M, F a manifold and I any set of indices. Such local models can be *glued* together in a (possibly) non-trivial way so that the topology of B may be fairly different from the topology of a global Cartesian product $M \times F$.

By *glueing together* we mean coherently choosing at each point of each intersection $U_{\alpha\beta} = U_\alpha \bigcap U_\beta \neq \varnothing$ some suitable diffeomorphism of the *standard fiber* F; this introduces the group Diff(F), in which we choose the ways of glueing patches together. In general it is an infinite dimensional group (in the above example of *Fig.1* we had $F = (-1, 1)$ and Diff(F) is thence the group of bidifferentiable maps of the interval into itself). However, it may happen (e.g. if we have to glue a finite number of patches) that the relevant diffeomorphisms are chosen just in a finite dimensional Lie subgroup $G \subset \text{Diff}(F)$. In these cases we say that G is a *structure group* for the bundle[2].

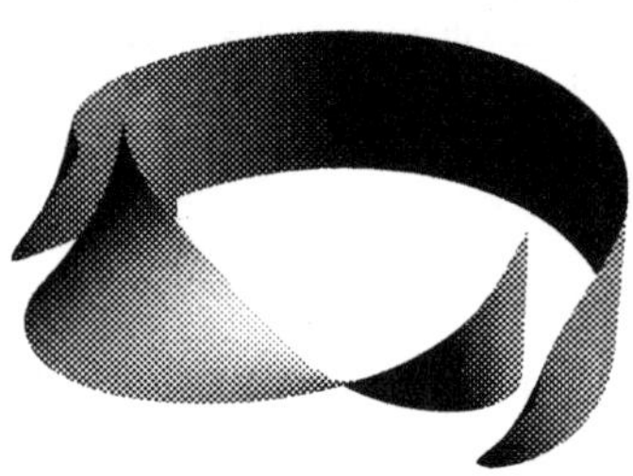

Fig. 2 – Construction of a non-trivial bundle.

Definition (1.1.1): let M and F be manifolds (possibly with boundary). A *trivial fiber bundle* is a quadruple $(M \times F, M, \text{pr}_1; F)$ in which the map $\text{pr}_1 : M \times F \longrightarrow M$ is the projection onto the first factor. ∎

Let us consider a simple example of trivial bundle. Let Π be a 2-plane and Λ a 1-line; they have to be thought as bare manifolds without further superimposed structures (such as affine, vector or group structure). Let us embed them into a 3-space and consider the projection π along a preferred direction, as shown in *Fig.* 3. Then $\mathcal{B} = (\Pi, \Lambda, \pi; \mathbb{R})$ is a trivial bundle.

[1] All manifolds and maps will be implicitly assumed to be smooth (i.e. differentiable in the C^∞ sense) unless explicitly stated.

[2] Notice that we say *"a" structure group*, since G is not uniquely defined. See Section 1.4 for the definition of *fiber bundles with structure group*.

We remark that the bundle structure of Π selects a preferred direction (the *vertical* fiber direction), leaving the complementary *horizontal* direction totally undefined. Accordingly, when considering fibered morphisms of the bundle $\mathcal{B} = (\Pi, \Lambda, \pi; \mathbb{R})$ the preferred direction is preserved, since a fibered morphism has to preserve fibers; contrarily, any possibly chosen *horizontal* direction may be twisted at will.

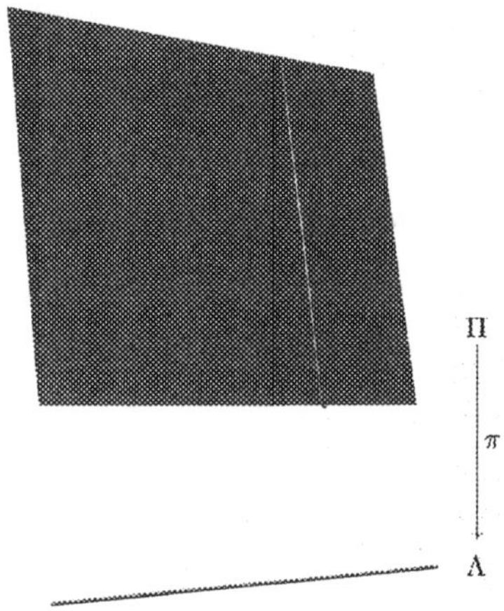

Fig. 3 – Trivial bundle.

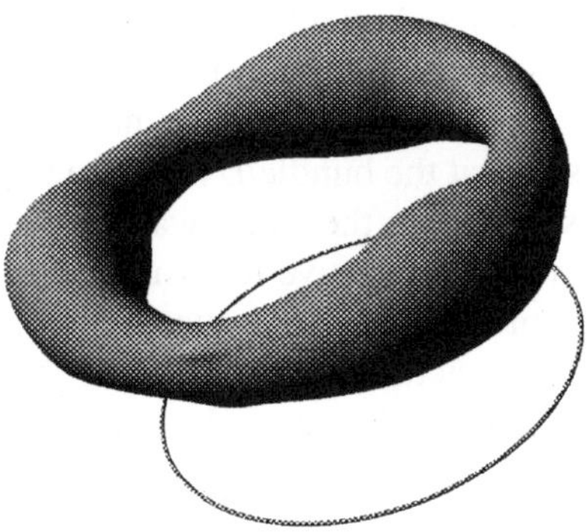

Fig. 4 – Trivial bundle.

Let us remark that no condition about the base manifold M or the standard fiber manifold F is required in trivial bundles. In particular we can consider trivial bundles over topologically non-trivial manifolds with topologically non-trivial standard fibers.

For example we can define the trivial bundle $(S^1 \times S^1, S^1, \mathrm{pr}_1; S^1)$ as shown in *Fig.* 4. It is called a 2-torus.

Definition (1.1.2): a *(geometric) fiber bundle* is a quadruple $\mathcal{B} = (B, M, \pi; F)$ such that:

(a) B, M, F are (paracompact) manifolds which are called *total space*, *base* and *standard fiber*, respectively. The map $\pi : B \longrightarrow M$ is surjective, it has maximal rank and is called the *projection*.

(b) there exists an open covering $\{U_\alpha\}_{\alpha\in I}$ of the base manifold M such that for all $\alpha \in I$ there exists a diffeomorphism[3] $t_{(\alpha)} : \pi^{-1}(U_\alpha) \longrightarrow U_\alpha \times F$ and a commutative diagram of spaces and arrows

$$\begin{array}{ccc} \pi^{-1}(U_\alpha) & \xrightarrow{t_{(\alpha)}} & U_\alpha \times F \\ & \searrow^{\pi} & \downarrow \mathrm{pr}_1 \\ & & U_\alpha \end{array} \qquad (1.1.3)$$

The pair $(U_\alpha, t_{(\alpha)})$ (or sometimes just the mapping $t_{(\alpha)}$) is called *a local trivialization* of the bundle $\mathcal{B}$. The given set of all local trivializations $\{(U_\alpha, t_{(\alpha)})\}_{\alpha\in I}$ is simply called *a trivialization* of $\mathcal{B}$. ∎

Sometimes a fiber bundle $\mathcal{B} = (B, M, \pi; F)$ will be also denoted by $\pi : B \longrightarrow M$ if there is no need to specify the standard fiber F. We remark that a trivialization is not *a priori* fixed on a fiber bundle, but it is simply required to exist. On the same bundle $\mathcal{B}$ there exist in fact (infinitely many) trivializations which are *equivalent*, in the sense that they define the same *bundle structure*. In other words one may say that the bundle $\mathcal{B}$ is the equivalence class of all these possible trivializations (in much the same way as a manifold admits several equivalent atlases). It is not restrictive to assume that $\{U_\alpha\}_{\alpha\in I}$ is an atlas of the differentiable structure of M. Then the family $\{\pi^{-1}(U_\alpha)\}_{\alpha\in I}$ forms an atlas of the differentiable structure of the manifold B, which is called a *fibered atlas* of the bundle $\mathcal{B}$. We stress however that trivialization domains need not to be coordinate domains. For example a cylinder $S^1 \times \mathbb{R}$ allows a global trivialization (so that one can fix $U \equiv S^1$) even if there is no global coordinate on S^1.

Let us fix a trivialization $\{(U_\alpha, t_{(\alpha)})\}_{\alpha\in I}$ of a fiber bundle $\mathcal{B}$ and let us denote by $U_{\alpha\beta} = U_\alpha \bigcap U_\beta$ the (possibly empty) overlaps of the open covering. Let $t_{(\alpha)}(x) : \pi^{-1}(x) \longrightarrow \{x\} \times F \simeq F$ be the restriction of $t_{(\alpha)}$ to the *fiber* $\pi^{-1}(x)$ over each point $x \in U_\alpha$. One can then define a family of maps $\hat{g}_{(\alpha\beta)} : U_{\alpha\beta} \longrightarrow$

3 $\pi^{-1}(U_\alpha)$ denotes the preimage of U_α with respect to the projection map π.

Diff(F) by setting $\hat{g}_{(\alpha\beta)}(x) = t_{(\alpha)}(x) \circ t_{(\beta)}^{-1}(x)$; the following properties, called *cocycle identities*, hold true

$$\begin{aligned} &\hat{g}_{(\alpha\alpha)}(x) = \mathrm{id}_F \\ &\hat{g}_{(\alpha\beta)}(x) = \left[\hat{g}_{(\beta\alpha)}(x)\right]^{-1} \\ &\hat{g}_{(\alpha\beta)}(x) \circ \hat{g}_{(\beta\gamma)}(x) \circ \hat{g}_{(\gamma\alpha)}(x) = \mathrm{id}_F \end{aligned} \tag{1.1.4}$$

where $\mathrm{id}_F \in$ Diff(F) is the identity diffeomorphism on F.

The maps $\hat{g}_{(\alpha\beta)}$ are called *transition functions* of the trivialization; one says that the whole collection $\{(U_{\alpha\beta}, \hat{g}_{(\alpha\beta)})\}$ forms a *cocycle* over M with values in Diff(F).

Definition (1.1.5): a *bundle morphism* between two bundles $\mathcal{B} = (B, M, \pi; F)$ and $\mathcal{B}' = (B', M', \pi'; F')$ is a pair of maps $\boldsymbol{\Phi} = (\Phi, \phi)$ where $\Phi : B \longrightarrow B'$ and $\phi : M \longrightarrow M'$ fill-in a commutative diagram:

$$\begin{array}{ccc} B & \xrightarrow{\ \Phi\ } & B' \\ {\scriptstyle\pi}\big\downarrow & & \big\downarrow{\scriptstyle\pi'} \\ M & \xrightarrow{\ \phi\ } & M' \end{array} \tag{1.1.6}$$

i.e., one has $\pi' \circ \Phi = \phi \circ \pi$. This property expresses the fact that Φ transports the fiber of $\mathcal{B}$ over a point $x \in M$ into the fiber of $\mathcal{B}'$ over the image point $\phi(x) \in M'$. A bundle morphism $\boldsymbol{\Phi} = (\Phi, \phi)$ is called *a strong morphism* if $\phi : M \longrightarrow M'$ is a diffeomorphism.

A *vertical morphism* $\boldsymbol{\Phi} = (\Phi, \mathrm{id}_M)$ is a bundle morphism between two fiber bundles $\mathcal{B} = (B, M, \pi; F)$ and $\mathcal{B}' = (B', M, \pi'; F')$ over the same base manifold M, which projects onto the identity over M. By an abuse of notation, if $M = M'$ and $\phi = \mathrm{id}_M$ a vertical bundle morphism $\boldsymbol{\Phi} = (\Phi, \mathrm{id}_M)$ is simply denoted by $\Phi : B \longrightarrow B'$. ∎

Due to the definitions (1.1.2) e (1.1.5), fiber bundles form a *category*[4]. All the standard terminology about morphisms applies to the category of fiber bundles. For example, a morphism $\boldsymbol{\Phi} = (\Phi, \phi)$ is called an *epimorphism* if $\Phi : B \longrightarrow B'$ is surjective, while if both $\Phi : B \longrightarrow B'$ and $\phi : M \longrightarrow M'$ are diffeomorphisms, then $\boldsymbol{\Phi}^{-1} = (\Phi^{-1}, \phi^{-1})$ is again a bundle morphism called *the inverse morphism*; in this case $\boldsymbol{\Phi} = (\Phi, \phi)$ is said to be an *isomorphism*. A bundle morphism $\boldsymbol{\Phi}$ from $\mathcal{B}$ into itself is called *endomorphism*, while it is called *automorphism* if it is an isomorphism of $\mathcal{B}$. The group of all automorphisms of $\mathcal{B}$ is denoted by Aut($\mathcal{B}$).

---

[4] Roughly speaking, a *category* is a class of *objects* together with a class of *morphisms* which is closed under composition and it includes the *identity morphisms*.

Proposition (1.1.7): let F and M be manifolds, $\{U_\alpha\}_{\alpha\in I}$ be an open covering of M and $\hat{g}_{(\alpha\beta)} : U_{\alpha\beta} \longrightarrow \text{Diff}(F)$ be a cocycle with values in Diff(F), i.e. a family of maps satisfying the cocycle identities (1.1.4).

Then there exists a bundle $\mathcal{B} = (B, M, \pi; F)$ which has $\hat{g}_{(\alpha\beta)}$ as transition functions; this bundle is unique modulo isomorphisms.

Sketch of the proof: the total space B of the bundle $\mathcal{B}$ is defined as the quotient space of the disjoint union $\coprod_{\alpha\in I}\left(U_\alpha \times F\right)$ with respect to the equivalence relation:

$$(\alpha, x, \varphi) \sim (\beta, x', \varphi') \Leftrightarrow \begin{cases} x = x' \\ \varphi = \left[\hat{g}_{(\alpha\beta)}(x)\right](\varphi') \end{cases} \tag{1.1.8}$$

We denote equivalence classes by $[x, \varphi]_{(\alpha)}$; local trivializations are then given by the maps $t_{(\alpha)} : \pi^{-1}(U_\alpha) \longrightarrow U_\alpha \times F$ if we set:

$$t_{(\alpha)} : [x, \varphi]_{(\alpha)} \mapsto (x, \varphi) \tag{1.1.9}$$

We leave to the reader the further necessary checks.

To show uniqueness, it is sufficient to notice that if two bundles $\mathcal{B}$ e $\mathcal{B}'$ having the same base M and the same standard fiber F have the same transition functions with respect to the same open covering, then they are (globally) isomorphic. ∎

Definition (1.1.10): let us consider a bundle $\mathcal{B} = (B, M, \pi; F)$. A map $\sigma : U \longrightarrow \pi^{-1}(U)$ such that $\pi \circ \sigma = \text{id}_U$ is called *a local section* on $U \subseteq M$. If $U = M$ then the map σ is called *a (global) section.* ∎

The existence of (infinitely many) local sections is guaranteed in any fiber bundle $\mathcal{B}$ by local trivializations; in fact, if we consider a local trivialization $t_{(\alpha)} : \pi^{-1}(U_\alpha) \longrightarrow U_\alpha \times F$ and any map $\varphi : U_\alpha \longrightarrow F$ (e.g. the constant maps) then we can define an *induced* local section σ_φ by setting:

$$\sigma_\varphi : x \mapsto t_{(\alpha)}^{-1}(x, \varphi(x)) = [x, \varphi(x)]_{(\alpha)} \tag{1.1.11}$$

Instead, global sections may not exist depending on the topology of the bundle under consideration (see below).

For any given open subset $U \subseteq M$ we denote by $\Gamma_U(\pi)$ the set of all local sections defined globally on U. If $D \subset M$ is a *domain* in M (i.e. a closed submanifold with a regular boundary $\partial D \subset M$) we denote by $\Gamma_D(\pi)$ the set of sections defined on D, i.e. the restrictions to D of (local) sections $\sigma \in \Gamma_U(\pi)$ defined on some open subset U such that $D \subset U \subseteq M$. The (possibly empty) set $\Gamma_M(\pi)$ of all global sections of a bundle $\mathcal{B} = (B, M, \pi; F)$ is simply denoted by $\Gamma(\mathcal{B})$, or equivalently by $\Gamma(\pi)$.

The local sections of a bundle $\mathcal{B} = (B, M, \pi; F)$ form a *sheaf*. In fact, let us consider two local sections $\sigma_{(\alpha)} : U_\alpha \longrightarrow B$ and $\sigma_{(\beta)} : U_\alpha \longrightarrow B$ defined on U_α and U_β respectively (and such that $U_{\alpha\beta}$ is non-empty); if these two local sections satisfy the following compatibility condition:

$$\sigma_{(\alpha)}(x)\Big|_{U_{\alpha\beta}} = \sigma_{(\beta)}(x)\Big|_{U_{\alpha\beta}} \tag{1.1.12}$$

then there exists a unique local section $\sigma_{(\alpha\beta)}$ defined on $U_\alpha \bigcup U_\beta$ which coincides with $\sigma_{(\alpha)}$ on U_α and with $\sigma_{(\beta)}$ on U_β, i.e. such that we have

$$\sigma_{(\alpha\beta)}(x)\Big|_{U_\alpha} = \sigma_{(\alpha)}(x) \qquad\qquad \sigma_{(\alpha\beta)}(x)\Big|_{U_\beta} = \sigma_{(\beta)}(x) \tag{1.1.13}$$

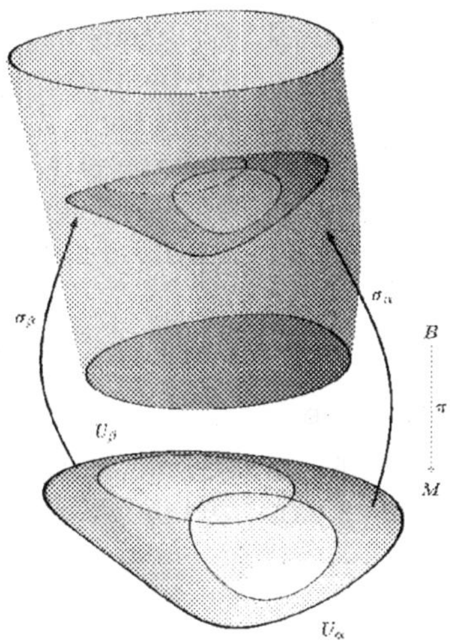

Fig. 5 – Sections.

Of course this is possible because of the compatibility condition (1.1.12). On the other hand let us assume that both U_α and U_β are trivialization domains for the local trivializations $t_{(\alpha)}$ and $t_{(\beta)}$, respectively. Let these two local sections be of the following form

$$\sigma_{(\alpha)}(x) = [x, f(x)]_{(\alpha)} \qquad\qquad \sigma_{(\beta)}(x) = [x, g(x)]_{(\beta)} \tag{1.1.14}$$

for two suitable maps $f : U_\alpha \longrightarrow F$ and $g : U_\beta \longrightarrow F$ which are thence called the *local representative of the local sections* with respect to the trivializations chosen. In this notation the compatibility condition (1.1.12) reads as

$$\hat{g}_{(\beta\alpha)}(x)[f(x)] = g(x) \tag{1.1.15}$$

Thence two local sections *glue together* if and only if their local representatives satisfy the compatibility condition $(1.1.15)$.

This can be generalized inductively to any number of patches. We can now understand why in general global sections may not exist. If a global section σ exists and $\{(U_\alpha, t_{(\alpha)})\}_{\alpha\in I}$ is a trivialization then one can define a family of local representatives $\sigma_{(\alpha)} : U_\alpha \longrightarrow B$. For any pair of such local representatives, they glue together, i.e. they satisfy the compatibility condition $(1.1.12)$ – or equivalently $(1.1.15)$. However, if we start from a family of local sections $\sigma_{(\alpha)} : U_\alpha \longrightarrow B$, they define a global section σ if and only if the compatibility conditions $(1.1.15)$ are satisfied. But depending on the particular form of the transition functions $\hat{g}_{(\beta\alpha)}$, satisfying such compatibility conditions may result to be impossible. We shall present later some examples in which no global section exists.

Trivial bundles always admit infinitely many global sections. In fact, on a trivial bundle one can fix a global trivialization by considering an open covering made of just one open set $\{M\}$ which covers the whole base manifold (transition functions are then trivial, being the open covering made of just one open set). This establishes through equation $(1.1.11)$ a one-to-one correspondence between global functions $\varphi : M \longrightarrow F$ and elements of $\Gamma(\pi)$. Changing the trivialization amounts then to perform just a global diffeomorphism in $\mathrm{Diff}(F)$.

Example: the Tangent Bundle

As a first important example, let us define the *tangent bundle* of a manifold M together with the tangent map of a differentiable map between two manifolds. Let M be a manifold and $\mathcal{F}$ denote the sheaf of (local) functions $f : U \longrightarrow \mathbb{R}$, where $U \subseteq M$ is any arbitrary open subset of M. When necessary, we denote by $\mathcal{F}(U)$ also the set of (global) functions over U, which can be identified with the (global) sections of the trivial bundle $U\times\mathbb{R} \subseteq M\times\mathbb{R}$. For any $U \subseteq M$ the set $\mathcal{F}(U)$ has a natural structure of algebra (i.e. it is a vector space with an internal multiplication defined pointwise by $(f \cdot g)(x) = f(x) \cdot g(x)$). We shall denote by $\mathcal{F}_{loc}(M)$ the class of all local functions on M, i.e. the functions defined on some unspecified open subset of M. When speaking of a local function $f \in \mathcal{F}_{loc}(M)$ we shall for simplicity write $f : M \longrightarrow \mathbb{R}$, having however in mind the local character of f.

Let us consider the set of (parametrized) curves[5] γ quotiented by the fol-

[5] By a *curve* we mean here a smooth map $\gamma : I \longrightarrow M$, where $I \subseteq \mathbb{R}$ is any open interval containing the zero. In principle (unless explicitly stated) we shall not be concerned with the problem of possibly extending a (local) curve $\gamma : I \longrightarrow M$ to a (global) curve $\gamma : \mathbb{R} \longrightarrow M$, so that when speaking generically of a curve we shall always have in mind a *local* one, but write $\gamma : \mathbb{R} \longrightarrow M$ for simplicity. We say that a curve $\gamma : I \longrightarrow M$ is *based* at a point $x \in M$ if and only if $\gamma(0) = x$.

lowing equivalence relation:

$$\begin{aligned} &\gamma \sim \gamma' \Leftrightarrow \forall f \in \mathcal{F}_{loc}(M),\\ &f \circ \gamma(0) = f \circ \gamma'(0) \quad \text{and} \quad \frac{\mathrm{d}}{\mathrm{d}t}(f \circ \gamma)\Big|_{t=0} = \frac{\mathrm{d}}{\mathrm{d}t}(f \circ \gamma')\Big|_{t=0} \end{aligned} \tag{1.1.16}$$

which requires that the two real functions $(f \circ \gamma) : \mathbb{R} \longrightarrow \mathbb{R}$ and $(f \circ \gamma') : \mathbb{R} \longrightarrow \mathbb{R}$ have the same first order Taylor expansion at $t = 0$. Equation (1.1.16) is meaningful provided $f \in \mathcal{F}_{loc}(M)$ is defined in an open set $U \subseteq M$ containing the base point $x = \gamma(0)$. The quotient set is denoted by TM and a canonical surjective map $\tau_M : TM \longrightarrow M : \dot{\gamma} \mapsto \gamma(0)$ is then well-defined[6].

We shall now prove that $T(M) = (TM, M, \tau_M; \mathbb{R}^m)$ is a fiber bundle. Let us fix an atlas $\{(U_\alpha, \varphi_{(\alpha)})\}$ on the manifold M. A local chart $(U_\alpha, \varphi_{(\alpha)})$ induces local coordinates $x^\mu_{(\alpha)}$ on M such that $\varphi_{(\alpha)} : U_\alpha \longrightarrow \mathbb{R}^m : x \mapsto (x^\mu_{(\alpha)})$. Equivalently, we shall write $(x^\mu)_{(\alpha)}$ to denote the point $x \in U_\alpha$ such that $\varphi_{(\alpha)}(x) = (x^\mu_{(\alpha)})$[7]. Let us consider the (local) curves $c^{(\alpha)}_\mu : \mathbb{R} \longrightarrow U_\alpha : t \mapsto (x^\nu_0 + \delta^\nu_\mu\, t)_{(\alpha)}$ based at $x_0 \in U_\alpha$ ($x_0 = (x^\nu_0)_{(\alpha)}$), which are called the *coordinate curves* of the chart.

Let $\partial^{(\alpha)}_\mu = [c^{(\alpha)}_\mu]$ denote the corresponding equivalence classes induced in TM. Of course, we have $\tau_M(\partial^{(\alpha)}_\mu) = x_0$. An arbitrary point in the preimage $\tau_M^{-1}(x_0)$ is a class $v = [c] \in TM$ which can be given in the following form:

$$c^{(\alpha)} : \mathbb{R} \longrightarrow U_\alpha : t \mapsto (x^\mu_0 + v^\mu\, t)_{(\alpha)} \tag{1.1.17}$$

Accordingly, $\tau_M^{-1}(x_0)$ inherits a natural structure of vector space and we have[8] $v = v^\mu\, \partial^{(\alpha)}_\mu$, i.e. $\partial^{(\alpha)}_\mu$ form a basis (called the *natural basis*) of $\tau_M^{-1}(x_0)$. We can then define the local trivializations (one for each α):

$$t_{(\alpha)} : \tau_M^{-1}(U_\alpha) \longrightarrow U_\alpha \times \mathbb{R}^m : v \mapsto (x, v^\mu_{(\alpha)}) \qquad v = v^\mu_{(\alpha)} \partial^{(\alpha)}_\mu,\ x = \tau_M(v) \tag{1.1.18}$$

We remark that the single local trivialization (1.1.18) depends on the chosen chart $(U_\alpha, \varphi_{(\alpha)})$ only. They are thence called *natural trivializations*. If we change

---

6 Here and in the sequel we shall adopt the following useful notation. If $\phi : X \longrightarrow Y$ is a map between two spaces acting as $\phi : x \mapsto \phi(x)$, we shall at once write

$$\phi : X \longrightarrow Y : x \mapsto \phi(x)$$

to denote both the map and its action on the generic point $x \in X$.

7 Whenever there is no need to specify the label (α) in the atlas, the local coordinates will be simply denoted by (x^μ).

8 Einstein convention on repeated indices is understood: whenever an index is repeated in a term once as a subscript and once as a superscript then the summation on that index on the appropriate range is understood. The trivialization labels are not indices and they do not obey Einstein convention on summation.

chart $(U_\beta, \varphi_{(\beta)})$ (and, of course, we assume $U_{\alpha\beta} \neq \varnothing$), the new chart induces a new local trivialization $t_{(\beta)} : \tau_M^{-1}(U_\beta) \longrightarrow U_\beta \times \mathbb{R}^m$; correspondingly, transition functions are thence determined by the transition functions $\varphi_{(\beta)} \circ \varphi_{(\alpha)}^{-1} : \mathbb{R}^m \longrightarrow \mathbb{R}^m$ of the manifold M, which are given by $x^\mu_{(\beta)} = x^\mu_{(\beta)}(x^\nu_{(\alpha)})$. In fact, the coordinate curve $c_\mu^{(\alpha)}$ identifies in $\tau_M^{-1}(x_0)$ the same equivalence class of the curve:

$$\gamma_\mu : t \mapsto (x_0^\nu + J^\nu_{\ \mu}\, t)_{(\beta)} \qquad\qquad J^\nu_{\ \mu} = \frac{\partial x^\nu_{(\beta)}}{\partial x^\mu_{(\alpha)}} \tag{1.1.19}$$

which is expressed in the new chart $(U_\beta, \varphi_{(\beta)})$.

Thence we have $\partial_\mu^{(\alpha)} = J^\nu_{\ \mu}\, \partial_\nu^{(\beta)}$ in $\tau_M^{-1}(U_{\alpha\beta})$. For any vector $v = v^\mu_{(\alpha)} \partial_\mu^{(\alpha)} = v^\mu_{(\beta)} \partial_\mu^{(\beta)}$ defined in $\tau_M^{-1}(U_{\alpha\beta})$ we have of course $v^\mu_{(\beta)} = J^\mu_{\ \nu}\, v^\nu_{(\alpha)}$. Transition functions on the bundle $T(M)$ are thence:

$$\hat{g}_{(\beta\alpha)}(x) : \mathbb{R}^m \longrightarrow \mathbb{R}^m : v^\nu_{(\alpha)} \mapsto v^\mu_{(\beta)} = J^\mu_{\ \nu}\, v^\nu_{(\alpha)} \tag{1.1.20}$$

Because of the chain rule the family $\{\hat{g}_{(\beta\alpha)}\}$ is then a cocycle over M with values in $\mathrm{Diff}(\mathbb{R}^m)$; actually it is valued into a copy of the linear group $\mathrm{GL}(\mathbb{R}^m)$ embedded into $\mathrm{Diff}(\mathbb{R}^m)$ since the maps (1.1.20) are linear.

We have proven thus that $T(M)$ is a fiber bundle, by directly relying on our definition (1.1.2). In many concrete applications this is hardly ever the way to define a bundle, since it is quite difficult to define the total space by a simple global definition as we did for TM. In these cases we shall frequently use proposition (1.1.7). For example, to define the tangent bundle $T(M)$ in an axiomatic way we can consider a manifold M together with an atlas $\{(U_\alpha, \varphi_{(\alpha)})\}$, we can choose the standard fiber $\mathbb{R}^m$ and consider the cocycle $\{\hat{g}_{(\beta\alpha)}\}$ defined axiomatically by (1.1.20). By proposition (1.1.7) there exists a (unique up to isomorphisms) fiber bundle $T(M) = (TM, M, \tau_M; \mathbb{R}^m)$ having these as transition functions.

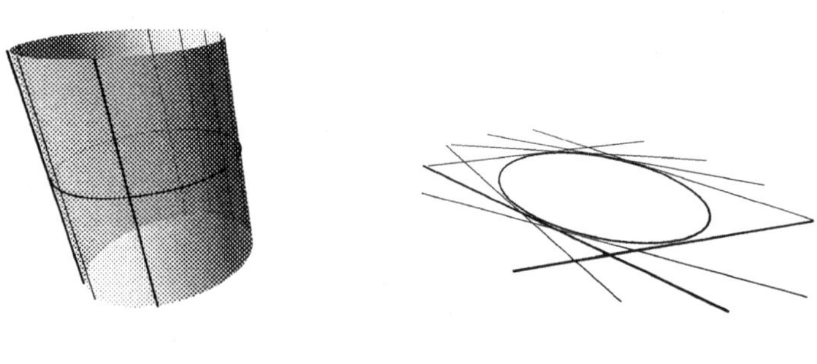

Fig. 6 – Tangent bundle $T(S^1) \simeq S^1 \times \mathbb{R}$.

Notice that the cocycle (1.1.20) is, by its very definition, completely determined by the atlas $\{(U_\alpha, \varphi_{(\alpha)})\}$ chosen in M. In other words, the tangent bundle encodes only information which is already encoded by the base manifold M. This is a key feature of the tangent bundle: in Chapter 4 we shall express this by saying that the tangent bundle is in fact a *natural bundle.*

Let now $\phi : M \longrightarrow N$ be a differentiable map; we can define the *tangent map of* ϕ by setting $T\phi : TM \longrightarrow TN : [\gamma] \mapsto [\phi \circ \gamma]$. It is evident that this definition does not depend on the representative chosen for the equivalence class. The pair $T(\phi) = (T\phi, \phi)$ is a fibered morphism between tangent bundles, i.e.:

$$\begin{array}{ccc} TM & \xrightarrow{T\phi} & TN \\ {\scriptstyle \tau_M}\downarrow & & \downarrow{\scriptstyle \tau_N} \\ M & \xrightarrow{\phi} & N \end{array} \tag{1.1.21}$$

is commutative, i.e. $\phi \circ \tau_M = \tau_N \circ T\phi$.

If we consider the composition of two differentiable maps:

$$M \xrightarrow{\phi} N \xrightarrow{\psi} Q \qquad (\psi \circ \phi : M \longrightarrow Q) \tag{1.1.22}$$

then we can easily prove that tangent maps preserve both composition and the identity, i.e.:

$$T(\psi \circ \phi) = T(\psi) \circ T(\phi) \ , \qquad T\mathrm{id}_M = \mathrm{id}_{TM} \tag{1.1.23}$$

Thus we see that $T(\cdot)$ is a *covariant functor*[9], which is called the *tangent functor*.

In the sequel, it will be often useful to regard a tangent vector $v = [\gamma] \in TM$ also as the *derivation* $v : \mathcal{F}_{loc}(M) \longrightarrow \mathbb{R}$ which associates to each $f \in \mathcal{F}_{loc}(M)$ (defined in a neighbourhood U of $x = \tau_M(v)$) the real number

$$v(f) = \frac{\mathrm{d}}{\mathrm{d}t}(f \circ \gamma)\Big|_{t=0} \tag{1.1.24}$$

The number $v(f)$ does not depend on the choice of the representative γ chosen for the vector v. In this equivalent perspective the tangent map $T\phi : TM \longrightarrow TN$ is then defined by $T\phi(v) : v \mapsto w$, with $w : \mathcal{F}_{loc}(N) \longrightarrow \mathbb{R}$ given by $w(g) = v(g \circ \phi)$, for any $g \in \mathcal{F}_{loc}(N)$ so that $g \circ \phi : M \longrightarrow \mathbb{R}$ is a local function on M.

---

[9] A *functor between two categories* is a map between the objects of the two categories together with a map of the morphisms of the two categories preserving composition and the identity. Thence a functor $F(\cdot)$ associates an object $F(A)$ of the target category to an object A of the category on which the functor is defined. A morphism $\phi : A \longrightarrow B$ of the starting category is mapped into a morphism $F(\phi) : F(A) \longrightarrow F(B)$ of the target category if the functor is *covariant* or into a "reverse" morphism $F(\phi) : F(B) \longrightarrow F(A)$ if the functor is *contravariant*.

If to any point $x \in M$ we associate a tangent vector $X(x)$ in the fiber $T_x M = \tau_M^{-1}(x)$ over that point, we obtain a *(global) vector field.* Vector fields are then (global) sections of $T(M)$; we usually assume these maps to be regular enough, e.g. smooth in the C^∞ sense. The set of all (global) smooth vector fields over M will be denoted by $\mathfrak{X}(M)$. Of course, local sections defined over $U \subseteq M$ provide *local vector fields.* The set of all local vector fields (globally) defined over $U \subseteq M$ will be denoted by $\mathfrak{X}(U)$. If $X \in \mathfrak{X}(M)$ is a vector field, we can define its action on functions $f \in \mathcal{F}(M)$. In fact, by evaluating X at each point $x \in M$ we obtain vectors $X_x = X(x) \in T_x M$. Thus we can define a new function $X(f) : M \longrightarrow \mathbb{R}$ by setting

$$X(f) : x \mapsto X_x(f) \tag{1.1.25}$$

In this way a vector field X may be regarded as a derivation[10] of the algebra $\mathcal{F}(M)$. Since vector fields are (differential) operators on $\mathcal{F}(M)$ one can define the *commutator* of two vector fields $X, Y \in \mathfrak{X}(M)$ by setting:

$$[X,Y](f) = X(Y(f)) - Y(X(f)) \qquad f \in \mathcal{F}(M) \tag{1.1.26}$$

It is easy to prove that $[X,Y]$ so defined is a linear (differential) operator on $\mathcal{F}(M)$. Then $[X,Y] \in \mathfrak{X}(M)$ and $(\mathfrak{X}(M), [\ ,\ ])$ is a Lie algebra since one can easily prove that the *Jacobi identity* holds true:

$$[X,[Y,Z]] + [Y,[Z,X]] + [Z,[X,Y]] = 0 \tag{1.1.27}$$

Example: the Hopf Bundle

As a further example let us introduce the Hopf bundle. It will be studied later in further detail as an example of principal bundle and it will provide an example of spin bundle over S^2.

Let us consider a 3-sphere S^3 embedded in $\mathbb{C}^2$ by the equation $|z_1|^2 + |z_2|^2 = 1$, and a 2-sphere S^2 embedded in $\mathbb{C} \times \mathbb{R}$ by the equation $|z_0|^2 + x^2 = 1$. Let us consider the map

$$\hat{\pi} : \mathbb{C}^2 \longrightarrow \mathbb{C} \times \mathbb{R} : (z_1, z_2) \mapsto (2\bar{z}_1 z_2, |z_1|^2 - |z_2|^2) \tag{1.1.28}$$

---

[10] A *derivation* of an algebra $\mathcal{A}$ is a linear map $D : \mathcal{A} \longrightarrow \mathcal{A}$ such that the Leibniz rule holds, i.e.

$$D(a \cdot b) = D(a) \cdot b + a \cdot D(b) \qquad \forall a, b \in \mathcal{A}$$

If $\mathcal{A} = \oplus \mathcal{A}_k$ is a graded algebra a *graded derivation* of degree n is a a linear map $D : \mathcal{A}_k \longrightarrow \mathcal{A}_{k+n}$ such that the graded Leibniz rule holds on homogenous terms, i.e.

$$D(a \cdot b) = D(a) \cdot b + (-1)^{kn} a \cdot D(b) \qquad \forall a \in \mathcal{A}_k,\ b \in \mathcal{A}_h$$

where $\bar{z}$ denotes the complex conjugate of $z \in \mathbb{C}$.

The map $\hat{\pi}$ restricts to a map $\pi : S^3 \longrightarrow S^2$, because of the constraint $|z_1|^2 + |z_2|^2 = 1$ which ensures that $\hat{\pi}$ maps the 3-sphere S^3 onto the 2-sphere S^2.

Let us now fix a point $q = (z_0, x) \in S^2$; because of the constraint $|z_0|^2 + x^2 = 1$ defining S^2 we have $z_0 = \sqrt{1-x^2}\, e^{i\alpha}$ and $-1 \le x \le 1$. Notice that the phase α is uniquely defined for $-1 < x < 1$ while it is undetermined when $x = \pm 1$. A point (z_1, z_2) is in the fiber $\pi^{-1}(q)$ if the following equations hold true

$$\begin{cases} |z_1|^2 + |z_2|^2 = 1 \\ |z_1|^2 - |z_2|^2 = x \\ 2\bar{z}_1 z_2 = z_0 \end{cases} \tag{1.1.29}$$

Then all points $(z_1, z_2) \in \pi^{-1}(q)$ are in the form

$$\left(\sqrt{\frac{1+x}{2}} e^{i\theta}, \sqrt{\frac{1-x}{2}} e^{i\alpha} e^{i\theta} \right) \tag{1.1.30}$$

for some $\theta \in \mathbb{R}$. One readily recognizes that each fiber is isomorphic to $U(1) \simeq S^1$, which is thence chosen as standard fiber. Let us now fix an open covering of the base manifold S^2 made of the open sets $S^2_N = \{q \in S^2 : q \ne S = (0, -1)\}$ and $S^2_S = \{q \in S^2 : q \ne N = (0, 1)\}$. We can define the following mappings:

$$\begin{cases} t_{(N)} : \pi^{-1}(S^2_N) \longrightarrow S^2_N \times U(1) : \\ \qquad : \left(\sqrt{\frac{1+x}{2}} e^{i\theta}, \sqrt{\frac{1-x}{2}} e^{i\alpha} e^{i\theta} \right) \mapsto (q, e^{i\theta}) \\ \\ t_{(S)} : \pi^{-1}(S^2_S) \longrightarrow S^2_S \times U(1) : \\ \qquad : \left(\sqrt{\frac{1+x}{2}} e^{-i\alpha} e^{i\theta'}, \sqrt{\frac{1-x}{2}} e^{i\theta'} \right) \mapsto (q, e^{i\theta'}) \end{cases} \tag{1.1.31}$$

where we set $\theta' = \alpha + \theta$.

These define a trivialization $\{(S^2_N, t_{(N)}), (S^2_S, t_{(S)})\}$ for the Hopf bundle $\mathcal{H} = (S^3, S^2, \pi; U(1))$. The transition function $\hat{g}_{(NS)} : S^2_{NS} \longrightarrow \mathrm{Diff}(U(1))$ is given by

$$\hat{g}_{(NS)}(q) : U(1) \longrightarrow U(1) : e^{i\theta} \mapsto e^{-i\alpha} e^{i\theta} \tag{1.1.32}$$

where we recall $q = (z_0, x)$ and $z_0 = \sqrt{1-x^2} e^{i\alpha}$.

The trivialization (1.1.31) allows us to define two local sections in the Hopf bundle:

$$\begin{cases} \sigma_{(N)} : q \mapsto \left(\sqrt{\frac{1+x}{2}}, \sqrt{\frac{1-x}{2}} e^{i\alpha} \right) \\ \\ \sigma_{(S)} : q \mapsto \left(\sqrt{\frac{1+x}{2}} e^{-i\alpha}, \sqrt{\frac{1-x}{2}} \right) \end{cases} \tag{1.1.33}$$

defined on S^2_N and S^2_S, respectively. They cannot patch together. In fact, at $N = (0,1)$, we have $x = 1$ and $z_0 = 0$ so that α is here undetermined. Nevertheless, $\sigma_{(N)}(N) = (1,0)$ is well-defined, while it fails at $S = (0,-1)$. Analogously, at $S = (0,-1)$, we have $x = -1$ and $z_0 = 0$ so that α is again undetermined. Nevertheless, $\sigma_{(S)}(S) = (0,1)$ is well-defined, while it fails at $N = (0,1)$. Accordingly, the local section $\sigma_{(N)}$, as well as $\sigma_{(S)}$, cannot be extended to a global section of the Hopf bundle.

Below in Section 1.3 we shall use the principal structure of the Hopf bundle to prove that it does not allow *any* global section.

2. Fibered Coordinates

Let $\mathcal{B} = (B, M, \pi; F)$ be a bundle and let us consider any point $b \in B$, fix a local trivialization $(U_\alpha, t_{(\alpha)})$ such that $t_{(\alpha)}(b) = (x, y)$ and denote briefly $b = [x, y]_{(\alpha)}$; we can choose the open set U_α to be the domain of a chart with local coordinates (x^μ) around the point x and we choose a chart (y^a) in some neighbourhood W of y in F. Hereafter, unless an explicit warning is given, Greek indices will be used to label coordinates in the base manifold M and Latin indices will label coordinates in the standard fiber F. In this way we have produced a local coordinate system $(x^\mu; y^a)$ on B, supported in the product $U_\alpha \times W$, using the local diffeomorphism between B and $M \times F$ induced by local trivializations, i.e. relying on the commutative diagram

$$\begin{array}{ccc} U_\alpha \times F & \xrightarrow{t^{-1}_{(\alpha)}} & B \\ {\scriptstyle \mathrm{pr}_1}\downarrow & & \downarrow{\scriptstyle \pi} \\ U_\alpha & \xrightarrow{i} & M \end{array} \tag{1.2.1}$$

Such local coordinates are called *fibered coordinates*.

A morphism $\boldsymbol{\Phi} = (\Phi, \phi)$ between two bundles $\mathcal{B}$ and $\mathcal{B}'$ has the following local expression[11] with respect to fibered coordinates:

$$\begin{cases} x'^\mu = \phi^\mu_{(\alpha)}(x) \\ y'^a = \Phi^a_{(\alpha)}(x,y) \end{cases} \tag{1.2.2}$$

If we consider a family $\{(\Phi_{(\alpha)}, \phi_{(\alpha)})\}_{\alpha\in I}$ of such local transformations, each defined on a trivialization neighbourhood U_α, and if the following compatibility condition is satisfied:

$$\begin{array}{ccc} [x,y]_{(\alpha)} & \xrightarrow{\quad\Phi\quad} & [\phi_{(\alpha)}(x), \Phi_{(\alpha)}(x,y)]_{(\alpha)} \\ \| & & \| \\ [x, \hat{g}_{(\beta\alpha)}(x,y)]_{(\beta)} & \xrightarrow{\quad\Phi\quad} & [\phi_{(\beta)}(x), \Phi_{(\beta)}(x, \hat{g}_{(\beta\alpha)}(x,y))]_{(\beta)} \end{array} \tag{1.2.3}$$

i.e.:

$$\begin{aligned} &\phi_{(\alpha)}(x) = \phi_{(\beta)}(x) =: x' \\ &\hat{g}_{(\beta\alpha)}\Big(x', \Phi_{(\alpha)}(x,y)\Big) = \Phi_{(\beta)}\Big(x, \hat{g}_{(\beta\alpha)}(x,y)\Big) \end{aligned} \tag{1.2.4}$$

then there exists a (unique) global morphism $\boldsymbol{\Phi} = (\Phi, \phi)$ having that family as its family of local expressions.

3. Particular Classes of Fiber Bundles

Within the class of fiber bundles we have to define some subclasses which will be used in the sequel. As we shall see below, in affine bundles one can identify a preferred class of trivializations the transition functions of which are not valued into the whole of Diff($\mathbb{A}$), but just into the subgroup of affine transformations on the affine space $\mathbb{A}$ which plays the role of standard fiber. Analogously, in vector bundles there exists a preferred class of trivializations the transition functions of which are valued into the group GL($\mathbb{V}$) of all linear transformations of a vector space V, while in a principal bundle $\mathcal{P} = (P, M, \pi; G)$ one can select a preferred class of trivializations the transition functions of which are valued into a Lie group G, represented on itself by left translations.

Of course, the requirement on the way transition functions act on the standard fiber may be regarded as a condition on local trivializations to be selected. For example, on a vector bundle one could define and use trivializations such

[11] Here and throughout this monograph we adopt as much as possible the following convenient notation. Whenever a coordinate change $x \mapsto x'$ is needed, instead of priming indices (and write as usual $x^\mu \mapsto x^{\mu'}(x)$) we just prime the coordinate and, if necessary or convenient, we use a different label for the index (writing, e.g., $x^\mu \mapsto x'^\nu(x)$).

that transition functions are not linear: in this way, however, one does not preserve the vector structure of the fibers. Analogously, on trivial bundles one can decide to use only global trivializations, so that transition functions take their values into the trivial group $\{e\}$, or to use more general trivializations, so that transition functions take their values into larger groups. This is a consequence of a general situation: the smallest structure group $G \subseteq \mathrm{Diff}(F)$ in which transition functions associated to some preferred atlas and trivialization take possibly their values is *not* uniquely determined by the bundle but it depends on the class of trivializations one chooses. In fact, if G is a structure group for a bundle $\mathcal{B}$ then any bigger subgroup $G' \supset G$ of $\mathrm{Diff}(F)$ is a structure group as well. Furthermore, even if G is minimal with respect to certain trivializations, it may happen that, by restricting the set of allowed trivializations, the structure group G reduces to a smaller group.

As a consequence, the notion of structure group (in the sense of Section 1.1) is not uniquely defined on fiber bundles. We shall give a rigorous framework for *fiber bundles with structure group* in Section 1.4.

Affine Bundles

Definition (1.3.1): a bundle $\mathcal{A} = (B, M, \pi; \mathbb{A})$ is an *affine bundle* if the standard fiber $\mathbb{A}$ is an affine space and there exists (at least) one trivialization the transition functions of which act on $\mathbb{A}$ in an affine way, i.e. they fill a subgroup of the affine group $\mathrm{GA}(\mathbb{A})$ canonically embedded into $\mathrm{Diff}(\mathbb{A})$ by the standard action. A morphism between affine bundles is called *affine* if, restricted to fibers, it is an affine map. ∎

Proposition (1.3.2): any affine bundle $\mathcal{A} = (B, M, \pi; \mathbb{A})$ allows global sections.

Proof: this can be shown by using a *partition of unity of* M, i.e. a family of positive C^∞-functions $\psi_{(\alpha)}$ having support contained in U_α (being $\{U_\alpha\}_{\alpha\in I}$ an open covering of the base manifold) and which sum up to unity at each point $x \in M$, i.e. $\sum_{\alpha\in I} \psi_{(\alpha)}(x) = 1$. A partition of unity always exists on M since M has been assumed to be paracompact. Then one can define local sections

$$\rho^{(\alpha)} : U_\alpha \longrightarrow B : x \mapsto [(x, P_{(\alpha)}(x))]_{(\alpha)} \qquad P_{(\alpha)} : U_\alpha \longrightarrow \mathbb{A} \tag{1.3.3}$$

Let us now fix a point $x \in M$. Whenever $x \in U_\alpha$ then the local section $\rho^{(\alpha)}$ selects a point $\rho^{(\alpha)}(x)$ in the fiber $\pi^{-1}(x)$. Since in any affine space (as $\pi^{-1}(x)$ is) the notion of weighted average makes sense, one can define a global section of $\mathcal{A}$ by setting

$$\rho : x \mapsto \sum_{\alpha\in I} \psi_{(\alpha)}(x) \cdot \rho^{(\alpha)}(x) \tag{1.3.4}$$

The section ρ is global; in fact the weighted average is not affected by locality of the local section $\rho^{(\alpha)}$ since where it does not exist (i.e. out of U_α) it has zero weight. ∎

Let us now consider any affine automorphism $\boldsymbol{\Phi} = (\Phi, \phi) \in \text{Aut}(\mathcal{A})$. Let $\{(U_\alpha, t_{(\alpha)})\}_{\alpha \in I}$ be a trivialization such that

$$t_{(\alpha)} : \pi^{-1}(U_\alpha) \longrightarrow U_\alpha \times \mathbb{A} : p \mapsto (x, P_{(\alpha)}) \tag{1.3.5}$$

Transition functions are then naturally seen as affine maps $g_{(\alpha\beta)}(x) : \mathbb{A} \longrightarrow \mathbb{A}$, i.e.

$$P_{(\alpha)} = [g_{(\alpha\beta)}(x)](P_{(\beta)}) = \tilde{g}_{(\alpha\beta)}(x) \cdot P_{(\beta)} + t_{(\alpha\beta)}(x) \tag{1.3.6}$$

where $\cdot$ denotes the linear action. The affine morphism Φ also acts by means of affine maps, i.e.

$$\left(x, P_{(\alpha)}\right) \mapsto \left(\phi(x), [\Phi_{(\alpha)}(x)](P_{(\alpha)})\right) \tag{1.3.7}$$

and the local representations $\Phi_{(\alpha)}$ are compatible in the sense of (1.2.4), i.e.

$$[\Phi_{(\beta)}(x)] = [g_{(\beta\alpha)}(x')] \circ [\Phi_{(\alpha)}(x)] \circ [g_{(\alpha\beta)}(x)] \tag{1.3.8}$$

so that they glue together to form a global morphism.

If we choose Cartesian coordinates P^i on the standard fiber $\mathbb{A}$, the local expression of a morphism (1.3.7) is

$$\begin{cases} x'^\mu = \phi^\mu(x) \\ P'^i = A^i_j(x)\, P^j + T^i(x) \end{cases} \tag{1.3.9}$$

Vector Bundles

Definition (1.3.10): a bundle $\mathcal{V} = (B, M, \pi; V)$ is a *vector bundle* if the standard fiber V is a vector space and there exists (at least) one trivialization the transition functions of which act linearly on V, i.e. they fill a subgroup of the linear group GL(V) canonically embedded into Diff(V) by the standard action. A morphism between vector bundles is *linear* if it sends fibers into fibers in a linear way. ■

If dim$V = k$ one usually says that $\mathcal{V} = (B, M, \pi; V)$ has *fiber dimension* k (or simply *rank* k). Of course, dim$B$ = dim$M$ + dim$V$. If $k = 1$ (resp. $k = 2$) one speaks of a *line* (resp. *plane*) *bundle* over M.

Proposition (1.3.11): any vector bundle $\mathcal{V} = (B, M, \pi; V)$ allows global sections.

Proof: in fact, one can define the local zero sections on any trivialization patch:

$$\rho_0^{(\alpha)} : x \mapsto t^{-1}_{(\alpha)}(x, 0) \tag{1.3.12}$$

These expressions glue together to give a global section ρ_0, called the *zero section of* $\mathcal{V}$, since transition functions are linear and thence preserve the zero of each fiber. Then one can define infinitely many global sections by deforming the zero section ρ_0 on any compact support. ■

Let now $\boldsymbol{\Phi} = (\Phi, \phi) \in \mathrm{Aut}(\mathcal{V})$ be a linear automorphism and $\{(U_\alpha, t_{(\alpha)})\}_{\alpha\in I}$ be a trivialization such that

$$t_{(\alpha)} : \pi^{-1}(U_\alpha) \longrightarrow U_\alpha \times V : p \mapsto (x, v_{(\alpha)}) \tag{1.3.13}$$

Transition functions are then seen as linear maps $g_{(\alpha\beta)}(x) : V \longrightarrow V$, i.e.

$$v_{(\alpha)} = [g_{(\alpha\beta)}(x)](v_{(\beta)}) = \tilde{g}_{(\alpha\beta)}(x) \cdot v_{(\beta)} \tag{1.3.14}$$

where $\cdot$ denotes the linear action.

The linear morphism Φ also acts by means of linear maps, i.e.

$$(x, v_{(\alpha)}) \mapsto \Big(\phi(x), [\Phi_{(\alpha)}(x)](v_{(\alpha)})\Big) \tag{1.3.15}$$

and the local representations $\Phi_{(\alpha)}$ are compatible in the sense of (1.2.4), i.e.

$$[\Phi_{(\beta)}(x)] = [g_{(\beta\alpha)}(x')] \circ [\Phi_{(\alpha)}(x)] \circ [g_{(\alpha\beta)}(x)] \tag{1.3.16}$$

In other words, the local maps $\{\Phi_{(\alpha)}\}_{\alpha\in I}$ glue together to define a global morphism if and only if they differ as in (1.3.16).

If we choose Cartesian coordinates v^i on the standard fiber V, the local expression of a morphism (1.3.15) is

$$\begin{cases} x'^\mu = \phi^\mu(x) \\ v'^i = A^i_j(x)\, v^j \end{cases} \tag{1.3.17}$$

Principal Bundles

Definition (1.3.18): a bundle $\mathcal{P} = (P, M, \pi; G)$ is a *principal bundle* if the standard fiber is a Lie group G and there exists (at least) one trivialization the transition functions of which act on G by left translations $L_g : G \longrightarrow G : h \mapsto g \cdot h$ (where $\cdot$ denotes here the group multiplication). ■

The principal morphisms between principal bundles will be defined below. Under this viewpoint principal bundles are slightly different from affine bundles and vector bundles. In fact, while in affine bundles the fibers $\pi^{-1}(x)$ have a canonical structure of affine spaces and in vector bundles the fibers $\pi^{-1}(x)$ have a canonical structure of vector spaces, in principal bundles the fibers have no canonical Lie group structure. This is due to the fact that, while in affine bundles transition functions act by means of affine transformations and in vector bundles transition functions act by means of linear transformations, in principal bundles transition functions act by means of left translations which are not group automorphisms. Thus the fibers of a principal bundle do not carry a canonical group structure, but rather many non-canonical (trivialization-depending) group structures. One can immediately see this by noticing, for example, that in the fibers of a vector bundle there exists a preferred element (the "zero") the definition of which does not depend on the local trivialization. On the contrary, in the fibers of a principal bundle there is no preferred point which is fixed by transition functions to be selected as an identity. Thus, while in affine bundles affine morphisms are those which preserve the affine structure of the fibers and in vector bundles linear morphisms are the ones which preserve the linear structure of the fibers, in a principal bundle $\mathcal{P} = (P, M, \pi; G)$ principal morphisms preserve instead a structure, the *right action* of G on P, which we are going to define hereafter.

Let $\mathcal{P} = (P, M, \pi; G)$ be a principal bundle and $\{(U_\alpha, t_{(\alpha)})\}_{\alpha\in I}$ a trivialization. We can locally consider the maps

$$R_g^{(\alpha)} : \pi^{-1}(U_\alpha) \longrightarrow \pi^{-1}(U_\alpha) : [x, h]_{(\alpha)} \mapsto [x, h \cdot g]_{(\alpha)} \tag{1.3.19}$$

Proposition (1.3.20): there exists a (global) right action R_g of G on P which is free, vertical and transitive on fibers[12]; the local expression in the given trivialization of this action is given by $R_g^{(\alpha)}$.

Proof: using the local trivialization we set $p = [x, h]_{(\alpha)} = [x, g_{(\beta\alpha)}(x) \cdot h]_{(\beta)}$; then the following diagram commutes:

$$\begin{array}{ccc} [x, h]_{(\alpha)} & \xrightarrow{\quad R_g^{(\alpha)} \quad} & [x, h \cdot g]_{(\alpha)} \\ \| & & \| \\ [x, g_{(\beta\alpha)}(x) \cdot h]_{(\beta)} & \xrightarrow{\quad R_g^{(\beta)} \quad} & [x, g_{(\beta\alpha)}(x) \cdot h \cdot g]_{(\beta)} \end{array} \tag{1.3.21}$$

which clearly shows that the local expressions agree on the overlaps $U_{\alpha\beta}$, to define a right action.

This is obviously a vertical action; it is free because of the following:

$$R_g p = p \quad \Rightarrow \quad [x, h \cdot g]_{(\alpha)} = [x, h]_{(\alpha)} \quad \Rightarrow \quad h \cdot g = h \quad \Rightarrow \quad g = e \tag{1.3.22}$$

[12] An action of G over a manifold M is *free* if whenever $g \cdot x = x$ for some $x \in M$ then $g = e$. It is *transitive* if for all $x, y \in M$ then there exists a $g \in G$ such that $g \cdot x = y$. An action on a bundle $\mathcal{B} = (B, M, \pi; F)$ is *vertical* if for all $b \in B$, $g \cdot b$ is in the same fiber of b itself. It is *transitive on fibers* if for all $b_1, b_2 \in B$ such that $\pi(b_1) = \pi(b_2)$ then there exists a $g \in G$ such that $g \cdot b_1 = b_2$.

Finally, if $p = [x, h_1]_{(\alpha)}$ and $q = [x, h_2]_{(\alpha)}$ are two points in the same fiber of $\mathcal{P}$, one can choose $g = h_2^{-1} \cdot h_1$ (where $\cdot$ denotes the group multiplication) so that $p = R_g q$. This shows that the right action is also transitive on the fibers. ∎

Notice, on the contrary, that a global left action cannot be defined by using the local maps

$$L_g^{(\alpha)} : \pi^{-1}(U_\alpha) \longrightarrow \pi^{-1}(U_\alpha) : [x, h]_{(\alpha)} \mapsto [x, g \cdot h]_{(\alpha)} \tag{1.3.23}$$

since these local maps do not satisfy a compatibility condition analogous to the condition (1.3.21). Some examples of principal bundles will be considered below.

Definition (1.3.24): let $\mathcal{P} = (P, M, \pi; G)$ and $\mathcal{P}' = (P', M', \pi'; G')$ be two principal bundles and $\theta : G \longrightarrow G'$ be a homomorphism of Lie groups. A bundle morphism $\boldsymbol{\Phi} = (\Phi, \phi) : \mathcal{P} \longrightarrow \mathcal{P}'$ is a *principal morphism with respect to* θ if the following diagram is commutative:

$$\begin{array}{ccc} P & \xrightarrow{\ \Phi\ } & P' \\ {\scriptstyle R_g}\downarrow & & \downarrow{\scriptstyle R_{\theta(g)}} \\ P & \xrightarrow{\ \Phi\ } & P' \end{array} \tag{1.3.25}$$

When $G = G'$ and $\theta = \mathrm{id}_G$ we just say that $\boldsymbol{\Phi}$ is a *principal morphism.* ∎

A trivial principal bundle $(M \times G, M, \pi; G)$ naturally admits the global *unity section* $\mathbb{I} \in \Gamma(M \times G)$, defined with respect to a global trivialization, $\mathbb{I} : x \mapsto (x, e)$, e being the unit element of G. Conversely, one can show that principal bundles allow global sections if and only if they are trivial. In fact, on principal bundles there is a canonical correspondence between local sections and local trivializations, due to the presence of the global right action defined above.

Proposition (1.3.26): any local section $\sigma^{(\alpha)} : U_\alpha \longrightarrow P$ of a principal bundle $\mathcal{P} = (P, M, \pi; G)$ induces a local trivialization $t_{(\alpha)} : \pi^{-1}(U_\alpha) \longrightarrow U_\alpha \times G$ and this correspondence is one-to-one.

Proof: let U_α be an open subset of M, $\sigma^{(\alpha)} : U_\alpha \longrightarrow P$ be a local section, $p \in \pi^{-1}(U_\alpha)$ (with $x = \pi(p)$) and $g \in G$ be uniquely chosen so that $p = \sigma^{(\alpha)}(x) \cdot g$. We define a local trivialization by setting:

$$t_{(\alpha)} : \pi^{-1}(U_\alpha) \longrightarrow U_\alpha \times G : p \mapsto (x, g) \qquad p = \sigma^{(\alpha)}(x) \cdot g \tag{1.3.27}$$

Conversely, let us choose a local trivialization $t_{(\alpha)}$ and define a local section $\sigma^{(\alpha)}$ as follows:

$$\sigma^{(\alpha)} : U_\alpha \longrightarrow P : x \mapsto [x, e]_{(\alpha)} \tag{1.3.28}$$

As one can now easily show, these correspondences are inverse to each other. ∎

By an abuse of language, a trivialization on a principal bundle is thence assigned by giving a family of local sections $\{\sigma^{(\alpha)}\}$ the domains of which cover the whole of the base manifold M.

Moreover, in virtue of proposition (1.3.26), we see that principal bundles can be completely characterized by their right action. We have in fact the following:

Proposition (1.3.29): if a Lie group G acts freely on the right onto a manifold P and the orbit space $M = P/G$ is a manifold, then P is the total space of a principal bundle over the orbit manifold M.

Proof: the canonical projection $\pi : P \longrightarrow M$ is defined and local trivializations are induced by local sections of π (which always exist) thanks to the fact that the action is transitive on the orbits. We can in fact choose a local section $\sigma^{(\alpha)} : U \longrightarrow P$ inducing the following local trivialization:

$$t_{(\alpha)} : \pi^{-1}(U) \longrightarrow U \times G : p \mapsto (x, g) \qquad p = \sigma^{(\alpha)} \cdot g \tag{1.3.30}$$

Changing local section we get another local trivialization $\sigma^{(\beta)}$ and accordingly there exists a (unique) function $g_{(\alpha\beta)} : U_{\alpha\beta} \longrightarrow G$ such that $\sigma^{(\beta)}(x) = \sigma^{(\alpha)}(x) \cdot g_{(\alpha\beta)}(x)$. It is easy to check that $g_{(\alpha\beta)}$ are transition functions (i.e. they form a cocycle in G) acting by left translation on the standard fiber G. ∎

Let us consider now a principal automorphism $\boldsymbol{\Phi} = (\Phi, \phi) \in \mathrm{Aut}(\mathcal{P})$. Let $\{(U_\alpha, \sigma^{(\alpha)})\}_{\alpha \in I}$ be a trivialization such that

$$t_{(\alpha)} : \pi^{-1}(U_\alpha) \longrightarrow U_\alpha \times G : p \mapsto (x, h_{(\alpha)}) \qquad p = \sigma^{(\alpha)}(x) \cdot h_{(\alpha)} \tag{1.3.31}$$

Transition functions $g_{(\alpha\beta)}(x) : G \longrightarrow G$ act by left translation, i.e.

$$h_{(\alpha)} = [g_{(\alpha\beta)}(x)] \cdot h_{(\beta)} \tag{1.3.32}$$

or, equivalently

$$\sigma_{(\beta)} = \sigma_{(\alpha)} \cdot [g_{(\alpha\beta)}(x)] \tag{1.3.33}$$

The principal morphism $\boldsymbol{\Phi} = (\Phi, \phi)$ acts through left translation too, i.e. there exist local maps $\varphi_{(\alpha)} : U_\alpha \longrightarrow G$ such that (see below (3.3.3))

$$(x, h_{(\alpha)}) \mapsto (\phi(x), [\varphi_{(\alpha)}(x)] \cdot h_{(\alpha)}) \tag{1.3.34}$$

or, equivalently

$$\Phi(\sigma_{(\alpha)}) = \sigma_{(\alpha)} \cdot \varphi_{(\alpha)} \tag{1.3.35}$$

The local representations $\varphi_{(\alpha)}$ are compatible in the sense of (1.2.4), i.e.

$$[\varphi_{(\beta)}(x)] = [g_{(\beta\alpha)}(x')] \cdot [\varphi_{(\alpha)}(x)] \cdot [g_{(\alpha\beta)}(x)] \tag{1.3.36}$$

where we set $x' = \phi(x)$.

Example: the Tangent Bundle

The tangent bundle $T(M)$ of a manifold M (already introduced above) is the first significant example of vector bundle. Let us, in fact, consider two local trivializations $t_{(\alpha)}$ and $t_{(\beta)}$ induced by two different coordinate systems $x^\mu_{(\alpha)}$ and $x^\mu_{(\beta)}$ defined on two intersecting coordinate patches U_α and U_β, respectively. According to equation (1.1.18), for an element $v \in \tau_M^{-1}(U_{\alpha\beta})$ we have:

$$v = v^\mu_{(\alpha)}\, \partial^{(\alpha)}_\mu = v^\mu_{(\beta)}\, \partial^{(\beta)}_\mu \qquad \partial^{(\alpha)}_\mu = J^\nu_{\ \mu}\, \partial^{(\beta)}_\nu, \qquad J^\nu_{\ \mu} = \frac{\partial x^\nu_{(\beta)}}{\partial x^\mu_{(\alpha)}} \tag{1.3.37}$$

One immediately concludes that $v^\nu_{(\beta)} = J^\nu_{\ \mu}\, v^\mu_{(\alpha)}$. These are the transition functions of a bundle and thence they form a cocycle with values in $\mathrm{GL}(m)$ (as one can directly verify by noticing that the Jacobian of the identity is the unit matrix and using the chain rule). Transition functions act then linearly on the fiber $\mathbb{R}^m$ by means of $J^\nu_{\ \mu} \in \mathrm{GL}(m)$ so that the tangent bundle is a vector bundle of rank $m = \dim(M)$.

If $\phi : M \longrightarrow N$ is a diffeomorphism, the local expression of the tangent map $T(\phi) : TM \longrightarrow TN$ in any system of natural fibered coordinates $(x^\mu; v^\mu)$ on TM and $(y^a; v^a)$ on TN is given by

$$T\phi : v^\mu \partial_\mu \mapsto \frac{\partial \phi^a}{\partial x^\nu} v^\nu \partial_a \tag{1.3.38}$$

so that it is a fiberwise linear map. The tangent functor is thence a covariant functor from the category of *manifolds* into the category of *vector bundles*.

We remind that one can use the tangent functor $T(\cdot)$ to define a cocycle $T\varphi_{(\alpha\beta)}$ with values in $\mathrm{GL}(m)$ associated to the transition functions $\varphi_{(\alpha\beta)}$ of any atlas on M. By theorem (1.1.7), one can use such a cocycle $T\varphi_{(\alpha\beta)}$ to define the tangent bundle TM axiomatically.

We stress again that the transition functions of $T(M)$ act linearly on fibers, *because* we restricted our attention to trivializations of the particular form (1.1.18). If one allows a larger set of trivializations this property fails of course down.

As an example, let us in fact consider the trivial bundle $(\mathbb{R}^2, \mathbb{R}, \mathrm{pr}_1; \mathbb{R})$ together with its natural trivialization

$$t : \mathbb{R}^2 \longrightarrow \mathbb{R} \times \mathbb{R} : (x_0, y_0) \mapsto (x_0; y_0) \tag{1.3.39}$$

For any point $(x_0, y_0) \in \mathbb{R}^2$, projecting over $x_0 \in \mathbb{R}$, there exists one and only one parabola of the form $y = a(x^2 + 1) + 1$ passing through (x_0, y_0). Then we can define a new global trivialization through the one-to-one mapping:

$$t' : \mathbb{R}^2 \longrightarrow \mathbb{R} \times \mathbb{R} : (x_0, y_0) \mapsto (x_0; a) \tag{1.3.40}$$

The transition function with respect to the trivializations t and t' is then

$$g(x_0) : \mathbb{R} \longrightarrow \mathbb{R} : a \mapsto a(x_0^2 + 1) + 1 \tag{1.3.41}$$

which is *not* linear in a since it does not fix $a = 0$ (see *Fig.* 7).

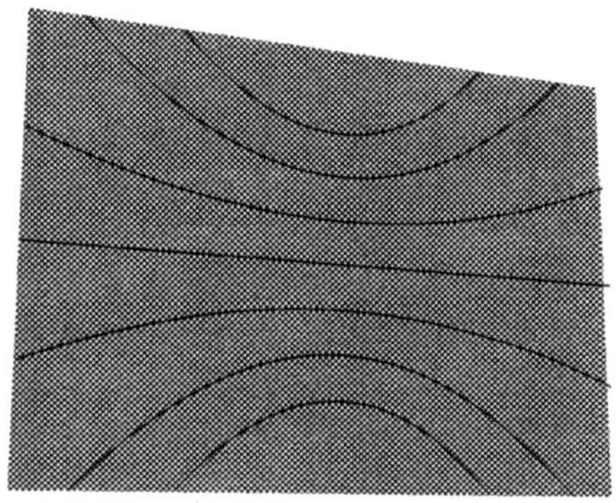

Fig. 7 – Non-linear transition functions on $(\mathbb{R}^2, \mathbb{R}, \mathrm{pr}_1; \mathbb{R})$.

Notice that transition functions (1.3.41) are affine maps. Accordingly, the use of the trivializations (1.3.40) corresponds to forgetting the structure of vector bundle of $(\mathbb{R}^2, \mathbb{R}, \mathrm{pr}_1; \mathbb{R})$, but still preserving its affine structure.

If we allow even more general trivializations, e.g.:

$$t'' : \mathbb{R}^2 \longrightarrow \mathbb{R} \times \mathbb{R} : (x_0, y_0) \mapsto (x_0; a^3 + a) \tag{1.3.42}$$

which correspond to transition functions between t'' and t' of the form:

$$g(x_0) : \mathbb{R} \longrightarrow \mathbb{R} : a \mapsto a^3 + a \tag{1.3.43}$$

then also the affine structure is given up. Then, as we noticed, the requirement on the way transition functions act on the standard fiber reverberates on the choice of a class of preferred trivializations.

Example: the Hopf Bundle

The Hopf bundle $\mathcal{H} = (S^3, S^2, \pi; U(1))$ introduced in Section 1.1 is an example of principal bundle. The local trivializations (1.1.31) induce the transition function $g_{(NS)}$ given by (1.1.32). This function acts on $U(1)$ by translation along the point-dependent element $e^{i\alpha} \in U(1)$ at the point $(z_0, x) \in S^2$ with $z_0 = \sqrt{1 - x^2} e^{i\alpha}$. Of course, being $U(1)$ a commutative group, it can be regarded as acting by left translation. Notice that the trivialization (1.1.31) is associated to the family of sections given by (1.1.33).

We can now also prove that the Hopf bundle does not allow global sections. In fact, if there existed a global section, by proposition (1.3.26) the bundle would be trivial. This is clearly false, since there is no global diffeomorphism between S^3 and $S^2 \times S^1$.

Example: the Frame Bundle

Let LM be the set of all pairs (x, e_a) with e_a any basis of the tangent space $T_xM = \tau_M^{-1}(x)$ to M at the point $x \in M$. Of course, one can define a natural projection $\tau : LM \longrightarrow M : (x, e_a) \mapsto x$. We shall set $L_xM = \tau^{-1}(x)$ for the fiber over the point $x \in M$. Consequently, LM coincides with the disjoint union $\coprod_{x\in M} (L_xM)$.

We can prove directly that $L(M) = (LM, M, \tau; \mathrm{GL}(m))$ is a principal bundle. Let us consider an atlas $\{(U_\alpha, \varphi^{(\alpha)})\}_{\alpha\in I}$ of M inducing local coordinates $x^\mu_{(\alpha)}$ on $U_\alpha \subseteq M$. As we saw in Section 1.1, we can define a fiberwise basis $\partial^{(\alpha)}_\mu$ of T_xM for each $x \in U_\alpha$. In other words, we can define a local map $V^{(\alpha)} : U_{(\alpha)} \longrightarrow LM : x \mapsto (x, \partial^{(\alpha)}_\mu)$. Any other element $(x, e_a) \in LM$ can be written as $(x, (e_{(\alpha)})^\mu_a\, \partial^{(\alpha)}_\mu)$ for some (uniquely determined) matrix $(e_{(\alpha)})^\mu_a \in \mathrm{GL}(m)$. We can thence define *natural local trivializations* of $L(M)$ as follows:

$$t_{(\alpha)} : \tau^{-1}(U_\alpha) \longrightarrow U_\alpha \times \mathrm{GL}(m) : (x, e_a) \mapsto (x, (e_{(\alpha)})^\mu_a) \tag{1.3.44}$$

As for the tangent bundle, such local trivializations depend only on the chart $(U_\alpha, \varphi^{(\alpha)})$. If we change chart we obtain transition functions in the following form:

$$\hat{g}_{(\beta\alpha)}(x) : \mathrm{GL}(m) \longrightarrow \mathrm{GL}(m) : (e_{(\alpha)})^\mu_a \mapsto J^\mu_\nu\, (e_{(\alpha)})^\nu_a \tag{1.3.45}$$

where we set J^μ_ν for the Jacobian of the transition functions between the charts $\varphi^{(\alpha)}$ and $\varphi^{(\beta)}$. Notice that transition functions (1.3.45) act on $\mathrm{GL}(m)$ by means of the left translation. The bundle $L(M) = (LM, M, \tau; \mathrm{GL}(m))$ is thence a principal bundle and it is called *the frame bundle* over the manifold M. As for the tangent bundle, the frame bundle $L(M)$ can be also built axiomatically as follows. Let us consider a manifold M together with an atlas $\{(U_\alpha, \varphi_{(\alpha)})\}$, choose the standard fiber $\mathrm{GL}(m)$ and consider the cocycle $\{\hat{g}_{(\beta\alpha)}\}$ defined by (1.3.45). By proposition (1.1.7) there exists a (unique up to isomorphisms) fiber bundle $L(M) = (LM, M, \tau; \mathrm{GL}(m))$ with the transition functions given by (1.3.45).

Finally, the frame bundle $L(M)$ can be defined also by means of proposition (1.3.29). In this way, we prove at once that $L(M)$ is a bundle and that it is principal. In fact, we can define a right action of $\mathrm{GL}(m)$ over $L(M)$ by setting

$$R_g : LM \longrightarrow LM : e_a \mapsto e_b\, g^b_{\ a} \qquad g = \|g^b_{\ a}\| \in \mathrm{GL}(m) \tag{1.3.46}$$

which turns out to be a well-defined, vertical with respect to the projection τ, free and fiber transitive right action. Then $L(M)$ is a principal bundle because of proposition (1.3.29). Trivializations are associated to local sections, i.e. to *(local) frames* $V^{(\alpha)}_{a} : U_\alpha \longrightarrow \pi^{-1}(U_\alpha) \subseteq LM$, and they are given by

$$t_{(\alpha)} : \pi^{-1}(U_\alpha) \longrightarrow U_\alpha \times \mathrm{GL}(m) : e_a \mapsto (x, g) \qquad e_a = V^{(\alpha)}_{b}(x)\, g^{b}_{\ a} \qquad (1.3.47)$$

The transition functions $g_{(\alpha\beta)} : U_{\alpha\beta} \longrightarrow \mathrm{GL}(m)$ are then given through transition matrices between local frames, i.e.

$$V^{(\beta)}_{a} = V^{(\alpha)}_{b}\, [g_{(\alpha\beta)}(x)]^{b}_{\ a} \qquad (1.3.48)$$

We remark that the natural trivializations (1.3.44) are a particular class of the trivializations (1.3.47) when the reference frame $V^{(\alpha)}_{a}$ is the natural frame $\partial^{(\alpha)}_{\mu}$ induced by a chart of the base manifold M.

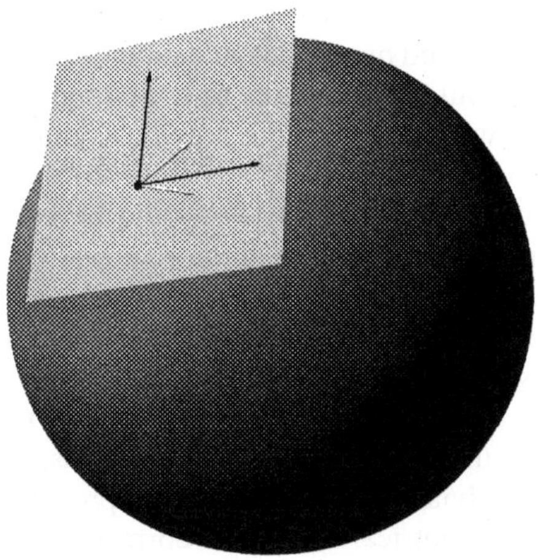

Fig. 8 – Local structure of $L(S^2)$.

Let now $\phi : M \longrightarrow N$ be a diffeomorphism. It induces a principal morphism $L(\phi) : L(M) \longrightarrow L(N)$ by setting

$$L(\phi) : (x, e_a) \mapsto (\phi(x), T\phi(e_a)) \qquad (1.3.49)$$

Since ϕ is a diffeomorphism, its tangent map is non-degenerate so that $T\phi(e_a)$ is a frame at $\phi(x)$. It is easy to see that $L(\cdot)$ is a covariant functor from the category of manifolds into the category of principal bundles; it is called the *frame functor*. Moreover, also the frame bundle, as well as we realized to be true for the tangent bundle, encodes information which is already encoded into the base manifold M. We shall see later that this is another example of *natural bundle*.

4. Fiber Bundles with Structure Group

In physical applications, fiber bundles often come with a *preferred group of transformations* (usually the symmetry group of the system). The modern attitude of physicists is to regard this group as a fundamental structure which should be implemented from the very beginning. We then enrich bundles with a further structure and define a new category.

A similar feature appears on manifolds as well: for example, on $\mathbb{R}^2$ one can restrict to Cartesian coordinates when we regard it just as a vector space endowed with a differentiable structure, but one can allow also translations if the "bigger" affine structure is considered. Moreover, coordinates can be chosen in much bigger sets: for instance one can fix the symplectic form $\omega = \mathrm{d}x \wedge \mathrm{d}y$ on $\mathbb{R}^2$ so that $\mathbb{R}^2$ is covered by an atlas of canonical coordinates (which include all Cartesian ones). But $\mathbb{R}^2$ also happens to be identifiable with the cotangent bundle $T^*\mathbb{R}$ so that we can restrict the previous symplectic atlas to allow only natural fibered coordinates. Finally, $\mathbb{R}^2$ can be considered as a bare manifold so that general curvilinear coordinates should be allowed accordingly; only if the full (i.e., unrestricted) manifold structure is considered one can use a full maximal atlas. Other choices define instead maximal atlases in suitably restricted sub-classes of allowed charts. As any manifold structure is associated with a maximal atlas, geometric bundles as defined in Section 1.1 are associated to "maximal trivializations". However, it may happen that one can restrict (or enlarge) the allowed local trivializations, so that the *same* geometrical bundle can be trivialized just using the appropriate smaller class of local trivializations. In geometrical terms this corresponds, of course, to impose a further structure on the bare bundle. Of course, this newly structured bundle is defined by the same basic ingredients, i.e. the same base manifold M, the same total space B, the same projection π and the same standard fiber F, but it is characterized by a new maximal trivialization where, however, *maximal* refers now to a smaller set of local trivializations

Examples have been already introduced in Section 1.3: vector bundles are characterized by *linear local trivializations*, affine bundles are characterized by *affine local trivializations*, principal bundles are characterized by *left translations* on the fiber group. Further examples come from Physics: gauge transformations are used as transition functions for the configuration bundles of any gauge theory. For these reasons we give the following definition of a *fiber bundle with structure group*.

Definition (1.4.1): a *fiber bundle with structure group* G is given by a sextuple $\mathcal{B} = (B, M, \pi; F; \lambda, G)$ such that:

(a) $(B, M, \pi; F)$ is a fiber bundle (see definition (1.1.2)). The *structure group* G is a Lie group (possibly a discrete one) and $\lambda : G \longrightarrow \text{Diff}(F)$ defines a left action of G on the standard fiber F.[13]

(b) There is a family of preferred trivializations $\{(U_\alpha, t_{(\alpha)})\}_{\alpha \in I}$ of $\mathcal{B}$ such that the following holds: let the transition functions be $\hat{g}_{(\alpha\beta)} : U_{\alpha\beta} \longrightarrow \text{Diff}(F)$ and let e_G be the neutral element of G. There exists a family of maps $g_{(\alpha\beta)} : U_{\alpha\beta} \longrightarrow G$ such that, for each $x \in U_{\alpha\beta\gamma} = U_\alpha \bigcap U_\beta \bigcap U_\gamma$

$$\begin{aligned} &g_{(\alpha\alpha)}(x) = e_G \\ &g_{(\alpha\beta)}(x) = \left[g_{(\beta\alpha)}(x)\right]^{-1} \\ &g_{(\alpha\beta)}(x) \cdot g_{(\beta\gamma)}(x) \cdot g_{(\gamma\alpha)}(x) = e_G \end{aligned} \tag{1.4.2}$$

and

$$\hat{g}_{(\alpha\beta)}(x) = \lambda(g_{(\alpha\beta)}(x)) \in \text{Diff}(F) \tag{1.4.3}$$

The maps $g_{(\alpha\beta)} : U_{\alpha\beta} \longrightarrow G$, which depend on the trivialization, are said to form a *cocycle with values in* G. They are called the *transition functions with values in* G (or also shortly the transition functions, if there is no danger of confusion). The preferred trivializations will be said to be *compatible with the structure.* Whenever dealing with fiber bundles with structure group the choice of a compatible trivialization will be implicitly assumed. ■

Fiber bundles with structure group provide the suitable framework to deal with bundles with a preferred group of transformations. To see this, let us begin by introducing the notion of *structure bundle* of a fiber bundle with structure group $\mathcal{B} = (B, M, \pi; F; \lambda, G)$.

Let $\mathcal{B} = (B, M, \pi; F; \lambda, G)$ be a bundle with structure group; let us fix a trivialization $\{(U_\alpha, t_{(\alpha)})\}_{\alpha \in I}$ and denote by $g_{(\alpha\beta)} : U_{\alpha\beta} \longrightarrow G$ its transition functions according to definition (1.4.1). By using the canonical left action $L : G \longrightarrow \text{Diff}(G)$ of G onto itself[14], let us define $\hat{g}_{(\alpha\beta)} : U_{\alpha\beta} \longrightarrow \text{Diff}(G)$ given by $\hat{g}_{(\alpha\beta)}(x) = L(g_{(\alpha\beta)}(x))$; they obviously satisfy the cocycle properties (1.1.4). Applying theorem (1.1.7) we can construct a (unique modulo isomorphisms)

[13] Recall that a left action $\lambda : G \times F \longrightarrow F$ defines uniquely a representation of G into $\text{Diff}(F)$, which, by an abuse of language, will be denoted by the same symbol λ; the definition is the standard one $\lambda(g) : F \longrightarrow F : f \mapsto \lambda(g, f)$. We shall also denote $\lambda(g, f)$ by $[\lambda(g)](f)$.

[14] The *canonical left action* is given by left multiplication on G by a element $g \in G$, namely $L(g) = L_g$ is the diffeomorphism over G given by $L_g : G \longrightarrow G : h \mapsto g \cdot h$.

principal bundle $\mathcal{P}_{\mathcal{B}} = \mathcal{P}(\mathcal{B})$ having G as structure group and $g_{(\alpha\beta)}$ as transition functions acting on G by left translation $L_g : G \longrightarrow G$.

Definition (1.4.4): the principal bundle $\mathcal{P}(\mathcal{B}) = (P, M, p; G)$ constructed above is called the *structure bundle of* $\mathcal{B} = (B, M, \pi; F; \lambda, G)$. ■

Notice that there is no similar canonical way of associating a structure bundle to a geometric bundle $\mathcal{B} = (B, M, \pi; F)$, since in that case the structure group G is at least partially undetermined.

Proposition (1.4.5): each automorphism of $\mathcal{P}(\mathcal{B})$ naturally acts over $\mathcal{B}$.

Proof: let, in fact, $\{\sigma^{(\alpha)}\}_{\alpha\in I}$ be a trivialization of $\mathcal{P}_{\mathcal{B}}$ together with its transition functions $g_{(\alpha\beta)} : U_{\alpha\beta} \longrightarrow G$ defined by $\sigma^{(\beta)} = \sigma^{(\alpha)} \cdot g_{(\alpha\beta)}$. Then any principal morphism $\boldsymbol{\Phi} = (\Phi, \phi)$ over $\mathcal{P}_{\mathcal{B}}$ is locally represented by local maps $\varphi_{(\alpha)} : U_\alpha \longrightarrow G$ such that

$$\Phi : [x, h]_{(\alpha)} \mapsto [\phi_{(\alpha)}(x), \varphi_{(\alpha)}(x) \cdot h]_{(\alpha)} \tag{1.4.6}$$

Since $\boldsymbol{\Phi}$ is a global automorphism of $\mathcal{P}_{\mathcal{B}}$, for the local expressions $(1.4.6)$ the following property holds true in $U_{\alpha\beta}$

$$\phi_{(\alpha)}(x) = \phi_{(\beta)}(x) \equiv x' \qquad\qquad \varphi_{(\alpha)}(x) = g_{(\alpha\beta)}(x') \cdot \varphi_{(\beta)}(x) \cdot g_{(\beta\alpha)}(x) \tag{1.4.7}$$

By using the family of maps $\{(\phi_{(\alpha)}, \varphi_{(\alpha)})\}$ one can thence define a family of global automorphisms of $\mathcal{B}$. In fact, using the trivialization $\{(U_\alpha, t_{(\alpha)})\}_{\alpha\in I}$ introduced above, one can define local automorphisms of $\mathcal{B}$ given by

$$\Phi_{\mathcal{B}}^{(\alpha)} : \Big(x, y\Big) \mapsto \Big(\phi_{(\alpha)}(x), [\lambda(\varphi_{(\alpha)}(x))](y)\Big) \tag{1.4.8}$$

These local maps glue together to give a global automorphism $\boldsymbol{\Phi}_{\mathcal{B}}$ of the bundle $\mathcal{B}$, due to $(1.4.7)$ and to the fact that $g_{(\alpha\beta)}$ are also transition functions of $\mathcal{B}$ with respect to its trivialization $\{(U_\alpha, t_{(\alpha)})\}_{\alpha\in I}$. ■

In this way $\mathcal{B}$ is endowed with a preferred group of transformations, namely the group $\mathrm{Aut}(\mathcal{P}_{\mathcal{B}})$ of automorphisms of the structure bundle $\mathcal{P}_{\mathcal{B}}$, represented on $\mathcal{B}$ by means of the canonical action $(1.4.8)$. These transformations are called *(generalized) gauge transformations*. Vertical gauge transformations, i.e. gauge transformations projecting over the identity, are also called *pure gauge transformations*.

Different Structure Groups over the Tangent Bundle

Once a geometric bundle is given one may try to enrich it with a suitable structure of bundle with structure group. However, as we shall see, the *same* geometric bundle may support different structures of bundle with structure group, with respect to *different* groups.

Affine bundles, vector bundles and principal bundles are typical examples of bundles which have been enriched to become bundles with structure group. In fact, affine bundles (vector bundles, resp.) are characterized by the affine (linear, resp.) structure group acting on the standard fiber $\mathbb{A}$ (V, resp.) according to the standard representation. Similarly, the structure group of a principal bundle $\mathcal{P} = (P, M, \pi; G)$ is the standard fiber G itself, acting by means of the left translation $L : G \longrightarrow \mathrm{Diff}(G)$. Of course these are canonical choices for the structure group, but, as we mentioned above, they do not represent a unique choice.

As a concrete example, let us consider again the tangent bundle $T(M)$. As we already did notice, it is a vector bundle having $\mathbb{R}^m$ as standard fiber ($m = \dim(M)$); it can be seen as a bundle with the structure group $\mathrm{GL}(m)$, acting on $\mathbb{R}^m$ by the standard representation. In fact, whenever we consider m pointwise independent local vector fields $e_a^{(\alpha)}$ ($a = 1, \dots, m$) over $U_\alpha \subseteq M$, they form a pointwise basis for vectors tangent to $U_\alpha \subseteq M$, i.e. a *(local) frame* of M over an open subset $U_\alpha \subseteq M$. Any other tangent vector $v \in \tau_M^{-1}(U_\alpha) \subseteq TM$ at $x \in U_\alpha$ may be uniquely written as $v = e_a^{(\alpha)} v_{(\alpha)}^a$. A local trivialization can be thence defined by

$$t_{(\alpha)} : \tau_M^{-1}(U_\alpha) \longrightarrow U_\alpha \times \mathbb{R}^m : v \mapsto (x, v_{(\alpha)}^a) \tag{1.4.9}$$

The local trivialization $(U_\alpha, t_{(\alpha)})$ is said to be *associated to the (local) frame* $e_a^{(\alpha)}$.

Let us consider two such trivializations $t_{(\alpha)}$ and $t_{(\beta)}$ defined over U_α and U_β and associated to the local frames $e_a^{(\alpha)}$ and $e_a^{(\beta)}$, respectively. Let $U_{\alpha\beta} = U_\alpha \bigcap U_\beta \neq \varnothing$ and let $g_{(\alpha\beta)} : U_{\alpha\beta} \longrightarrow \mathrm{GL}(m)$ be defined by the relation

$$e^{(\beta)}_a = e^{(\alpha)}_b \, [g_{(\alpha\beta)}(x)]^b{}_a \tag{1.4.10}$$

These are the transition functions which act by the standard representation of $\mathrm{GL}(m)$ onto $\mathbb{R}^m$; in fact we have

$$v = e_a^{(\beta)} v_{(\beta)}^a = e_b^{(\alpha)} [g_{(\alpha\beta)}(x)]^b{}_a v_{(\beta)}^a \qquad \Rightarrow v_{(\alpha)}^b = [g_{(\alpha\beta)}(x)]^b{}_a v_{(\beta)}^a \tag{1.4.11}$$

Notice that the trivializations of $T(M)$ introduced in (1.1.18) are particular cases of this construction for the local frame $\partial_\mu^{(\alpha)}$ there defined.

Since according to (1.4.11) each $g_{(\alpha\beta)}$ is given by a *transition matrix* between the local frames $e_a^{(\beta)}$ and $e_a^{(\alpha)}$, the corresponding transition functions of $T(M)$ take their values into the linear group $\mathrm{GL}(m)$. In this way we have endowed the tangent bundle $T(M)$ with a canonical structure of (vector) bundle with structure group $\mathrm{GL}(m)$. The compatible trivializations are all those trivializations which are constructed out of a family of local frames, i.e. out of local sections of $L(M)$, by means of (1.4.10). Also in view of proposition (1.3.26), we have established a sort of correspondence between trivializations of $L(M)$, defined by local frames, and *compatible* trivializations of $T(M)$ seen as a bundle with structure group $\mathrm{GL}(m)$. Of course, the *structure bundle* of $T(M) = (TM, M, \tau_M; \mathbb{R}^m; \lambda, \mathrm{GL}(m))$ so constructed is nothing but the principal bundle $L(M) = (LM, M, \tau; \mathrm{GL}(m))$ itself.

Notice that, since frames are for the moment completely arbitrary, transition functions can take *any* value in $\mathrm{GL}(m)$ which seems, in that sense, a *minimal* structure group for the tangent bundle. However, let us stress once again that this is not true. Let us in fact consider a (strictly) Riemannian metric g over M (we recall that Riemannian metrics always exist, due to a partition of unity argument; see example 5.7 of [KN63] Vol.1, page 60). Let $\{U_\alpha\}_{\alpha\in I}$ be a covering of M and $e_a^{(\alpha)}$ a (smooth) family of local frames over U_α. Grahm-Schmidt procedure produces a well-known algorithm to associate a g-orthonormal frame $\hat{e}_a^{(\alpha)}$ to the frame $e_a^{(\alpha)}$. We have then produced a family of g-orthonormal frames $\hat{e}_a^{(\alpha)}$ over M and we can consider the induced trivializations

$$\hat{t}_{(\alpha)} : \tau_M^{-1}(U_\alpha) \longrightarrow U_\alpha \times \mathbb{R}^m : v \mapsto (x, \hat{v}^a_{(\alpha)}) \qquad v = \hat{e}_a^{(\alpha)} \hat{v}^a_{(\alpha)} \tag{1.4.12}$$

The corresponding transition functions are given by $\hat{g}_{(\alpha\beta)} : U_{\alpha\beta} \longrightarrow \mathrm{GL}(m)$

$$\hat{e}_a^{(\beta)} = \hat{e}_b^{(\alpha)} \left[\hat{g}_{(\alpha\beta)}(x)\right]^b{}_a \tag{1.4.13}$$

But now each $\hat{g}_{(\alpha\beta)}$ takes its values into the proper subgroup $\mathrm{O}(m) \subseteq \mathrm{GL}(m)$, being $\hat{e}_a^{(\alpha)}$ and $\hat{e}_a^{(\beta)}$ two g-orthonormal frames.

Thus if one restricts the set of allowed local trivializations from trivializations associated to general frames (1.4.9) to the family (1.4.12), the structure group of $T(M)$ may be *reduced* from $\mathrm{GL}(m)$ to the smaller group $\mathrm{O}(m)$ in the sense used in [KN63] (see Vol.1, page 53; see also example 5.7 at page 60).

In other words, the same geometric fiber bundle $(TM, M, \tau_M; \mathbb{R}^m)$ has been endowed with two different structures of bundle with structure group, namely $(TM, M, \tau_M; \mathbb{R}^m; \lambda, \mathrm{GL}(m))$ and $(TM, M, \tau_M; \mathbb{R}^m; \hat{\lambda}, \mathrm{O}(m))$, being λ and $\hat{\lambda}$ the standard actions of $\mathrm{GL}(m)$ and $\mathrm{O}(m)$ onto $\mathbb{R}^m$, respectively.

Notice that each different Riemannian metric g over M gives a different reduction to the orthogonal group and a different set of compatible local trivializations on the tangent bundle $T(M)$. More precisely, there exists a one-to-one correspondence between Riemannian metrics and reductions of the structure group to the orthogonal group. Thus, at least in this case, the structure group is a truly additional structure superimposed to the geometric fiber bundle. Similarly, if M is orientable[15] one can use an orientation in order to produce a further reduction of the structure group from $O(m)$ to the special orthogonal group $SO(m)$.

5. Canonical Constructions of Bundles

There are lots of ways of constructing bundles out of other bundles. In this Section we are going to analyze some of these constructions; many more do exist but we limit ourselves to define only those which shall be useful later for our purposes.

The Fibered Product Bundle

Let $\mathcal{B} = (B, M, \pi; F)$ and $\mathcal{B}' = (B', M, \pi'; F')$ be two bundles on the same base M. We can then define the manifold:

$$B \times_M B' = \{(b, b') \in B \times B' \mid \pi(b) = \pi'(b')\} \tag{1.5.1}$$

The projection $\pi_\times : B \times_M B' \longrightarrow M : (b, b') \mapsto \pi(b) = \pi'(b')$ is well-defined. One can easily check that $\mathcal{B} \times_M \mathcal{B}' = (B \times_M B', M, \pi_\times; F \times F')$ is canonically a bundle called the *(fibered) product bundle*. Given in fact two local trivializations $t_{(\alpha)}$ and $t'_{(\alpha)}$ of $\mathcal{B}$ and $\mathcal{B}'$ over the same open subset $U_\alpha \subset M$, respectively, we can define a local trivialization of the product bundle as follows:

$$\begin{gathered} t_{(\alpha)} \times_M t'_{(\alpha)} : \pi_\times^{-1}(U_\alpha) \longrightarrow U_\alpha \times (F \times F') : (b, b') \mapsto (x; y, y') \\ x = \pi(b) = \pi'(b'),\ b = [x, y]_{(\alpha)},\ b' = [x, y']_{(\alpha)} \end{gathered} \tag{1.5.2}$$

Further checks are left to the reader.

The Pull-back Bundle

Let $\mathcal{B} = (B, N, \pi; F)$ be a bundle and $\phi : M \longrightarrow N$ be a local diffeomorphism. Let us define the manifold:

$$\phi^* B = \{(x, b) \in M \times B \mid \phi(x) = \pi(b)\} \tag{1.5.3}$$

The projection $\pi^* : \phi^* B \longrightarrow M : (x, b) \mapsto x$ is well-defined. By this definition $\phi^* \mathcal{B} = (\phi^* B, M, \pi^*; F)$ is a bundle; in fact, if we choose a trivialization of $\mathcal{B}$ on

[15] A manifold M of dimension m is *orientable* if and only if it admits a global volume form, i.e. an everywhere non-zero m-form.

the covering $\{V_\alpha\}_{\alpha\in I}$ of N, we can define the covering $\{U_\alpha = \phi^{-1}(V_\alpha)\}_{\alpha\in I}$ of M, and then the trivialization:

$$t^*_{(\alpha)} : (\pi^*)^{-1}(U_\alpha) \longrightarrow U_\alpha \times F : (x,b) \mapsto (x,y) \qquad b = [\phi(x), y]_{(\alpha)} \tag{1.5.4}$$

The pull-back bundle is thus obtained together with the fiber bundle morphism $\boldsymbol{\Phi} = (\Phi, \phi) : \phi^*\mathcal{B} \longrightarrow \mathcal{B}$, defined by:

$$\Phi : \phi^* B \longrightarrow B : (x,b) \mapsto b \tag{1.5.5}$$

We have the following result (see [S51]):

Theorem (1.5.6): let $\mathcal{B} = (B, M, \pi; F)$ be a fiber bundle and ϕ, $\psi : N \longrightarrow M$ be two homotopic maps[16]. Then the bundles $\phi^*\mathcal{B} \simeq \psi^*\mathcal{B}$ are isomorphic. ∎

We remark that the converse is not true. If we fix an arbitrary bundle $\mathcal{B} = (B, M, \pi; F)$ and consider two maps $\phi, \psi : N \longrightarrow M$, then the pull-back bundles $\phi^*\mathcal{B}$ and $\psi^*\mathcal{B}$ may be isomorphic even if the maps are not homotopic. For example if $\mathcal{B}$ is trivial, the pull-back bundle $\phi^*\mathcal{B}$ along any map $\phi : N \longrightarrow M$ is a trivial bundle.

However, one can define a particular bundle $\mathcal{U} = (U, M, \pi; F)$, called the *universal bundle*, which classifies homotopic maps, i.e. for which the converse of theorem (1.5.6) holds. One obtains that any bundle $\mathcal{C} = (C, N, \pi; F)$ can be obtained as a pull-back of the universal bundle $\mathcal{U}$ along some map $\phi : N \longrightarrow M$, i.e. $\mathcal{B} \simeq \phi^*\mathcal{U}$. Furthermore, the pull-backs of the universal bundle along two maps $\phi, \psi : N \longrightarrow M$ are two isomorphic bundles $\phi^*\mathcal{U}$, $\psi^*\mathcal{U}$ if and only if the maps $\phi, \psi : N \longrightarrow M$ are homotopic. This is also the starting point for the theory of characteristic classes. For greater details see [MM92], [MS63], [S51].

As a consequence, as long as the base manifold M is contractible, i.e. the identity map id_M is homotopic to a constant map $c_{x_0} : M \longrightarrow M : x \mapsto x_0$ for some $x_0 \in M$, the following result holds:

Corollary (1.5.7): if M is contractible, then any bundle over M is trivial.

Proof: since M is contractible the two maps $\mathrm{id}_M : M \longrightarrow M$ and $c_{x_0} : M \longrightarrow M : x \mapsto x_0$ are homotopic. The bundle $c^*_{x_0}\mathcal{B}$ is trivial and $\mathrm{id}^*_M\mathcal{B} \simeq \mathcal{B}$; $\mathcal{B}$ is then trivial thanks to the theorem (1.5.6). ∎

16 Two maps $\phi, \psi : N \longrightarrow M$ are said to be *homotopic* if there exists a continuous map $F : [0,1] \times N \longrightarrow M$ such that $\phi = F_0$ and $\psi = F_1$, where $F_0 : N \longrightarrow M : x \mapsto F(0,x)$ and $F_1 : N \longrightarrow M : x \mapsto F(1,x)$. We denote by F_t the maps $F(t,\cdot) : N \longrightarrow M$ for each $t \in [0,1]$.

Tensor Product of Vector Bundles

Let $\mathcal{V} = (E, M, \pi; V)$ and $\mathcal{V}' = (E', M, \pi'; V')$ be two vector bundles on the same base manifold M.

Definition (1.5.8): the *tensor product* of the two vector bundles $\mathcal{V}$ and $\mathcal{V}'$ is a vector bundle $\mathcal{V} \otimes_M \mathcal{V}'$ together with a vector bundle morphism $\otimes : \mathcal{V} \times_M \mathcal{V}' \longrightarrow \mathcal{V} \otimes_M \mathcal{V}'$ which satisfies the following universal property:
for any vector bundle $\mathcal{W}$ and for any vector bundle morphism $\Phi : \mathcal{V} \times_M \mathcal{V}' \longrightarrow \mathcal{W}$, there exists a unique vector bundle morphism $\bar{\Phi} : \mathcal{V} \otimes_M \mathcal{V}' \longrightarrow \mathcal{W}$ such that the following diagram is commutative:

$$\begin{array}{ccc} \mathcal{V} \times_M \mathcal{V}' & \xrightarrow{\ \otimes\ } & \mathcal{V} \otimes_M \mathcal{V}' \\ {\scriptstyle \Phi}\big\downarrow & \swarrow {\scriptstyle \bar{\Phi}} & \\ \mathcal{W} & & \end{array} \qquad (1.5.9)$$

One can easily prove that if such pair $(\mathcal{V} \otimes_M \mathcal{V}', \otimes)$ exists, it is unique. ∎

This property is completely analogous to the property which characterizes the tensor product of two vector spaces (see [AMD69], page 24); from a practical viewpoint $\mathcal{V} \otimes_M \mathcal{V}'$ is a vector bundle on M having as standard fiber the tensor product $V \otimes V'$. Local trivializations of $\mathcal{V} \otimes_M \mathcal{V}'$ are the *tensor products* of local trivializations on $\mathcal{V}$ and $\mathcal{V}'$, i.e. we have:

$$\begin{aligned} t_{(\alpha)} \otimes t'_{(\alpha)} :& \pi^{-1}(U_\alpha) \otimes \pi'^{-1}(U_\alpha) \longrightarrow U_\alpha \times (V \otimes V') \\ :& v \otimes v' \mapsto (x, t_{(\alpha)}(v) \otimes t'_{(\alpha)}(v')) \end{aligned} \qquad (1.5.10)$$

where $x = \pi(v) = \pi'(v')$. Equivalently, the tensor product bundle is a bundle which has as fibers the fiberwise tensor products of the fibers of $\mathcal{V}$ and $\mathcal{V}'$ and as transition functions the tensor products of transition functions of $\mathcal{V}$ and $\mathcal{V}'$. Whenever there is no ambiguity about the base manifold, the tensor product $\mathcal{V} \otimes_M \mathcal{V}'$ will be simply denoted by $\mathcal{V} \otimes \mathcal{V}'$.

Sometimes it is necessary to consider the tensor product of two vector bundles $\mathcal{V} = (E, M, \pi; V)$ and $\mathcal{V}' = (E', M', \pi'; V')$ on two different base manifolds M and M' along a map $\phi : M \longrightarrow M'$. This is by definition the bundle $\mathcal{V} \otimes_\phi \mathcal{V}' = \mathcal{V} \otimes \phi^*(\mathcal{V}')$ over M. When there is no possible confusion about the map to be used, by a strong abuse of language $\mathcal{V} \otimes_\phi \mathcal{V}'$ may be simply denoted by $\mathcal{V} \otimes_M \mathcal{V}'$ (or even by $\mathcal{V} \otimes \mathcal{V}'$).

All the other elementary linear constructions existing in the category of vector spaces easily extend to the category of vector bundles, by the principle of extending the linear construction to the relevant cocycles of transition functions. In this way, one easily defines the *dual bundle* $\mathcal{V}^*$ of a vector bundle $\mathcal{V}$

by using the transpose of the transition functions of $\mathcal{V}$. Analogously, one can define the *direct sum* $\mathcal{V} \oplus_M \mathcal{V}'$ of two vector bundles, as well as the *kernel bundle* $\mathrm{Ker}(\boldsymbol{\Phi})$ (*image bundle* $\mathrm{Im}(\boldsymbol{\Phi})$, resp.) of a vector bundle morphism $\boldsymbol{\Phi} : \mathcal{V} \longrightarrow \mathcal{V}'$.

For example, let us consider the dual bundle $T^*(M) = (T^*M, M, \tau^*_M; (\mathbb{R}^m)^*)$ to the tangent bundle $T(M) = (TM, M, \tau_M; \mathbb{R}^m)$. An element ω in a fiber $T^*_x M$ is then a linear form over the vector space $T_x M$, i.e. a linear map $\omega \in \mathrm{Hom}(T_x M, \mathbb{R})$. The bundle $T^*(M)$ is called the *cotangent bundle over* M and its elements are called *covectors over* M.

If ∂_μ is the natural basis of $T_x M$ associated to some chart, the dual basis will be denoted by $\mathrm{d}x^\mu$, so that one has $\mathrm{d}x^\mu(\partial_\nu) = \delta^\mu_\nu$ and each $\omega \in T^*_x M$ is of the form $\omega = \omega_\mu \, \mathrm{d}x^\mu$. We can then define the duality between the tangent and the cotangent bundle by setting

$$< \omega \mid v >= \omega(v) = \omega_\mu v^\mu \in \mathbb{R} \qquad \forall v = v^\mu \partial_\mu \in T_x M \tag{1.5.11}$$

If we choose a new chart x'^μ in M and denote by $J^\mu_\nu = \partial_\nu x'^\mu$ the Jacobian of the transition function $x'^\mu = x'^\mu(x)$, we have

$$\begin{array}{rclcrcl} \partial_\mu &=& J^\nu_\mu \, \partial'_\nu & \Leftrightarrow & \partial'_\nu &=& \bar{J}^\mu_\nu \, \partial_\mu \\ \mathrm{d}x^\mu &=& \bar{J}^\mu_\nu \, \mathrm{d}x'^\nu & \Leftrightarrow & \mathrm{d}x'^\nu &=& J^\nu_\mu \, \mathrm{d}x^\mu \end{array} \tag{1.5.12}$$

where $\|\bar{J}^\mu_\nu\|$ is the inverse matrix of $\|J^\nu_\mu\|$, i.e. $\bar{J}^\mu_\nu = \frac{\partial x^\mu}{\partial x'^\nu}$.

Consequently, transition functions of the cotangent bundle $T^*(M)$ are:

$$\omega_\mu = \omega'_\nu \, J^\nu_\mu \tag{1.5.13}$$

Notice that they act *dually* with respect to the transition functions of the tangent bundle (1.4.10).

If $\phi : M \longrightarrow N$ is a smooth map (not necessarily locally invertible) we can define linear maps $T^*_{\phi(x)}\phi : T^*_{\phi(x)}N \longrightarrow T^*_x M : \omega' \mapsto \omega$ by setting:

$$\omega(v) = \omega'(T_x\phi(v)) \qquad \forall v \in T_x M \tag{1.5.14}$$

where $T_x\phi$ denotes the tangent map (at $x \in M$) defined by (1.1.21). Then there is a unique map $T^*\phi : T^*N \longrightarrow T^*M$, called the *cotangent map*, which, when restricted to a fiber, coincides with the map (1.5.14) defined above. If ϕ is a diffeomorphism then $T^*(\phi) = (T^*\phi, \phi^{-1})$ is a linear bundle morphism.

If we consider the composition of two diffeomorphisms (i.e. $M \xrightarrow{\phi} N \xrightarrow{\psi} Q$) then we have $T^*(Q) \xrightarrow{T^*(\psi)} T^*(N) \xrightarrow{T^*(\phi)} T^*(M)$ and $T^*(\mathrm{id}_M) = \mathrm{id}_{T^*(M)}$. Consequently, if we work with all manifolds but we restrict morphisms to be diffeomorphisms (when they exist), $T^*(\cdot)$ is a *contravariant functor* from this smaller category of manifolds into the category of vector bundles, since it reverts compositions. This functor is called the *cotangent functor*.

Locally, let x^μ and x'^λ be coordinates on M and N respectively, while (x^μ, ω_μ) and (x'^μ, ω'_μ) are coordinates on T^*M and T^*N respectively. If the map ϕ has local expression $x'^\lambda = \phi^\lambda(x)$ then the cotangent morphism $T^*(\phi)$ has the following expression:

$$\begin{cases} x^\mu = (\phi^{-1})^\mu(x') \\ \omega_\mu = \left(\partial_\mu \phi^\lambda\right) \, \omega'_\lambda \end{cases} \tag{1.5.15}$$

Let us now consider a map $\omega : M \longrightarrow T^*M$ which (smoothly) associates to each point $x \in M$ a covector $\omega_x = \omega(x) \in T^*_x M$. We can define the canonical action of the 1-form ω on a vector field $X \in \mathfrak{X}(M)$ by setting $\omega(X) = \omega_x(X_x) \equiv < \omega_x \mid X_x >$. Then ω is called a *(differential) 1-form* over M and it can be identified with a global section of $T^*(M)$. The set of all (global) 1-forms will be denoted by $\Omega^1(M)$.

We can easily construct tensor bundles $T^p_q(M)$ as the tensor products of p copies of the tangent bundle $T(M)$ and q copies of the cotangent bundle $T^*(M)$, i.e.:

$$T^p_q(M) = \overbrace{T(M) \otimes \ldots \otimes T(M)}^{p \text{ times}} \otimes \overbrace{T^*(M) \otimes \ldots \otimes T^*(M)}^{q \text{ times}} \tag{1.5.16}$$

This definition trivially extends to the cases $p = 0$ or $q = 0$.

A section of $T^p_q(M)$ is a *tensor field* of rank (p, q). Any other operation of tensor algebra easily extends to these vector bundles; for example one can consider the exterior bundle $A(M) = \oplus_{k \leq m} A_k(M)$, where $A_k(M)$ is the bundle of skew-symmetric covariant tensors. Global sections of $A_k(M)$ are called *(differential) k-forms* over M. The set of all k-forms is denoted by $\Omega^k(M)$. In particular, we have that $A_0(M) = M \times \mathbb{R}$ is the trivial bundle and $A_1(M) = T^*(M)$ coincides with the cotangent bundle. Accordingly, the algebra of functions $\mathcal{F}(M)$ is identified with $\Omega^0(M)$.

The set $\Omega(M) = \oplus_{k \leq m} \Omega^k(M)$ of non-homogeneous (differential) forms over M, i.e. sections of the *exterior bundle* $A(M)$, is endowed with a ($\mathbb{Z}$-graded) Lie algebra structure induced by pointwise exterior product.

Bundles Associated to a Principal Bundle

Let $\mathcal{P} = (P, M, \pi; G)$ be a principal bundle, F be any manifold and let $\lambda : G \times F \longrightarrow F$ be a left action of G on F. We define a total space $P \times_\lambda F$ as the quotient of $P \times F$ with respect to the equivalence relation:

$$(p, y) \sim (p', y') \Leftrightarrow \exists g \in G \mid R_g p = p' \text{ and } y = [\lambda(g)](y') \tag{1.5.17}$$

Equivalence classes will be denoted by $[p, y]_\lambda$. We define a (smooth) projection $\pi_\lambda : P \times_\lambda F \longrightarrow M$ by $\pi_\lambda : [p, y]_\lambda \mapsto \pi(p)$; then $\mathcal{P} \times_\lambda F = (P \times_\lambda F, M, \pi_\lambda; F)$ is a fiber bundle, called the *bundle associated to $\mathcal{P}$ through λ* (shortly an *associated bundle*).

If we fix a (local) trivialization $\sigma^{(\alpha)}$ of $\mathcal{P}$ we can define a corresponding (local) trivialization of $\mathcal{P}\times_\lambda F$ as follows:

$$t_\lambda^{(\alpha)}:\pi_\lambda^{-1}(U_\alpha)\longrightarrow U_\alpha\times F:[p,y]_\lambda\mapsto\Big(x,[\lambda(g)](y)\Big)\qquad p=\sigma^{(\alpha)}(x)\cdot g \quad (1.5.18)$$

The correspondence between $\sigma^{(\alpha)}$ and $t_\lambda^{(\alpha)}$ is one-to-one thanks to equation (1.5.17).

Equivalently, by using proposition (1.1.7), one can define the associated bundle $\mathcal{P}\times_\lambda F$ as the unique (up to isomorphisms) bundle having F as standard fiber and as transition functions the representations through λ of the transition functions of $\mathcal{P}$ with values in G, as given by (1.4.2).

Let now $\mathcal{B}=(B,M,\pi;F;\lambda,G)$ be a fiber bundle with structure group together with its structure bundle $\mathcal{P}=(P,M,\pi;G)$ (see definition (1.4.4)). If we consider $\mathcal{P}$, choose F as standard fiber and construct the associated bundle with respect to the action $\lambda:G\times F\longrightarrow F$, we obtain again the bundle $\mathcal{B}$ (modulo isomorphisms). Thus any bundle with structure group can be thought as a bundle associated to its own structure bundle.

In the sequel we shall often produce a fiber bundle $\mathcal{B}=\mathcal{P}\times_\lambda F$ as an associated bundle to some principal bundle $\mathcal{P}=(P,M,\pi;G)$ with respect to an action λ of G over F. It is understood in this case that $\mathcal{B}=(P\times_\lambda F,M,\pi;F)$ is a bundle with structure group, namely $\mathcal{B}=(P\times_\lambda F,M,\pi;F;\lambda,G)$.

Structure Bundles for the Tangent Bundle

Let us recall the definition of the frame bundle $L(M)$ as given in Section 1.3. Then, according to definition (1.4.4) together with the form (1.4.10) of the transition functions of the tangent bundle $T(M)$, we see that the frame bundle is the structure bundle of the tangent bundle when the latter is thought as a bundle with structure group GL(m).

It is also easy to check that the tangent bundle $T(M)$ is canonically isomorphic to the bundle $L(M)\times_\lambda\mathbb{R}^m$, where λ denotes the standard representation of GL(m) onto $\mathbb{R}^m$. In fact, a point in $LM\times_\lambda\mathbb{R}^m$ is in the form $[e_b^{(\alpha)}\alpha^b{}_a,v^a]_\lambda=[e_b^{(\alpha)},\alpha^b{}_a\cdot v^a]_\lambda$ and we can define the local isomorphisms:

$$\psi_{(\alpha)}:LM\times_\lambda\mathbb{R}^m\longrightarrow TM:[e_a^{(\alpha)},v^a_{(\alpha)}]_\lambda\mapsto e_a^{(\alpha)}\,v^a_{(\alpha)} \qquad (1.5.19)$$

These local morphisms glue together on the overlaps $U_{\alpha\beta}$ hence inducing a (unique) global isomorphism $\psi:LM\times_\lambda\mathbb{R}^m\longrightarrow TM$. The local morphisms $\psi_{(\alpha)}$ satisfy the compatibility conditions (1.2.4). In fact, if we choose another local trivialization associated to another local frame $e_b^{(\beta)}$, we have:

$$e_a^{(\alpha)}=e_b^{(\beta)}\cdot J^b_{\ a} \qquad (1.5.20)$$

We then have the compatibility condition:

$$
\begin{array}{ccc}
[e_a^{(\alpha)}, v^a_{(\alpha)}] & \xrightarrow{\psi_{(\alpha)}} & e_a^{(\alpha)}\, v^a_{(\alpha)} \\
\Big\| & & \Big\| \\
[e_b^{(\beta)}, v^b_{(\beta)}] & \xrightarrow{\psi_{(\beta)}} & e_b^{(\beta)}\, v^b_{(\beta)}
\end{array}
\qquad
\begin{cases} e_b^{(\alpha)} = J^a_b e_a^{(\beta)} \\ v^b_{(\beta)} = J^b{}_a v^a_{(\alpha)} \end{cases}
\tag{1.5.21}
$$

The isomorphism $T(M) \simeq L(M) \times_\lambda \mathbb{R}^m$ defined in this way does not depend on any further structure added to the bundles so that it is a canonical isomorphism.

Let us now fix a Riemannian metric g on M, so that we can uniquely select a sub-bundle $\mathrm{O}(M, g)$ of the frame bundle $L(M)$, i.e. the bundle of all g-orthonormal frames, called the *orthonormal frame bundle of the metric g*. We can consider the canonical right action of the group $\mathrm{O}(m)$ over $\mathrm{O}(M, g)$ given by

$$
\hat{e}'_a = \hat{e}_b\, \alpha^b{}_a \qquad\qquad \|\alpha^b{}_a\| \in \mathrm{O}(m) \tag{1.5.22}
$$

i.e. the restriction of the canonical action (1.3.46).

It is a vertical, free and fiber transitive action of $\mathrm{O}(m)$. Thus $\mathrm{O}(M, g)$ is a principal bundle. Let us finally consider a trivialization of $\mathrm{O}(M, g)$ induced by g-orthonormal frames $\hat{e}_a^{(\alpha)}$. Transition functions $\hat{g}_{(\alpha\beta)} : U_{\alpha\beta} \longrightarrow \mathrm{O}(m)$ are then given through transition matrices between local g-orthonormal frames, i.e.

$$
\hat{e}_a^{(\beta)} = \hat{e}_b^{(\alpha)}\, [\hat{g}_{(\alpha\beta)}(x)]^b{}_a \qquad\qquad \hat{g}_{(\alpha\beta)}(x) \in \mathrm{O}(m) \tag{1.5.23}
$$

According to definition (1.4.4) together with the form (1.4.12) of transition functions of the tangent bundle, we see that the g-orthonormal frame bundle $\mathrm{O}(M, g)$ is the structure bundle of the tangent bundle when the last one is thought as a bundle with structure group $\mathrm{O}(m)$.

This simple example shows that a geometric bundle may be considered as a bundle with different structure groups (just by restricting the suitable preferred classes of compatible trivializations); accordingly, the same geometric bundle may support more than one structure of bundle with structure group (with different structure groups). In the sequel, bundles with structure group will be particularly important because gauge transformations on them will be regarded as fundamental symmetries of Nature. Under this viewpoint, it will be particularly important to distinguish among possible different structure groups on the same geometric bundle because this will end up in different groups of symmetries for fields having, *a priori*, the same set of configurations. To distinguish among different structure groups of a given bundle it will be enough to think of the bundle $\mathcal{B}$ as the associated bundle to some suitable principal bundle $\mathcal{P} = (P, M, \pi; G)$, meaning that G is the structure group selected on

$\mathcal{B}$. For example, one can consider the tangent bundle $T(M)$ both as a bundle associated to the frame bundle $L(M)$ as well as a bundle associated to the smaller g-orthonormal frame bundle $\mathrm{O}(M,g)$ of some fixed metric g, meaning the structure group to be $\mathrm{GL}(m)$ or the smaller group $\mathrm{O}(m)$, respectively.

6. Vector Fields on a Bundle

Let us consider a fiber bundle $\mathcal{B} = (B, M, \pi; F)$. Of course, one can consider the tangent bundle $T(B) = (TB, B, \tau_B; \mathbb{R}^{m+k})$ to the total space B, where $\dim(M) = m$ and $\dim(F) = k$, so that $\dim(B) = m + k$. A (global) section Ξ of this bundle is a (global) vector field over B. Clearly, the flow of such a vector field does not preserve in general the bundle structure of B, which is here regarded as a bare manifold. In fact, being (x^μ, y^i) local coordinates on B, the generic vector field $\Xi = \xi^\mu(x,y)\partial_\mu + \xi^i(x,y)\partial_i$ over B is not compatible with the projection, meaning that $T_{[x,y]_{(\alpha)}}\pi : T_{[x,y]_{(\alpha)}}B \longrightarrow T_xM$ does not allow to define a vector field ξ over M, since for two different points $[x,y]_{(\alpha)}$ and $[x,y']_{(\alpha)}$ in the fiber over x one has in general $T_{[x,y]_{(\alpha)}}\pi(\Xi) \neq T_{[x,y']_{(\alpha)}}\pi(\Xi)$. If the equality holds for all $x \in M$ and $y \in F$ then we say that Ξ is *projectable*. In other words:

Definition (1.6.1): a *projectable vector field over* $\mathcal{B}$ is a vector field $\Xi \in \mathfrak{X}(B)$ over B such that there exists a vector field $\xi \in \mathfrak{X}(M)$ over M satisfying $\pi_*\Xi = \xi$. The vector field ξ is called the *projection* of Ξ. ∎

The set of all projectable vector fields of $\mathcal{B}$ is denoted by $\mathfrak{X}_{proj}(\mathcal{B})$ (or sometimes $\mathfrak{X}_{proj}(\pi)$). It is a Lie sub-algebra of the Lie algebra $\mathfrak{X}(B)$. We remark that, if Ξ is a projectable vector field on $\mathcal{B} = (B, M, \pi; F)$ and $U \subseteq M$ is a sub-manifold, by an abuse of language we say that the vector field Ξ *vanishes over* $U \subseteq M$ if it vanishes in $\pi^{-1}(U) \subseteq B$. In this case we shall write $\Xi\big|_U = 0$.

A vector field Ξ is projectable if and only if in each local trivialization it has the form

$$\Xi = \xi^\mu(x)\partial_\mu + \xi^i(x,y)\partial_i \tag{1.6.2}$$

so that it projects over $\xi = \xi^\mu(x)\partial_\mu$, which is a well-defined vector field over M, as the components ξ^μ transform into $\xi'^\mu = J^\mu_\nu \xi^\nu$ under any change of local trivialization $\Phi : (x^\mu, y^i) \mapsto (x'^\mu(x), y'^i(x,y))$. The flow of a projectable vector field Ξ is in the form

$$\begin{cases} x' = \phi_s(x) \\ y' = Y_s(x,y) \end{cases} \qquad \begin{aligned} \xi^\mu(x) &= \frac{\mathrm{d}}{\mathrm{d}s}\phi^\mu_s(x)\Big|_{s=0} \\ \xi^i(x,y) &= \frac{\mathrm{d}}{\mathrm{d}s}Y^i_s(x,y)\Big|_{s=0} \end{aligned} \tag{1.6.3}$$

with $s \in \mathbb{R}$. Thence the flow of a projectable vector field is a flow of (local) fibered automorphisms of $\mathcal{B}$.

Definition (1.6.4): the *vertical bundle* $V(\mathcal{B}) = (VB, B, \tau_V; \mathbb{R}^k)$ *over* $\mathcal{B}$ is the vector sub-bundle of the tangent bundle $T(B)$ of all tangent vectors $v \in TB$ such that $T\pi(v) = 0$, i.e. it is the bundle Ker$(T\pi)$. Here $\tau_V : VB \longrightarrow B$ is the induced projection which associates to each vertical vector its application point. The elements of VB are called *vertical vectors*; a section of $V(\mathcal{B})$ over B is called a *vertical vector field over* $\mathcal{B}$. The set of all vertical vector fields is denoted by $\mathfrak{X}_{(V)}(\mathcal{B})$ (or sometimes by $\mathfrak{X}_{(V)}(\pi)$). It is a Lie sub-algebra of $\mathfrak{X}(\pi)$ (as well as a Lie sub-algebra of $\mathfrak{X}_{proj}(\pi)$ since any vertical vector field always projects onto the zero vector field of M). ∎

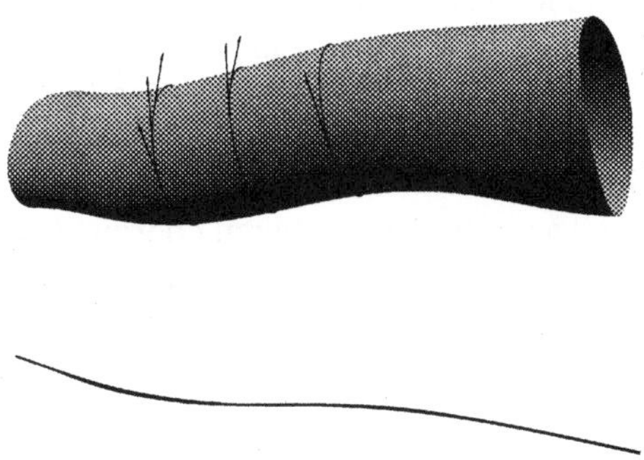

Fig. 9 – Vertical vectors.

As it easily follows from (1.6.2), once a trivialization of $\mathcal{B}$ is chosen a vertical vector is necessarily of the form $\Xi = \xi^i(x,y)\partial_i$, so that $(x^\mu, y^i; \xi^i)$ are natural fibered coordinates in $V(\mathcal{B})$ associated to the given trivialization. Vertical vectors may thence be regarded as vectors which are tangent to the fibers of $\mathcal{B}$. Since a vertical vector field Ξ projects over the zero vector field of M, then its flow is made of vertical automorphisms.

Of course, one can endow the total space VB with a bundle structure over M by composition with the bundle structure of $\mathcal{B}$, namely defining the projection $\pi_V = \pi \circ \tau_V : VB \longrightarrow M$. In this way, a new bundle over M is obtained and denoted by $V_M(\mathcal{B}) = (VB, M, \pi_V; F \times \mathbb{R}^k)$; this is not in general a vector bundle, unless $\mathcal{B}$ is. In fact, if $t_{(\alpha)} : \pi^{-1}(U_\alpha) \longrightarrow U_\alpha \times F$ is a local trivialization of $\mathcal{B}$ and $(x^\mu; y^i)$ are fibered coordinates around a point $b \in \pi^{-1}(U_\alpha)$, then one can define a local trivialization of $V(\mathcal{B})$ over the open subset $\hat{U}_\alpha = \pi^{-1}(U_\alpha)$ by

$$Vt_{(\alpha)} : \xi^i\partial_i \mapsto (b, \xi^i) \qquad \xi^i\partial_i \in \mathrm{Ker}_b(T\pi) \equiv V_b(\mathcal{B}) \subseteq T_bB \tag{1.6.5}$$

If $g_{(\beta\alpha)}$ are the transition functions of $\mathcal{B}$ over the trivialization atlas $\{U_\alpha\}_{\alpha\in I}$, the change of fibered local coordinates on $\mathcal{B}$ is of the form:

$$\begin{cases} x'^\mu = \phi^\mu(x) \\ y'^i = Y^i(x, y) \end{cases} \tag{1.6.6}$$

Let us set $J^i_j(x, y) = \partial_j Y^i(x, y)$. The transition functions of $V(\mathcal{B})$ on the atlas $\{\pi^{-1}(U_\alpha)\}_{\alpha\in I}$ are then in the following form

$$\begin{cases} x'^\mu = \phi^\mu(x) \\ y'^i = Y^i(x, y) \\ \xi'^i = J^i_j(x, y)\, \xi^j \end{cases} \tag{1.6.7}$$

The transition functions (1.6.7) of $V(\mathcal{B})$ are then linear transformations of the fiber coordinates ξ^i with coefficients depending on the point $b = (x^\mu, y^i)$; thence $V(\mathcal{B})$, when considered as a bundle over B, is a vector bundle. When considering the bundle $V_M(\mathcal{B})$ over M the equations (1.6.7) define also the transition functions of a trivialization over the atlas $\{U_\alpha\}_{\alpha\in I}$ of M. In this case they are interpreted as a change of fibered coordinates from $(x^\mu; y^i, \xi^i)$ to new coordinates $(x'^\mu; y'^i, \xi'^i)$, with $x'^\mu(x)$ given by transition functions on M. In general, these transition functions are not linear in (y^i, ξ^i) and still depend on (x^μ). However, if $\mathcal{B}$ is a vector bundle over M, then its transition functions $g_{(\beta\alpha)}$ are linear, i.e. the functions $Y^i(x, y)$ are linear in y so that $Y^i(x, y) = J^i_j(x) y^j$ and J^i_j does no longer depend on y. Thence, under these stronger hypotheses the transition functions (1.6.7) become linear transformations on (y^i, ξ^i) with coefficients depending on (x^μ) and the bundle $V_M(\mathcal{B})$ is a vector bundle also when regarded as a bundle over M.

Notice that sections of the two fiber bundles

$$VB \xrightarrow{\ \tau_V\ } B \qquad\qquad VB \xrightarrow{\ \pi_V\ } M \tag{1.6.8}$$

have a completely different meaning.

In fact, a section $\Xi : B \longrightarrow VB$ of $V(\mathcal{B})$ associates to any point $b \in B$ a vertical vector $\Xi \in V_b B$, i.e. it is a vertical vector field over B in the sense of definition (1.6.4). On the contrary, a section $\hat{\Xi} : M \longrightarrow VB$ over $V_M(\mathcal{B})$ has a local expression

$$\hat{\Xi} : x \mapsto (x, y^i(x), \xi^i(x)) \tag{1.6.9}$$

We can thence uniquely associate to $\hat{\Xi}$ a section of $\mathcal{B}$, given by $\sigma_{\hat{\Xi}} = \tau_V \circ \hat{\Xi} : x \mapsto (x; y^i(x))$; the vertical vector $\xi^i(x)\partial_i$ is applied to the point $\sigma_{\hat{\Xi}}(x)$ belonging to the fiber $\pi^{-1}(x)$. We stress that out of the image of the section $\sigma_{\hat{\Xi}}$ no vertical vector is defined. The section $\hat{\Xi}$ is then called a *vertical vector field defined on the section* $\sigma_{\hat{\Xi}}$.

This is an example of a fairly general situation which will be commonly encountered in our applications. It is the case of a *double fibration*. By this we mean a double bundle structure $C \xrightarrow{\pi} B \xrightarrow{\tau} M$. Any section $\Xi : M \longrightarrow C$ of the composed projection $\tau \circ \pi$ defines uniquely a section $\sigma_\Xi = \pi \circ \Xi$ of the projection τ. In this case one says as before that Ξ is a section of C over the section σ_Ξ of B. Obviously, these definitions and properties extend immediately to *multiple fibrations* $B_n \xrightarrow{\pi_n} \ldots \xrightarrow{\pi_2} B_1 \xrightarrow{\pi_1} B_0 \xrightarrow{\tau} M$. The easy details are left to the reader.

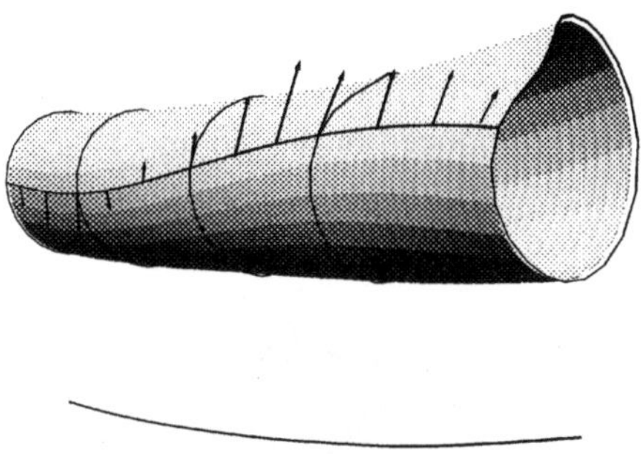

Fig. 10 – Vertical vectors on a section.

Finally let us consider the dual bundle $V^*(\mathcal{B}) = (V^*B, B, \tau_V^*; (\mathbb{R}^k)^*)$. The bundle $V^*(\mathcal{B})$ is a vector bundle which is fiberwise dual to $V(\mathcal{B})$. An element $\omega \in V_b^*(\mathcal{B})$ is then a linear form over the vector space $V_b(\mathcal{B})$. If $\Xi \in V_b(\mathcal{B})$ we can evaluate $< \omega \mid \Xi > \equiv \omega(\Xi) \in \mathbb{R}$ and the duality $< \ \mid \ >$ is a bilinear form. Thence if $\{\rho_i\}$ is a fiberwise basis of vertical vector fields, we can define the *dual* basis $\{\theta^i\}$ such that $< \theta^i \mid \rho_i > \equiv \theta^i(\rho_j) = \delta^i_j$. Then we obtain $\omega = \omega_i \, \theta^i$.

In a fibered chart $(x^\mu; y^i)$ of B, if $\{\partial_i\}$ is the natural local frame in $V(\mathcal{B})$, then we denote the dual basis by $\{\overline{\mathrm{d}}y^i\}$. Notice that $V^*(\mathcal{B})$ is not a sub-bundle of $T^*\mathcal{B}$. In fact, if we consider a new fibered chart $(x'^\mu(x); y'^i(x,y))$, we have

$$\partial'_i = J^j_i \, \partial_j \qquad\qquad \overline{\mathrm{d}}y^j = \overline{\mathrm{d}}y'^i \, J^j_i \tag{1.6.10}$$

where we set

$$J^j_i = \frac{\partial y'^j}{\partial y^i}(x, y) \tag{1.6.11}$$

Consequently, the components of an element $\omega = \omega_i \, \overline{\mathrm{d}}y^i$ transform according to the rule

$$\omega'_i = \omega_j \, J^j_i \tag{1.6.12}$$

On the contrary, for an element $\tilde{\omega} = \omega_\mu \, \mathrm{d}x^\mu + \omega_i \, \mathrm{d}y^i$ of T^*B we obtain, by specializing the transition functions (1.5.13) of the cotangent bundle:

$$\begin{cases} \omega'_\mu = \omega_\nu \, J^\nu_\mu + \omega_i J^i_\mu \\ \omega'_i = \omega_j \, J^j_i \end{cases} \tag{1.6.13}$$

where we set

$$\begin{aligned} J^{\nu}_{\ \mu} &= \frac{\partial x'^{\nu}}{\partial x^{\mu}}(x) \\ J^{i}_{\ \mu} &= \frac{\partial y'^{i}}{\partial x^{\mu}}(x, y) \end{aligned} \tag{1.6.14}$$

A direct comparison between (1.6.12) and (1.6.13) shows that elements in V^*B are not elements in T^*B constrained by the condition $\omega_\nu = 0$, since this constraint has only a local meaning. In fact if we set the constraint $\omega_\nu = 0$ in the first chart, the same constraint is violated in the second chart, i.e. $\omega'_{\ \mu} \neq 0$.

Notice that equation (1.6.12) and (1.6.13), resp., are to be understood as the transition functions of the bundle $V^*(\mathcal{B})$ and $T^*(\mathcal{B})$, resp., with respect to the canonical trivialization induced on $V^*(\mathcal{B})$ and $T^*(\mathcal{B})$, resp., by a trivialization of $\mathcal{B}$.

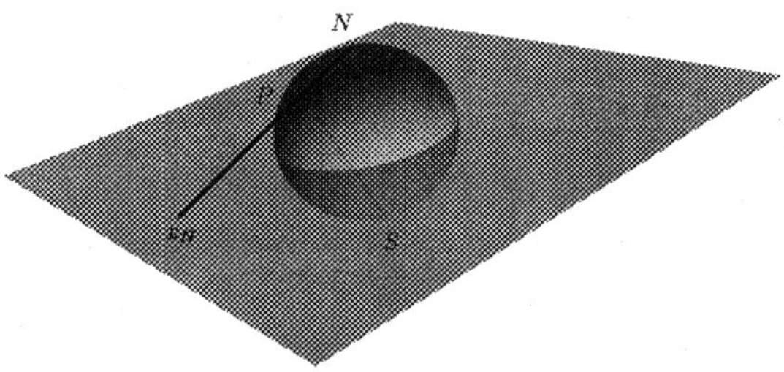

Fig. 11 – Stereographic coordinates on the 2-sphere.

Exercises

1- Show that the tangent bundle $T(M)$ is associated to $\mathrm{O}(M, g)$. In particular do this by explicitly producing a canonical isomorphism between $T(M)$ and an associated bundle to $\mathrm{O}(M, g)$.

Hint: a tangent vector can be written as a linear combination of an orthonormal basis, can't it?

2- Consider the 2-sphere S^2 with the stereographic charts shown in *Fig.11*. Consider a natural trivialization of the tangent bundle $T(S^2)$ as in (1.1.18) and compute explicitly its transition functions.

Hint: transition functions of natural trivializations are the Jacobian of coordinate changes $x'^{\mu} = x'^{\mu}(x^{\nu})$.

3- Consider the projective real space $\mathbb{PR}^m$ (the space of all lines in $\mathbb{R}^{m+1}$), i.e. the manifold obtained by quotienting $\mathbb{R}^{m+1} - \{\overline{0}\}$ with respect to the following equivalence relation

$$\overline{x} \sim \overline{y} \Leftrightarrow \exists \lambda \neq 0 \,|\, \overline{x} = \lambda \overline{y} \tag{1.6.15}$$

Consider the *tautologic line bundle* $\tau(L^m_{\mathbb{R}})$, i.e. the line bundle in which the fiber over a point $[\overline{x}] \in \mathbb{PR}^m$ is the whole line through $\overline{x}$ and $\overline{0}$ in $\mathbb{R}^{m+1}$. Define a trivialization and compute its transition functions. In particular, show that $\mathbb{PR}^1$ is a circle S^1 and that $\tau(L^1_{\mathbb{R}})$ is a Moebius strip.

Hint: write down proper coordinates on all spaces and write down the local expression on all maps w.r.t. those coordinates.

4- Consider an orientable manifold M and show that the group $\mathrm{SO}(m)$ is a structure group of the tangent bundle $T(M)$. Construct the corresponding structure bundle and show that $T(M)$ is isomorphic to a vector bundle associated to $\mathrm{SO}(M, g)$.

Hint: orientability provides a coherent choice of Jacobians with the same sign. The structure bundle is called the bundle of oriented g-orthonormal frames and it is denoted by $\mathrm{SO}(M, g)$.

Chapter 2

JET BUNDLES

Abstract *Jet bundles* are introduced and their structure is investigated in some detail. Some of the canonical structures defined on jet bundles, e.g. *contact forms, holonomic sections* and *prolongations of strong fibered morphisms, projectable vector fields and sections* are also introduced. The *infinite jet bundle* is introduced as the inverse limit of k-order jet bundles as k tends to infinity. *Lie derivatives* of sections are also defined and some of their properties are analyzed. As an application, jet bundles are used to define differential equations and the *s-frame bundle*.

References

The main references about jet bundles are [MM92], [S89], [S95], [V96]. Some further details can be found in [KMS93], [F91]. As long as differential operators are concerned we refer to [AKO93], [CBWM82], [W80].

1. The Jet Bundles

Let $\mathcal{B} = (B, M, \pi; F)$ be a bundle over M and $\Gamma_x(\mathcal{B})$ be the set of all local sections of $\mathcal{B}$ defined in a neighbourhood of a point $x \in M$. Let $k \geq 1$ be any positive integer and let us consider the equivalence relation $\sim^k_x$ defined on $\Gamma_x(\mathcal{B})$ by

$$\begin{aligned}\rho \sim^k_x \sigma \Leftrightarrow &\forall f : B \longrightarrow \mathbb{R} \text{ and } \forall \gamma : \mathbb{R} \longrightarrow M \text{ such that } \gamma(0) = x,\\ &\text{the functions } (f \circ \rho \circ \gamma) \text{ and } (f \circ \sigma \circ \gamma) \text{ have}\\ &\text{the same } k\text{-order Taylor expansion at } t = 0\end{aligned} \tag{2.1.1}$$

If we choose a trivialization $\{(U_\alpha, t_{(\alpha)})\}_{\alpha \in I}$ of $\mathcal{B}$ and fibered coordinates $(x^\mu; y^i)$ in each patch U_α, then a local section $\rho : U_\alpha \longrightarrow \pi^{-1}(U_\alpha)$ has the local expression $\rho : x \mapsto [x, \rho_{(\alpha)}(x)]_{(\alpha)}$, where $\rho_{(\alpha)} : U_\alpha \longrightarrow F$. The equivalence relation (2.1.1) reduces to require that the local expressions $\rho_{(\alpha)}$ and $\sigma_{(\alpha)}$ of the local sections have the same k-order Taylor expansion near $x \in M$. This equivalence relation is obviously independent of the choice of the local trivialization and the fibered coordinates chosen on B. Of course formula (2.1.1) is meaningful also for $k = 0$. In this case it reduces to require that σ and ρ pass through the same point $b \in B_x = \pi^{-1}(x)$ at $x \in M$.

Let us denote by $j^k_x\rho$ the equivalence class identified by the representative $\rho \in \Gamma_x(\mathcal{B})$. If $\phi : M \longrightarrow N$ is a map between manifolds, then we denote by $j^k_x\phi$ the equivalence class of the induced section $\hat{\phi} : M \longrightarrow M \times N$ of the trivial bundle $(M \times N, M, \mathrm{pr}_1; N)$. Each section $\hat{\psi}$ of the bundle $M \times N$ may be regarded as a map $\psi : M \longrightarrow N$ so that any element in $j^k_x\phi$ may be equivalently interpreted as manifold map.

Let us denote by J^k_xB the quotient space $\Gamma_x(\mathcal{B})/\sim^k_x$ of all equivalence classes $j^k_x\rho$ and by $J^kB = \coprod_{x\in M}(J^k_xB)$ the disjoint union of all J^k_xB. We define a map $\pi^k : J^kB \longrightarrow M$ by setting $\pi^k(j^k_x\rho) = x$ and a map $\pi^k_0 : J^kB \longrightarrow B$ by setting $\pi^k_0(j^k_x\rho) = \rho(x)$. Both these maps are well defined since they do not depend on the representative chosen for the class $j^k_x\rho$.

Using a trivialization $\{(U_\alpha, t_{(\alpha)})\}_{\alpha\in I}$ of $\mathcal{B}$, with fibered coordinates $(x^\mu; y^i)$, the coefficients of the Taylor expansion of ρ (i.e. the partial derivatives of $\rho(x)$ with respect to x^μ up to order k, evaluated at the point $x \in M$) define in J^kB a natural set of coordinates, denoted by $(x^\mu, y^i, y^i_{\mu_1}, \ldots, y^i_{\mu_1\ldots\mu_k})$ which are called *natural coordinates on* J^kB. Notice that the coordinates $y^i_{\mu_1\ldots\mu_h}$ are meant to be symmetric in the indices $(\mu_1 \ldots \mu_h)$ because they represent the values of partial derivatives of local sections.

As we shall immediately see below, changing local fibered coordinates in B by means of the transition functions $g_{(\alpha\beta)}$ of the chosen trivialization, produces (by Taylor expansion) a family of transition functions in J^kB which define uniquely a manifold structure (a bundle over B, in fact, with respect to the projection π^k_0). The dimension of this manifold is $\dim(J^kB) = m + n \cdot \sum_{h=0}^k D(h, m)$, with $m = \dim(M)$, $n = \dim(F)$ and $D(h, m)$ the number of independent partial derivatives of order h of one function of m variables. The numbers $D(h, m)$ satisfy the recurrence relation

$$D(h, m) = \sum_{l=0}^{m-1} D(h-1, m-l) \qquad\qquad D(0, m) = 1 \tag{2.1.2}$$

For example, $D(1, m) = m$ and $D(2, m) = \frac{m(m+1)}{2}$.

Let us also introduce for any given trivialization a whole family of (local) operators d_μ, called *formal derivatives*, which take (local) functions of J^kB to give (local) functions on $J^{k+1}B$ according to the rule:

$$\begin{aligned} &\mathrm{d}_\mu F(x^\mu, y^i, y^i_{\mu_1}, \ldots, y^i_{\mu_1\ldots\mu_k}) = \\ &\quad = \partial_\mu F + y^i_\mu\,(\partial_i F) + y^i_{\mu\mu_1}\,(\partial_i^{\mu_1} F) + \ldots + y^i_{\mu\mu_1\ldots\mu_k}\,(\partial_i^{\mu_1\ldots\mu_k} F) \end{aligned} \tag{2.1.3}$$

where ∂_μ denotes the partial derivative with respect to x^μ, ∂_i denotes the partial derivative with respect to y^i, $\partial_i^{\mu_1}$ the partial derivative with respect to $y^i_{\mu_1}$ and so on. For the moment, formal derivatives are considered as local operators (although, as we shall see below, they have a global character). They will be useful to write in a more compact form some of the following formulae.

If we consider the case $k = 1$, we can easily compute the transition functions of J^1B induced by the local trivialization chosen in $\mathcal{B}$. In the new coordinates in $\pi^{-1}(U_{\alpha\beta})$, the local expression for the same representative ρ is given by $\rho_{(\beta)}$ such that

$$\rho_{(\alpha)}(x) = [g_{(\alpha\beta)}(x)]\big(\rho_{(\beta)}(x)\big) \tag{2.1.4}$$

where $g_{(\alpha\beta)}$ are the transition functions of the local trivialization considered in $\mathcal{B}$. The transition functions on J^1B are then obtained just by taking the derivatives of the new local expressions of the representative, i.e.

$$\begin{cases} y'^i = [g_{(\alpha\beta)}(x)]^i(y) \equiv Y^i(x,y) \\ y'^i_\mu = \bar{J}^\nu_\mu\big(\partial_\nu Y^i(x,y) + y^j_\nu\, \partial_j Y^i(x,y)\big) = \bar{J}^\nu_\mu \mathrm{d}_\nu Y^i(x,y) \end{cases} \tag{2.1.5}$$

Notice that these transition functions are explicitly given in the affine form[1] $y'^i_\mu = A^i_\mu(x,y) + B^{i\ \nu}_{j\mu}(x,y)\, y^j_\nu$. Let us also write the explicit form of the non-trivial part of the transition functions of J^2B, i.e.:

$$\begin{aligned} y'^i_{\mu\nu} =& \bar{J}^\rho_\mu\big(\partial_{\rho\nu}Y^i(x,y) + y^j_\nu\, \partial_{\rho j}Y^i(x,y) + y^j_\rho\, \partial_{\nu j}Y^i(x,y) + \\ &+ y^j_\rho\, y^k_\nu\, \partial_{jk}Y^i(x,y) + y^j_{\rho\nu}\, \partial_j Y^i(x,y)\big) + \\ &+ \bar{J}^\rho_{\mu\nu}\big(\partial_\rho Y^i(x,y) + y^j_\rho\, \partial_j Y^i(x,y)\big) = \\ =& \bar{J}^\rho_\mu \mathrm{d}_\rho\, \big(\bar{J}^\sigma_\nu \mathrm{d}_\sigma Y^i(x,y)\big) \end{aligned} \tag{2.1.6}$$

Again the general structure is

$$y'^i_{\mu\nu} = A^i_{\mu\nu}(x^\lambda, y^j, y^j_\lambda) + B^{i\ \alpha\beta}_{j\mu\nu}(x,y)\, y^j_{\alpha\beta} \tag{2.1.7}$$

where the coefficients $A^i_{\mu\nu}$ are polynomial functions of the second degree in y^j_λ.

This general structure is preserved in the k-order case, being explicitly:

$$y'^i_{\mu_1\ldots\mu_k} = A^i_{\mu_1\ldots\mu_k}(x^\lambda, y^j, y^j_\lambda, \ldots, y^j_{\lambda_1\ldots\lambda_{k-1}}) + B^{i\ \nu_1\ldots\nu_k}_{j\mu_1\ldots\mu_k}(x,y)\, y^j_{\nu_1\ldots\nu_k} \tag{2.1.8}$$

where $A^i_{\mu_1\ldots\mu_k}$ are polynomial functions of kth degree in $y^i_{\lambda_1\ldots\lambda_{k-1}}$.

---

[1] The notation $(x^\mu; y^i) \mapsto (x'^\mu; y'^i)$ for coordinate changes is thoroughly used according to the note after (1.2.1). All derived notation is coherent with this choice.

In fact, in the general case we have recursively:

$$
\begin{cases}
x'^{\mu} = \phi^{\mu}(x) & J^{\mu}_{\nu} = \partial_{\nu}\phi^{\mu}(x) \\
y'^{i} = Y^{i}(x,y) & \bar{J}^{\nu}_{\mu} = \partial_{\mu}(\phi^{-1})^{\nu}(x) \\
y'^{i}_{\mu_1} = \bar{J}^{\nu_1}_{\mu_1}\mathrm{d}_{\nu_1}Y^{i}(x,y) & \\
y'^{i}_{\mu_1\mu_2} = \bar{J}^{\nu_2}_{\mu_2}\mathrm{d}_{\nu_2}\left(\bar{J}^{\nu_1}_{\mu_1}\mathrm{d}_{\nu_1}Y^{i}(x,y)\right) = & \\
\qquad = \bar{J}^{\nu_2}_{\mu_2}\mathrm{d}_{\nu_2}y'^{i}_{\mu_1} & \\
\dots & \\
y'^{i}_{\mu_1\mu_2\dots\mu_k} = \bar{J}^{\nu_k}_{\mu_k}\mathrm{d}_{\nu_k}y'^{i}_{\mu_1\mu_2\dots\mu_{k-1}} &
\end{cases}
\tag{2.1.9}
$$

As we said, these define a C^{∞}-atlas which provides the topological and differentiable structure to the set J^kB, making it a manifold. For notational convenience we set $J^0B = B$ (since they are canonically isomorphic) and assume, from now on, that k is any integer $k \geq 0$. Moreover, we can define projections $\pi^{k+1}_k : J^{k+1}B \longrightarrow J^kB$ by setting

$$
\pi^{k+1}_k : (x^{\mu}, y^i, y^i_{\mu_1}, \dots; y^i_{\mu_1\dots\mu_{k+1}}) \mapsto (x^{\mu}, y^i, y^i_{\mu_1}, \dots, y^i_{\mu_1\dots\mu_k}) \tag{2.1.10}
$$

Since the transition functions (2.1.8) are affine maps in the fiber variables $y^i_{\mu_1\dots\mu_k}$, these define an infinite family of affine bundles $J^{k+1}B \longrightarrow J^kB$ having π^{k+1}_k as projections. We have thence a sequence of bundles:

$$
M \xleftarrow{\ \pi\ } B \xleftarrow{\ \pi^1_0\ } J^1B \xleftarrow{\ \pi^2_1\ } J^2B \xleftarrow{\ \pi^3_2\ } \dots \tag{2.1.11}
$$

each one of which (except π) is affine. It is a multiple fibration in the sense of Section 1.6. We then define, by composition, the *"intermediate" projections* $\pi^{k+h}_k : J^{k+h}B \longrightarrow J^kB$ (for any $h \geq 2$) and we denote simply by $\pi^k : J^kB \longrightarrow M$ the composed projections $\pi^k = \pi \circ \pi^k_0$ onto the base manifold M. This notation is made coherent by setting $\pi^0 \equiv \pi$.

Let us finally denote by $J^k_0(\mathbb{R}^m \times F)$ the fiber at $0 \in \mathbb{R}^m$ of the k-order jet prolongation of the trivial bundle $(\mathbb{R}^m \times F, \mathbb{R}^m, \mathrm{pr}_1; F)$. We shall then denote by $J^k\mathcal{B}$ the bundle $(J^kB, M, \pi^k; J^k_0(\mathbb{R}^m \times F))$. The bundle structure $\pi^k_0 : J^kB \longrightarrow B$ will be instead denoted by $J^k_0\mathcal{B}$ (as well as, more generally, $J^{k+h}_k\mathcal{B}$ will denote the bundle structures associated to π^{k+h}_k). Coherently, we set $J^0\mathcal{B} \simeq \mathcal{B}$.

As for the case of the vertical bundle, the total space J^kB has many bundle structures over different bases, one for each projection π^k_h $(h < k)$ and one for π^k. While the projection π^{k+1}_k defines an affine bundle for each k, the other projections define bundles which are not affine.

2. Prolongation of Morphisms and Sections

Let $\mathcal{B}$ and $\mathcal{B}'$ be two bundles and $\boldsymbol{\Phi} = (\Phi, \phi) : \mathcal{B} \longrightarrow \mathcal{B}'$ be a strong bundle morphism. We define the *k-order prolongation of the strong morphism* $\boldsymbol{\Phi}$ (or *lift*) by the following

$$J^k\Phi : J^kB \longrightarrow J^kB' : j^k_x\rho \mapsto j^k_{\phi(x)}(\Phi \circ \rho \circ \phi^{-1}) \tag{2.2.1}$$

which defines a bundle morphism from $J^k\mathcal{B}$ into $J^k\mathcal{B}'$, denoted by $J^k\boldsymbol{\Phi} = (J^k\Phi, \phi)$.

The following properties hold true:

Proposition (2.2.2): let $\boldsymbol{\Phi} : \mathcal{B} \longrightarrow \mathcal{C}$ and $\boldsymbol{\Psi} : \mathcal{C} \longrightarrow \mathcal{D}$ be two strong bundle morphisms. We have:

$$J^k(\boldsymbol{\Psi} \circ \boldsymbol{\Phi}) = J^k(\boldsymbol{\Psi}) \circ J^k(\boldsymbol{\Phi}) \qquad\qquad J^k(\mathrm{id}_B) = \mathrm{id}_{J^kB} \tag{2.2.3}$$

so that $J^k(\cdot)$ is a *covariant functor*, called the *functor of k-order jet prolongation.* ■

If the local representation of the morphism $\boldsymbol{\Phi}$ is

$$\begin{cases} x'^\mu = \phi^\mu(x) \\ y'^i = Y^i(x, y) \end{cases} \qquad\qquad \begin{matrix} J^\mu_\nu = \partial_\nu \phi^\mu(x) \\ \bar{J}^\nu_\mu = \partial_\mu(\phi^{-1})^\nu(x) \end{matrix} \tag{2.2.4}$$

then the k-order prolongation of $\boldsymbol{\Phi}$ is given by

$$\begin{cases} y'^i_{\mu_1} = \bar{J}^{\nu_1}_{\mu_1} \mathrm{d}_{\nu_1} Y^i(x, y) \\ y'^i_{\mu_1\mu_2} = \bar{J}^{\nu_2}_{\mu_2} \mathrm{d}_{\nu_2} \Big(\bar{J}^{\nu_1}_{\mu_1} \mathrm{d}_{\nu_1} Y^i(x, y)\Big) = \bar{J}^{\nu_2}_{\mu_2} \mathrm{d}_{\nu_2} y'^i_{\mu_1} \\ \dots \\ y'^i_{\mu_1\mu_2\dots\mu_k} = \bar{J}^{\nu_k}_{\mu_k} \mathrm{d}_{\nu_k} y'^i_{\mu_1\mu_2\dots\mu_{k-1}} \end{cases} \tag{2.2.5}$$

Notice the complete analogy between the local expression of the k-order prolongation of $\boldsymbol{\Phi}$ and the transition functions of $J^k\mathcal{B}$ as given by (2.1.9). This is a general feature due to the fact that the transition functions $g_{(\alpha\beta)}$ of a bundle $\mathcal{B}$ may be regarded as (local) fibered morphisms between the trivial bundles $U_\alpha \times F$ and $U_\beta \times F$, i.e. between the local models of $\mathcal{B}$. Consequently, the transition functions in $J^k\mathcal{B}$ are one-to-one related to the k-order prolongation of these (local) fibered morphisms. As a consequence, any result about morphisms may be regarded as a result about transition functions and vice versa. The first viewpoint is often called the *active viewpoint* since morphisms actually move points in the bundle, while the second one is called the *passive viewpoint* since trivializations just change the point labels without moving anything. As in this case, the same formula can be interpreted in both a passive and an active way. In particular this holds for $k = 0$, giving rise to the active and passive view about bundle mappings.

We remark also that the notion of k-order prolongation allows us to provide an alternative definition of $J^k\mathcal{B}$. Consider in fact a trivialization $\{(U_\alpha, t_{(\alpha)})\}_{\alpha\in I}$ of $\mathcal{B}$, together with its transition functions $g_{(\alpha\beta)}$. For any $k \geq 0$, the prolongations $J^k g_{(\alpha\beta)}$ form a cocycle, due to (2.2.3). The bundle $J^k\mathcal{B}$ is thence uniquely defined (up to isomorphisms) by proposition (1.1.7).

Using the lift of strong morphisms, the prolongation of projectable vector fields (see definition (1.6.1)) can be also defined. Recall, in fact, that the flow of a projectable vector field Ξ over a bundle $\mathcal{B}$ is made of (strong) fibered morphisms so that it can be prolonged to $J^k\mathcal{B}$ by using (2.2.5). Thus we obtain a flow on J^kB. Denote by $j^k\Xi$ its infinitesimal generator, which is a vector field over J^kB. This vector field $j^k\Xi$ is projectable with respect to all projections π^k_h $(h < k)$ as well as with respect to π^k and it is called the *k-order prolongation of* Ξ. Locally, if $\Xi = \xi^\mu(x)\partial_\mu + \xi^i(x,y)\partial_i$, its k-order prolongation is of the form $j^k\Xi = \xi^\mu\partial_\mu + \xi^i\partial_i + \xi^i_{\mu_1}\partial_i^{\mu_1} + \ldots + \xi^i_{\mu_1\ldots\mu_k}\partial_i^{\mu_1\ldots\mu_k}$ where we set recursively:

$$\begin{cases} \xi^i_{\mu_1} = \mathrm{d}_{\mu_1}\xi^i - y^i_\nu\, \mathrm{d}_{\mu_1}\xi^\nu \\ \ldots \\ \xi^i_{\mu_1\ldots\mu_k} = \mathrm{d}_{\mu_k}\xi^i_{\mu_1\ldots\mu_{k-1}} - y^i_{\mu_1\ldots\mu_{k-1}\nu}\, \mathrm{d}_{\mu_k}\xi^\nu \end{cases} \tag{2.2.6}$$

This is also a general feature. Whenever a functor is defined on a category in which vector fields and their flows have a meaning, the same functor may be used to induce a flow also in the target category. In this case the infinitesimal generator of the target flow is thought as a vector field induced by the vector field we started from.

Let now ρ be a section of $\mathcal{B}$. We can associate to it a bundle morphism

$$\begin{array}{ccc} M & \xrightarrow{\ \rho\ } & B \\ \| & & \downarrow{\scriptstyle \pi} \\ M & = & M \end{array} \tag{2.2.7}$$

and apply to it the functor $J^k(\cdot)$ having then:

$$\begin{array}{ccc} M & \xrightarrow{\ \rho\ } & B \\ \| & & \downarrow{\scriptstyle \pi} \\ M & = & M \end{array} \quad \overset{J^k(\cdot)}{\rightsquigarrow} \quad \begin{array}{ccc} M & \xrightarrow{\ J^k\rho\ } & J^kB \\ \| & & \downarrow{\scriptstyle \pi^k} \\ M & = & M \end{array} \tag{2.2.8}$$

The map $J^k\rho$ may be interpreted as a section of the bundle $\pi^k : J^kB \longrightarrow M$ and it is called the *k-order jet prolongation of* ρ. Usually this section is denoted by $j^k\rho$ rather than by $J^k\rho$. If ρ is a section of $\mathcal{B}$ locally represented by $\rho_{(\alpha)} : U_\alpha \longrightarrow F : x^\mu \mapsto \rho^i_{(\alpha)}(x)$, then its jet prolongation is given by:

$$j^k\rho : x^\mu \mapsto \left(x^\mu, \rho^i_{(\alpha)}(x), \partial_{\mu_1}\rho^i_{(\alpha)}(x), \ldots, \partial_{\mu_1\ldots\mu_k}\rho^i_{(\alpha)}(x)\right) \tag{2.2.9}$$

as one can immediately check. We stress the difference between $j^k\rho$ (which is a section of $J^k\mathcal{B}$) and $j^k_x\rho$ (which is a point in J^kB).

Among all sections Σ of $J^k\mathcal{B}$, those sections which can be obtained as jet prolongations $j^k\sigma$ of some section σ of $\mathcal{B}$ are called *holonomic sections*. For example, if $\Sigma : M \longrightarrow J^1B : x^\mu \mapsto (x^\mu, \Sigma^i(x), \Sigma^i_\mu(x))$ is a section, it is holonomic if and only if $\Sigma^i_\mu(x) = \partial_\mu\Sigma^i(x)$. In this case $\Sigma = j^1\sigma_\Sigma$, where $\sigma_\Sigma : M \longrightarrow B$ is the section of the double fibration $J^1B \longrightarrow B \longrightarrow M$ uniquely induced by Σ. We leave to the reader the easy task of understanding the higher order situation.

We can now provide a global meaning for the local operators d_μ introduced by (2.1.3). Let $\theta \in \Omega^p(J^kB)$ be any p-form on J^kB and let $\sigma \in \Gamma(\mathcal{B})$ be any local section of π; the pull-back $(j^k\sigma)^*\theta$ is a p-form of $\Omega^p(M)$.

Definition (2.2.10): on a bundle $\mathcal{B} = (B, M, \pi; F)$, a p-form $\theta \in \Omega^p(B)$ is called *horizontal* with respect to π if one has

$$\theta(\Xi_1, \ldots, \Xi_p) = 0 \tag{2.2.11}$$

if at least one of the vector fields $(\Xi_1, \ldots, \Xi_p)$ is vertical. ∎

In local coordinates a horizontal p-form is of the form:

$$\theta = \frac{1}{p!}\theta_{\mu_1\ldots\mu_p}(x, y)\, \mathrm{d}x^{\mu_1} \wedge \ldots \wedge \mathrm{d}x^{\mu_p} \tag{2.2.12}$$

i.e. it has no components along fiber coordinates y^i. The set of horizontal p-forms on a bundle $\mathcal{B}$ is denoted by $\Omega^p_{hor}(\mathcal{B})$ and the set of all (non-homogeneous) horizontal forms is denoted by $\Omega_{hor}(\mathcal{B}) = \oplus_p \Omega^p_{hor}(\mathcal{B})$. One can immediately verify that $\Omega_{hor}(\mathcal{B})$ is a graded subalgebra of the graded algebra of forms $\Omega(\mathcal{B})$ over $\mathcal{B}$. In particular for each k we can consider horizontal forms on $J^k\mathcal{B}$, which are accordingly denoted by $\Omega_{hor}(J^k\mathcal{B})$.

Definition (2.2.13): the *formal differential* of a horizontal p-form $\theta \in \Omega^p_{hor}(J^k\mathcal{B})$ is the horizontal $(p+1)$-form $D\theta \in \Omega^{p+1}_{hor}(J^{k+1}\mathcal{B})$ uniquely defined by:

$$(j^{k+1}\sigma)^* D\theta = \mathrm{d}((j^k\sigma)^*\theta) \qquad \forall\sigma \in \Gamma(\mathcal{B}) \tag{2.2.14}$$

where $\mathrm{d} : \Omega(M) \longrightarrow \Omega(M)$ is the exterior differential in M. ∎

Expanding (2.2.14) for the formal differential of a 0-form $F \in \Omega^0_{hor}(J^k\mathcal{B})$ (i.e. a function $F \in \mathcal{F}(J^k\mathcal{B})$) in local coordinates one finds:

$$\begin{aligned}(j^{k+1}\sigma)^* DF =& (j^{k+1}\sigma)^* \left[\mathrm{d}_\mu F\, \mathrm{d}x^\mu\right] = \\ =& \left(\partial_\mu F + (\partial_i F)\, \partial_\mu\sigma^i + \ldots \left(\partial_i^{\mu_1\ldots\mu_k} F\right) \partial_{\mu\mu_1\ldots\mu_k}\sigma^i\right) \mathrm{d}x^\mu\end{aligned} \tag{2.2.15}$$

with $\mathrm{d}_\mu F$ given by (2.1.3). Notice that the formal differential of F is thence given by:

$$DF = (\mathrm{d}_\mu F)\, \mathrm{d}x^\mu = \Big(\partial_\mu F + (\partial_i F)\, y^i_\mu + \ldots (\partial_i^{\mu_1 \ldots \mu_k} F)\, y^i_{\mu\mu_1\ldots\mu_k}\Big)\, \mathrm{d}x^\mu \quad (2.2.16)$$

3. Contact Forms

Definition (2.3.1): a 1-form ω on J^kB is called a *contact* 1*-form* of $J^k\mathcal{B}$ if it vanishes on all holonomic sections, i.e.:

$$(j^k\rho)^*\omega = 0 \qquad \forall \rho \in \Gamma(\mathcal{B}) \quad (2.3.2)$$

The set of contact 1-forms over J^kB will be denoted by $\Omega^1_c(J^k\mathcal{B})$. ■

A basis for contact 1-forms of $J^k\mathcal{B}$ is

$$\begin{aligned} \omega^i &= \mathrm{d}y^i - y^i_\nu\, \mathrm{d}x^\nu \\ \omega^i_\mu &= \mathrm{d}y^i_\mu - y^i_{\mu\nu}\, \mathrm{d}x^\nu \\ &\ldots \\ \omega^i_{\mu_1\ldots\mu_{k-1}} &= \mathrm{d}y^i_{\mu_1\ldots\mu_{k-1}} - y^i_{\mu_1\ldots\mu_{k-1}\nu}\, \mathrm{d}x^\nu \end{aligned} \quad (2.3.3)$$

i.e. any contact 1-form $\theta \in \Omega^1_c(J^k\mathcal{B})$ can be expanded as a linear combination of the basic 1-forms (2.3.3), having then

$$\theta = \theta_i\, \omega^i + \theta^\mu_i\, \omega^i_\mu + \ldots + \theta_i^{\mu_1\ldots\mu_{k-1}}\, \omega^i_{\mu_1\ldots\mu_{k-1}} \quad (2.3.4)$$

For example, one immediately sees that $\Sigma : M \longrightarrow J^1B$ is holonomic (i.e. $\Sigma = j^1\sigma$ for some section $\sigma : M \longrightarrow B$) if and only if $\Sigma^*(\omega^i) = 0$. Analogously, $\Sigma : M \longrightarrow J^2B$ is holonomic (i.e. $\Sigma = j^2\sigma$ for some section $\sigma : M \longrightarrow B$) if and only if $\Sigma^*(\omega^i) = 0$ and $\Sigma^*(\omega^i_\mu) = 0$ and so on for holonomic sections of higher order prolongations $J^k\mathcal{B}$ $(k \geq 3)$.

The *ideal of contact forms* $\Omega_c(J^k\mathcal{B})$ is the bilateral ideal of $\Omega(J^kB)$ generated by contact 1-forms (2.3.3); it is in fact a graded ideal $\Omega_c(J^k\mathcal{B}) = \oplus_p \Omega^p_c(J^k\mathcal{B}) \hookrightarrow \Omega(J^kB)$, i.e. any product $\alpha \wedge \beta$ is in $\Omega_c(J^k\mathcal{B})$ whenever at least one of the two forms α and β is in $\Omega_c(J^k\mathcal{B})$. A p-form $\omega \in \Omega^p_c(J^k\mathcal{B})$ is called a *(pure) contact p-form.*

The *differential ideal of extended contact forms* $\Omega^*_c(J^k\mathcal{B})$ is the bilateral ideal of $\Omega(J^kB)$ generated by contact 1-forms (2.3.3) together with their exterior differentials. Notice that just the differentials of the highest basic forms $\omega^i_{\mu_1\ldots\mu_{k-1}}$

are actually involved in the construction because of the recurrence relations

$$\begin{aligned}
&\mathrm{d}\omega^i = -\mathrm{d}y^i_\nu \wedge \mathrm{d}x^\nu = -\omega^i_\nu \wedge \mathrm{d}x^\nu\\
&\mathrm{d}\omega^i_\mu = -\mathrm{d}y^i_{\mu\nu} \wedge \mathrm{d}x^\nu = -\omega^i_{\mu\nu} \wedge \mathrm{d}x^\nu\\
&\dots\\
&\mathrm{d}\omega^i_{\mu_1\dots\mu_{k-2}} = -\mathrm{d}y^i_{\mu_1\dots\mu_{k-2}\nu} \wedge \mathrm{d}x^\nu = -\omega^i_{\mu_1\dots\mu_{k-2}\nu} \wedge \mathrm{d}x^\nu
\end{aligned} \tag{2.3.5}$$

These relations express the fact that the differentials of ω^i, ω^i_μ, ..., $\omega^i_{\mu_1\dots\mu_{k-2}}$ are already contained in the (bilateral) ideal generated by the contact 1-forms (2.3.3). On the contrary, the differential $\mathrm{d}\omega^i_{\mu_1\dots\mu_{k-1}}$ of the highest basic form is not contained in the ideal $\Omega_c(J^k\mathcal{B})$. In fact an analogous relation holds only *formally* for $\mathrm{d}\omega^i_{\mu_1\dots\mu_{k-1}}$, i.e.

$$\mathrm{d}\omega^i_{\mu_1\dots\mu_{k-1}} = -\mathrm{d}y^i_{\mu_1\dots\mu_{k-1}\nu} \wedge \mathrm{d}x^\nu = -\omega^i_{\mu_1\dots\mu_{k-1}\nu} \wedge \mathrm{d}x^\nu \tag{2.3.6}$$

but this is however impossible in $\Omega(J^kB)$ since $\omega^i_{\mu_1\dots\mu_{k-1}\nu}$ is not a form over J^kB, being it defined over $J^{k+1}B$. Of course a suitable equivalent of (2.3.6) has a genuine meaning in $\Omega(J^{k+1}\mathcal{B})$; see however (2.3.15) below.

As a consequence of definition (2.3.1), if $\omega \in \Omega^*_c(J^k\mathcal{B})$ is an (extended) contact p-form then it vanishes along holonomic sections, i.e.

$$(j^k\rho)^*\omega = 0 \tag{2.3.7}$$

We thence have the following inclusions

$$\Omega_c(J^k\mathcal{B}) \subset \Omega^*_c(J^k\mathcal{B}) \subset \Omega(J^kB) \tag{2.3.8}$$

For example, for $k = 2$ the pure contact forms are generated by

$$\omega^i = \mathrm{d}y^i - y^i_\nu\,\mathrm{d}x^\nu \qquad \omega^i_\mu = \mathrm{d}y^i_\mu - y^i_{\mu\nu}\,\mathrm{d}x^\nu \tag{2.3.9}$$

the extended contact forms are generated by

$$\omega^i = \mathrm{d}y^i - y^i_\nu\,\mathrm{d}x^\nu \qquad \omega^i_\mu = \mathrm{d}y^i_\mu - y^i_{\mu\nu}\,\mathrm{d}x^\nu \qquad \mathrm{d}\omega^i_\mu = -\mathrm{d}y^i_{\mu\nu} \wedge \mathrm{d}x^\nu \tag{2.3.10}$$

For example, if $\dim(M) = 2$ and $(x^1, x^2; y^i)$, with $i = 1, \dots, n$, are fibered coordinates on B, the forms $\mathrm{d}x^1 \wedge \mathrm{d}x^2 \wedge \mathrm{d}y^i$ and $\mathrm{d}x^1 \wedge \mathrm{d}x^2 \wedge \mathrm{d}y^i_\mu$ are in $\Omega_c(J^2\mathcal{B})$ while the forms $\mathrm{d}x^1 \wedge \mathrm{d}x^\nu \wedge \mathrm{d}y^i_{\mu\nu}$ are all in $\Omega^*_c(J^k\mathcal{B})$ since

$$\begin{aligned}
&\mathrm{d}x^1 \wedge \mathrm{d}x^2 \wedge \mathrm{d}y^i = \mathrm{d}x^1 \wedge \mathrm{d}x^2 \wedge (\mathrm{d}y^i - y^i_\mu \mathrm{d}x^\mu) = \mathrm{d}x^1 \wedge \mathrm{d}x^2 \wedge \omega^i\\
&\mathrm{d}x^1 \wedge \mathrm{d}x^2 \wedge \mathrm{d}y^i_\mu = \mathrm{d}x^1 \wedge \mathrm{d}x^2 \wedge (\mathrm{d}y^i_\mu - y^i_{\mu\nu} \mathrm{d}x^\nu) = \mathrm{d}x^1 \wedge \mathrm{d}x^2 \wedge \omega^i_\mu\\
&\mathrm{d}x^1 \wedge \mathrm{d}x^\nu \wedge \mathrm{d}y^i_{\mu\nu} = \mathrm{d}x^1 \wedge \mathrm{d}\omega^i_\mu
\end{aligned} \tag{2.3.11}$$

Let us finally consider a form $\omega \in \Omega(J^kB)$ over J^kB and for $h \geq 0$ the projection $\pi^{k+h}_k : J^{k+h}B \longrightarrow J^kB$. We can consider the pull-back of ω along π^{k+h}_k which is a form $(\pi^{k+h}_k)^*\omega$ over $J^{k+h}B$. Thus any form in $\Omega(J^kB)$ canonically defines a form $(\pi^{k+h}_k)^*\omega$ on any higher order prolongation $J^{k+h}B$. Because of the particular form of the projection π^{k+h}_k in local natural coordinates, the local expressions of $(\pi^{k+h}_k)^*\omega$ are obtained from the local expressions of ω simply by interpreting the natural coordinates in J^kB as a (subset) of the natural coordinates of $J^{k+h}B$. For this reason and by an abuse of language sometimes the form ω is identified with its pull-backs $(\pi^{k+h}_k)^*\omega \in \Omega(J^{k+h}B)$. As we shall see in the next Section, this is also justified by the fact that all $(\pi^{k+h}_k)^*\omega$ $(h \geq 0)$ identify the same form in the infinite jet bundle $J^\infty\mathcal{B}$.

Using the above definitions we can define two projectors

$$H : \Omega^p(J^kB) \longrightarrow \Omega^p_{hor}(J^{k+1}\mathcal{B}) \qquad K : \Omega^p(J^kB) \longrightarrow \Omega^p_c(J^{k+1}\mathcal{B}) \tag{2.3.12}$$

which can be regarded as a formal splitting of the forms $\omega \in \Omega^p(J^kB)$. We remark that a general p-form in $\Omega^p(J^kB)$ cannot be split into a horizontal part and a contact part both remaining in the algebra $\Omega^p(J^kB)$. To see this, let us first discuss the example of $k = 0$, $p = 1$. If $\theta \in \Omega^1(B)$ has local expression $\theta = \theta_\mu\, \mathrm{d}x^\mu + \theta_i\, \mathrm{d}y^i$, one can formally try to split it as follows

$$\theta = (\theta_\mu + \theta_i\, y^i_\mu)\, \mathrm{d}x^\mu \oplus \theta_i\, \omega^i = \theta_{(H)} \oplus \theta_{(c)} \tag{2.3.13}$$

but that splitting (2.3.13) has a meaning in $\Omega^1(J^1B)$, only. In fact, $\theta_{(H)} = (\theta_\mu + \theta_i\, y^i_\mu)\, \mathrm{d}x^\mu$ and $\theta_{(c)} = \theta_i\, \omega^i$ are both well-defined 1-forms in $\Omega^1(J^1B)$ so that one can in fact split $(\pi^1_0)^*\theta$ in $\Omega^1(J^1B)$ as follows

$$(\pi^1_0)^*\theta = (\theta_\mu + \theta_i\, y^i_\mu)\, \mathrm{d}x^\mu \oplus \theta_i\, \omega^i = \theta_{(H)} \oplus \theta_{(c)} \tag{2.3.14}$$

Accordingly, despite there is no canonical splitting of $\Omega^p(J^kB)$, one can canonically split $(\pi^{k+1}_k)^*\Omega^p(J^kB)$. The possibility of canonically splitting (the pull-back of) forms into a horizontal and a contact part is a characteristic feature of jet prolongation bundles.

Thence the pull-backs along the projection π^{k+1}_k of the forms in $\Omega^p(J^kB)$ identify a sub-set $(\pi^{k+1}_k)^*\Omega^p(J^kB)$ of $\Omega^p(J^{k+1}B)$. The forms in $(\pi^{k+1}_k)^*\Omega^p(J^kB)$ can be canonically decomposed as $(\pi^{k+1}_k)^*\Omega^p(J^kB) = \Omega^p_{hor}(J^{k+1}B) \oplus \Omega^p_c(J^{k+1}B)$. The projectors associated to this splitting are the projectors H and K defined above.

One can immediately see that for each $h \geq 0$ the pull-back $(\pi^{k+h}_k)^*\omega$ of a (pure, extended or generalized) contact form ω (i.e. a form ω in $\Omega_c(J^k\mathcal{B})$, $\Omega^*_c(J^k\mathcal{B})$ or in $\bar{\Omega}^*_c(J^k\mathcal{B})$) is still a contact form (of the same type). As a consequence, if we consider the pull-back of the form $\mathrm{d}\omega^i_{\mu_1\dots\mu_{k-1}}$ over $J^{k+1}B$, we obtain the correct version of the formal equation (2.3.6), namely:

$$(\pi^{k+1}_k)^*(\mathrm{d}\omega^i_{\mu_1\dots\mu_{k-1}}) = -\mathrm{d}y^i_{\mu_1\dots\mu_{k-1}\nu} \wedge \mathrm{d}x^\nu = -\omega^i_{\mu_1\dots\mu_{k-1}\nu} \wedge \mathrm{d}x^\nu \tag{2.3.15}$$

We stress that equation (2.3.15) is now written in $J^{k+1}B$ where it is completely meaningful.

We can now extend the definition of formal differential to (not necessarily horizontal) forms in $\Omega(J^kB)$. Let $\omega \in \Omega^p(J^kB)$ be any p-form on J^kB and $\theta_{(H)} = H(\omega) \in \Omega^p_{hor}(J^{k+1}\mathcal{B})$ be the horizontal part of the pull-back $(\pi^{k+1}_k)^*\omega$. Then the formal differential $D\omega$ of ω is, by definition, the formal differential $D\theta_{(H)}$ as defined in definition (2.2.13). Equivalently, $D\omega$ is the horizontal part $H(\mathrm{d}\omega)$ of the differential $\mathrm{d}\omega$, where $d : \Omega(J^kB) \longrightarrow \Omega(J^kB)$ is the exterior differential for forms on J^kB (see Exercise 7 below).

4. The Infinite Jet Bundle

The *infinite jet bundle* $J^\infty\mathcal{B}$ is the *inverse limit* of the family of all jet bundles $J^k\mathcal{B}$. Sometimes a clearer formulation of some properties can be obtained by stating them on the infinite jet bundle. Though the infinite jet bundle can be rigorously introduced as an infinite dimensional manifold (and bundle), it is usually used whenever one does not want to (or cannot *a priori*) keep track of the order of prolongation. Roughly speaking, one can regard objects in the infinite jet bundle as objects defined on *some* finite jet bundle for some (high enough) prolongation order. Moreover, the contact structure on infinite jet bundles is much simpler than the contact structure on finite jet bundles, basically because of our remarks above, which among other things tell us that $\Omega_c(J^\infty\mathcal{B}) \equiv \Omega^*_c(J^\infty\mathcal{B})$ since the differential of a contact 1-form is always in the ideal of contact forms $\Omega_c(J^\infty\mathcal{B})$.

Direct and Inverse Limit

Let $(I, \geq)$ be an ordered set and let $\{V_i\}_{i\in I}$ be an indexed family of objects in a category (e.g. topological vector spaces, groups, manifolds, fiber bundles, etc.), together with a family of morphisms $p^i_j : V_i \longrightarrow V_j$ $(i \geq j)$, called *projections*, such that $p^k_j \circ p^i_k = p^i_j$. The collection $\{V_i, p^i_j\}$ is called an *inverse family*.

Definition (2.4.1): we say that the category under consideration *allows inverse limits* if for any inverse family there is an object V_∞ together with a family of projection morphisms $p^\infty_j : V_\infty \longrightarrow V_j$ such that:

(i) $p^k_j \circ p^\infty_k = p^\infty_j$

(ii) for any object W and any family of morphisms $f_i : W \longrightarrow V_i$ such that $p_i^k \circ f_k = f_i$, there exists a unique morphism $f_\infty : W \longrightarrow V_\infty$ which makes the following diagram commutative for all $i \in I$

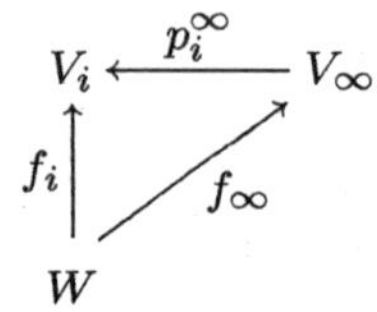

(2.4.2)

The family $\{V_\infty, p_j^\infty\}$ is then called the *inverse limit of the family* $\{V_i, p_j^i\}$. ∎

As an example, let us consider the inverse family $\{\mathbb{R}^n, p_m^n\}$ formed by the topological vector spaces $\mathbb{R}^n$ together with the projections $p_m^n : \mathbb{R}^n \longrightarrow \mathbb{R}^m$ $(n \geq m)$ over the first m factors. We can define $\mathbb{R}^\infty$ to be the vector space of infinite sequences of real numbers and $p_m^\infty : \mathbb{R}^\infty \longrightarrow \mathbb{R}^m$ the projection over the first m factors. The family $\{\mathbb{R}^n, p_m^n\}$ is an inverse family which allows $\{\mathbb{R}^\infty, p_m^\infty\}$ as inverse limit. The topology of $\mathbb{R}^\infty$ is defined to be the weakest topology for which the maps p_m^∞ are all continuous, so that open sets in $\mathbb{R}^\infty$ are exactly the preimages through p_m^∞ of open sets in $\mathbb{R}^m$, for each m. Let now W be a topological vector space and let us consider a family of morphisms $f_m : W \longrightarrow \mathbb{R}^m$, such that $p_m^n \circ f_n = f_m$. We can define a map $f_\infty : W \longrightarrow \mathbb{R}^\infty$ which acts on $w \in W$ to give $f_\infty(w) \in \mathbb{R}^\infty$. The first m elements in $f_\infty(w)$, i.e. $p_m^\infty \circ f_\infty(w)$, are defined to be $f_m(w) \in \mathbb{R}^m$ for each m, so to have commutativity of the diagram (2.4.2). Clearly, f_∞ is well-defined, unique, linear and continuous. Thus $\{\mathbb{R}^\infty, p_m^\infty\}$ is the inverse limit of the family $\{\mathbb{R}^n, p_m^n\}$.

Let us now recall the definition of *direct limit* which in a suitable sense is the dual with respect to the previous one. Let $\{V_k\}_{k \in I}$ be an indexed family of objects in a category (e.g. topological vector spaces, groups, manifolds, fiber bundles, etc.), together with a family of morphisms $i_h^k : V_h \longrightarrow V_k$ $(k \geq h)$, called *injections*, such that $i_j^k \circ i_h^j = i_h^k$. The collection $\{V_k, i_h^k\}$ is called a *direct family*.

Definition (2.4.3): we say that the category under consideration *allows direct limits* if for any direct family there is an object V_∞ together with a family of morphisms $i_h^\infty : V_h \longrightarrow V_\infty$ such that:

(i) $i_h^\infty \circ i_k^h = i_k^\infty$

(ii) for any object W and any family of morphisms $f_k : V_k \longrightarrow W$ such that $f_h \circ i_k^h = f_k$, there exists a unique morphism $f_\infty : V_\infty \longrightarrow W$ which makes the following diagram commutative for all $i \in I$

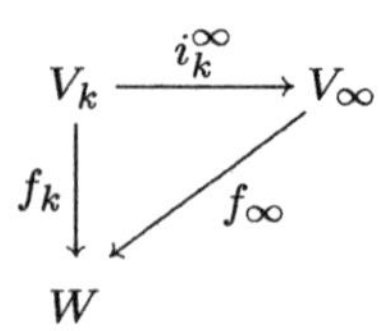

(2.4.4)

The family $\{V_\infty, i_h^\infty\}$ is then called the *direct limit of the family* $\{V_k, i_h^k\}$. ■

As an example let us now consider the direct family $\{\mathbb{R}^n, i_m^n\}$ formed by the topological vector spaces $\mathbb{R}^n$ together with the canonical injections $i_m^n : \mathbb{R}^m \longrightarrow \mathbb{R}^n$ $(n \geq m)$ into the first m factors. We can define $\mathbb{R}_0^\infty$ to be the vector space of infinite sequences of real numbers with just a finite sequence of non-zero elements. Let $i_m^\infty : \mathbb{R}^m \longrightarrow \mathbb{R}_0^\infty$ be the injections into the first m factors. The family $\{\mathbb{R}^n, i_m^n\}$ is a direct family which allows $\{\mathbb{R}_0^\infty, i_n^\infty\}$ as the direct limit. The topology of $\mathbb{R}_0^\infty$ is defined to be the strongest topology for which the maps i_m^∞ are all continuous, so that open sets in $\mathbb{R}_0^\infty$ are exactly the images through i_m^∞ of open sets in $\mathbb{R}^m$, for any m. Let now W be a topological vector space and let us consider a family of morphisms $f_m : \mathbb{R}^m \longrightarrow W$, such that $f_n \circ i_m^n = f_m$. We can define a map $f_\infty : \mathbb{R}_0^\infty \longrightarrow W$ which acts on $a \in \mathbb{R}_0^\infty$ to give $f_\infty(a) \in W$. Let a be $a = i_m^\infty(\bar{a})$ for some $\bar{a} \in \mathbb{R}^m$, and $f_\infty(a)$ is defined to be $f_m(\bar{a}) \in W$, so to have commutativity of the diagram (2.4.4). Clearly, f_∞ is well-defined, unique, linear and continuous. Thus $\{\mathbb{R}_0^\infty, i_m^\infty\}$ is the inverse limit of the family $\{\mathbb{R}^n, i_m^n\}$. Clearly, $\mathbb{R}_0^\infty$ may be identified with a subspace of $\mathbb{R}^\infty$.

Proposition (2.4.5): the dual space of $\mathbb{R}^\infty$ is isomorphic to $\mathbb{R}_0^\infty$.

Proof: let $\alpha : \mathbb{R}^\infty \longrightarrow \mathbb{R}$ be a point in the dual of $\mathbb{R}^\infty$, i.e. a linear continuous map.

In order to be continuous, it has to depend just on finitely many elements of its argument. If this were not the case, let us consider, for each n, two points, x_n and y_n, in $\mathbb{R}^\infty$ which agree up to the nth element, i.e. $p_n^\infty(x_n) = p_n^\infty(y_n)$; since α depends on infinitely many elements of its argument, the points x_n and y_n may be always chosen so that $\alpha(x_n) \neq \alpha(y_n)$.

Let us then consider the sequence

$$z_n = \frac{x_n - y_n}{\alpha(x_n - y_n)} \tag{2.4.6}$$

The sequence $\{z_n\}$ is convergent and its limit is $0 \in \mathbb{R}^\infty$.

However, the sequence $\alpha(z_n)$ is constant and $\alpha(z_n) = 1$ so that it converges to $1 \in \mathbb{R}$; then α cannot be linear and continuous. ■

One can also prove the opposite (see [S89] page 255):

Proposition (2.4.7): the dual space of $\mathbb{R}_0^\infty$ is isomorphic to $\mathbb{R}^\infty$. ■

Analogously, whenever one considers a function f_∞ on the inverse limit, it depends just on the first i elements in its argument, being then the pull-back of a function f_i of V_i, i.e. $f_\infty = f_i \circ p_i^\infty$, for some arbitrary but finite $i \in I$.

Basically, that is why the infinite jet bundle we are going to define is used when one does not want to keep track of the order of prolongation.

One can extend the standard category of manifolds by allowing manifolds modelled on $\mathbb{R}^\infty$, which are a particular class of infinite dimensional manifolds. Transition functions are in this case required to be smooth in $\mathbb{R}^\infty$, i.e. to be the pull-back over $\mathbb{R}^\infty$ of smooth functions of some $\mathbb{R}^n$.

The Infinite Jet Bundle

The family $\{J^k B, \pi^k\}$ is an inverse family of (finite dimensional) manifolds. Of course, the standard category does not allow inverse limits unless infinite dimensional manifolds are considered. In the extended category of infinite dimensional manifolds, one can define the inverse limit $\{J^\infty B, \pi^\infty\}$ of the family $\{J^k B, \pi^k\}$; $J^\infty \mathcal{B} = (J^\infty B, M, \pi^\infty; \mathbb{R}^\infty)$ is a bundle over the base manifold M.

The infinite jet prolongation $\{J^\infty \mathcal{B}, \pi^\infty\}$ will be also denoted by $\{J\mathcal{B}, \pi^\infty\}$ to remember that basically everything works as if it were a jet bundle of a *"suitable finite order"*.

A point in $J^\infty B$ will be denoted by $j^\infty_x \sigma$ (or simply by $j_x\sigma$) and it is the infinite string $(j^1_x\sigma, j^2_x\sigma, \ldots)$. Notice that $j^1_x\sigma = \pi^2_1(j^2_x\sigma) = \ldots$, i.e. any entry of the string $j_x\sigma$ projects over the previous ones.

The manifold $J^\infty B$ has a bundle structure also over each finite jet extension $J^k B$ with a suitably defined projection π^∞_k. Such a structure will be denoted by $J^\infty_k \mathcal{B}$ or simply by $J_k\mathcal{B}$.

One can extend the prolongation of morphisms and sections of $\mathcal{B}$ to $J^\infty\mathcal{B}$. If $\boldsymbol{\Phi} = (\Phi, \phi)$ is a fibered morphism with $\Phi : B \longrightarrow B'$, then the map Φ can be prolonged to any order $J^k\Phi : J^k B \longrightarrow J^k B'$, inducing an infinite prolongation

$$J\Phi : JB \longrightarrow JB' : (j^1_x\sigma, j^2_x\sigma, \ldots) \mapsto (J^1\Phi(j^1_x\sigma), J^2\Phi(j^2_x\sigma), \ldots) \tag{2.4.8}$$

Analogously, if $\sigma : M \longrightarrow B$ is a section of $\mathcal{B}$, then it can be prolonged to the infinite jet bundle by

$$j\sigma : M \longrightarrow JB : x \mapsto j_x\sigma \tag{2.4.9}$$

A projectable vector field $\Xi = \xi^\mu \partial_\mu + \xi^i \partial_i$ can also be prolonged to $J\mathcal{B}$ obtaining a *"vector field"* $j\Xi = \xi^\mu\partial_\mu + \xi^i\partial_i + \xi^i_\mu \partial^\mu_i + \ldots$. This object $j\Xi$ may have infinitely many non-zero components, but each one of them depends just on finitely many coordinates of JB; namely

$$\begin{cases} \xi^i_{\mu_1} = \mathrm{d}_{\mu_1}\xi^i - y^i_\nu \,\mathrm{d}_{\mu_1}\xi^\nu \\ \xi^i_{\mu_1\mu_2} = \mathrm{d}_{\mu_1}\xi^i_{\mu_2} - y^i_{\nu\mu_2}\,\mathrm{d}_{\mu_1}\xi^\nu \\ \ldots \end{cases} \tag{2.4.10}$$

The *"vector field"* $j\Xi$ projects over the prolongations $j^k\Xi$, for each k. Analogously, for each r, r-forms over $J\mathcal{B}$ are pull-back of r-forms over $J^k\mathcal{B}$, for some arbitrary k. Notice that, despite $j\Xi$ has infinitely many components, the interior product $i_{j\Xi}\omega$ is well defined for any r-form ω over JB, because of the fact that ω has just finitely many non-zero components. Actually, one has that ω is the pull-back of an r-form $\omega^{(k)}$ over $J^k\mathcal{B}$, i.e. $\omega = (\pi^\infty_k)^*\omega^{(k)}$, so that $i_{j\Xi}\omega = (\pi^\infty_k)^*(i_{j^k\Xi}\omega^{(k)})$.

Contact forms over $J\mathcal{B}$ have a simpler structure, basically due to the fact that there is no *"last"* contact 1-form, so that for any basic contact 1-form $\omega^i_{\mu_1\dots\mu_k}$ we have in $J^\infty\mathcal{B}$:

$$\mathrm{d}\omega^i_{\mu_1\dots\mu_k} = -\mathrm{d}y^i_{\mu_1\dots\mu_k\sigma}\wedge \mathrm{d}x^\sigma = -\omega^i_{\mu_1\dots\mu_k\sigma}\wedge \mathrm{d}x^\sigma \tag{2.4.11}$$

which is thence in $\Omega_c(J\mathcal{B})$ automatically.

Notice once again that also this property can be obtained in terms of finite jet bundles. As we noticed above, one can in fact say that for any contact 1-form $\omega^i_{\mu_1\dots\mu_{k-1}}$ over $J^k\mathcal{B}$, the pull-back over $J^{k+1}\mathcal{B}$ of its differential, namely $(\pi^{k+1}_k)^*d\omega^i_{\mu_1\dots\mu_{k-1}}$, belongs to the ideal of contact forms $\Omega_c(J^{k+1}\mathcal{B})$ defined over $J^{k+1}\mathcal{B}$. Taking the inverse limit of equation (2.3.15) gives then the appropriate equation (2.4.11) in $J^\infty\mathcal{B}$. Of course, this is somehow cumbersome to be stated and proved; that is why the infinite jet bundle provides a good language to deal at least formally with these properties. For greater details see for example [S89].

5. Generalized Vector Fields

Generalized vector fields over a bundle are not vector fields on the bundle in the standard sense; nevertheless, one can drag sections along them and thence define their Lie derivative. The formal Lie derivative on a bundle may be seen as a generalized vector field. Furthermore, generalized vector fields are objects suitable to describe generalized symmetries.

Let $\mathcal{B} = (B, M, \pi; F)$ be a bundle, with local fibered coordinates $(x^\mu; y^i)$. Let us consider the pull-back of the tangent bundle $\tau_B : TB \longrightarrow B$ along the map $\pi^k_0 : J^kB \longrightarrow B$:

$$\begin{array}{ccc}
(\pi^k_0)^*TB & \xrightarrow{\ i\ } & TB \\
{\scriptstyle \pi^*}\downarrow & & \downarrow{\scriptstyle \tau_B} \\
J^kB & \xrightarrow{\ \pi^k_0\ } & B \\
 & & \downarrow{\scriptstyle \pi} \\
 & & M
\end{array} \tag{2.5.1}$$

Definition (2.5.2): a *generalized vector field (of order k) over $\mathcal{B}$* is a section Ξ of the fiber bundle $\pi^* : (\pi_0^k)^*T\mathcal{B} \longrightarrow J^k\mathcal{B}$, i.e.:

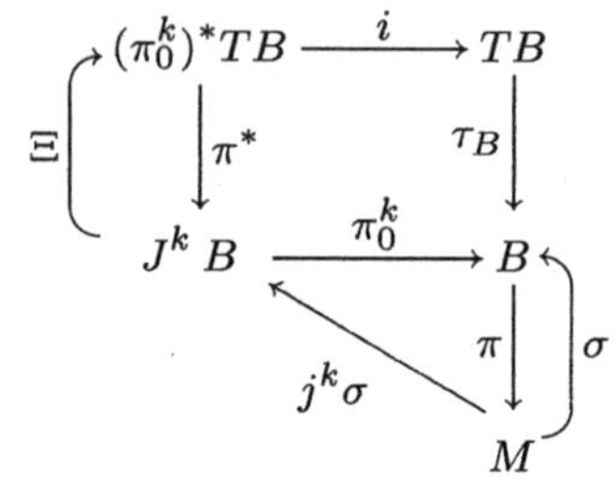

(2.5.3)

For each section $\sigma : M \longrightarrow \mathcal{B}$ one can define $\Xi_\sigma = i \circ \Xi \circ j^k\sigma : M \longrightarrow T\mathcal{B}$ which is a vector field over the section σ. ∎

Clearly, generalized vector fields of order $k = 0$ are ordinary vector fields over $\mathcal{B}$. Locally, $\Xi(x^\mu, y^i, \dots, y^i_{\mu_1\dots\mu_k})$ is given in the following form:

$$\Xi = \xi^\mu(x^\mu, y^i, \dots, y^i_{\mu_1\dots\mu_k})\partial_\mu + \xi^i(x^\mu, y^i, \dots, y^i_{\mu_1\dots\mu_k})\partial_i \tag{2.5.4}$$

which, for $k \neq 0$, is not an ordinary vector field on $\mathcal{B}$ due to the dependence of the components (ξ^μ, ξ^i) on the derivatives of fields. Once one computes it on a section σ then the pulled-back components depend just on the base coordinates (x^μ), so that Ξ_σ is a vector field over the section σ, in the standard sense.

Clearly, generalized vector fields over $\mathcal{B}$ do not preserve the fiber structure of $\mathcal{B}$.

Definition (2.5.5): a *generalized projectable vector field (of order k) over the bundle $\mathcal{B}$* is a generalized vector field Ξ over $\mathcal{B}$ which projects onto an ordinary vector field $\xi = \xi^\mu(x)\partial_\mu$ on the base. ∎

Locally, a generalized projectable vector field over $\mathcal{B}$ is in the form:

$$\Xi = \xi^\mu(x^\mu)\partial_\mu + \xi^i(x^\mu, y^i, \dots, y^i_{\mu_1\dots\mu_k})\partial_i \tag{2.5.6}$$

As a particular case, one can define *generalized vertical vector fields (of order k) over $\mathcal{B}$*, which are locally of the form:

$$\Xi = \xi^i(x^\mu, y^i, \dots, y^i_{\mu_1\dots\mu_k})\partial_i \tag{2.5.7}$$

In particular, for any section σ of $\mathcal{B}$ and any generalized vertical vector field Ξ over $\mathcal{B}$, one can define a vertical vector field over σ given by:

$$\Xi_\sigma = \xi^i(x^\mu, \sigma^i(x), \dots, \partial_{\mu_1\dots\mu_k}\sigma^i(x))\partial_i \tag{2.5.8}$$

which is a section of the bundle $\pi_V : VB \longrightarrow M$, as noticed in Section 1.6.

Proposition (2.5.9): if $\Xi = \xi^\mu \partial_\mu + \xi^i \partial_i$ is a generalized projectable vector field, then $\Xi_{(V)} = (\xi^i - y^i_\mu \xi^\mu)\partial_i = \xi^i_{(V)} \partial_i$ is a generalized vertical vector field. The generalized vector field $\Xi_{(V)}$ is called the *vertical part of* Ξ.

Proof: the transformation law of the components of Ξ with respect to new fibered coordinates is:

$$\begin{cases} x' = \phi(x) \\ y' = Y(x,y) \end{cases} \qquad \begin{cases} \xi'^\mu = J^\mu_\nu \xi^\nu \\ \xi'^i = J^i_\nu \xi^\nu + J^i_j \xi^j \end{cases} \tag{2.5.10}$$

where we set $J^\mu_\nu = \partial_\nu \phi^\mu$, $J^i_\nu = \partial_\nu Y^i$ and $J^i_j = \partial_j Y^i$. The induced transformation law on J^1B is given by

$$y'^i_\mu = \bar{J}^\nu_\mu (J^i_\nu + J^i_j\, y^j_\nu) \tag{2.5.11}$$

Thence the transformation rule for $\xi^i_{(V)}$ is:

$$\xi'^i_{(V)} = \xi'^i - y'^i_\mu\, \xi'^\mu = J^i_\nu \xi^\nu + J^i_j \xi^j - \bar{J}^\nu_\mu (J^i_\nu + J^i_j\, y^j_\nu) J^\mu_\lambda \xi^\lambda = J^i_j\, \xi^j_{(V)} \tag{2.5.12}$$

which agrees with the transformation rule for the components of a vertical vector field. ∎

Let now $\bar{\sigma} : \mathbb{R} \times M \longrightarrow B$ be a smooth map such that for any fixed $s \in \mathbb{R}$ $\sigma_s(x) = \bar{\sigma}(s,x) : M \longrightarrow B$ is a (global) section of $\mathcal{B}$. The map $\bar{\sigma}$, as well as the family $\{\sigma_s\}$, is then called a 1*-parameter family of sections*. In other words, (a suitable restriction of) the family σ_s is a homotopic deformation with $s \in \mathbb{R}$ of the *"central"* section $\sigma = \sigma_0$. Often one restricts it to a finite (open) interval, conventionally $(-1, 1)$ (or $(-\epsilon, \epsilon)$ if *"small"* deformations are considered). Analogous definitions are given for the homotopic families of sections over a fixed open subset $U \subseteq M$ or on some domain $D \subset M$ (possibly with values fixed at the boundary ∂D, together with any number of their derivatives). This will be helpful in the Calculus of Variations.

Definition (2.5.13): a 1-parameter family of sections σ_s is *Lie-dragged* along a generalized projectable vector field Ξ if and only if one has

$$(\Xi_{(V)})_{\sigma_s} = \frac{\mathrm{d}}{\mathrm{d}s}\sigma_s \tag{2.5.14}$$

∎

In this way, a generalized projectable vector field Ξ can drag a section.

Proposition (2.5.15): a 0-order generalized projectable vector field Ξ, i.e. an ordinary projectable vector field over $\mathcal{B}$, drags sections along its flow $\boldsymbol{\Phi}_s = (\Phi_s, \phi_s)$, according to the rule

$$\sigma_s = \Phi_s \circ \sigma \circ \phi_s^{-1} \tag{2.5.16}$$

Proof: we have

$$\frac{\mathrm{d}}{\mathrm{d}s}\sigma_s = \xi^i \circ \sigma_s - T\sigma_s(\xi) = (\Xi_{(V)})_{\sigma_s} \tag{2.5.17}$$

where $\xi = \xi^\mu \partial_\mu$ is the projection of Ξ on the base manifold M. ∎

6. Lie Derivative of Sections

Definition (2.6.1): let $\sigma : M \longrightarrow B$ be a section of $\mathcal{B}$, Ξ be a projectable vector field and ξ be its projection on M. We define the *Lie derivative of a section σ along the vector field* Ξ by:

$$\pounds_\Xi \sigma = T\sigma(\xi) - \Xi \circ \sigma \tag{2.6.2}$$

Then $\pounds_\Xi \sigma$ is a vertical vector field defined over σ, i.e. the Lie derivative of a section may be interpreted as a section of the bundle $\pi_V : VB \longrightarrow M$. ∎

Locally, if the section σ is given by the map $x \mapsto (x^\mu; \sigma^i(x))$ and the vector field Ξ is given by the expression $\Xi = \xi^\mu(x)\partial_\mu + \xi^i(x,y)\partial_i$, then the Lie derivative is

$$\pounds_\Xi \sigma = (\xi^\mu \, \partial_\mu \sigma^i - \xi^i)\partial_i \equiv \left(\pounds_\Xi \sigma^i\right)\partial_i \tag{2.6.3}$$

One can also consider the generalized vertical vector field (of order 1) defined by:

$$\pounds_\Xi = (\xi^\mu \, y^i_\mu - \xi^i)\partial_i \equiv -\Xi_{(V)} \tag{2.6.4}$$

It is called the *formal Lie derivative on $\mathcal{B}$*. If one evaluates it along a section $\sigma : M \longrightarrow B$ one produces the Lie derivative $\pounds_\Xi \sigma$ of the section σ.

Let now σ be a section of $\mathcal{B}$, Ξ a projectable vector field and $\boldsymbol{\Phi}_s = (\Phi_s, \phi_s)$ its flow. The dragging of σ along Ξ is given by $\sigma_s = \Phi_s \circ \sigma \circ \phi_s^{-1}$ (see proposition (2.5.15)); then the following holds:

Proposition (2.6.5): $\pounds_\Xi \sigma = -\left.\frac{\mathrm{d}}{\mathrm{d}s}\sigma_s\right|_{s=0}$

Proof: see definition (2.5.13). ∎

Thence the Lie derivative of a section is related to the infinitesimal generator of the Lie-dragging along the projectable vector field Ξ. Since one can Lie-drag sections along generalized projectable vector fields, one can also define the Lie derivative of a section (as well as the formal Lie derivative) along a generalized projectable vector field Ξ by the rule:

$$\pounds_\Xi \sigma = -(\Xi_{(V)})_\sigma \,, \qquad \pounds_\Xi = -(\Xi_{(V)}) \tag{2.6.6}$$

This will turn out to be important for generalized symmetries.

Let us now prove some of the fundamental properties of Lie derivatives which shall be extensively used below.

Proposition (2.6.7): let Ξ be a (projectable) vector field over $\mathcal{B}$, σ be a section and k be any positive integer. Then the following holds:

$$j^k(\pounds_\Xi \sigma) = \pounds_{j^k\Xi}(j^k\sigma) \tag{2.6.8}$$

Proof: the left hand side reads as:

$$j^k(\pounds_\Xi \sigma) = (\xi^\nu\, \partial_\nu \sigma^i - \xi^i)\partial_i + \mathrm{d}_\mu(\xi^\nu\, \partial_\nu \sigma^i - \xi^i)\partial_i^\mu + \dots \tag{2.6.9}$$

On the other hand, $j^k\sigma = (x^\mu, \sigma^i(x), \partial_\mu\sigma^i(x), \dots)$ and $j^k\Xi = \xi^\mu\partial_\mu + \xi^i\partial_i + (\mathrm{d}_\mu\xi^i - y^i_\nu \mathrm{d}_\mu\xi^\nu)\partial_i^\mu + \dots$, so that the right hand side reads as:

$$\pounds_{j^k\Xi}(j^k\sigma) = (\xi^\nu\, \partial_\nu\sigma^i - \xi^i)\partial_i + (\xi^\nu\, \partial_{\mu\nu}\sigma^i - (\mathrm{d}_\mu\xi^i - \partial_\nu\sigma^i\, \mathrm{d}_\mu\xi^\nu))\partial_i^\mu + \dots \tag{2.6.10}$$

which perfectly agrees with equation (2.6.9). ∎

Proposition (2.6.11): let $(x^\mu; y^i, X^i)$ be natural fibered coordinates on the vertical bundle $\pi_V : VB \longrightarrow M$. Then we have:

$$[\pounds_{\Xi_1}, \pounds_{\Xi_2}] = (\pounds_{\Xi_1} - \pounds_{\Xi_2})\partial_i + (\pounds_{[\Xi_1,\Xi_2]})\hat\partial_i \tag{2.6.12}$$

where we denoted by $\hat\partial_i$ the partial derivatives with respect to X^i.

Proof: as we noticed above the Lie-derivative $\pounds_{\Xi_2}\sigma$ may be regarded as a section

$$\pounds_{\Xi_2}\sigma : M \longrightarrow VB : x \mapsto \Big(x^\mu; \sigma^i(x), \xi^\nu_{(2)}(x)\partial_\nu\sigma^j(x) - \xi^j_{(2)}(x, \sigma(x))\Big) \tag{2.6.13}$$

Thence the formal Lie derivative reads as

$$\pounds_{\Xi_2} = y^i\partial_i + (\xi^\nu_{(2)} y^j_\nu - \xi^j_{(2)})\, \hat\partial_i \tag{2.6.14}$$

The vector field Ξ_1 induces a vector field on VB by:

$$\Xi_1^* = \xi^\mu_{(1)}\partial_\mu + \xi^i_{(1)}\partial_i + \partial_j\xi^i_{(1)} X^j\hat\partial_i \tag{2.6.15}$$

It is then meaningful to define the composite Lie derivative and get:

$$\begin{aligned}\pounds_{\Xi_1}\circ\pounds_{\Xi_2} =& [\xi^\nu_{(1)} y^i_\nu - \xi^i_{(1)}]\partial_i + \\ &+ [\xi^\mu_{(1)}(\partial_\mu\xi^\rho_{(2)}\, y^i_\rho + \xi^\rho_{(2)}\, y^i_{\rho\mu} - \mathrm{d}_\mu\xi^i_{(2)}) - \partial_j\xi^i_{(1)}\, \xi^j_{(2)}]\hat\partial_i\end{aligned} \tag{2.6.16}$$

Then the commutator of Lie derivatives is

$$\begin{aligned}[\pounds_{\Xi_1}, \pounds_{\Xi_2}]\sigma =& (\xi^\nu_{(1)} y^i_\nu - \xi^i_{(1)})\partial_i - (\xi^\nu_{(2)} y^i_\nu - \xi^i_{(2)})\partial_i + \\ &+ \Big((\xi^\mu_{(1)}\partial_\mu\xi^\rho_{(2)} - \xi^\mu_{(2)}\partial_\mu\xi^\rho_{(1)})y^i_\rho - \\ &+ \xi^\mu_{(1)}\mathrm{d}_\mu\xi^i_{(2)} + \partial_j\xi^i_{(1)}\, \xi^j_{(2)} - \xi^\mu_{(2)}\mathrm{d}_\mu\xi^i_{(1)} - \partial_j\xi^i_{(2)}\, \xi^j_{(1)}\Big)\hat\partial_i = \\ =& (\pounds_{\Xi_1} - \pounds_{\Xi_2})\partial_i + (\pounds_{[\Xi_1,\Xi_2]})\hat\partial_i\end{aligned} \tag{2.6.17}$$

which proves our claim. ∎

7. Examples

Let us apply the jet bundle techniques to some relevant examples.

The Phase Bundle for Mechanics

Let us consider the trivial bundle $(\mathbb{R} \times Q, \mathbb{R}, \mathrm{pr}_1; Q)$, where Q is a manifold having (q^i) as local coordinates. Using the preferred global trivialization induced by the choice of the (Newtonian absolute) time t, sections of $\mathbb{R} \times Q$ are given by:

$$\rho : \mathbb{R} \longrightarrow \mathbb{R} \times Q : t \mapsto (t, q(t)) \qquad q : \mathbb{R} \longrightarrow Q \tag{2.7.1}$$

so that they identify curves in Q, the *configuration space* of a mechanical system.

To define $J^1(\mathbb{R} \times Q)$ using relation (2.1.1) we have to identify sections if they have the same first-order Taylor expansion. In this case our relation essentially coincides with the one which is used to define vectors tangent to the manifold Q. Then we have natural coordinates (t, q^i, u^i) on $J^1(\mathbb{R} \times Q)$ such that (q^i, u^i) can be also seen as coordinates on TQ. In other words, we have an isomorphism between $J^1(\mathbb{R} \times Q)$ and $\mathbb{R} \times TQ$; notice that this isomorphism depends on the trivialization chosen (i.e. it depends on the choice of a time function).

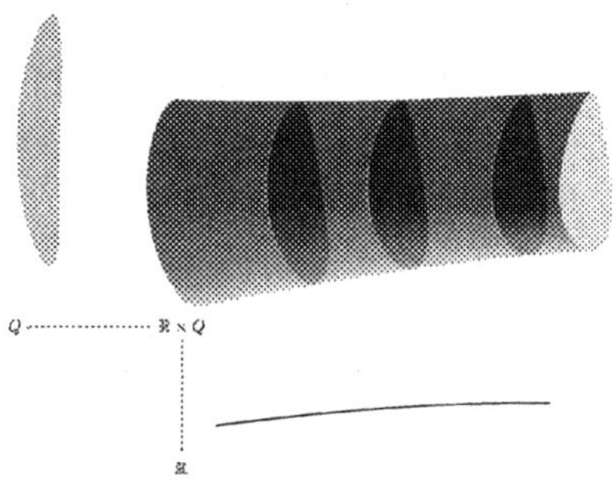

Fig. 12 – Configuration bundle in Mechanics.

This latter space is the space in which one works to describe first-order (possibly time-dependent) mechanical systems having Q as configuration manifold. Thus the framework we are going to define for field theories in Part II encompasses, as a particular case, the case of (time-dependent) Mechanics. We refer the reader to [MS00] and [FR] for further details.

The Higher Order Frame Bundles

To investigate a further example of principal bundles we consider the manifolds

$$L^s(M) = \{j^s_0\epsilon \mid \epsilon : \mathbb{R}^m \longrightarrow M, \text{ loc. invertible around the origin}\} \qquad (2.7.2)$$

for all integers $s \geq 1$. The projection $\pi : L^s(M) \longrightarrow M : j^s_0\epsilon \mapsto \epsilon(0)$ is surjective and of maximal rank.

As a standard fiber let us define:

$$\begin{aligned}\mathrm{GL}^s(m) =\{j^s_0\alpha \mid \alpha : \mathbb{R}^m \longrightarrow \mathbb{R}^m, &\text{ loc. invertible at the origin} \\ &\text{and with } \alpha(0) = 0\}\end{aligned} \qquad (2.7.3)$$

together with the group product:

$$j^s_0\alpha \cdot j^s_0\beta := j^s_0(\alpha \circ \beta) \qquad (2.7.4)$$

Then $(L^s(M), M, \pi; \mathrm{GL}^s(m))$ is a principal bundle. We can in fact define the global right action:

$$j^s_0\epsilon \cdot j^s_0\alpha := j^s_0(\epsilon \circ \alpha) \qquad (2.7.5)$$

which turns out to be a free action.

If we consider a diffeomorphism $\phi : M \longrightarrow M'$, it induces a map:

$$L^s(\phi) : L^s(M) \longrightarrow L^s(M') : j^s_0\epsilon \mapsto j^s_0(\phi \circ \epsilon) \qquad (2.7.6)$$

It can be shown that $(L^s(\phi), \phi)$ is a principal morphism so that $L^s(\cdot)$ is a covariant functor. The bundle $L^s(M)$ is called the *s-frame bundle* and $L^s(\phi)$ is called the *prolongation of ϕ to s-frames*. In the particular case $s = 1$ we have a canonical isomorphism $L^1(M) \simeq L(M)$ as defined in Section 1.3.

Lie Derivative of a Vector Field over M

Let M be a manifold and $T(M) = (TM, M, \tau_M; \mathbb{R}^m)$ be its tangent bundle. Let $X : M \longrightarrow TM$ be a section, i.e. a vector field over M, locally given by $X = X^\mu(x)\partial_\mu$. Let us choose $(x^\mu; v^\mu)$ as fibered coordinates on TM; locally the expression of the section is $X : x^\mu \mapsto (x^\mu; X^\mu(x))$. Let $\xi = \xi^\mu \partial_\mu$ be another vector field over M and ϕ_s its flow. As required in Exercise 1 below, one can define a lifted vector field $\hat{\xi}$ over TM, which is given by:

$$\hat{\xi} = \xi^\mu\, \partial_\mu + v^\nu\, (\partial_\nu \xi^\mu)\, \hat{\partial}_\mu \qquad (2.7.7)$$

where by $\hat{\partial}_\mu$ we denoted the partial derivatives with respect to v^μ.

The Lie derivative of the section X with respect to $\hat{\xi}$ is given by:

$$\pounds_{\hat{\xi}} X = (\xi^\mu \, \partial_\mu X^\nu - X^\mu \partial_\mu \xi^\nu) \, \hat{\partial}_\nu = [\xi, X]^\nu \, \hat{\partial}_\nu \tag{2.7.8}$$

which reproduces the well known fact that the Lie derivative of a vector field X with respect to a vector field ξ is the commutator of the two vector fields.

Actually, $\pounds_{\hat{\xi}} X$ is not yet a vector field on M as in the classical sense of Differential Geometry. To obtain the standard result, we still have to introduce the so-called *swap map*. It is an automorphism of the tangent bundle $T(TM)$ of TM locally defined by

$$s : TTM \longrightarrow TTM : (x^\mu, v^\mu, X^\mu) \mapsto (x^\mu, X^\mu, v^\mu) \tag{2.7.9}$$

It is easy to prove that it is a global map. If we compose it with the Lie derivative $\pounds_{\hat{\xi}} X$ and project onto TM we obtain a genuine vector field over M.

The property (2.7.8) is in fact also a general feature. If one considers a tensor field t as a section of the appropriate tensor bundle over M and calculates its Lie derivative with respect to $\Xi = \hat{\xi}$ given by an equation analogous to (2.7.7), the result is the ordinary Lie derivative defined by (see Exercise 8 below):

$$\pounds_\xi t = \frac{\mathrm{d}}{\mathrm{d}s} (\phi_s)_* t \Big|_{s=0} \tag{2.7.10}$$

8. Differential Equations

Jet bundles provide a general framework to describe systems of partial differential equations in a geometric way. Let us in fact consider a bundle $\mathcal{B} = (B, M, \pi; F)$.

Definition (2.8.1): a *system of (partial) differential equations* of order k on $\mathcal{B}$ is a closed embedded sub-manifold S of the jet prolongation $J^k\mathcal{B}$. ■

Of course, it is understood that S *really* seats in the total space J^kB. In fact, if $S^k \subset J^kB$ is a system of partial differential equations of order k on $\mathcal{B}$ and $h \geq 1$ is a positive integer, then we can define $S^{k+h} = (\pi^{k+h}_k)^{-1}(S^k) \subset J^{k+h}B$ which strictly speaking defines a system of (partial) differential equations of order $k + h$ on $\mathcal{B}$. Of course S^{k+h} does not really depend on derivatives of order greater than k. The two systems are thence said to be *equivalent*. In a sense one should remark that the *order* of a differential equation is sharp (i.e. minimal). The way of requiring it is to require that if one has a system of (partial) differential equations $S^k \subset J^k\mathcal{B}$ of order k on $\mathcal{B}$ and it is projected down onto $J^{k-h}\mathcal{B}$ by setting $S^{k-k} = \pi^k_{k-h}(S^k)$ then the result is not equivalent to the system S^k, i.e.

$$(\pi^k_{k-h})^{-1}(\pi^k_{k-h}(S^k)) \neq S^k \tag{2.8.2}$$

If this case we say that $S^k \subset J^k\mathcal{B}$ is a system of (partial) differential equations *strictly of order* k on $\mathcal{B}$.

Definition (2.8.3): a *solution* of a system of (partial) differential equations $S \subset J^k\mathcal{B}$ is a section $\sigma : M \longrightarrow B$ such that its kth-order prolongation $j^k\sigma$ takes its value in S, i.e. $j^k\sigma(x) \in S$ for all $x \in M$. ∎

For example, the equation for the harmonic oscillator

$$\ddot{q} + \omega^2 q = 0 \tag{2.8.4}$$

may be regarded as a differential equation in the sense of (2.8.1). Let us in fact consider the trivial bundle $(\mathbb{R}^2, \mathbb{R}, \mathrm{pr}_1; \mathbb{R})$ with (t, q) as fibered coordinates on $\mathbb{R}^2$. The second jet bundle $J^2\mathbb{R}^2$ has fibered coordinates (t, q, u, a). One can consider the function $F(t, q, u, a) = a + \omega^2 q$. Let us consider the preimage of $0 \in \mathbb{R}$ with respect to F which is a hypersurface $S \subset J^2\mathbb{R}^2$. It represents the equation (2.8.4) and a section $\gamma : t \mapsto (t, q(t))$ is a solution of S if and only if $j^2\gamma : t \mapsto (t, q(t), \dot{q}(t), \ddot{q}(t)) \in S$, i.e. $\ddot{q}(t) + \omega^2 q(t) = 0$.

We remark that in general a sub-manifold $S \subset J^kB$ (of codimension h) can be locally identified with the set of all zeros of h functions $F_i : B \longrightarrow \mathbb{R}$ $(i = 1, \ldots, h)$. From a global viewpoint, one can often identify S with the kernel of some bundle morphism $\mathbb{E} : J^k\mathcal{B} \longrightarrow \mathcal{V}$ valued into some suitable vector bundle $\mathcal{V} = (V, M, \tau; \mathbb{R}^h)$. In this way one automatically takes into account the behavior with respect to changes of trivialization. The Euler-Lagrange morphism will define field equations according to this rule.

When a differential operator $S \subset J^k\mathcal{B}$ is identified as the kernel of a bundle morphism $\mathbb{E} : J^kB \longrightarrow V$ into some suitable vector bundle $\mathcal{V} = (V, M, \tau; \mathbb{R}^h)$, then one can define a differential operator $D : \Gamma(\mathcal{B}) \longrightarrow \Gamma(\mathcal{V})$ as follows

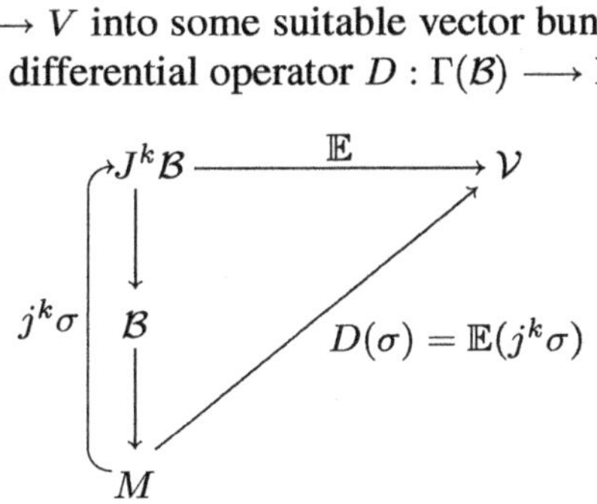

(2.8.5)

The operator D acts on sections of $\mathcal{B}$ to give sections of $\mathcal{V}$ by $D(\sigma) = \mathbb{E}(j^k\sigma)$. If $\mathcal{B}$ (and thence $J^k\mathcal{B}$) is a vector bundle and $\mathbb{E}$ is a linear morphism then the underlying differential operator is called a *linear operator*. The operator D is called *quasi-linear* if $\mathbb{E}$ is an affine morphism over the bundle $J^k\mathcal{B} \longrightarrow J^{k-1}\mathcal{B}$. We remark that both in Mechanics and in Field Theory the Euler-Lagrange morphism will define field equations which are generically (i.e. except for a number of particularly degenerate Lagrangians) quasi-linear.

Let us fix a system of natural fibered coordinates $(x^\mu; y^i, y^i_\mu, \dots, y^i_{\mu_1\dots\mu_k})$ on $J^k\mathcal{B}$ and fibered coordinates $(x^\mu; v^A)$ on $\mathcal{V}$. The local expression of a linear differential operator is

$$v^A = a^A_i(x)y^i + a^{A\mu}_i(x)y^i_\mu + \dots + a^{A\mu_i\dots\mu_k}_i(x)y^i_{\mu_1\dots\mu_k} \tag{2.8.6}$$

Let n be the rank of $\mathcal{B}$ (i.e. the dimension of its standard fiber F) and n' be the rank of $\mathcal{V}$. We can define the *symbol* of the operator D to be the map[2] $p: M \times \mathbb{R}^m \longrightarrow M(n, n')$ defined as follows

$$p(x,\xi) = a^A_i(x) + a^{A\mu}_i(x)\xi_\mu + \dots + a^{A\mu_i\dots\mu_k}_i(x)\xi_{\mu_1}\dots\xi_{\mu_k} \tag{2.8.7}$$

We have to stress that in general the symbol depends on the fiber coordinates chosen. Thence let us define the *principal symbol* of the operator D to be the highest order term of the symbol, i.e.

$$\mathfrak{P}(x,\xi) = a^{A\mu_i\dots\mu_k}_i(x)\xi_{\mu_1}\dots\xi_{\mu_k} \tag{2.8.8}$$

which turns out to be an invariant polynomial of degree k associated to D.

Let us now restrict to the case $n = n'$, i.e. when one has the same number of equations and unknown functions. An operator is *elliptic* if and only if the principal symbol is valued in $\mathrm{GL}(n) \subset M(n,n)$, i.e. if and only if $\mathfrak{P}(x,\xi)$ is invertible for all $\xi \neq 0$ and all $x \in M$.

As a further example let us consider the wave equation in $\mathbb{R}^m$. We choose the trivial configuration bundle $\mathcal{B} = (\mathbb{R}^{m+1} \times \mathbb{R}, \mathbb{R}^{m+1}, \mathrm{pr}_1; \mathbb{R})$ with coordinates $(x^\mu; \varphi)$ with $\mu = 0, \dots, m$; we set $t = x^0$ and $(x^\mu) = (t, x^i)$ with $i = 1, \dots, m$. The wave equation

$$\partial_t\partial_t\varphi - \sum_i \partial_i\partial_i\varphi = 0 \tag{2.8.9}$$

is of the second order; as a consequence it can be seen as a sub-manifold of $J^2\mathcal{B}$ having natural coordinates $(x^\mu; \varphi, \varphi_\mu, \varphi_{\mu\nu})$. We see that (2.8.9) corresponds to the sub-manifold $S \subset J^2\mathcal{B}$ given by the constraint $F = \varphi_{00} - \sum_i \varphi_{ii} = 0$. A section $\sigma : \mathbb{R}^{m+1} \longrightarrow \mathbb{R}^{m+1} \times \mathbb{R} : (t, x^i) \mapsto (t, x^i; \varphi(t,x))$ is a solution of the wave equation (2.8.9) if and only if $j^2\sigma(t,x) \in S \subset J^2\mathcal{B}$. The principal symbol is defined as

$$\mathfrak{P}(x,\xi) = (\xi_0)^2 - \sum_i (\xi_i)^2 \tag{2.8.10}$$

[2] $M(n,n')$ denotes the set of all $n \times n'$ matrices.

so that the operator underlying equation (2.8.9) is not elliptic since $\mathfrak{P}(x, \xi)$ vanishes on the *light cone*

$$(\xi_0)^2 = \sum_i (\xi_i)^2 \tag{2.8.11}$$

Analogously, the so-called *heat kernel* (or *heat equation*)

$$\partial_t \varphi + \sum_i \partial_i \partial_i \varphi = 0 \tag{2.8.12}$$

has the following principal symbol

$$\mathfrak{P}(x, \xi) = \sum_i (\xi_i)^2 \tag{2.8.13}$$

so that the operator is not elliptic since it vanishes on the 0-axe $(\xi_0, 0, \dots, 0)$ for all ξ_0.

Finally, the *Laplacian operator*

$$\partial_t \partial_t \varphi + \sum_i \partial_i \partial_i \varphi = 0 \tag{2.8.14}$$

is elliptic since its principal symbol is

$$\mathfrak{P}(x, \xi) = (\xi_0)^2 + \sum_i (\xi_i)^2 \tag{2.8.15}$$

which is an invertible (1 by 1) matrix, i.e. it is a non-zero number for all $(\xi_0, \xi_i) \neq (0, 0)$.

Exercises

1- Use the tangent functor to define the prolongation (2.7.7) of a vector field $\xi = \xi^\mu \partial_\mu$ over a manifold M to a vector field $\hat{\xi} \equiv T(\xi)$ over the tangent bundle TM.
Hint: the lift $\hat{\xi}$ is the infinitesimal generator of the prolongation of the flow of ξ.

2- Describe the first order prolongation of the Hopf bundle $\mathcal{H}$ introduced in Section 1.1 and the prolongation of the right action $R_g : \mathcal{H} \longrightarrow \mathcal{H}$.
Hint: choose fibered coordinates $(x^\mu; \theta)$ on the Hopf bundle and express everything locally in those coordinates.

3- a) Prove the isomorphism $J^k(V(\mathcal{B})) \simeq V(J^k\mathcal{B})$ and discuss the canonicity of this isomorphism. (Both $V(\mathcal{B})$ and $J^k(\mathcal{B})$ are to be considered as bundles over M). Consider in particular the case $\mathcal{B} = (\mathbb{R} \times Q, \mathbb{R}, \mathrm{pr}_1; Q)$ and show that the above isomorphism reduces to a canonical isomorphism of TTQ into itself (the *swap map* of equation (2.7.9)).

b) Define finally a canonical isomorphism $T^*(TQ) \simeq T(T^*Q)$.

Hint: choose natural fibered coordinates and compute transition functions.

4- Determine the constraints which identify $J^2\mathcal{B}$ as a sub-bundle of $J^1\left(J^1\mathcal{B}\right)$. Generalize this to $J^{k+1}\mathcal{B}$ and $J^1\left(J^k\mathcal{B}\right)$ for each $k \geq 2$.

Hint: the coordinate $y^i_{\mu\nu}$ on $J^2\mathcal{B}$ is symmetric w.r.t. (μ,ν). Use induction.

5- Describe the differential equation of the mathematical pendulum as a submanifold of $J^2(\mathbb{R}\times S^1)$.

Hint: the configuration space is S^1; choose fibered coordinates (t,θ) on $\mathbb{R}\times Q$.

6- Calculate the Lie derivative of a section of T^*M, i.e. of a 1-form over M.

Hint: the standard result is $\pounds_\xi\omega = (i_\xi\, \mathrm{d} + \mathrm{d}\, i_\xi)\omega$, where i_ξ is contraction.

7- Check that $H(\mathrm{d}\omega) = D(H\omega)$ where $\omega \in \Omega(J^kB)$.

Hint: expand both sides modulo contact terms.

8- Check that the Lie derivative of a tensor field t defined as a section of the relevant tensor bundle $T^p_q(M)$ is equivalent to the ordinary Lie derivative defined by (2.7.10).

Hint: which is the relation between the Lie-dragging of t along $\hat{\xi}$ and $(\phi_s)_*t$?

Chapter 3

PRINCIPAL BUNDLES AND CONNECTIONS

Abstract The notion of *principal bundle* as introduced in Section 1.3 is recalled. *Invariant vector fields* are then introduced as well as the *bundle of vertical automorphisms* and the *bundle of infinitesimal generators of vertical principal automorphisms.* Lie derivatives for principal automorphisms and infinitesimal generators of vertical automorphisms are considered in detail. *Bundle connections* and *principal connections* are then introduced in many equivalent ways and Lie derivatives of principal connections are calculated. The *induced connections* on bundles which are associated to a principal bundle are introduced. As examples and applications, the explicit expression for principal connections is obtained on the frame bundle and on higher order frame bundles. The induced connection on the bundle of automorphisms and infinitesimal generators of vertical automorphisms are also calculated. Finally, we introduce the connection-dependent structures on bundles, such as *parallel transport, covariant derivatives of a section* and the *curvature* of a principal connection.

References

Some good references about principal bundles are [KN63], [MM92], [A79]. A fairly didactical introduction is [GS87], while one of the most complete and updated is [KMS93]. Some information can also be found in [CBWM82]. For the applications to Physics we refer to [BM94], [B81], [N90].

Some interesting result can be found also in the original papers [GFF88].

1. Principal Bundles and Their Right Action

We shall briefly recall the definition and properties of principal bundles we gave in Section 1.3.

A principal bundle is a bundle $\mathcal{P} = (P, M, \pi; G)$ which has a Lie group G as standard fiber and such that there exists a trivialization the transition functions of which act on G by left translations $L_g : G \longrightarrow G : h \mapsto g \cdot h$. A fibered morphism $\boldsymbol{\Phi} = (\Phi, \phi)$ between two principal bundles $\mathcal{P} = (P, M, \pi; G)$ and $\mathcal{P}' = (P', M', \pi'; G')$ is a *principal morphism with respect to a Lie group homomorphisms* $\theta : G \longrightarrow G'$ if the following diagram commutes:

$$\begin{array}{ccc} P & \xrightarrow{\Phi} & P' \\ {\scriptstyle R_g}\downarrow & & \downarrow{\scriptstyle R_{\theta(g)}} \\ P & \xrightarrow{\Phi} & P' \end{array} \tag{3.1.1}$$

We say that Φ is *equivariant* with respect to θ, i.e. $\Phi \circ R_g = R_{\theta(g)} \circ \Phi$. If $G = G'$ and $\theta = \mathrm{id}_G$ then Φ is simply called a *principal morphism*.

On any principal bundle $\mathcal{P} = (P, M, \pi; G)$ there is a canonical right action $R_g : P \longrightarrow P$ of G which is free, vertical and transitive on fibers; see proposition (1.3.20). Because of the right action there exists a correspondence between local sections and local trivializations. A trivialization of a principal bundle $\mathcal{P} = (P, M, \pi; G)$ corresponds to a family of local sections $\{(U_\alpha, \sigma^{(\alpha)})\}_{\alpha \in I}$ such that $M = \cup_{\alpha \in I} U_\alpha$. To a local section $\sigma^{(\alpha)}$ there corresponds the local trivialization:

$$t_{(\alpha)} : \pi^{-1}(U_\alpha) \longrightarrow U_\alpha \times G : p \mapsto (x, g) \tag{3.1.2}$$

where $x = \pi(p)$ and $g = g(x, p) \in G$ is uniquely determined by the condition $R_g \sigma^{(\alpha)}(x) = p$. Since p and $\sigma^{(\alpha)}(x)$ belong to the same fiber $\pi^{-1}(x)$, the group element g exists as the right action R_g is transitive on fibers and it is uniquely determined (the action is free). As usual let us denote $p = t_{(\alpha)}^{-1}(x, g)$ by $p = [x, g]_{(\alpha)}$. The local expression of the right action is thence

$$R_g : [x, h]_{(\alpha)} \mapsto [x, h \cdot g]_{(\alpha)} \tag{3.1.3}$$

Let $\{(U_\alpha, \sigma^{(\alpha)})\}_{\alpha \in I}$ be a trivialization of $\mathcal{P} = (P, M, \pi; G)$ and let U_α, U_β be two trivialization patches such that $U_{\alpha\beta} = U_\alpha \bigcap U_\beta \neq \varnothing$. Let $x \in U_{\alpha\beta}$ be a point in their intersection. Then we have:

$$\sigma^{(\alpha)}(x) = \sigma^{(\beta)}(x) \cdot g_{(\beta\alpha)}(x) \tag{3.1.4}$$

and $g_{(\beta\alpha)} : U_{\alpha\beta} \longrightarrow G$ are the transition functions since for any point $p \in \pi^{-1}(x)$ one has:

$$p = [x, h]_{(\alpha)} \qquad p = \sigma^{(\alpha)} \cdot h = \sigma^{(\beta)} \cdot g_{(\beta\alpha)} \cdot h = [x, g_{(\beta\alpha)} \cdot h]_{(\beta)} \tag{3.1.5}$$

2. Invariant Vector Fields

Let $\mathcal{P} = (P, M, \pi; G)$ be a principal bundle. One can consider the following local maps defined over a trivialization patch U_α:

$$\begin{aligned} R_g^{(\alpha)} &: \pi^{-1}(U_\alpha) \longrightarrow \pi^{-1}(U_\alpha) : & L_p^{(\alpha)} &: G \longrightarrow \pi^{-1}(U_\alpha) : \\ &: [x,h]_{(\alpha)} \mapsto [x, h\cdot g]_{(\alpha)} & &: g \mapsto R_g^{(\alpha)}(p) \\ L_g^{(\alpha)} &: \pi^{-1}(U_\alpha) \longrightarrow \pi^{-1}(U_\alpha) : & R_p^{(\alpha)} &: G \longrightarrow \pi^{-1}(U_\alpha) : \\ &: [x,h]_{(\alpha)} \mapsto [x, g\cdot h]_{(\alpha)} & &: g \mapsto L_g^{(\alpha)}(p) \end{aligned} \tag{3.2.1}$$

We remark that the maps $R_g^{(\alpha)}$ (and consequently $L_p^{(\alpha)}$) glue together to define a global map, i.e. the right action R_g, while on the contrary $L_g^{(\alpha)}$ (and consequently $R_p^{(\alpha)}$) do not.

Let us then fix a basis T_A ($A = 1, \dots, p = \dim(G)$) of the Lie algebra[1] $\mathfrak{g} \simeq T_eG$ of G and a system of coordinates g^a ($a = 1, \dots, p$) in a neighbourhood of the identity in G (which in turn define another basis $\partial_a \in T_eG$ in the Lie algebra). By considering now the right (resp., left) translation $\tilde{R}_g : G \longrightarrow G$ (resp., $\tilde{L}_g : G \longrightarrow G$) of G onto itself, one can define the *right* (resp., *left*) *invariant vector fields* $r_A(g)$ (resp., $l_A(g)$) on the Lie group G:

$$r_A(g) = T_e\tilde{R}_g(T_A) = R^a_A(g)\partial_a \qquad \left(l_A(g) = T_e\tilde{L}_g(T_A) = L^a_A(g)\partial_a\right) \tag{3.2.2}$$

Here we also introduced the invertible matrix $R^a_A(g)$ (resp., $L^a_A(g)$) which is the tangent map at $e \in G$ of the right (resp., left) translation, which is in fact invertible.

By using $L_p^{(\alpha)}$ (resp., $R_p^{(\alpha)}$) as defined in (3.2.1), we can define a pointwise basis of vertical right invariant (left invariant, respectively) vector fields on P as follows:

$$\rho_A(p) = T_eR_p^{(\alpha)}(T_A) \qquad \left(\lambda_A(p) = T_eL_p^{(\alpha)}(T_A)\right) \tag{3.2.3}$$

Proposition (3.2.4): the local vector fields ρ_A (resp., λ_A) are right (resp., left) invariant vector fields on the principal bundle $\mathcal{P}$.

Proof:

$$\begin{aligned} T_pR_g\rho_A(p) &= T_pR_g \circ T_eR_p^{(\alpha)}(T_A) = T_e(R_g \circ R_p^{(\alpha)})(T_A) = \\ &= T_eR^{(\alpha)}_{p\cdot g}(T_A) = \rho_A(p\cdot g) \end{aligned} \tag{3.2.5}$$

---

[1] The set of left invariant vector fields on a Lie group G is called the *Lie algebra of* G and it is denoted by $\mathfrak{g}$. As a vector space, it is isomorphic to T_eG which is thence endowed with a Lie algebra structure and it is identified with $\mathfrak{g}$.

Analogously, $T_p L_g^{(\alpha)} \lambda_A(p) = \lambda_A(L_g^{(\alpha)} p)$. ∎

We recall that the *structure constants* of the group G (also called the structure constants of its Lie algebra $\mathfrak{g}$) are defined by:

$$[l_B, l_C] = c^A_{\cdot BC}\, l_A \tag{3.2.6}$$

or equivalently by:

$$[r_B, r_C] = -c^A_{\cdot BC}\, r_A \tag{3.2.7}$$

Since the push-forward commutes with commutators, at the bundle level we have the following:

$$[\rho_B, \rho_C] = -c^A_{\cdot BC}\, \rho_A \qquad \Big([\lambda_B, \lambda_C] = c^A_{\cdot BC}\, \lambda_A\Big) \tag{3.2.8}$$

3. Local Expression of Principal Automorphisms

Let us fix as usual a local trivialization $\sigma^{(\alpha)}$ on $\mathcal{P}$ and a bundle endomorphism $\boldsymbol{\Phi} = (\Phi, \phi)$. We have:

$$\forall p = [x, h]_{(\alpha)} \in P,\ \exists \varphi(x, h) \in G \quad \Phi([x, h]_{(\alpha)}) = [\phi(x), h \cdot \varphi(x, h)]_{(\alpha)} \tag{3.3.1}$$

Here we have tacitly assumed that $\Phi(p)$ is again in $\pi^{-1}(U_\alpha)$. In general this is not true. However, this is true if $\boldsymbol{\Phi}$ is *"near"* the identity, which will be often the case. Anyway, one can consider the general case by choosing a local trivialization near the image point $\Phi(p)$. In this case nothing relevant changes but the expressions turn out to be more cumbersome. We leave to the reader the task of outlining the more general expressions.

Requiring $\boldsymbol{\Phi}$ to be a principal morphism we obtain:

$$\begin{aligned}
&\Phi([x, h\cdot g]_{(\alpha)}) = \Phi\left([x, h]_{(\alpha)} \cdot g\right) = \Phi\left([x, h]_{(\alpha)}\right) \cdot g \quad \Rightarrow \\
\Rightarrow \quad & [\phi(x), h \cdot g \cdot \varphi(x, h\cdot g)]_{(\alpha)} = [\phi(x), h \cdot \varphi(x, h) \cdot g]_{(\alpha)} \quad \Rightarrow \\
\Rightarrow \quad & g \cdot \varphi(x, h \cdot g) = \varphi(x, h) \cdot g \quad \Rightarrow \\
\Rightarrow \quad & \varphi(x, h\cdot g) = g^{-1} \cdot \varphi(x, h) \cdot g \quad \Rightarrow \\
\Rightarrow \quad & \varphi(x, g) = g^{-1} \cdot \varphi(x) \cdot g \qquad\qquad \varphi(x) := \varphi(x, e)
\end{aligned} \tag{3.3.2}$$

Then any principal endomorphism on P (in particular all automorphisms) can be locally written as:

$$\Phi([x, h]_{(\alpha)}) = [\phi(x), \varphi(x) \cdot h]_{(\alpha)} \qquad \varphi : U_\alpha \longrightarrow G \tag{3.3.3}$$

Accordingly, any global automorphism of $\mathcal{P}$ can be locally represented by a family of maps $\varphi^{(\alpha)} : U_\alpha \longrightarrow G$ which satisfy the *glueing condition*:

$$\varphi^{(\alpha)}(x) = g_{(\alpha\beta)}(x') \cdot \varphi^{(\beta)}(x) \cdot g_{(\beta\alpha)}(x) = g_{(\alpha\beta)}(x') \cdot \varphi^{(\beta)}(x) \cdot g^{-1}_{(\alpha\beta)}(x) \qquad (3.3.4)$$

as one can easily prove by setting $x' = \phi(x)$ and by considering the following diagram:

$$\begin{array}{ccc}
U_\alpha \times G & \xrightarrow{\quad\Phi\quad} & U_\alpha \times G \\
[x, h]_{(\alpha)} & & [\phi(x), \varphi^{(\alpha)}(x) \cdot h]_{(\alpha)} \\
\Big\downarrow t_{(\beta)} \circ t^{-1}_{(\alpha)} & & \Big\downarrow t_{(\beta)} \circ t^{-1}_{(\alpha)} \\
 & & [\phi(x), g_{(\beta\alpha)}(x') \cdot \varphi^{(\alpha)}(x) \cdot h]_{(\beta)} \\
U_\beta \times G & \xrightarrow{\quad\Phi\quad} & U_\beta \times G \\
[x, g_{(\beta\alpha)}(x) \cdot h]_{(\beta)} & & [\phi(x), \varphi^{(\beta)}(x) \cdot g_{(\beta\alpha)}(x) \cdot h]_{(\beta)}
\end{array} \qquad (3.3.5)$$

which is commutative if and only if equation (3.3.4) holds.

Since we can describe a vertical automorphism by the local objects $\varphi^{(\alpha)}$ and we know that these local objects glue together by means of the adjoint action

$$\mathrm{ad}_g : G \longrightarrow G : h \mapsto g \cdot h \cdot g^{-1} \qquad (3.3.6)$$

we can construct a fiber bundle such that its sections correspond to vertical automorphisms of $\mathcal{P}$. Let us in fact consider the associated bundle $\mathcal{P} \times_{\mathrm{ad}} G$, defined by the prescription of Section 1.5, i.e. the space of orbits of the action $\overline{\mathrm{ad}}$ on $P \times G$ obtained by extending the adjoint action on G by setting $\overline{\mathrm{ad}}_g(p, h) = (p \cdot g^{-1}, g \cdot h \cdot g^{-1})$.

Proposition (3.3.7): there is a correspondence between sections of $\mathcal{P} \times_{\mathrm{ad}} G$ and vertical automorphisms of $\mathcal{P}$.

Proof: Let $\boldsymbol{\Phi} = (\Phi, \mathrm{id}_M)$ be a vertical automorphism of $\mathcal{P}$. Locally, it can be represented by the expression

$$\Phi\left([x, h]_{(\alpha)}\right) = [x, \varphi^{(\alpha)}(x) \cdot h]_{(\alpha)} \qquad (3.3.8)$$

so that it defines a family of maps $\varphi^{(\alpha)} : U_\alpha \longrightarrow G$. Since Φ is global the elements of this family glue together by

$$\varphi^{(\alpha)}(x) = g_{(\alpha\beta)}(x) \cdot \varphi^{(\beta)}(x) \cdot g^{-1}_{(\alpha\beta)}(x) \qquad (3.3.9)$$

which is exactly the equation which provides globality for the section on $\mathcal{P}\times_{\mathrm{ad}} G$ locally given by the expressions

$$\varphi : M \longrightarrow P \times_{\mathrm{ad}} G : x \mapsto [x, \varphi^{(\alpha)}(x)]_{\mathrm{ad}} \tag{3.3.10}$$

Conversely, any section $\varphi : M \longrightarrow \mathcal{P}\times_{\mathrm{ad}} G$ has local expressions

$$\varphi : M \longrightarrow P \times_{\mathrm{ad}} G : x \mapsto [x, \varphi^{(\alpha)}(x)]_{\mathrm{ad}} \tag{3.3.11}$$

and the maps $\varphi^{(\alpha)} : U_\alpha \longrightarrow G$ glue together, i.e. equation (3.3.9) holds true. Thence one can define a global vertical automorphism of $\mathcal{P}$ locally given by

$$\Phi\left([x,h]_{(\alpha)}\right) = [x, \varphi^{(\alpha)}(x)\cdot h]_{(\alpha)} \tag{3.3.12}$$

These two correspondences are one the inverse of the other. ∎

Let us consider a flow of automorphisms of a principal fiber bundle $\mathcal{P} = (P, M, \pi; G)$; according to equation (3.3.3), it is locally given by:

$$\begin{cases} x' = \phi_s(x) \\ g' = \varphi_s(x)\cdot g \qquad\qquad \varphi_s : U_\alpha \longrightarrow G \end{cases} \tag{3.3.13}$$

The infinitesimal generator of this 1-parameter subgroup of automorphisms is $\Xi = \xi^\mu\, \partial_\mu + \xi^A\, \rho_A$, where we set:

$$\begin{cases} \xi^\mu = \dfrac{\mathrm{d}}{\mathrm{d}s}\phi^\mu_s\Big|_{s=0} \\ \xi^A = \bar{R}^A_a(g)\, \dfrac{\mathrm{d}}{\mathrm{d}s}(\varphi_s\cdot g)^a\Big|_{s=0} \end{cases} \tag{3.3.14}$$

We fixed $\rho_A = R^a_A(g)\, \partial_a$ to be a basis for vertical right invariant vector fields and we denoted by $\bar{R}^A_a(g)$ the inverse matrix of $R^a_A(g)$.

Since the flow is formed by principal automorphisms, then ξ^A is independent of g and it depends just on x. In fact, φ_s depends just on x and we have:

$$\begin{aligned} \xi^A =&\bar{R}^A_a(g)\, \frac{\mathrm{d}}{\mathrm{d}s}(\varphi_s\cdot g)^a\Big|_{s=0} = \\ =&\bar{R}^A_a(g)\, R^a_B(g)\, \frac{\mathrm{d}}{\mathrm{d}s}(\varphi_s)^b\Big|_{s=0}\, \bar{T}^B_b = \frac{\mathrm{d}}{\mathrm{d}s}(\varphi_s)^a\Big|_{s=0}\, \bar{T}^A_a \end{aligned} \tag{3.3.15}$$

where $\bar{T}^A_a$ is the inverse matrix of T^a_A, which is in turn given by $\rho_A(e) = T^a_A\partial_a \in V_p(\mathcal{P})$ or, equivalently, by $T_A = T^a_A\, \partial_a \in T_eG$.

On the bundle $\mathcal{P}\times_{\mathrm{ad}} G$ we have (local) fibered coordinates $(x^\mu; h^a)$. Any principal automorphism $\boldsymbol{\Phi} = (\Phi, \phi)$ on $\mathcal{P}$, locally expressed by

$$\begin{cases} x' = \phi(x) \\ g' = \varphi(x)\cdot g \end{cases} \tag{3.3.16}$$

canonically acts on $\mathcal{P} \times_{\text{ad}} G$. The induced automorphism on $\mathcal{P} \times_{\text{ad}} G$ is locally given by:

$$\begin{cases} x' = \phi(x) \\ h' = \varphi(x) \cdot h \cdot \varphi^{-1}(x) \end{cases} \tag{3.3.17}$$

Thence, we can define the Lie derivative of a section h of $\mathcal{P} \times_{\text{ad}} G$ with respect to the infinitesimal generator $\Xi = \xi^\mu \, \partial_\mu + \xi^A \, \rho_A$ of automorphisms on $\mathcal{P}$. The vector field Ξ induces a vector field $\hat{\Xi} = \xi^\mu \, \partial_\mu + \xi^a \, \partial_a$ on $\mathcal{P} \times_{\text{ad}} G$ locally given by

$$\xi^a(x, h) = \Big(R^a_A(h) - L^a_A(h) \Big) \xi^A(x) \tag{3.3.18}$$

Then, the Lie derivative of a section $h : x^\mu \mapsto (x^\mu; h^a(x))$ of the fiber bundle $\mathcal{P} \times_{\text{ad}} G$ is:

$$\pounds_\Xi h := \pounds_{\hat{\Xi}} h = \Big(\xi^\mu \text{d}_\mu h^a - \big(R^a_A(h) - L^a_A(h)\big) \xi^A \Big) \partial_a \tag{3.3.19}$$

We stress that, from a local viewpoint, Lie derivatives encode information about transformation rules of sections of the bundle. Being such transformation rules the transition functions of the bundles itself, Lie derivatives characterize in fact the global properties of the bundle considered.

4. Infinitesimal Generators of Automorphisms

Let us investigate the properties of infinitesimal generators of automorphisms on $\mathcal{P}$.

Theorem (3.4.1): if any infinitesimal generator of automorphisms is in the form

$$\Xi = \xi^\mu(x) \partial_\mu + \xi^A(x) \hat{\rho}_A \tag{3.4.2}$$

then $\hat{\rho}_A$ are a local basis of right invariant vertical vector fields.

Proof: infinitesimal generators of principal automorphisms are right invariant, then:

$$\begin{aligned} &T_p R_g \Xi_p = \Xi_{p \cdot g} \quad \Rightarrow \xi^A(x) T_p R_g \hat{\rho}_A(p) = \xi^A(x) \hat{\rho}_A(p \cdot g) \quad \Rightarrow \\ &\quad \Rightarrow T_p R_g \hat{\rho}_A(p) = \hat{\rho}_A(p \cdot g) \end{aligned} \tag{3.4.3}$$

which tells us that $\hat{\rho}_A$ are right invariant. They are trivially vertical and they are local generators of right invariant vertical vector fields. Then they are a local basis of right invariant vertical vector fields. ∎

Let us now consider another local trivialization $\sigma^{(\beta)}$ of $\mathcal{P}$ on U_β ($U_{\alpha\beta} \neq \varnothing$) with $\sigma^{(\beta)} = \sigma^{(\alpha)} \cdot \varphi(x)$. We have two local fibered coordinate systems $(x^\mu_{(\alpha)}, g^a_{(\alpha)})$

and $(x^\mu_{(\beta)}, g^a_{(\beta)})$ and then two local bases, $\{\partial^{(\alpha)}_\mu, \rho^{(\alpha)}_A\}$ and $\{\partial^{(\beta)}_\mu, \rho^{(\beta)}_A\}$, respectively, of vector fields on $\mathcal{P}$. The two local expressions $\Xi = \xi^\mu_{(\alpha)}(x)\partial^{(\alpha)}_\mu + \xi^A_{(\alpha)}(x)\rho^{(\alpha)}_A = \xi^\mu_{(\beta)}(x)\partial^{(\beta)}_\mu + \xi^A_{(\beta)}(x)\rho^{(\beta)}_A$ of an infinitesimal generator of automorphisms on $\mathcal{P}$ are then related by the equations

$$\begin{cases} \xi^\mu_{(\beta)} = J^\mu_\nu\, \xi^\nu_{(\alpha)} \\ \xi^A_{(\beta)} = \mathrm{Ad}^A_B(\varphi)\, \xi^B_{(\alpha)} + \xi^\mu_{(\alpha)}\, \partial_\mu \varphi^b \bar{R}^A_b(\varphi) \end{cases} \tag{3.4.4}$$

where we set

$$\mathrm{Ad}_g : \mathfrak{g} \longrightarrow \mathfrak{g} : T_A \mapsto \mathrm{Ad}^B_A(g) T_B \qquad\qquad \mathrm{Ad}^B_A(g) = \bar{R}^B_a(g)\, L^a_A(g) \tag{3.4.5}$$

for the adjoint representation of the group G onto its Lie algebra $\mathfrak{g}$. Namely, $\mathrm{Ad}_g \equiv T_e \mathrm{ad}_g : \mathfrak{g} \longrightarrow \mathfrak{g}$ is the tangent representation of the adjoint action $\mathrm{ad}_g : G \longrightarrow G$, given by equation (3.3.6), on the group G itself. We remark that one has

$$\lambda_A(p) = \mathrm{Ad}^B_A(g) \rho_B(p) \tag{3.4.6}$$

for the left invariant vector fields at $p = [(x, g)]_{(\alpha)}$.

Let us now consider the right action TR_g of G on TP. Locally, we can choose local natural coordinates $(x^\mu, h^a, \xi^\mu, \xi^a)$ on TP to represent the vector $\xi^\mu\, \partial_\mu + \xi^a\, \partial_a$ at $(x^\mu; h^a)_{(\alpha)} \in P$; then the action TR_g has the following expression:

$$(g; x^\mu, h^a, \xi^\mu, \xi^a) \mapsto (x^\mu, (h \cdot g)^a, \xi^\mu, R^a_b(h)\, \xi^b) \tag{3.4.7}$$

Then in each orbit $[x^\mu, h^a, \xi^\mu, \xi^b]$ there exists a representative of the form $(x^\mu, (e)^a, \xi^\mu, \bar{T}^A_a\, R^a_b(h^{-1})\, \xi^b)$. By setting $\xi^A = \bar{T}^A_a\, \bar{R}^a_b(h)\, \xi^b$, we define local coordinates (x^μ, ξ^μ, ξ^A) on the orbit space TP/G. It is fibered on M and we set $T\mathcal{P}/G = (TP/G, M, \pi; \mathbb{R}^m \oplus \mathfrak{g})$ for the bundle structure over M. If we now change local trivialization in $\mathcal{P}$ and consequently we choose a different set of natural fibered coordinates $(x'^\mu, h'^a, \xi'^\mu, \xi'^a)$ on TP, the induced coordinates on TP/G then change according to the rule:

$$\begin{cases} x' = x'(x) & \qquad\qquad J^\mu_\nu = \partial_\nu x'^\mu \\ \xi'^\mu = J^\mu_\nu\, \xi^\nu \\ \xi'^A = \mathrm{Ad}^A_B(\varphi)\, \xi^B + \xi^\mu\, \bar{R}^A_b(\varphi)\, \partial_\mu \varphi^b \end{cases} \tag{3.4.8}$$

which completely agrees with (3.4.4). Then infinitesimal generators of automorphisms are in one-to-one correspondence with global sections of $\pi : TP/G \longrightarrow M$.

Let us remark that we are not yet ready to fully exploit and understand the structure bundle of the bundle $T\mathcal{P}/G$ of the infinitesimal generators of automorphisms on $\mathcal{P}$. The structure bundle should in fact have the same transition functions of $T\mathcal{P}/G$, i.e. (3.4.4), but represented by left translations; the transition functions (3.4.4) depend on J^μ_ν (i.e. the transition functions of $L(M)$), on φ (i.e. the transition functions of $\mathcal{P}$) and also on $\partial_\mu \varphi^b$, something which we are not yet ready to interpret correctly. This will be done below, after gauge natural bundles will have been introduced. When we shall construct the structure bundle of $T\mathcal{P}/G$, we shall automatically endow it with a structure group.

For the moment let us thence restrict our attention to infinitesimal generators of *vertical* automorphisms, i.e. those in the local form $\Xi = \xi^A(x)\rho_A$ (i.e., $\xi^\mu = 0$). Of course, they are sections of the sub-bundle $V\mathcal{P}/G$ of $T\mathcal{P}/G$ which has local coordinates $(x^\mu; \xi^A)$ and transition functions

$$\xi'^A = \mathrm{Ad}^A_B(\varphi)\, \xi^B \tag{3.4.9}$$

In this case we can immediately show that $\mathcal{P}$ is the structure bundle of $V\mathcal{P}/G$. In fact, the components ξ^A transform as in equation (3.4.9) where φ are the transition functions of $\mathcal{P}$. Consequently, we can define the associated bundle $\mathcal{P} \times_{\mathrm{Ad}} \mathfrak{g}$ which is called *the bundle of vertical infinitesimal automorphisms of $\mathcal{P}$*, because of the following result:

Proposition (3.4.10): there is a one-to-one correspondence between vertical infinitesimal generator of automorphisms and sections of $\mathcal{P} \times_{\mathrm{Ad}} \mathfrak{g}$. As a consequence there is also a canonical isomorphism between $\mathcal{P} \times_{\mathrm{Ad}} \mathfrak{g}$ and $V\mathcal{P}/G$.

Proof: let us use the trivialization $\sigma^{(\alpha)}$ and the invariant vector fields $\rho^{(\alpha)}_A$ which correspond to some basis $T_A = T^a_A\, \partial_a \in \mathfrak{g}$ in the Lie algebra $\mathfrak{g}$.

Let us define local maps:

$$\psi_{(\alpha)} : \xi^A \rho^{(\alpha)}_A(p) \mapsto [\sigma^{(\alpha)}(x), \xi^A T_A]_{\mathrm{Ad}} \qquad p = [x, h]_{(\alpha)} \tag{3.4.11}$$

Let now $\sigma^{(\beta)}$ be another local trivialization such that $\sigma^{(\alpha)} = \sigma^{(\beta)} \cdot g_{(\beta\alpha)}$ and $\rho^{(\beta)}_A$ the basis of vertical right invariant vector fields which correspond to the same basis T_A in the Lie algebra $\mathfrak{g}$. The compatibility condition is then:

$$\begin{array}{ccc}
\xi^A \rho^{(\alpha)}_A(p) & \xrightarrow{\quad\psi_{(\alpha)}\quad} & [\sigma^{(\alpha)}(x), \xi^A T_A]_{\mathrm{Ad}} \\
\Big\| & & \Big\| \\
\xi^A \mathrm{Ad}^B_A(g_{(\beta\alpha)}) \rho^{(\beta)}_B(p) & \xrightarrow{\psi_{(\beta)}} & [\sigma^{(\beta)}(x), \xi^A \mathrm{Ad}^B_A(g_{(\beta\alpha)}(x)) T_B]_{\mathrm{Ad}}
\end{array} \tag{3.4.12}$$

The maps $\psi_{(\alpha)}$ are thence independent of the choice of the local trivialization; consequently, there is a canonical and one-to-one correspondence between vertical infinitesimal generators of automorphisms and sections of $\mathcal{P} \times_{\mathrm{Ad}} \mathfrak{g}$, as claimed.

As far as the isomorphism between $\mathcal{P} \times_{\mathrm{Ad}} \mathfrak{g}$ and $V\mathcal{P}/G$ is concerned, do simply consider the local maps $\Phi_{(\alpha)} : [\sigma^{(\alpha)}, \xi^A]_\lambda \mapsto [\xi^A\, \rho_A]_G$. They glue together since $\mathcal{P} \times_{\mathrm{Ad}} \mathfrak{g}$ and $V\mathcal{P}/G$ have the same transition functions. ■

On the bundle of infinitesimal generators of vertical automorphisms we have fibered coordinates $(x^\mu; \xi^A)$. Any infinitesimal generator $\Theta = \theta^\mu\, \partial_\mu + \theta^A\, \rho_A$ of automorphisms on $\mathcal{P}$ induces a vector field $\hat\Theta = \theta^\mu\, \partial_\mu + \hat\theta^A\, \partial_A$ on $\mathcal{P} \times_{\mathrm{Ad}} \mathfrak{g}$ locally given by:

$$\hat\theta^A = c^A_{\cdot\, BC}\, \theta^B \xi^C \tag{3.4.13}$$

We remark that in order to obtain the expression above we used the general identity:

$$R^a_A(g)\, \partial_a \mathrm{Ad}^E_B(g) = c^E_{\cdot\, AC}\, \mathrm{Ad}^C_B(g) \tag{3.4.14}$$

which is easy to prove.

As for sections of $\mathcal{P} \times_{\mathrm{ad}} G$, we can define the Lie derivative of sections of $\mathcal{P} \times_{\mathrm{Ad}} \mathfrak{g}$ with respect to infinitesimal generators of vertical automorphisms on $\mathcal{P}$, which is locally expressed by:

$$\pounds_\Theta \Xi := \pounds_{\hat\Theta} \Xi = \left(\theta^\mu \mathrm{d}_\mu \xi^A - c^A_{\cdot\, BC}\, \theta^B \xi^C\right) \partial_A \tag{3.4.15}$$

5. Bundle Connections

We shall here introduce bundle connections in several equivalent ways. The different definitions are all important because they are all used in the literature and because each one of them is well suited to deal with some specific problem.

Connections and Principal Connections

Let $\mathcal{B} = (B, M, \pi; F)$ be a fiber bundle, $t_{(\alpha)}$ be a local trivialization of $\mathcal{B}$ and $(x^\mu; y^i)$ be a system of fibered coordinates over $\mathcal{B}$. Fibered coordinates induce a natural local basis $\{\partial^{(\alpha)}_\mu, \partial^{(\alpha)}_i\}$ for tangent vectors to B.

Notice that vertical vectors are intrinsically defined on $\mathcal{B}$, i.e. without using trivializations or any other additional structure on $\mathcal{B}$. On any fiber bundle $\mathcal{B}$, in fact, $\ker(T\pi)$ is a well defined sub-bundle of TB; we have already introduced it (see Section 1.6) and we have denoted it by $V(\pi)$. On the contrary, there is no canonical notion of horizontal vectors. In fact, if we consider vectors in the form $\xi = \xi^\mu_{(\alpha)}\, \partial^{(\alpha)}_\mu$ and then change fibered coordinate systems from $(x^\mu_{(\alpha)}; y^i_{(\alpha)})$ to $(x^\mu_{(\beta)}; y^i_{(\beta)})$, the vector field ξ gains a "vertical" component, i.e. we have:

$$\xi = \xi^\nu_{(\alpha)} \frac{\partial x^\mu_{(\beta)}}{\partial x^\nu_{(\alpha)}} \partial^{(\beta)}_\mu + \xi^\nu_{(\alpha)} \frac{\partial y^i_{(\beta)}}{\partial x^\nu_{(\alpha)}} \partial^{(\beta)}_i \tag{3.5.1}$$

In other words, such vectors (which may be said to be *horizontal with respect to the trivialization*) depend on the local trivialization and, as such, are not canonical (nor, in general, global).

Since there is no canonical choice, connections provide a definition which is at least global.

Definition (3.5.2): we call *connection on $\mathcal{B}$* a family of subspaces (also called a *distribution*) $H : b \mapsto H_b \subset T_bB$ such that:

(a) it is a *smooth* family with respect to the point b, i.e. $\mathcal{H} = \{v \in TB \mid b = \tau_B(v),\ v \in H_b\}$ is a sub-bundle of TB

(b) $\forall b \in B,\ T_bB = H_b \oplus V_b(\pi)$ where $V_b(\pi)$ is the sub-space of vertical vectors at $b \in B$

A vector $v \in T_bB$ is *horizontal with respect to the connection* H if $v \in H_b$. ∎

Let now H be a connection over $\mathcal{B}$. Since $T_bB = H_b \oplus V_b(\pi)$ is a direct sum, then for any $v \in TM$ there exists a unique $\hat{v} \in H_b$ such that $T\pi(\hat{v}) = v$. The vector $\hat{v}$ is called the *horizontal lift* of $v \in TM$. If $u \in T_bB$ is a vector on the bundle which projects over $v = T_b\pi(u) \in TM$, then we shall call $u_{(H)} = \hat{v} \in H_b$ the *horizontal part of* $u \in T_bB$. Consequently, the vector $u_{(V)} = u - u_{(H)} \in V_b(\pi)$ is called the *vertical part of* $u \in T_bB$. Notice that the horizontal lift of $v \in TM$, the horizontal part and the vertical part of $u \in TB$ all explicitly depend on the choice of the connection H.

On a principal bundle $\mathcal{P} = (P, M, \pi; G)$ there are particular connections which preserve the right action.

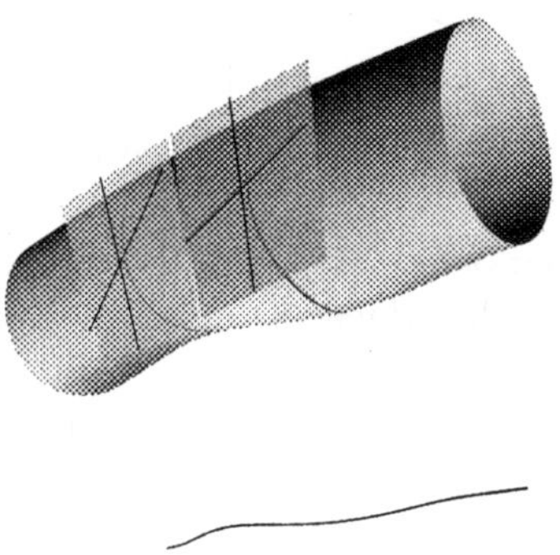

Fig. 13 – The splitting of TB induced by a connection.

Definition (3.5.3): a *principal connection* on a principal bundle $\mathcal{P}$ is a connection $H : p \mapsto H_p \subset T_pP$ such that:

(c) $\forall p \in P$, $\forall g \in G$, we have $\left(T_pR_g\right)H_p = H_{p\cdot g}$. ∎

Notice that (c) completely determines $H_{p\cdot g}$ once H_p is given. Since the right action is transitive on fibers, this means that in order to define a principal connection H it is sufficient to define H_p (in a smooth manner) at a family of points $p \in \pi^{-1}(x)$, one for each single fiber.

Example

Consider the (trivial) principal bundle $(\mathbb{R} \times \mathbb{R}, \mathbb{R}, p_1; \mathbb{R})$ where the fiber $\mathbb{R}$ is provided together with the Abelian group structure induced by the sum of real numbers. The right action is given by translations

$$R_a : (x, r) \mapsto (x, r + a) \tag{3.5.4}$$

Since the group is commutative, this is also a left action but this is of no interest to us.

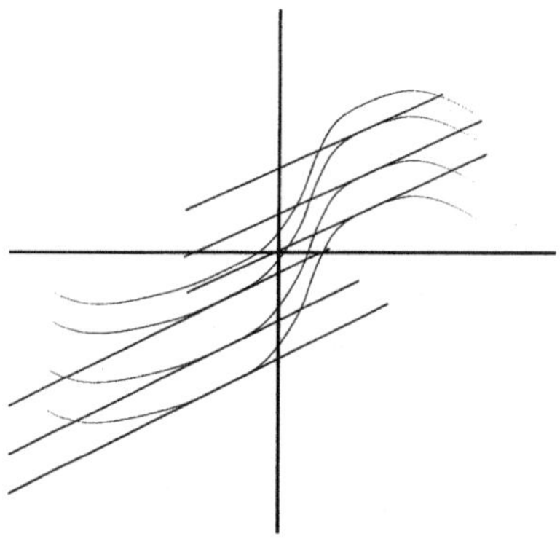

Fig. 14 – Principal connection on $\mathbb{R} \times \mathbb{R}$.

Consider a $\mathcal{C}^\infty$-function $f : \mathbb{R} \longrightarrow \mathbb{R}$ (the graph of which is also a global section of the bundle). It induces a slicing of $\mathbb{R} \times \mathbb{R}$ given by all translated graphs $F_a = \{(x, f(x) + a)\} \subset \mathbb{R} \times \mathbb{R}$.

Let $(x, r) \in F_a$ and let us define $H_{(x,r)} = \{\text{tangent line to } F_a \text{ at } (x, r)\}$. Any point $(x, r) \in \mathbb{R} \times \mathbb{R}$ belongs to one and only one F_a, the family $H_{(x,r)}$ is defined anywhere on $\mathbb{R} \times \mathbb{R}$ and it is a principal connection.

Notice however that this prescription to construct principal connections is not general as it requires a global section, which exists only on trivial principal bundles (because of corollary (1.3.26)).

Connections as Splittings

The geometric definition (3.5.2) of connection is simple and beautiful but it is somehow unsuited for applications. However, let H be a connection over the fiber bundle $\mathcal{B} = (B, M, \pi; F)$. The connection defines the map $\omega_H(b) : T_{\pi(b)}M \longrightarrow T_bB$ (or equivalently $\omega_H : B \longrightarrow T^*M \otimes TB$) such that $T_b\pi \circ \omega_H(b) = \mathrm{id}_{TM}$. The map $\omega_H(b)$ associates to a vector $v \in T_{\pi(b)}M$ its horizontal lift $\omega_H^b(v) = \hat{v} \in T_bB$. Conversely, this map ω_H contains all the information about the connection H. In fact, if one defines such a map ω, a connection H_ω can be obtained by setting

$$(H_\omega)_b \equiv \{u \in T_bB : \; u \in \mathrm{Im}(\omega(b))\} \tag{3.5.5}$$

The maps $H \mapsto \omega_H$ and $\omega \mapsto H_\omega$ are one the inverse to the other.

Thence, the connection on the bundle $\mathcal{B}$ may be seen as a splitting of the exact sequence of vector spaces:

$$0 \longrightarrow V_b(\pi) \longrightarrow T_bB \underset{\omega(b)}{\overset{T_b\pi}{\rightleftarrows}} T_{\pi(b)}M \longrightarrow 0 \tag{3.5.6}$$

Locally, in a system of fibered coordinates $(x^\mu; y^i)$ on $\mathcal{B}$, the map $\omega(b) : T_xM \longrightarrow T_bB$ is given by:

$$\omega_b(\xi) = \omega(\xi^\mu \partial_\mu) = \xi^\mu (\partial_\mu - \omega^i{}_\mu(b)\, \partial_i) \tag{3.5.7}$$

We remark that the minus sign in equation (3.5.7) has been introduced for later notational convenience. The map ω itself can now be seen as a map (for which we use the same symbol) $\omega : B \longrightarrow T^*M \otimes TB$, i.e.

$$\omega(b) = \mathrm{d}x^\mu \otimes (\partial_\mu - \omega^i{}_\mu(b)\, \partial_i) \tag{3.5.8}$$

We remark that, though $T^*(M) \otimes T(B)$ is a vector bundle over B and the map $\omega : B \longrightarrow T^*M \otimes TB$ may be seen as a section, the set of all such maps ω is not endowed with a vector space structure. In fact, the maps ω associated to connections on $\mathcal{B}$ are not general sections of $T^*(M) \otimes T(B)$, due to the fact that $T\pi \circ \omega = \mathrm{id}_{TM}$. In general, the sum of two maps ω_1 and ω_2 associated to two connections on $\mathcal{B}$ is in the form:

$$\omega_1 + \omega_2 = \mathrm{d}x^\mu \otimes (2\partial_\mu - ((\omega_1)^i_\mu(b) + (\omega_2)^i_\mu(b))\, \partial_i) \tag{3.5.9}$$

which is not associated to any connection on $\mathcal{B}$, not being in the same form (3.5.8). This remark will turn out to be very important from a physical viewpoint. In fact, as we shall see below, connections on a bundle are not sections of a vector bundle, but rather of an affine bundle. Then, even if there are always global connections (see theorem (1.3.2)), there is no preferred section (as the zero section on vector bundles). Consequently, if we choose a connection as a physical field, there is no preferred configuration to be chosen as a *vacuum state.* This is not a characteristic of connections, but it is quite a general feature of fields which are not sections of a vector bundle (e.g. metric fields); it is a feature which relies on the very basic structure of field theory and we shall later learn how to cope with it.

However, the local expression (3.5.8) for connections on $\mathcal{B}$ turns out to be important in order to analyze the behavior of connections with respect to bundle automorphisms. In fact, if $\boldsymbol{\Phi} = (\Phi, \phi)$ is a bundle automorphism of $\mathcal{B}$ locally given by

$$\begin{cases} x' = \phi(x) \\ y' = Y(x, y) \end{cases} \qquad \begin{array}{ll} J^\nu_\mu = \partial_\mu \phi^\nu & \bar{J}^\mu_\nu = \partial_\nu (\phi^{-1})^\mu \\ J^i_\nu = \partial_\nu Y^i & J^i_j = \partial_j Y^i \end{array} \tag{3.5.10}$$

and recalling the transformation rules (2.5.10) for the components of a vector field, we obtain immediately:

$$\omega'^i_\mu = \bar{J}^\nu_\mu \left(J^i_j \, \omega^j_\nu - J^i_\nu \right) \tag{3.5.11}$$

As usual, these are also the expression for transition functions. It is then evident that transition functions are affine and not linear.

Before looking for a characterization of connections as general sections of a suitable bundle let us specialize the characterization (3.5.8) to principal connections. Let $\mathcal{P} = (P, M, \pi; G)$ be a principal bundle, $\sigma^{(\alpha)}$ a local trivialization and ρ_A a local basis for vertical right invariant vector fields.

Proposition (3.5.12): a connection ω over $\mathcal{P}$ is principal if and only if it is locally given by $\omega = dx^\mu \otimes (\partial_\mu - \omega^A_\mu(x)\rho_A)$, i.e. if and only if the coefficients ω^A_μ are fiberwise constant.

Proof: if ω is in the required form, then $\omega(\xi)$ is right invariant and thus the connection is principal.

Conversely, if the connection $\omega = dx^\mu \otimes (\partial_\mu - \omega^A_\mu(x, g)\rho_A)$ is principal, then $\omega^A_\mu(x, g)$ is constant on fibers so that it depends just on x. ■

If we choose a trivialization $\sigma^{(\alpha)}$ of $\mathcal{P}$, a natural trivialization $\partial^{(\alpha)}_\mu$ of the frame bundle $L(M)$ of the basis M and we denote by $dx^\mu_{(\alpha)}$ the local dual basis of 1-forms over M, then the local expression of a principal connection is the following:

$$\omega = dx^\mu_{(\alpha)} \otimes (\partial^{(\alpha)}_\mu - (\omega^{(\alpha)})^A_\mu(x)\rho^{(\alpha)}_A) \tag{3.5.13}$$

If other local trivializations $\sigma^{(\beta)} = \sigma^{(\alpha)} \cdot g_{(\alpha\beta)}$ and $\partial^{(\beta)}_\mu = \partial^{(\alpha)}_\nu J^\nu_\mu$ are chosen, we get:

$$(\omega^{(\beta)})^A_\mu = \bar{J}^\nu_\mu \left(\mathrm{Ad}^A_B(g_{(\beta\alpha)})(\omega^{(\alpha)})^B_\nu - \bar{R}^A_a(g_{(\beta\alpha)})\partial_\nu g^a_{(\beta\alpha)} \right) \tag{3.5.14}$$

since we have:

$$\begin{cases} \partial_\mu^{(\alpha)} = J^\nu_{\ \mu}\partial_\nu^{(\beta)} + \bar{R}^A_{\ a}(g_{(\beta\alpha)})\partial_\mu g^a_{(\beta\alpha)}\rho_A^{(\beta)} \\ \rho_A^{(\alpha)} = \mathrm{Ad}^B_{\ A}(g_{(\beta\alpha)})\rho_B^{(\beta)} \\ \mathrm{d}x^\mu_{(\alpha)} = J^\mu_{\ \nu}\,\mathrm{d}x^\mu_{(\beta)} \end{cases} \tag{3.5.15}$$

Connections as Sections of the Jet Bundle

The representation (3.5.8) of connections on a fiber bundle $\mathcal{B} = (B, M, \pi; F)$ is unsatisfactory under some viewpoints. It is in fact good for transformation rules and quite effective when dealing with horizontal lifts; but it is cumbersome if connections have to be seen as sections of a bundle, which is particularly important, for example, to define Lie derivatives of a connection.

Let us then consider a local trivialization of $\mathcal{B}$ and the local expression $\omega = \mathrm{d}x^\mu \otimes (\partial_\mu - (\omega^{(\alpha)})^i_{\ \mu}\,\partial_i)$ of a connection on $\mathcal{B}$.

Proposition (3.5.16): there is a one-to-one correspondence between connections on $\mathcal{B}$ and sections of $\pi^1_0 : J^1B \longrightarrow B$.

Proof: we can associate to a connection H a section $\sigma : B \longrightarrow J^1B$ of the bundle $\pi^1_0 : J^1B \longrightarrow B$. Let us in fact consider a point $b \in B$ and the sub-space $H_b \subset T_bB$. Then there exists a local section $\rho_b : M \longrightarrow B$ defined over a neighbourhood of $x = \pi(b) \in M$ such that $\rho_b(x) = b$ and the tangent space $\mathrm{Im}(T_x\rho_b) \subset T_bB$ at the point b coincides with H_b. This can always be done because H_b does not contain vertical vectors other than the zero vector. Then we can associate to the connection ω the section

$$\sigma : B \longrightarrow J^1B : b \mapsto j^1_x\rho_b \tag{3.5.17}$$

Of course, the class $j^1_x\rho_b$ does not depend on the representative chosen for ρ_b, thus the map is well defined and global. Conversely, given a global section $\sigma : B \longrightarrow J^1B$, the connection can be defined by setting

$$H_b = \mathrm{Im}(T_x\rho_b) \tag{3.5.18}$$

where ρ_b is a representative of the class $j^1_x\rho_b = \sigma(b)$. ∎

The *bundle of connections* over $\mathcal{B}$ defined by the projection $\pi^1_0 : J^1B \longrightarrow B$ will be denoted by $\mathrm{Con}(\mathcal{B})$.

We can also prove the proposition (3.5.16) by a local argument. Let us choose local fibered coordinates $(x^\mu, y^i; y^i_\mu)$ on $J^1\mathcal{B}$. The local expression of the section σ induced by a connection $\omega = \mathrm{d}x^\mu \otimes (\partial_\mu - (\omega^{(\alpha)})^i_{\ \mu}\,\partial_i)$ is then:

$$\sigma^{(\alpha)} : (x^\mu, y^i) \mapsto (x^\mu, y^i; -(\omega^{(\alpha)})^i_{\ \mu}) \tag{3.5.19}$$

A brief comparison between transformation rules of connections $(3.5.11)$ and transition functions $(2.1.5)$ of $J^1\mathcal{B}$ shows that the globality of the connection ensures the glueing properties for the local sections $\sigma^{(\alpha)}$ and vice versa.

Thence generic bundle connections can be represented as generic sections of the bundle $\pi^1_0 : J^1B \longrightarrow B$. Of course, principal connections are not generic sections of $J^1\mathcal{P}$, due to the equivariance property (c) they have to satisfy. However, they can be represented as generic sections of another suitable bundle. First of all, let us notice that the right action of G defined on $\mathcal{P}$ induces a right action $J^1R_g : J^1P \longrightarrow J^1P$ of the same group G on the 1-jet prolongation $J^1\mathcal{P}$. We can consider the orbit space $J^1\mathcal{P}/G$ with respect to this action. If (x^μ, h^a, h^a_μ) are local natural coordinates on $J^1\mathcal{P}$, we can choose in each orbit $[x^\mu, h^a, h^a_\mu]_G$ a representative in the form:

$$[x^\mu, e^a, \theta^A_\mu]_G \tag{3.5.20}$$

where e is the identity element in G and $\theta^A_\mu = \bar{R}^A_a(h)h^a_\mu$. Then (x^μ, θ^A_μ) are local coordinates on $J^1\mathcal{P}/G$ which is a bundle over M with respect to the projection $\pi' : (x^\mu, \theta^A_\mu) \mapsto (x^\mu)$; the standard fiber is $\mathfrak{g} \otimes \mathbb{R}^m$. Its transition functions are given by:

$$(\theta^{(\beta)})^A_\mu = \bar{J}^\nu_\mu \Big(\mathrm{Ad}^A_B(g_{(\beta\alpha)})(\theta^{(\alpha)})^B_\nu + \bar{R}^A_a(g_{(\beta\alpha)})\partial_\nu g^a_{(\beta\alpha)}\Big) \tag{3.5.21}$$

Notice that $J^1\mathcal{P}/G$ is an affine bundle being transition functions affine maps.

Proposition (3.5.22): there is a one-to-one correspondence between principal connections on $\mathcal{P}$ and sections of the affine bundle $\pi' : J^1P/G \longrightarrow M$.

Proof: the transition functions $(3.5.21)$ of $J^1\mathcal{P}/G$ correspond to the transformation rules $(3.5.14)$ for the coefficients of the principal connection. This provides the globality of the section $\theta : x \mapsto (x; \theta^A_\mu(x))$ from the globality of the connection $\omega = \mathrm{d}x^\mu(\partial_\mu + \theta^A_\mu\, \rho_A)$ and vice versa. ∎

We shall denote by $\mathcal{C}_\mathcal{P}$ (or, by an abuse of language, by Con($\mathcal{P}$)) the bundle of principal connections over $\mathcal{P}$ given by the projection $\pi' : J^1P/G \longrightarrow M$. We stress that principal connections are sections of a bundle over M while generic connections are sections of a bundle over B.[2] When we try to set up a field theory, fields have to be sections of a bundle over spacetime M. In this perspective, principal connections are well suited to be considered as physical

[2] As we saw above, this is due to equivariance with respect to the vertical right action of G on P, which allows to reduce from $\mathcal{P}$ to the base manifold M.

fields, while generic connections usually are not. In fact, principal connections when regarded as fields, provide a geometric framework which is suited to deal with gauge theories.

Any infinitesimal generator $\Xi = \xi^\mu\,\partial_\mu + \xi^A\,\rho_A$ of automorphisms of $\mathcal{P}$ induces a vector field $\hat{\Xi} = \xi^\mu\,\partial_\mu + \xi^A_\mu\,\partial^\mu_A$ over $J^1\mathcal{P}/G$ (here ∂^μ_A denotes the partial derivative with respect to θ^A_μ) which is locally given by the following expression:

$$\xi^A_\mu = \mathrm{d}_\mu \xi^A + c^A_{\cdot BC}\xi^B\theta^C_\mu - \mathrm{d}_\mu\xi^\nu\,\theta^A_\nu \tag{3.5.23}$$

where we used again identity (3.4.14). The Lie derivative of the principal connection ω is thence:

$$\pounds_\Xi\omega := \pounds_{\hat{\Xi}}\omega = \left(\xi^\nu \mathrm{d}_\nu\omega^A_\mu + \mathrm{d}_\mu\xi^A - c^A_{\cdot BC}\xi^B\omega^C_\mu + \mathrm{d}_\mu\xi^\nu\,\omega^A_\nu\right)\partial^\mu_A \tag{3.5.24}$$

We shall see another expression for the Lie derivative of ω as soon as we introduce the curvature of ω and the notion of covariant derivative. The systematic approach to the way in which infinitesimal generators of automorphisms of $\mathcal{P}$ induce vector fields on bundles such as $J^1\mathcal{P}/G$ will be given in Chapter 5, where we shall deal with gauge natural bundles.

Connections as Forms Valued in the Lie Algebra

Finally, principal connections may be regarded as 1-forms valued in the Lie algebra $\mathfrak{g}$ of the structure group G. As already noticed, once a principal connection ω is fixed on a principal bundle $\mathcal{P} = (P, M, \pi; G)$, each vector $\Xi \in T_pP$ can be split into a horizontal part $\Xi_{(H)} = \omega_p(\Xi) \in H_p$ and a vertical part $\Xi_{(V)} \in V_p(\pi)$. Even if vertical vectors are defined regardless to any connection, both the horizontal and the vertical part depend on the connection. We also notice that $T_eL_p : \mathfrak{g} \longrightarrow V_p(\pi) : T_A \mapsto \lambda_A(p)$ is an isomorphism, for all $p \in P$. In this way, one can define a form $\bar{\omega}$ over P with values in the Lie algebra $\mathfrak{g}$ such that $\bar{\omega}_p(\Xi) \in \mathfrak{g}$ is the element such that $T_eL_p(\bar{\omega}_p(\Xi)) = \Xi_{(V)}(p)$ corresponding, via $(T_eL_p)^{-1}$, to the vertical vector $\Xi_{(V)}$.

The form $\bar{\omega}$ satisfies the following properties:

$$\begin{aligned} &\text{(a)}\ \ \bar{\omega}_p(H_p) = 0\\ &\text{(b)}\ \ \forall\xi \in \mathfrak{g},\ \bar{\omega}_p(T_eL_p(\xi)) = \xi\\ &\text{(c)}\ \ \tilde{R}^*_\gamma\bar{\omega}_{p\cdot\gamma} = \mathrm{Ad}_{\gamma^{-1}}\,\bar{\omega}_p \end{aligned} \tag{3.5.25}$$

Conversely, if a 1-form $\bar{\omega}$ over P with values in $\mathfrak{g}$ satisfies conditions (b) and (c) of (3.5.25) then property (a) defines a family of sub-spaces $H_p \subset TP$ and thence a principal connection which is the connection inducing $\bar{\omega}$. Locally, let us fix $p = [x, g]_{(\alpha)}$ and denote by $\theta^A_{(L)} = \bar{L}^A_a(g)\,\mathrm{d}g^a$ the local basis of left invariant 1-forms which is the dual basis to $\lambda_A = L^a_A(g)\,\partial_a$ (here the matrix $\bar{L}^A_a$ is the inverse of L^a_A); then the 1-form $\bar{\omega}$ has the following expression:

$$\bar{\omega}_p = \left[\theta^A_{(L)}(p) + \mathrm{Ad}^A_B(g^{-1})\omega^B_\mu(x)\mathrm{d}x^\mu\right] \otimes T_A \tag{3.5.26}$$

The components $\omega^A_\mu(x)$ are usually called *vector potentials*. If the connection form $\bar{\omega}$ is given by (3.5.26) then the induced connection is in the form:

$$\omega = \mathrm{d}x^\mu \otimes (\partial_\mu - \omega^A_\mu(x)\,\rho_A) \tag{3.5.27}$$

If we consider a local section $\sigma : x^\mu \mapsto (x^\mu, g^a(x))$ of the principal bundle $\mathcal{P}$ and a connection 1-form $\bar{\omega} = (\theta^A_{(L)} + \mathrm{Ad}^A_B\omega^B_\mu\mathrm{d}x^\mu) \otimes T_A$, we can define the pull-back of $\bar{\omega}$ along σ obtaining a local 1-form in M valued in the Lie algebra $\mathfrak{g}$, given by:

$$\sigma^*\,\bar{\omega} = \left[\bar{L}^A_a(g)\,\partial_\mu g^a(x) + \mathrm{Ad}^A_B(g^{-1})\omega^B_\mu(x)\right]\,\mathrm{d}x^\mu \otimes T_A \tag{3.5.28}$$

The local section σ is called a *(local) gauge fixing* (or just a *gauge*) and the form $\sigma^*\,\bar{\omega}$ is called the *vector potential* with respect to that (local) gauge fixing. We remark that the local gauge σ also induces a local trivialization of $\mathcal{P}$. In the induced local trivialization, the section σ has the expression $\sigma : x^\mu \mapsto (x^\mu, e)$ and the vector potential is of the form

$$\sigma^*\,\bar{\omega} = \omega^A_\mu(x)\,\mathrm{d}x^\mu \otimes T_A = \mathrm{d}x^\mu \otimes \omega_\mu = \omega^A \otimes T_A \qquad \begin{cases} \omega_\mu = \omega^A_\mu\,T_A \\ \omega^A = \omega^A_\mu\,\mathrm{d}x^\mu \end{cases} \tag{3.5.29}$$

which justifies the abuse of language.

The vector potentials are often used in physical applications. Of course vector potentials are local (unless the bundle $\mathcal{P}$ is trivial and a global gauge fixing is chosen). Let now G be a (real) matrix group, i.e. a subgroup of some linear group $\mathrm{GL}(k,\mathbb{R})$ for some integer k; if one changes the local gauge fixing by $\sigma' : x^\mu \mapsto (x^\mu, g'^a(x))$ with $g' = g\cdot\gamma$ where $\gamma(x)$ is a local function, so that $\sigma' = \sigma\cdot\gamma(x)$, then the vector potential $\omega_\mu = \omega^A_\mu\,T_A$ changes according to the transformation law:

$$\omega'_\mu = \gamma^{-1}\cdot(\omega_\mu\cdot\gamma + \partial_\mu\gamma) \tag{3.5.30}$$

which is obtained by specializing equation (3.5.14). If $G \subseteq \mathrm{GL}(k,\mathbb{R})$ then the inclusion $i : G \longrightarrow \mathrm{GL}(k,\mathbb{R})$ induces an inclusion of the corresponding Lie algebras $T_e i : \mathfrak{g} \longrightarrow \mathfrak{gl}(k,\mathbb{R})$. Let $\{T_A\}$ be a basis of $\mathfrak{g}$; each generator T_A can be regarded as a matrix $T_A = (T_A)^i{}_j$ with $i,j = 1,\dots,k$ in $\mathfrak{gl}(k,\mathbb{R})$ by means of the inclusion $T_e i$. Then the coordinate expression of (3.5.30) on G reads as follows in $\mathrm{GL}(k,\mathbb{R})$:

$$\omega'^i{}_{j\mu} = \bar{\gamma}^i_k\,\omega^k{}_{h\mu}\,\gamma^h_j + \bar{\gamma}^i_k\,\partial_\mu\gamma^k_j \tag{3.5.31}$$

where $\bar{\gamma}^i_k$ denotes the inverse of the matrix $\gamma^k{}_i$ corresponding, via $T_e i$, to the transition functions on $\mathcal{P}$. Clearly, the matrix entries $\gamma^k{}_i$ are not coordinates on G; they are called *overdetermined coordinates* and they are always meant to be constrained by the equations identifying out G into $\mathrm{GL}(k, \mathbb{R})$.

In general let us denote by $\Omega^k(P, \mathfrak{g})$ the set of all k-forms on P valued in the Lie algebra $\mathfrak{g}$, i.e. the space of all sections of the vector bundle $\Lambda^k T^*(P) \otimes \mathfrak{g} \equiv \Lambda^k T^*(P) \otimes_P (P \times \mathfrak{g})$, where Λ^k denotes exterior power. The set $\Omega(P, \mathfrak{g}) = \oplus_k \Omega^k(P, \mathfrak{g})$ of all forms on P valued in the Lie algebra $\mathfrak{g}$ is a graded Lie algebra with the commutator defined by:

$$[\theta, \omega] = \theta^A \wedge \omega^B \otimes [T_A, T_B] = c^C_{\cdot AB}\, \theta^A \wedge \omega^B \otimes T_C \tag{3.5.32}$$

for all $\theta = \theta^A\, T_A$ and $\omega = \omega^A\, T_A \in \Omega^k(P, \mathfrak{g})$.

We can define the *horizontal exterior differential* D_ω (also called *exterior covariant differential*) which acts on any k-form $\theta \in \Omega^k(\mathcal{P}, \mathfrak{g})$ on P to give a horizontal $(k+1)$-form $D_\omega \theta \in \Omega^{k+1}_{\mathrm{hor}}(\mathcal{P}, \mathfrak{g})$ on $\mathcal{P}$, according to the following rule:

$$\mathrm{D}_\omega \theta(\Xi^{(1)}, \ldots, \Xi^{(k+1)}) := d\theta(\Xi^{(1)}_{(H)}, \ldots, \Xi^{(k+1)}_{(H)}) \qquad \forall \theta \in \Omega^k(P, \mathfrak{g}) \tag{3.5.33}$$

For example, let $\theta : P \longrightarrow \mathfrak{g}$ be a 0-form valued in the Lie algebra $\mathfrak{g}$ (i.e. a $\mathfrak{g}$-valued function). The horizontal exterior differential is then:

$$\mathrm{D}_\omega \theta(\Xi) = \mathrm{d}\theta(\Xi_{(H)}) = \xi^\mu (\partial_\mu \theta^A - \partial_b \theta^A\, R^b_B\, \omega^B_\mu) \otimes T_A \tag{3.5.34}$$

where $\Xi = \xi^\mu \partial_\mu + \xi^A\, \rho_A$ and $\omega = \mathrm{d}x^\mu \otimes (\partial_\mu - \omega^A_\mu\, \rho_A)$. Then we obtain

$$\mathrm{D}_\omega \theta = (\partial_\mu \theta^A - \partial_b \theta^A\, R^b_B\, \omega^B_\mu)\, \mathrm{d}x^\mu \otimes T_A \tag{3.5.35}$$

Analogously, for a 1-form $\theta = (\theta^A_\mu \mathrm{d}x^\mu + \theta^A_a \mathrm{d}g^a) \otimes T_A$ valued in the Lie algebra $\mathfrak{g}$, the horizontal exterior differential is given by:

$$\begin{aligned} \mathrm{D}_\omega \theta =& (\partial_\mu \theta^A_\nu - \partial_\mu \theta^A_a\, R^a_B\, \omega^B_\nu - \partial_a \theta^A_\nu\, R^a_B\, \omega^B_\mu + \\ & + \partial_b \theta^A_c\, R^b_B\, R^c_C\, \omega^B_\mu\, \omega^C_\nu)\, \mathrm{d}x^\mu \wedge \mathrm{d}x^\nu \otimes T_A \end{aligned} \tag{3.5.36}$$

Notice that if θ is a connection 1-form (which locally amounts to require that $\theta^A_\nu = \mathrm{Ad}^A_B(g^{-1})\theta^B_\nu(x)$ and $\theta^A_a = \bar{L}^A_a(g)$) this expression simplifies to

$$\begin{aligned} \mathrm{D}_\omega \theta =& \mathrm{Ad}^A_E(g^{-1}) \Big(\partial_\mu \theta^E_\nu + c^E_{\cdot BC}\, \theta^B_\mu\, \omega^C_\nu + \\ & - \tfrac{1}{2} c^E_{\cdot BC}\, \omega^B_\mu\, \omega^C_\nu \Big)\, \mathrm{d}x^\mu \wedge \mathrm{d}x^\nu \otimes T_A \end{aligned} \tag{3.5.37}$$

For example, if we consider the horizontal differential $\Omega = \mathrm{D}_\omega \bar{\omega}$ of the connection 1-form $\bar{\omega}_p = (\mathrm{Ad}^A_B(g^{-1})\omega^B_\mu(x)\, \mathrm{d}x^\mu + \theta^A_{(L)}) \otimes T_A$ at $p = [x, g]_{(\alpha)}$, we obtain:

$$\Omega = \tfrac{1}{2} \mathrm{Ad}^A_D(g^{-1}) \Big(\partial_\mu \omega^D_\nu - \partial_\nu \omega^D_\mu + c^D_{\cdot BC}\, \omega^B_\mu\, \omega^C_\nu \Big)\, \mathrm{d}x^\mu \wedge \mathrm{d}x^\nu \otimes T_A \tag{3.5.38}$$

where we used for the structure constants of G the identity:

$$c^{A}_{\cdot\,BC} = \bar{L}^{A}_{b}\,(L^{a}_{B}\,\partial_a L^{b}_{C} - L^{a}_{C}\,\partial_a L^{b}_{B}) \tag{3.5.39}$$

Using (3.5.32) we see that (3.5.38) is the local expression of the following identity

$$\Omega = \mathrm{d}\bar{\omega} + \tfrac{1}{2}[\bar{\omega},\bar{\omega}] \tag{3.5.40}$$

which is called the *Maurer-Cartan equation.*

Definition (3.5.41): the 2-form $\Omega = \frac{1}{2}\mathrm{Ad}^{A}_{B}(g^{-1})\, F^{B}_{\mu\nu}\, \mathrm{d}x^{\mu} \wedge \mathrm{d}x^{\nu} \otimes T_A$ the coefficients of which are given by:

$$F^{B}_{\mu\nu} = \partial_\mu \omega^{B}_{\ \nu} - \partial_\nu \omega^{B}_{\ \mu} + c^{B}_{\cdot\,CD}\, \omega^{C}_{\ \mu}\, \omega^{D}_{\ \nu} \tag{3.5.42}$$

is called the *curvature of the connection* ω. ∎

6. Induced Connections on Associated Bundles

We shall now investigate how a principal connection on $\mathcal{P}$ induces a connection on each bundle associated to $\mathcal{P}$. Let us then choose a principal connection $\omega = dx^\mu \otimes (\partial_\mu - \omega^{A}_{\mu}\rho_A)$ on a principal bundle $\mathcal{P} = (P, M, \pi; G)$ and let us proceed to define a canonical connection over any fiber bundle $(\mathcal{P}\times_\lambda F)$ associated to $\mathcal{P}$. That connection will be called the *induced connection.*

If we fix a point $\varphi \in F$, we can define a map

$$\Phi_\varphi : P \longrightarrow P \times_\lambda F : p \mapsto [p,\varphi]_\lambda \tag{3.6.1}$$

Let (p,φ) be a representative for the point $[p,\varphi]_\lambda \in P \times_\lambda F$. Consider the tangent map $T_p\Phi_\varphi : T_pP \longrightarrow T_{[p,\varphi]_\lambda}(P \times_\lambda F)$ which defines horizontal subspaces in the associated bundle:

$$\hat{H}_{[p,\varphi]_\lambda} := T_p\Phi_\varphi(H_p) \tag{3.6.2}$$

Proposition (3.6.3): $\hat{H}_{[p,\varphi]_\lambda}$ is well defined and it does not depend on the representative chosen for the point $[p,\varphi]_\lambda \in P \times_\lambda F$.

Proof: in fact, let $(p',\varphi') = (p\cdot g, \lambda(g^{-1},\varphi))$ be another representative of the same point $[p,\varphi]_\lambda$ in the associated bundle. We have:

$$\begin{aligned}\hat{H}'_{[p',\varphi']_\lambda} &:= T_{p'}\Phi_{\varphi'}(H_{p'}) = T_{p'}\Phi_{\varphi'} \circ T_pR_g(H_p) = \\ &= T_p\Big(\Phi_{\varphi'} \circ R_g\Big)(H_p) = T_p\Phi_\varphi(H_p) = \hat{H}_{[p,\varphi]_\lambda}\end{aligned} \tag{3.6.4}$$

Thus $\hat{H}'_{[p',\varphi']_\lambda} = \hat{H}_{[p,\varphi]_\lambda}$ is well defined. ∎

Proposition (3.6.5): $\hat{H} : [p,\varphi]_\lambda \mapsto \hat{H}_{[p,\varphi]_\lambda}$ is a connection over $\mathcal{P} \times_\lambda F$.

Proof: for each $\xi \in T_xM$ we define a vector $\hat{\omega}(\xi) \in \hat{H}_{[p,\varphi]_\lambda}$ (which projects over ξ) in the following manner:

$$\hat{\omega}(\xi) = T_p\Phi_\varphi\omega_p(\xi) \tag{3.6.6}$$

This vector is well defined; in fact, if a different representative $(p',\varphi') = (p\cdot g,\ \lambda(g^{-1},\varphi))$ is chosen for the same point $[p,\varphi]_\lambda \in P\times_\lambda F$ we have

$$T_{p'}\Phi_{\varphi'}\omega_{p'}(\xi) = T_p\Big(\Phi_{\varphi'}\circ R_g\Big)\omega_p(\xi) = T_p\Phi_\varphi\omega_p(\xi) = \hat{\omega}(\xi) \tag{3.6.7}$$

Choosing fibered coordinates $(x^\mu;\varphi^i)$ in $P\times_\lambda F$, we have the following local expressions for $\hat{\omega}$:

$$\hat{\omega} = \mathrm{d}x^\mu \otimes \Big(\partial_\mu - \omega^i{}_\mu(x,\varphi)\partial_i\Big) \qquad \omega^i{}_\mu(x,\varphi) = T^a_A\,\partial_a\lambda^i(e,\varphi)\,\omega^A{}_\mu(x) \tag{3.6.8}$$

If $v \in \hat{H}_{[p,\varphi]_\lambda}$ and $\xi = T\pi(v)$, then $v = \hat{\omega}(\xi)$; thus $\hat{H}_{[p,\varphi]_\lambda} = \hat{\omega}(T_xM)$. Accordingly, one easily obtains $T_{[p,\varphi]_\lambda}(\mathcal{P}\times_\lambda F) = V_{[p,\varphi]_\lambda}(\pi)\oplus\hat{H}_{[p,\varphi]_\lambda}$, i.e. $\hat{H}$ is a connection on $\mathcal{P}\times_\lambda F$.
∎

As already stressed above, in the sequel we shall mainly use bundles with a structure group, i.e. bundles $\mathcal{B}$ obtained as bundles associated to some principal bundle $\mathcal{P}$. Accordingly, we shall use connections on $\mathcal{B}$ which are induced by principal connections on $\mathcal{P}$.

7. Examples

We shall here provide some examples of connections on principal bundles and associated bundles.

Principal Connections on the Frame Bundle

Let us consider the frame bundle $L(M) = (LM, M, \tau; \mathrm{GL}(m))$, which as we saw is a principal bundle having $\mathrm{GL}(m)$ as structure group[3]. A point in $L(M)$ is a basis e_a of the tangent space T_xM for some $x \in M$. Each local section $V^{(\alpha)}_b$ of $L(M)$ induces a local trivialization given by (see Section 1.3):

$$\begin{aligned} t_{(\alpha)} : \tau^{-1}(U_\alpha) \longrightarrow U_\alpha \times \mathrm{GL}(m) : e_a \mapsto (x, \overset{(\alpha)}{e}{}^b{}_a) \\ \text{where } x = \tau(e_a) \text{ and } e_a = \overset{(\alpha)}{V}_b\, \overset{(\alpha)}{e}{}^b{}_a \end{aligned} \tag{3.7.1}$$

[3] The Lie algebra of $\mathrm{GL}(m)$ is $\mathfrak{gl}(m) \simeq \mathbb{R}^{m^2}$.

Then $(x^\mu; (e^{(\alpha)})^b_{\ a})$ are fibered coordinates[4] on $L(M)$. The right action is expressed by:

$$R_\alpha\, e_a = e_b\ \alpha^b_{\ a} \qquad\qquad \alpha = \|\alpha^b_{\ a}\| \tag{3.7.2}$$

A local basis for vertical right invariant vector fields is then given by:

$$\overset{(\alpha)}{\rho}{}^a_{\ c} = \overset{(\alpha)}{e}{}^a_{\ b}\ \frac{\partial}{\partial e^{(\alpha)}{}^c_{\ b}} \equiv \overset{(\alpha)}{e}{}^a_{\ b}\ \overset{(\alpha)}{\partial}{}^b_{\ c} \tag{3.7.3}$$

and a principal connection on the frame bundle is locally given by:

$$\omega = dx^\mu \otimes \left(\partial^{(\alpha)}_\mu - \overset{(\alpha)}{\Gamma}{}^c_{\ a\mu}(x)\ \overset{(\alpha)}{\rho}{}^a_{\ c}\right) \tag{3.7.4}$$

where $(\Gamma^{(\alpha)})^c_{\ a\mu}$ are the coefficients of the connection, i.e. the vector potentials, over U_α. Principal connections of $L(M)$ will be also called *linear connections over* M.

In particular, we can choose $V^{(\alpha)}_b$ to be a natural frame $\partial^{(\alpha)}_\mu$. Accordingly, we have $e_a = (e^{(\alpha)})^\mu_{\ a}\ \partial^{(\alpha)}_\mu$, natural fibered coordinates are $(x^\mu; (e^{(\alpha)})^\mu_{\ a})$ and the local basis for vertical right invariant fields is

$$\overset{(\alpha)}{\rho}{}^\lambda_{\ \sigma} = \overset{(\alpha)}{e}{}^\lambda_{\ a}\ \frac{\partial}{\partial e^{(\alpha)}{}^\sigma_{\ a}} \equiv \overset{(\alpha)}{e}{}^\lambda_{\ a}\ \overset{(\alpha)}{\partial}{}^a_{\ \sigma} \tag{3.7.5}$$

In this case, the expression for the principal connection reproduces the classical form of linear connections in classical Differential Geometry:

$$\omega = dx^\mu \otimes \left(\partial^{(\alpha)}_\mu - \overset{(\alpha)}{\Gamma}{}^\sigma_{\ \lambda\mu}(x)\ \overset{(\alpha)}{\rho}{}^\lambda_{\ \sigma}\right) \tag{3.7.6}$$

where $(\Gamma^{(\alpha)})^\sigma_{\ \lambda\mu}$ are the coefficients of the connection, i.e. the vector potentials, in the natural trivialization of $L(M)$ over U_α. This justifies our terminology.

Let us now consider other fibered coordinates $(x'^\mu; (e^{(\beta)})^b_{\ a})$ induced by another local trivialization $V^{(\beta)}_b = V^{(\alpha)}_c\ g^c_b$. We then have transition functions:

$$\overset{(\alpha)}{e}{}^c_{\ a} = g^c_{\ d}\ \overset{(\beta)}{e}{}^d_{\ a} \tag{3.7.7}$$

which act on the left, as it must be, being $L(M)$ a principal bundle. In the tangent space of $L(M)$ we have:

$$\begin{cases} \overset{(\alpha)}{\partial}_\mu = J^\nu_\mu\ \overset{(\beta)}{\partial}_\nu + (\partial_\mu \bar{g}^d_{\ c})\ g^c_b\ \overset{(\beta)}{\rho}{}^b_{\ d} \\ \overset{(\alpha)}{\rho}{}^a_{\ c} = g^a_b\ \overset{(\beta)}{\rho}{}^b_{\ d}\ \bar{g}^d_{\ c} \end{cases} \qquad\qquad J^\nu_\mu = \frac{\partial x'^\nu}{\partial x^\mu}(x) \tag{3.7.8}$$

[4] Here x^μ is a shorthand for $x^\mu_{(\alpha)}$ on U_α.

and, consequently, for the components of a vector field $\Xi = \xi^\mu\, \partial_\mu + \xi^a_b\, \rho^b_a$ we have the transformation rule:

$$\begin{cases} \overset{(\beta)}{\xi}{}^\nu = J^\nu_\mu\, \overset{(\alpha)}{\xi}{}^\mu \\ \overset{(\beta)}{\xi}{}^d_b = \bar{g}^d_c\, \overset{(\alpha)}{\xi}{}^c_a\, g^a_b + \overset{(\alpha)}{\xi}{}^\mu\, \partial_\mu \bar{g}^d_c\, g^c_b \end{cases} \tag{3.7.9}$$

Accordingly, for the coefficients of the connection ω we have:

$$\overset{(\beta)}{\Gamma}{}^d_{b\mu} = \bar{J}^\nu_\mu \bar{g}^d{}_c \Big(\overset{(\alpha)}{\Gamma}{}^c_{a\nu} g^a_b + \partial_\nu g^c_b \Big) \tag{3.7.10}$$

In the natural trivialization, we have the Jacobian $\bar{J}^\nu_\mu$ in place of the transition functions $g^a{}_b$, because of $\partial^{(\beta)}_\mu = \bar{J}^\nu_\mu\, \partial^{(\alpha)}_\nu$. In that case, the transformation rules (3.7.10) reproduce the well known classical transformation rules for the coefficients of a linear connection over M, namely:

$$\overset{(\beta)}{\Gamma}{}^\sigma_{\lambda\mu} = J^\sigma_\rho \Big(\overset{(\alpha)}{\Gamma}{}^\rho_{\epsilon\nu} \bar{J}^\epsilon_\lambda \bar{J}^\nu_\mu + \bar{J}^\rho_{\lambda\mu} \Big) \tag{3.7.11}$$

If we fix a connection ω on $L(M)$ and we consider a natural local trivialization ∂_μ together with a general local trivialization $V_a = V^\mu_a\, \partial_\mu$, then the coefficients of the connection with respect to the frame V_a are related to the natural coefficients by the following expression:

$$\Gamma^a_{b\mu} = \bar{V}^a_\sigma \Big(\Gamma^\sigma_{\lambda\mu} V^\lambda_b + \partial_\mu V^\sigma_b \Big) \tag{3.7.12}$$

where we set $\bar{V}^a_\lambda$ for the inverse of the matrix V^λ_b.

To any connection $\omega = \mathrm{d}x^\mu \otimes (\partial_\mu - \Gamma^\sigma_{\lambda\mu}\, \rho^\lambda_\sigma)$ on $L(M)$ we can associate the so-called *transpose connection* ${}^t\omega$ which, in a natural trivialization, is expressed by:

$${}^t\omega = \mathrm{d}x^\mu \otimes (\partial_\mu - {}^t\Gamma^\sigma_{\lambda\mu}\, \rho^\lambda_\sigma) \qquad \text{where } {}^t\Gamma^\sigma_{\lambda\mu} = \Gamma^\sigma_{\mu\lambda} \tag{3.7.13}$$

Proposition (3.7.14): the difference between a connection ω on the frame bundle $L(M)$ and its transposed ${}^t\omega$ can be identified with a tensor on M, $T \in \Gamma(T^1_2(M))$

$$T = T^\lambda_{\mu\nu}\, \partial_\lambda \otimes \mathrm{d}x^\mu \otimes \mathrm{d}x^\mu \qquad T^\lambda_{\mu\nu} = \Gamma^\lambda_{\mu\nu} - {}^t\Gamma^\lambda_{\mu\nu} = \Gamma^\lambda_{\mu\nu} - \Gamma^\lambda_{\nu\mu} \tag{3.7.15}$$

called the *torsion of* ω.

Proof: in a natural trivialization of $L(M)$ the coefficients of the connections transform according to equation (3.7.11). Consequently, the coefficients of the torsion transform as:

$$\overset{(\beta)}{T}{}^\lambda_{\mu\nu} = J^\lambda_\epsilon \overset{(\alpha)}{T}{}^\epsilon_{\rho\sigma} \bar{J}^\rho_\mu \bar{J}^\sigma_\nu \tag{3.7.16}$$

Thus they transform as a tensor of $T^1_2(M)$. ∎

These are general features of objects defined on associated bundles to $L(M)$. This bundles are called *'natural bundles'* and they will be studied in detail in Chapter 4. Trivializations on natural bundles are induced by trivializations on $L(M)$. If we consider natural trivializations on $L(M)$, the induced trivializations are also called *'natural trivializations'*. In such trivializations transition functions depend just on the atlas chosen on the base manifold M and consequently sections may be regarded as global objects on M. In particular the torsion may be regarded as a tensor field on M. Analogously, this is the key feature which allows us to interpret the objects of Differential Geometry in terms of bundle theory.

We can also interpret proposition (3.7.14) in a different way. We proved in fact that the difference of two particular linear connections is a tensor field over M, i.e. a section of the tensor bundle $T^1_2(M)$ over M. This fact can be immediately generalized to each pair of linear connections on $L(M)$. Actually, we can define a (vertical) bundle morphism Φ

$$\Phi : \left(\frac{J^1L(M)}{\mathrm{GL}(m)} \times_M \frac{J^1L(M)}{\mathrm{GL}(m)}\right) \longrightarrow T^1_2(M) \tag{3.7.17}$$

If $(x^\mu, \Gamma^\lambda_{\mu\nu}, \Gamma'^\lambda_{\mu\nu})$ are natural coordinates on $\left(\frac{J^1L(M)}{\mathrm{GL}(m)} \times_M \frac{J^1L(M)}{\mathrm{GL}(m)}\right)$ and $(x^\mu, T^\lambda_{\mu\nu})$ are natural coordinates in $T^1_2(M)$, then the morphism has local expression $T^\lambda_{\mu\nu} = \Gamma^\lambda_{\mu\nu} - \Gamma'^\lambda_{\mu\nu}$. This bundle morphism Φ allows us to associate to each local section of the bundle $\left(\frac{J^1L(M)}{\mathrm{GL}(m)} \times_M \frac{J^1L(M)}{\mathrm{GL}(m)}\right)$, i.e. to each pair of linear connections of $L(M)$, a section of $T^1_2(M)$, i.e. a tensor field over M. If the two connections are transpose to each other, the induced tensor field is the torsion.

Finally, proposition (3.7.14) proves that the affine bundle $\frac{J^1L(M)}{\mathrm{GL}(m)}$ of linear connections is *modelled* on the vector bundle $T^1_2(M)$.

If a Riemannian metric $g = g_{\mu\nu}(x)\, dx^\mu \otimes dx^\nu$ is given on M, then one can define uniquely the so-called *Levi-Civita connection of* g, the natural coefficients of which are the Christoffel symbols:

$$\{^{\sigma}_{\lambda\mu}\}_g = \frac{1}{2} g^{\sigma\nu}\Big(- \partial_\nu g_{\lambda\mu} + \partial_\lambda g_{\mu\nu} + \partial_\mu g_{\nu\lambda}\Big) \tag{3.7.18}$$

As is well known, the Levi-Civita connection of g is torsionless, i.e.

$$T^\lambda_{\mu\nu} = \{^{\lambda}_{\mu\nu}\}_g - \{^{\lambda}_{\nu\mu}\}_g = 0 \tag{3.7.19}$$

and it is characterized by the compatibility condition:

$$\nabla_\mu g_{\lambda\sigma} = d_\mu g_{\lambda\sigma} - \{^{\epsilon}_{\lambda\mu}\}_g\, g_{\epsilon\sigma} - \{^{\epsilon}_{\sigma\mu}\}_g\, g_{\lambda\epsilon} = 0 \tag{3.7.20}$$

where $\nabla_\mu g_{\lambda\sigma}$ is the *covariant derivative* of the metric g with respect to its Levi-Civita connection (see equation (3.8.7) below for a general definition of covariant derivative).

For each metric on M there exists in fact a unique linear connection on $L(M)$ which satisfies the properties (3.7.19) and (3.7.20). The converse is not true. In fact, if we consider a new metric $g' = \lambda^2 g$ conformally equivalent to g with a constant conformal factor λ, then $\left\{{}^{\sigma}_{\mu\nu}\right\}_g = \left\{{}^{\sigma}_{\mu\nu}\right\}_{g'}$, i.e. they induce the same Levi-Civita connection. The curvature of a linear connection will be considered later.

Invariant Vector Fields on Higher Order Frame Bundles

This notion can be easily worked out in all frame bundles $L^s(M)$ of any order $s \geq 1$. We shall show how to construct it in the case $s = 3$. The generalization to order $s > 3$ is left to the reader, while the cases $s = 1$ and $s = 2$ can be naturally considered as sub-cases of our present discussion.

Let us now consider the bundle $L^3(M)$ of 3-frames. A point in $L^3(M)$ is in the form $j^3_x\epsilon$ with $\epsilon : \mathbb{R}^m \longrightarrow M$ locally invertible around $\overline{0} \in \mathbb{R}^m$ and $\epsilon(\overline{0}) = x$. Let us choose local coordinates (y^i) in a neighbourhood of $\overline{0} \in \mathbb{R}^m$ and coordinates (x^μ) near $x \in M$. The map $\epsilon : \mathbb{R}^m \longrightarrow M$ reads as $\epsilon : y^i \mapsto \epsilon^\mu(y)$ and the class $j^3_x\epsilon$ is represented by the coefficients of the Taylor expansion

$$\epsilon^\mu(y) = x^\mu + \partial_i\epsilon^\mu(0)\, y^i + \frac{1}{2!}\partial_{ij}\epsilon^\mu(0)\, y^i y^j + \frac{1}{3!}\partial_{ijk}\epsilon^\mu(0)\, y^i y^j y^k + \dots \tag{3.7.21}$$

i.e. $(x^\mu, \epsilon^\mu_i, \epsilon^\mu_{ij}, \epsilon^\mu_{ijk})$ are fibered coordinates on $L^3(M)$.

A diffeomorphism $\phi : M \longrightarrow M$ on the base manifold naturally induces an automorphism of $L^3(M)$ which is locally:

$$\begin{cases} x'^\mu = \phi^\mu(x) & J^\mu_\nu = \partial_\nu\phi^\mu \\ \epsilon'^\mu_i = J^\mu_\nu\, \epsilon^\nu_i & J^\mu_{\nu\lambda} = \partial_{\nu\lambda}\phi^\mu \\ \epsilon'^\mu_{ij} = J^\mu_{\alpha\beta}\, \epsilon^\alpha_i \epsilon^\beta_j + J^\mu_\nu\, \epsilon^\nu_{ij} & J^\mu_{\nu\lambda\sigma} = \partial_{\nu\lambda\sigma}\phi^\mu \\ \epsilon'^\mu_{ijk} = J^\mu_{\alpha\beta\gamma}\, \epsilon^\alpha_i \epsilon^\beta_j \epsilon^\gamma_k + J^\mu_{\alpha\beta}\left(\epsilon^\alpha_{ij}\epsilon^\beta_k + \epsilon^\alpha_{jk}\epsilon^\beta_i + \epsilon^\alpha_{ki}\epsilon^\beta_j\right) + J^\mu_\nu\, \epsilon^\nu_{ijk} & \end{cases} \tag{3.7.22}$$

Then any vector field $\xi = \xi^\mu\, \partial_\mu$ on the base manifold M induces a projectable vector field $\hat{\xi} = \xi^\mu\, \partial_\mu + \xi^\mu_i\, \partial^i_\mu + \xi^\mu_{ij}\, \partial^{ij}_\mu + \xi^\mu_{ijk}\, \partial^{ijk}_\mu$ on $L^3(M)$ which is locally given by:

$$\begin{cases} \xi^\mu_i = \mathrm{d}_\nu\xi^\mu\, \epsilon^\nu_i \\ \xi^\mu_{ij} = \mathrm{d}_{\alpha\beta}\xi^\mu\, \epsilon^\alpha_i\epsilon^\beta_j + \mathrm{d}_\nu\xi^\mu\, \epsilon^\nu_{ij} \\ \xi^\mu_{ijk} = \mathrm{d}_{\alpha\beta\gamma}\xi^\mu\, \epsilon^\alpha_i\epsilon^\beta_j\epsilon^\gamma_k + \mathrm{d}_{\alpha\beta}\xi^\mu\, \left(\epsilon^\alpha_{ij}\epsilon^\beta_k + \epsilon^\alpha_{jk}\epsilon^\beta_i + \epsilon^\alpha_{ki}\epsilon^\beta_j\right) + \\ \qquad + \mathrm{d}_\nu\xi^\mu\, \epsilon^\nu_{ijk} \end{cases} \tag{3.7.23}$$

The same vector field $\hat{\xi}$ can be also written as $\hat{\xi} = \xi^\mu\, \partial_\mu + \xi_\nu^{\ \mu}\, \rho^{\ \nu}_\mu + \xi^{\ \mu}_{\alpha\beta}\, \rho^{\ \alpha\beta}_\mu + \xi^{\ \mu}_{\alpha\beta\gamma}\, \rho^{\ \alpha\beta\gamma}_\mu$ where we set $\xi^\mu_\nu = \mathrm{d}_\nu\xi^\mu$, $\xi^{\ \mu}_{\alpha\beta} = \mathrm{d}_{\alpha\beta}\xi^\mu$ and $\xi^{\ \mu}_{\alpha\beta\gamma} = \mathrm{d}_{\alpha\beta\gamma}\xi^\mu$ and we defined:

$$\begin{cases} \rho_\mu^\nu = \epsilon_i^\alpha \, \partial_\mu^i + \epsilon_{ij}^\alpha \, \partial_\mu^{ij} + \epsilon_{ijk}^\alpha \, \partial_\mu^{ijk} \\ \rho_\mu^{\alpha\beta} = \epsilon_i^\alpha \epsilon_j^\beta \, \partial_\mu^{ij} + (\epsilon_{ij}^\alpha \epsilon_k^\beta + \epsilon_{jk}^\alpha \epsilon_i^\beta + \epsilon_{ki}^\alpha \epsilon_j^\beta) \partial_\mu^{ijk} \\ \rho_\mu^{\alpha\beta\gamma} = \epsilon_i^\alpha \epsilon_j^\beta \epsilon_k^\gamma \, \partial_\mu^{ijk} \end{cases} \tag{3.7.24}$$

The vector fields $(\rho_\mu^\nu, \rho_\mu^{\alpha\beta}, \rho_\mu^{\alpha\beta\gamma})$ defined above fiberwise generate vertical vector fields on $L^3(M)$ and, since $(\xi_\nu^\mu, \xi^\mu_{\alpha\beta}, \xi^\mu_{\alpha\beta\gamma})$ are independent at each point, they form a basis for right invariant vertical vector fields on $L^3(M)$ (see proposition (3.4.1)). Notice that ρ_μ^ν reproduce the basis of right invariant vertical vector fields on $L(M)$ already defined in (3.7.5).

The principal connections on $L^3(M)$ are thence in the following form:

$$\hat{\omega} = \mathrm{d}x^\mu \otimes \left(\partial_\mu - \Gamma^\lambda_{\nu\mu}(x) \, \rho_\lambda^\nu - \Gamma^\lambda_{\alpha\beta\mu}(x) \, \rho_\lambda^{\alpha\beta} - \Gamma^\lambda_{\alpha\beta\gamma\mu}(x) \, \rho_\lambda^{\alpha\beta\gamma} \right) \tag{3.7.25}$$

Notice that (3.7.6) is a particular case of (3.7.25).

Connections on the Tangent Bundle

The tangent bundle $T(M) = (TM, M, \tau_M; \mathbb{R}^m)$ is associated to the principal bundle $L(M)$. Using fibered coordinates $(x^\mu; v^a)$ we have for a vector $v = V_a \, v^a$, where $\{V_a\}$ is a frame on M. Therefore, we have the following local expression for the induced connection

$$\hat{\omega} = \mathrm{d}x^\mu \otimes \left(\partial_\mu - \Gamma^a_{b\mu} \, v^b \, \partial_a \right) \tag{3.7.26}$$

Connections on the Bundle of Vertical Automorphisms

On the bundle $P \times_{\mathrm{ad}} G$ of vertical automorphisms of $\mathcal{P}$ we have local fibered coordinates $(x^\mu; h^a)$. By taking the derivative of the adjoint representation:

$$T_A^a \partial_a \mathrm{ad}^b(e, h) = R_A^b(h) - L_A^b(h) \tag{3.7.27}$$

we can apply formula (3.6.8) for the induced connection and obtain

$$\hat{\omega} = \mathrm{d}x^\mu \otimes \left(\partial_\mu - (R_A^b(h) - L_A^b(h)) \omega_\mu^A \, \partial_b \right) \qquad \partial_b = \frac{\partial}{\partial h^b} \tag{3.7.28}$$

Connections on the Bundle of Infinitesimal Vertical Automorphisms

On the bundle $\mathcal{P} \times_{\mathrm{Ad}} \mathfrak{g}$ we use fibered coordinates $(x^\mu; \xi^A)$. The connection $\omega = \mathrm{d}x^\mu \otimes (\partial_\mu - \omega^A_\mu \rho_A)$ over $\mathcal{P}$ induces a connection on $\mathcal{P} \times_{\mathrm{Ad}} \mathfrak{g}$ of the following form:

$$\hat{\omega} = \mathrm{d}x^\mu \otimes \left(\partial_\mu - c^A_{\cdot BC}\, \omega^B_\mu\, \xi^C\, \partial_A\right) \tag{3.7.29}$$

where $c^A_{\cdot BC}$ are as usual the structure constants of G.

8. Structures Induced by a Connection

Bundle connections define a rich structure on a bundle $\mathcal{B}$ and they solve some classical problems of differential geometry.

Parallel Transport

Let $\mathcal{B} = (B, M, \pi; F)$ be a fiber bundle and $\omega = \mathrm{d}x^\mu \otimes (\partial_\mu - \omega^i_{\ \mu}\, \partial_i)$ be a connection on $\mathcal{B}$. Let $\gamma : [0,1] \longrightarrow M$ be a curve in M based at any point $x = \gamma(0) \in M$. Of course, there exist infinitely many curves $\tilde{\gamma}$ on $\mathcal{B}$ which project over γ, i.e. such that $\pi \circ \tilde{\gamma} = \gamma$. Even if we fix a point $b \in \pi^{-1}(x)$ in the fiber over x, still there exist infinitely many curves $\tilde{\gamma}$ on $\mathcal{B}$ which project over γ and such that $\tilde{\gamma}(0) = b$. Thus curves in M cannot be canonically lifted to the bundle $\mathcal{B}$.

However, they can be lifted with respect to the connection ω. We have in fact:

Proposition (3.8.1): let us consider a curve $\gamma : [0,1] \longrightarrow M$ based at $x = \gamma(0) \in M$ and let $b \in \pi^{-1}(x)$ be any point in the fiber over x. There exists a unique curve $\hat{\gamma} : [0,1] \longrightarrow B$ based at b which projects over γ, i.e. $\pi \circ \hat{\gamma} = \gamma$, and such that its tangent vector $\tau_{\hat{\gamma}} \in TB$ is the horizontal lift of the tangent vector $\tau_\gamma \in TM$ for any $t \in [0,1]$.

Proof: Consider the pull-back bundle[5]

$$\begin{array}{ccc} \gamma^*(B) & \xrightarrow{\ i\ } & B \\ \big\downarrow & & \big\downarrow \pi \\ [0,1] & \xrightarrow{\ \gamma\ } & M \end{array} \tag{3.8.2}$$

It is a sub-bundle $\gamma^*(\mathcal{B})$ of $\mathcal{B}$ so that a connection $\gamma^*(\omega)$ is induced on it. The tangent vector τ_γ may be regarded as a vector field in $[0,1]$ and one can consider the horizontal lift $(\gamma^*\omega)(\tau_\gamma)$ which can be regarded as a vector field over $\gamma^*(\mathcal{B})$. Let us denote by Φ_s its flow.

---

[5] The interval $[0,1]$ is a manifold with boundary. The notion we are using here extends in fact from the category of manifolds to the category of manifolds with boundary.

We can then define a curve $\hat{\gamma}' : [0,1] \longrightarrow \gamma^*(\mathcal{B}) : s \mapsto \Phi_s\left(i^{-1}(b)\right)$ which in turn induces a curve $\hat{\gamma} = i \circ \hat{\gamma}'$ in B. The curve $\hat{\gamma}$ is based at b and it projects on γ by construction. Notice that to determine the curve $\hat{\gamma}$ we fix the initial condition $\hat{\gamma}(0) = b$ and its tangent vector $\omega(\tau_\gamma)$, thence the curve $\hat{\gamma}$ is unique because of Cauchy theorem on differential equations. ■

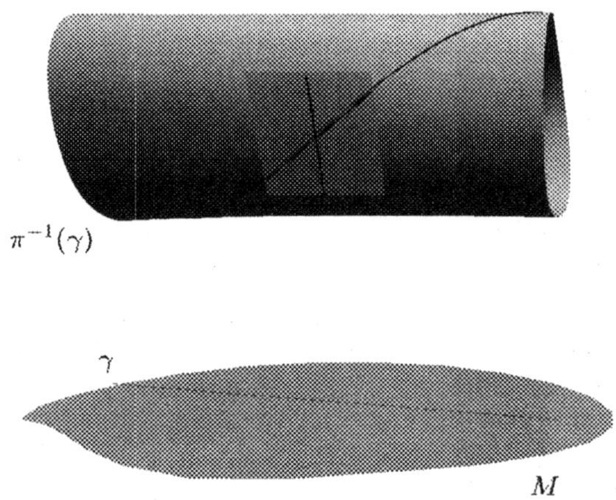

Fig. 15 – The horizonal lift of a curve.

Proposition (3.8.3): each curve $\gamma : [0,1] \longrightarrow M$ such that $x = \gamma(0)$ and $y = \gamma(1)$ induces a bijection $\Phi_\gamma : \pi^{-1}(x) \longrightarrow \pi^{-1}(y)$. The map Φ_γ is defined by setting $\Phi_\gamma(b) = \hat{\gamma}(1)$, where $\hat{\gamma}$ is the (unique) horizontal lift of γ based at $b \in \pi^{-1}(x)$.

Proof: the map Φ_γ so defined is surjective. If not, there would exist a point $b' \in \pi^{-1}(y)$ which is not in the image of Φ_γ. Let us then consider the curve $\gamma'(t) = \gamma(1-t)$ which is based at y. It induces a map $\Phi_{\gamma'} : \pi^{-1}(y) \longrightarrow \pi^{-1}(x)$ and let us consider $b'' = \Phi_{\gamma'}(b') \in \pi^{-1}(x)$. It is easy to check that $\Phi_\gamma(b'') = b'$ which contradicts the assumption that b' is not in the image of Φ_γ.

The map Φ_γ is also injective. If not, there would exist two different points $b, \bar{b} \in \pi^{-1}(x)$ such that $\Phi_\gamma(b) = \Phi_\gamma(\bar{b}) = b' \in \pi^{-1}(y)$. One can easily prove that $\Phi_{\gamma'}(b') = b$ and $\Phi_{\gamma'}(b') = \bar{b}$. Thus $b = \bar{b}$ follows, which is against the assumption $b \neq \bar{b}$.

In other words, the map $\Phi_{\gamma'} : \pi^{-1}(y) \longrightarrow \pi^{-1}(x)$ is the inverse of the map $\Phi_\gamma : \pi^{-1}(x) \longrightarrow \pi^{-1}(y)$ which is thence invertible. ■

Definition (3.8.4): the bijection $\Phi_\gamma : \pi^{-1}(x) \longrightarrow \pi^{-1}(y)$ is called the *parallel transport along the curve* γ. ■

We remark that the parallel transport along a curve γ depends just on the homotopy class of γ.[6] If the bundle is the tangent bundle $T(M)$, points in $\tau_M^{-1}(x)$ are tangent vectors at $x \in M$. The parallel transport Φ_γ along a curve γ in M then induces an isomorphism between the tangent spaces $T_x M$ and $T_y M$ $(y = \gamma(1))$. The vector $\Phi_\gamma(v)$ will be the parallel transport of the vector $v \in T_x M$ along the curve γ and with respect to the connection $\hat{\omega}$ given by (3.7.26).

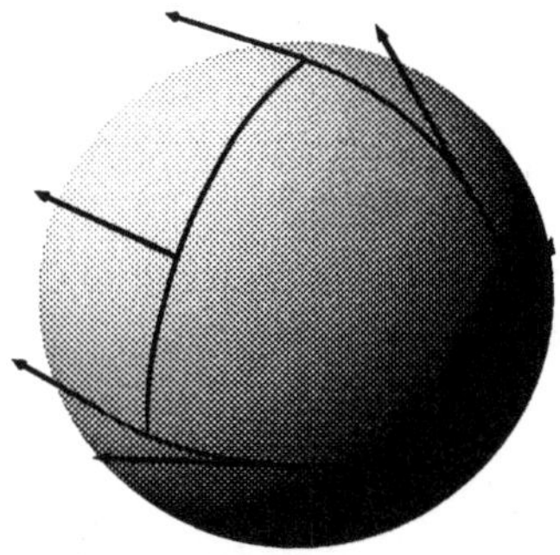

Fig. 16 – Parallel transport along a path.

The Covariant Derivative of a Section

Let $\mathcal{B} = (B, M, \pi; F)$ be a fiber bundle and ω be a connection over $\mathcal{B}$.

Definition (3.8.6): let ξ be a vector field over the base manifold M and $\rho \in \Gamma(\pi)$ be a (local) section of $\mathcal{B}$. We call *covariant derivative of ρ along ξ* the object defined by the following:

$$\nabla_\xi \rho(x) = T_x\rho(\xi(x)) - \omega_{\rho(x)}(\xi(x)) \tag{3.8.7}$$

If we choose a chart on M and the natural basis $\partial_\mu^{(\alpha)}$ of tangent vectors, then let us denote by $\nabla_\mu^{(\alpha)}\rho(x)$ the covariant derivative of ρ along $\partial_\mu^{(\alpha)}$.

---

[6] In Section 1.5 we already defined homotopy. Here we mean homotopy with fixed extrema. Two curves γ and $\gamma' : I \longrightarrow \mathbb{R}$ are homotopic if and only if there exists a continuous map $F : [0,1] \times I \longrightarrow M$ such that

$$\begin{cases} F(0,t) = \gamma(t) \\ F(1,t) = \gamma'(t) \\ F(s,0) = \gamma(0) = \gamma'(0) \qquad F(s,1) = \gamma(1) = \gamma'(1) \end{cases} \tag{3.8.5}$$

Proposition (3.8.8): the vector $\nabla_\xi \rho(x)$ is a vertical vector field defined on the section $\rho(x)$.

Proof: both $T_x\rho(\xi(x))$ and $\omega_{\rho(x)}(\xi(x))$ are tangent vectors to B at the point $\rho(x)$ and both of them project over $\xi(x)$. ∎

If $(x^\mu; y^i)$ are fibered coordinates on B, $\omega = \mathrm{d}x^\mu \otimes \left(\partial_\mu - \omega^i_{\ \mu}(x,y)\partial_i\right)$ is the connection over $\mathcal{B}$, $\xi = \xi^\mu \partial_\mu$ is the vector field over the base M and $\rho : x \mapsto (x, \rho^i(x))$ is the section, then the covariant derivative is given by

$$\nabla_\mu \rho = \left(\partial_\mu \rho^i(x) - \omega^i_{\ \mu}(x, \rho(x))\right)\partial_i \equiv (\nabla_\mu \rho^i)\, \partial_i \tag{3.8.9}$$

As for Lie derivatives, one can associate to covariant derivatives a generalized vertical vector field in the form:

$$\nabla_\mu = \left(y^i_\mu - \omega^i_{\ \mu}(x,y)\right)\partial_i \equiv (\nabla_\mu y^i)\, \partial_i \tag{3.8.10}$$

which will be called *formal covariant derivative*.

If the formal covariant derivative is calculated along the section ρ we immediately obtain

$$(\nabla_\mu) \circ j^1\rho = \left(\partial_\mu \rho^i - \omega^i_{\ \mu}(x,\rho)\right)\partial_i = \nabla_\mu \rho \tag{3.8.11}$$

Notice in fact that $\nabla = \nabla_\mu \otimes \mathrm{d}x^\mu$ may be regarded as a global section of a bundle, as shown by the following commutative diagram

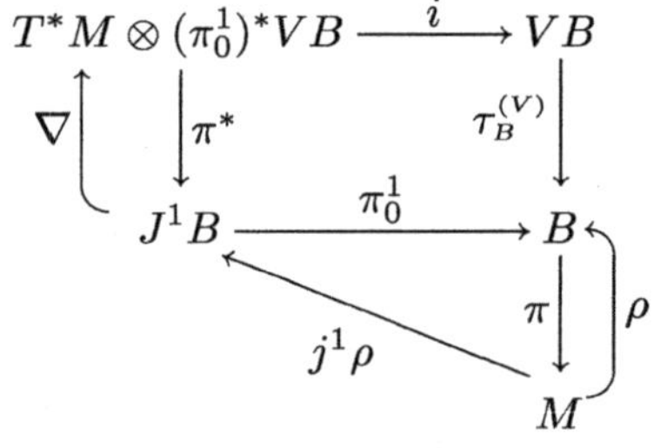

(3.8.12)

Consequently, if we evaluate ∇ along the section ρ and contract the result with a vector field $\xi = \xi^\mu \partial_\mu$ we obtain the covariant derivative of the section ρ along ξ, i.e.:

$$\nabla_\xi \rho = < \nabla \circ j^1\rho \,|\, \xi > = \xi^\mu\, \nabla_\mu \rho \tag{3.8.13}$$

where $< \ | \ >$ denotes the standard duality between T^*M and TM.

Curvature of a Principal Connection

Let $\mathcal{P} = (P, M, \pi; G)$ be a principal bundle and ω be a principal connection on $\mathcal{P}$. Let $\bar{\omega} = (\theta^A_{(L)} + \mathrm{Ad}^A_B(g^{-1})\omega^B_\mu \, \mathrm{d}x^\mu) \otimes T_A$ be the connection 1-form associated to ω. We recall that for each vector field Ξ on $\mathcal{P}$ the contraction $\bar{\omega}(\Xi) \in \mathfrak{g}$ is the element in the Lie algebra which corresponds to the vertical part $\Xi_{(V)}$. The horizontal exterior differential of $\bar{\omega}$ is a horizontal 2-form, called the *curvature 2-form*; according to equations (3.5.40) and (3.5.42) it reads as:

$$F_\omega := \mathrm{D}_\omega \bar{\omega} = \tfrac{1}{2}\mathrm{Ad}^A_B(g^{-1}) F^B_{\mu\nu} \, \mathrm{d}x^\mu \wedge \mathrm{d}x^\nu \otimes T_A = \mathrm{d}\bar{\omega} + \tfrac{1}{2}[\bar{\omega}, \bar{\omega}] \tag{3.8.14}$$

The curvature coefficients $F^A_{\mu\nu} = \mathrm{d}_\mu \omega^A_\nu - \mathrm{d}_\nu \omega^A_\mu + c^A{}_{\cdot BC}\omega^B_\mu \omega^C_\nu$ transform as follows

$$F'^A_{\mu\nu} = \mathrm{Ad}^A_B(\varphi) \, F^B_{\rho\sigma} \, \bar{J}^\rho_\mu \bar{J}^\sigma_\nu \tag{3.8.15}$$

We remark that if the group G is Abelian we have $\mathrm{Ad}^A_B(\varphi) = \delta^A_B$ and the curvature 2-form is an ordinary 2-form on the base manifold M. We also remark that the Lie derivative of the connection, expressed by (3.5.24), can now be written in a more elegant form:

$$\pounds_\Xi \omega = \left(\xi^\nu F^A_{\nu\mu} + \nabla_\mu \xi^A_{(V)}\right) \frac{\partial}{\partial \omega^A_\mu} \tag{3.8.16}$$

where we set $\xi^A_{(V)} = \xi^A + \omega^A_\nu \, \xi^\nu$ for the vertical part of Ξ with respect to ω.

The curvature form can be seen as a section of a bundle associated to the principal bundle $\mathcal{P} \times_M L(M)$. That bundle has $G \times \mathrm{GL}(m)$ as structure group, which is represented on the vector space $V = \mathfrak{g} \otimes \mathbb{R}^{\frac{m(m-1)}{2}}$ by the representation

$$\lambda : G \times \mathrm{GL}(m) \times V \longrightarrow V : (\varphi, J, F^A_{\mu\nu}) \mapsto F'^A_{\mu\nu} = \mathrm{Ad}^A_B(\varphi) \, F^B_{\rho\sigma} \, \bar{J}^\rho_\mu \bar{J}^\sigma_\nu \tag{3.8.17}$$

where as usual the bar denotes the inverse matrix. Then there is a one-to-one correspondence between curvature 2-forms and sections of the bundle $\mathcal{F} = (\mathcal{P} \times_M L(M)) \times_\lambda V$. Local coordinates on $\mathcal{F}$ will be denoted by $(x^\mu; F^A_{\mu\nu})$.

If we fix a principal connection $\omega = \mathrm{d}x^\mu \otimes (\partial_\mu - \omega^A_\mu \, \rho_A)$ on $\mathcal{P}$ and a linear connection $\Gamma = \mathrm{d}x^\mu \otimes (\partial_\mu - \Gamma^\lambda_{\nu\mu} \, \rho^\nu_\lambda)$ on $L(M)$, then a principal connection is induced on $\mathcal{P} \times_M L(M)$ by the rule:

$$\hat{\Omega} = \mathrm{d}x^\mu \otimes \left(\partial_\mu - \omega^A_\mu \, \rho_A - \Gamma^\lambda_{\nu\mu} \, \rho^\nu_\lambda\right) \tag{3.8.18}$$

A connection Ω is also induced on $\mathcal{F}$ and it reads as:

$$\Omega = \mathrm{d}x^\mu \otimes \left(\partial_\mu - (\Gamma^\rho_{\lambda\mu} \, F^A_{\rho\nu} + \Gamma^\rho_{\nu\mu} \, F^A_{\lambda\rho} + c^A{}_{\cdot BC}\omega^B_\mu \, F^C_{\lambda\nu}) \, \partial^{\lambda\nu}_A\right) \tag{3.8.19}$$

with $\partial^{\lambda\nu}_A = \frac{\partial}{\partial F^A_{\lambda\nu}}$. Consequently, one can define the covariant derivative of the curvature by

$$\nabla_\rho F^A_{\mu\nu} = \mathrm{d}_\rho F^A_{\mu\nu} - F^A_{\lambda\nu}\Gamma^\lambda_{\mu\rho} - F^A_{\mu\lambda}\Gamma^\lambda_{\nu\rho} + c^A{}_{\cdot BC}\omega^B_\rho \, F^C_{\mu\nu} \tag{3.8.20}$$

Notice that the general framework to deal with induced connections, covariant derivatives and Lie derivatives will be discussed in Chapter 5 when speaking of gauge natural bundles. Our definitions above have thence, for the moment, just a heuristic meaning.

One can easily prove the following identities

$$\nabla_\rho F^A_{\ \mu\nu} + \nabla_\mu F^A_{\ \nu\rho} + \nabla_\nu F^A_{\ \rho\mu} = F^A_{\ \lambda\nu}\, T^\lambda_{\ \rho\mu} + F^A_{\ \lambda\mu}\, T^\lambda_{\ \nu\rho} + F^A_{\ \lambda\rho}\, T^\lambda_{\ \mu\nu} \qquad (3.8.21)$$

which are the so-called *generalized Bianchi's identities*. They are the local expression of the following global identity:

$$\mathrm{D}_\omega \Omega = 0 \qquad (3.8.22)$$

In the case of a linear connection, i.e. a connection on the frame bundle $L(M)$, we choose $\rho^a_{\ b}$ as a basis of vertical right invariant so that a linear connection is expressed by

$$\omega = \mathrm{d}x^\mu \otimes (\partial_\mu - \omega^a_{\ b\mu}\rho^b_{\ a}) \qquad (3.8.23)$$

The curvature is nothing but the Riemann tensor of classical Differential Geometry

$$R^a_{\ b\mu\nu} = \mathrm{d}_\mu \Gamma^a_{\ b\nu} - \mathrm{d}_\nu \Gamma^a_{\ b\mu} + \Gamma^a_{\ c\mu}\Gamma^c_{\ b\nu} - \Gamma^a_{\ c\nu}\Gamma^c_{\ b\mu} \qquad (3.8.24)$$

and the generalized Bianchi's identities reproduce the standard Bianchi's identities:

$$\nabla_\rho R^a_{\ b\mu\nu} + \nabla_\mu R^a_{\ b\nu\rho} + \nabla_\nu R^a_{\ b\rho\mu} = R^a_{\ b\lambda\nu}\, T^\lambda_{\ \rho\mu} + R^a_{\ b\lambda\mu}\, T^\lambda_{\ \nu\rho} + R^a_{\ b\lambda\rho}\, T^\lambda_{\ \mu\nu} \qquad (3.8.25)$$

which for a torsionless connection reduce to:

$$\nabla_\rho R^a_{\ b\mu\nu} + \nabla_\mu R^a_{\ b\nu\rho} + \nabla_\nu R^a_{\ b\rho\mu} = 0 \qquad (3.8.26)$$

Any infinitesimal generator $\Xi = \xi^\mu\,\partial_\mu + \xi^A\,\rho_A + \xi^a_{\ b}\,\rho^b_{\ a}$ of automorphisms of $\mathcal{P}\times_M L(M)$ induces on the bundle $\mathcal{F}$ a vector field $\hat{\Xi} = \xi^\mu\,\partial_\mu + \xi^A_{\ \mu\nu}\,\partial^{\mu\nu}_A$ locally given by:

$$\xi^A_{\ \mu\nu} = -\partial_\mu\xi^\rho\, F^A_{\ \nu\rho} - \partial_\nu\xi^\rho\, F^A_{\ \mu\rho} + c^A_{\cdot\, BC}\,\xi^B F^C_{\ \mu\nu} \qquad (3.8.27)$$

Then we can define the Lie derivative of the curvature F by:

$$\pounds_\Xi F := \pounds_{\hat{\Xi}} F = (\xi^\rho\nabla_\rho F^A_{\ \mu\nu} + \nabla_\mu\xi^\rho F^A_{\ \rho\nu} + \nabla_\nu\xi^\rho F^A_{\ \mu\rho} - c^A_{\cdot\, BC}\xi^B_{(V)}\, F^C_{\ \mu\nu})\partial^{\mu\nu}_A \quad (3.8.28)$$

Let now $\Xi = \Xi^A(x)\,\rho_A$ be an infinitesimal generator of vertical automorphisms on a principal bundle $\mathcal{P} = (P, M, \pi; G)$. As we have already shown in Section 3.4, it corresponds to a section of $\mathcal{P}\times_{\mathrm{Ad}}\mathfrak{g}$. If we fix a principal connection ω on $\mathcal{P}$ we can thence consider the covariant derivative of that section along a vector field $\zeta = \zeta^\mu(x)\, \partial_\mu$ with respect to the induced connection $\hat\omega$ on $\mathcal{P} \times_{\mathrm{Ad}} \mathfrak{g}$. This is a section of the bundle $\tau_{(V)} : V(\mathcal{P} \times_{\mathrm{Ad}} \mathfrak{g}) \longrightarrow M$, which is again associated to $\mathcal{P}$. Then we can calculate the second covariant derivative along a vector field $\xi = \xi^\mu(x)\, \partial_\mu$ with respect to the connection induced on $\tau_{(V)} : V(\mathcal{P} \times_{\mathrm{Ad}} \mathfrak{g}) \longrightarrow M$ by ω.

Proposition (3.8.29): the following identity holds true:

$$[\nabla_\xi, \nabla_\zeta]\, \Xi^A = \nabla_{[\xi,\zeta]}\, \Xi^A + \xi^\mu\, \zeta^\nu\, (c^A_{\cdot BC} F^B_{\mu\nu} \Xi^C) \tag{3.8.30}$$

Proof: the local expression of the section representing Ξ is : $x \mapsto (x, \Xi^A(x))$.

The covariant derivative of this section is thence:

$$\nabla_\zeta \Xi = \zeta^\mu \left(\partial_\mu \Xi^A - c^A_{\cdot BC} \omega^B_\mu \Xi^C \right) \partial_A \tag{3.8.31}$$

The object $\nabla_\zeta \Xi$ is a section of the bundle $\tau_{(V)} : V(\mathcal{P} \times_{\mathrm{Ad}} \mathfrak{g}) \longrightarrow M$ which has local coordinates $(x^\mu; \Xi^A, \bar\Xi^A)$. Changing local trivialization on $\mathcal{P}$ one obtains:

$$\begin{cases} x'^\mu = x'^\mu(x) \\ \Xi'^A = \mathrm{Ad}^A_B(\varphi)\, \Xi^B \\ \bar\Xi'^A = \mathrm{Ad}^A_B(\varphi)\, \bar\Xi^B \end{cases} \tag{3.8.32}$$

In other words, the transformation rules of the covariant derivative $\bar\Xi'^A = \nabla_\zeta \Xi^A$ coincide with those of infinitesimal generators Ξ^a of automorphisms. Thence, we can define a connection on $V(\mathcal{P} \times_{\mathrm{Ad}} \mathfrak{g})$ by

$$\hat\omega = dx^\mu \otimes (\partial_\mu - c^A_{\cdot BC} \omega^B_\mu\, \Xi^C\, \partial_A - c^A_{\cdot BC} \omega^B_\mu\, \bar\Xi^C\, \bar\partial_A) \tag{3.8.33}$$

where we set $\bar\partial_A = \partial / \partial \bar\Xi^A$. The second covariant derivative of the section Ξ is:

$$\begin{aligned} \nabla_\xi \nabla_\zeta \Xi^A =& \xi^\nu \partial_\nu \zeta^\mu \partial_\mu \Xi^A + \xi^\nu \zeta^\mu \partial_{\mu\nu} \Xi^A + \xi^\nu \zeta^\mu c^A_{\cdot BC} \partial_\nu \omega^B_\mu\, \Xi^C + \\ &+ \xi^\nu \zeta^\mu c^A_{\cdot BC} \omega^B_\mu\, \partial_\nu \Xi^C + \xi^\nu \partial_\nu \zeta^\mu c^A_{\cdot BC} \omega^B_\mu\, \Xi^C + \\ &+ \xi^\nu \zeta^\mu c^A_{\cdot BC} \omega^B_\nu\, \partial_\mu \Xi^C + \xi^\nu \zeta^\mu c^A_{\cdot BC} \omega^B_\nu\, c^C_{\cdot DE} \omega^D_\mu \Xi^E \end{aligned} \tag{3.8.34}$$

and thence the commutator is in the following form

$$[\nabla_\xi, \nabla_\zeta]\, \Xi^A = \nabla_{[\xi,\zeta]}\, \Xi^A + \xi^\mu\, \zeta^\nu\, (c^A_{\cdot BC} F^B_{\mu\nu} \Xi^C) \tag{3.8.35}$$

as we claimed. ∎

Proposition (3.8.36): if ξ, $\zeta \in \mathfrak{X}(M)$ and $\omega = dx^\mu \otimes (\partial_\mu - \omega^A_\mu\, \rho_A)$ is a principal connection, then we have:

$$[\omega(\xi), \omega(\zeta)] = \omega([\xi, \zeta]) - \xi^\mu \zeta^\nu\, F^A_{\mu\nu}\, \rho_A \tag{3.8.37}$$

Proof: the commutator of $\omega(\xi) = \xi^\mu(\partial_\mu - \omega^A_\mu \rho_A)$ and $\omega(\zeta) = \zeta^\mu(\partial_\mu - \omega^A_\mu \rho_A)$ is given by

$$\begin{aligned}
[\omega(\xi),&\omega(\zeta)] = (\xi^\lambda \partial_\lambda \zeta^\mu - \zeta^\lambda \partial_\lambda \xi^\mu)\, \partial_\mu +\\
&+ \Big(- \xi^\lambda \partial_\lambda (\zeta^\mu \omega^A_\mu) + \zeta^\lambda \partial_\lambda (\xi^\mu \omega^A_\mu) - \xi^\mu \zeta^\nu \omega^B_\mu \omega^C_\nu \, c^A{}_{BC} \Big)\, \rho_A =\\
=&(\xi^\lambda \partial_\lambda \zeta^\mu - \zeta^\lambda \partial_\lambda \zeta^\mu)\, \partial_\mu - (\xi^\lambda \partial_\lambda \zeta^\mu - \zeta^\lambda \partial_\lambda \xi^\mu) \omega^A_\mu \, \rho_A\\
&- \xi^\mu \zeta^\nu (\partial_\mu \omega^A_\nu - \partial_\nu \omega^A_\mu + c^A{}_{BC} \omega^B_\mu \omega^C_\nu)\, \rho_A =\\
=&\omega([\xi,\zeta]) - \xi^\mu \zeta^\nu \, F^A_{\mu\nu}\, \rho_A
\end{aligned} \tag{3.8.38}$$

■

Exercises

1- Find out the form of the most general principal connection on the Hopf bundle.
Hint: when one finds a basis for vertical right invariant vector fields then principal connections are in the form (3.5.12).

2- Generalize the construction of the trivial principal connection on the bundle $(\mathbb{R}\times\mathbb{R}, \mathbb{R}, p_1; \mathbb{R})$ given in Section 3.5 to the case of a general trivial principal bundle $(M \times G, M, p_1; G)$.
Hint: if a principal bundle is trivial it admits a global section.

3- Compute the covariant derivative of a vector field and of a 1-form on M with respect to the connection induced by a linear connection on $L(M)$.
Hint: compare with the classical prescriptions in Differential Geometry.

4- Show that parallel transport on the tangent bundle induced by a metric connection (i.e. by the Levi-Civita connection of a Riemannian metric g) preserves the norm of vectors.
Hint: write down the relevant parallel transport in its infinitesimal form.

5- Prove the following identities (see equation (3.4.14))

$$\begin{cases} R^a_A(g)\, \partial_a \mathrm{Ad}^E_B(g) = c^E_{\cdot\, AC}\, \mathrm{Ad}^C_B(g) \\ L^a_A(g)\, \partial_a \mathrm{Ad}^E_B(g) = \mathrm{Ad}^E_C(g)\, c^C_{\cdot\, AB} \end{cases}$$

Hint: the adjoint representation transforms left invariant vector fields in the corresponding right invariant vector fields. Play with commutators.

6- Prove the following identities (see equation (3.5.39))

$$\begin{cases} c^A_{\cdot\, BC} = -\left(R^a_B(g)\, \partial_a R^b_C(g) - R^a_C(g)\, \partial_a R^b_B(g)\right)\, \bar{R}^A_b(g) \\ c^A_{\cdot\, BC} = \bar{L}^A_b(g)\, \left(L^a_B(g)\, \partial_a L^b_C(g) - L^a_C(g)\, \partial_a L^b_B(g)\right) \end{cases}$$

Hint: the structure constants are related to the commutators of invariant vector fields.

Chapter 4

NATURAL BUNDLES

Abstract We shall here define *natural bundles* together with the structures defined on them, e.g. the *lift of diffeomorphisms and vector fields*. Both the categorial and the constructive definitions are given and compared. The *tangent and cotangent bundles* are again considered as examples of natural bundles. The bundles of metrics of signature (r, s) and the bundle of linear connections on M are considered as further examples of natural bundles. Finally, the bundles of tensors of rank (p, q) and tensor densities of rank (p, q) and weight k are also introduced.

References

Some information about natural bundles can be found in [KMS93], [F91], [T72], [F88]. Further material can be found in [CB68], [CBWM82]. The original paper about naturality and finite order theorem is [PT77].

1. Natural Bundles

Natural bundles are a sub-category of bundles with structure group. They are important for Differential Geometry (for which they provide a modern perspective) as well as for Physics (they are the mathematical framework in which General Relativity is defined together with all *generally covariant* theories).

As we already mentioned, natural bundles encode the properties of the base manifold M since their trivializations and transition functions are *canonically* constructed out of atlases of M. We already encountered a few examples of natural bundles: the tangent bundle, the cotangent bundle, the (higher order) frame bundles and all tensor bundles on M. They all allow so-called *natural trivializations* in which transition functions are constructed out of the Jacobians of the transition functions of an atlas on M. As we shall see this is the simplest case of a general feature of natural bundles.

We shall provide two equivalent definitions of natural bundles. The first definition is functorial and it is well suited for understanding the general properties of natural bundles which are then used in applications to Physics. The second definition is constructive and it provides the appropriate framework to actually apply the general framework to particular cases and situations. All along the rest of this monograph we shall build examples and applications to Physics of the theory of natural (as well as gauge natural) bundles.

Let us consider a covariant functor $\mathfrak{B}$ from the category of manifolds (and local diffeomorphisms[1]) into the category of bundles with structure group (and local automorphisms). The functor $\mathfrak{B}$ then maps each manifold M into a bundle $\mathfrak{B}(M) = (BM, M, \pi; F)$ and each local diffeomorphism $\phi : M \longrightarrow N$ into a bundle morphism $\mathfrak{B}(\phi) = (B\phi, \phi)$ with $B\phi : BM \longrightarrow BN$. Being $\mathfrak{B}$ a covariant functor it preserves compositions, i.e. for each pair of local diffeomorphisms $M \xrightarrow{\phi} N \xrightarrow{\psi} Q$ one has $\mathfrak{B}(\phi \circ \psi) = \mathfrak{B}(\phi) \circ \mathfrak{B}(\psi)$ and $\mathfrak{B}(\mathrm{id}_M) = \mathrm{id}_{BM}$.

Definition (4.1.1): a covariant functor $\mathfrak{B}$ from the category of manifolds into the category of bundles with structure group is a *natural functor* if the following properties hold true:

(a) $\mathfrak{B}(M) = (BM, M, \pi; F)$ is a bundle over M;

(b) if $i : U \longrightarrow M$ is the inclusion of an open submanifold $U \subseteq M$ then $\mathfrak{B}(i) = (Bi, i)$ defined by $Bi : BU \longrightarrow BM$ is the inclusion of the sub-bundle $\mathfrak{B}(U) = \pi^{-1}(U)$ into $\mathfrak{B}(M)$;

(c) let Σ be a parameter manifold and $\phi : \Sigma \times M \longrightarrow N$ a smooth family of local diffeomorphisms $\phi_s : M \longrightarrow N$ ($s \in \Sigma$), then $\mathfrak{B}(\phi_s) = (B\phi_s, \phi_s)$ is a smooth family of fibered local diffeomorphisms, i.e. the map $B\phi : \Sigma \times BM \longrightarrow BN : (s, b) \mapsto B\phi_s(b)$ is smooth. ∎

We already encountered examples of natural functors. Let us consider the tangent functor $T(\cdot)$ introduced in Section 1.1. It associates to any manifold M of dimension $\dim(M) = m$ its tangent bundle $T(M) = (TM, M, \tau_M; \mathbb{R}^m)$ and to any local diffeomorphism $\phi : M \longrightarrow N$ its tangent map $T(\phi) = (T\phi, \phi)$ given by equation (1.3.38), which is a local automorphism. The properties

[1] Here a local diffeomorphism $\phi : M \longrightarrow N$ is a *global* map such that for any $x \in M$ there exists a neighbourhood $U \subseteq M$ such that the restriction $\phi_U : U \longrightarrow \phi(U)$ of the map ϕ is a diffeomorphism onto the image $\phi(U) \subseteq N$.

From an equivalent viewpoint, a local diffeomorphism $\phi : M \longrightarrow N$ is a *global* map such that the tangent maps $T_x\phi : T_xM \longrightarrow T_{\phi(x)}N$ are isomorphisms for all $x \in M$. Analogously for the notion of local automorphism.

required in definition (1.3.38) hold true. In fact: (a), the tangent bundle $T(M)$ is a bundle over M; (b), the tangent map of the inclusion $i : U \longrightarrow M$ (which is a local diffeomorphism locally expressed by the identity) is the inclusion $Ti : TU \longrightarrow TM$ since the tangent map of the identity is the identity; (c), finally smooth families of local automorphisms are associated to smooth families of diffeomorphisms since the Jacobian commutes with the derivative with respect to the family parameter s, as already used in Exercise 1 of Chapter 2.

Another, more tricky, example can be obtained by the cotangent functor $T^*(\cdot)$ introduced in Section 1.5. Of course the cotangent functor $T^*(\cdot)$ itself is not a natural functor because it is contravariant. However, by restricting to local diffeomorphisms we can associate to any local diffeomorphism $\phi : M \longrightarrow N$ the local automorphism $\bar{T}^*(\phi) = T^*(\phi^{-1}) : T^*(M) \longrightarrow T^*(N)$ defined, according to equation (1.5.15), by setting

$$\begin{cases} x'^\mu = \phi^\mu(x) \\ \omega'_\mu = \partial_\mu(\phi^{-1})^\lambda \, \omega_\lambda \end{cases} \tag{4.1.2}$$

This is a natural functor, which is still denoted by $T^*(\cdot)$ by an abuse of language.

Let us finally consider a further example which will be important in the sequel. Let us fix an integer s, a manifold F and an action λ of the group $\mathrm{GL}^s(m)$ on the manifold F, i.e. $\lambda : \mathrm{GL}^s(m) \times F \longrightarrow F$. Then to any manifold M we can associate the bundle $M_\lambda = M_\lambda(F, s) = L^s(M) \times_\lambda F$ which is associated to the s-order frame bundle over M. As long as morphisms are concerned, if $\phi : M \longrightarrow N$ is a local diffeomorphism let us denote by $L^s(\phi) : L^s(M) \longrightarrow L^s(N)$ the local morphism induced on the relevant s-frame bundles which is given by equation (2.7.6). Since a point in $L^s(M) \times_\lambda F$ (resp., $L^s(N) \times_\lambda F$) is of the form $[j^s_0\epsilon, \varphi]_\lambda$ (resp., $[j^s_0\epsilon', \varphi']_\lambda$) where $j^s_0\epsilon \in L^s(M)$ and $\varphi \in F$ (resp., $j^s_0\epsilon' \in L^s(N)$ and $\varphi' \in F$) we can define a bundle morphism

$$\phi_\lambda : L^s(M) \times_\lambda F \longrightarrow L^s(N) \times_\lambda F : [j^s_0\epsilon, \varphi]_\lambda \mapsto [L^s(\phi)(j^s_0\epsilon), \varphi]_\lambda \tag{4.1.3}$$

Then it is immediate to check that the functor $(\cdot)_\lambda$ is a natural functor.

Definition (4.1.4): the *category of natural bundles* is the sub-category of the category of fiber bundles with structure group in which objects are the bundles, called *natural bundles*, which are in the image of some natural functor $\mathfrak{B}$; morphisms, called *natural morphisms*, are fibered morphisms obtained by the lift $\mathfrak{B}(\phi)$ of some local diffeomorphism ϕ. Then natural functors restrict to functors from the category of manifold (and local diffeomorphisms) to the category of natural bundles. Sections of natural bundles are called *natural objects* over M. ∎

In the classical literature natural bundles are also called *bundles of geometric objects* and natural objects are also called *geometric objects*. We stress that the construction of a natural bundle out of its base manifold M must be a *canonical* construction. Consequently, a natural bundle is never defined on its own. On the contrary we always define a whole family of natural bundles, one for each base manifold, as we did for the tangent bundle, the cotangent bundle, the tensor bundles, etc.

The fiber bundles $M_\lambda \equiv M_\lambda(F, s)$ canonically associated as above to some frame bundle (with structure group $\mathrm{GL}^s(m)$) are not just an example of natural bundles. One can prove (see e.g. [KMS93] where Chapter *V* is devoted to prove this theorem, or [PT77] for a brief proof) that *any* natural bundle is associated to some s-frame bundle in the above sense; the minimal integer $s \geq 1$ is called the *order of the natural bundle*. This theorem is usually called the *finite order theorem* for natural bundles. Here we shall not prove the theorem since in the sequel we shall always directly define natural bundles as associated bundles to some suitable $L^s(M)$. In this way they are always of finite order. To be more rigorous we can consider the following definition:

Definition (4.1.5): a *natural bundle (of finite order s)* is a bundle $\mathcal{B}$ which is canonically associated to $L^s(M)$ in such a way that s is minimal, i.e. $\mathcal{B}$ is not associated canonically to any other $L^{s'}(M)$, with $s' < s$. ■

Then the finite order theorem for natural bundles prove that any natural bundle in the sense of (4.1.4) is also a natural bundle of finite order in the sense of definition (4.1.5).

However, definition (4.1.1) can be very useful in some application; e.g., the following holds:

Proposition (4.1.6): if $\mathcal{B} = L^s(M) \times_\lambda F$ is a natural bundle then $J^k\mathcal{B}$ is a natural bundle.

Sketch of the proof: let us consider the functor $J^k(\cdot) \circ (\cdot)_\lambda$ and prove that it is a natural functor. Property (a) in definition (4.1.1) still holds true.

Property (b) is still true because if $i : C \longrightarrow B$ is the inclusion of a sub-bundle $\mathcal{C}$ of a bundle $\mathcal{B}$ then $j^k i : J^k C \longrightarrow J^k B$ is the inclusion of $J^k\mathcal{C}$ into $J^k\mathcal{B}$.

Property (c) is still true since if $\Phi_s : C \longrightarrow B$ is a 1-parameter smooth family of bundle morphisms then $j^k\phi_s : J^k C \longrightarrow J^k B$ is again a 1-parameter smooth family of bundle morphisms. ■

We remark that for any bundle $\mathcal{B} = (B, M, \pi; F)$ we have an exact short sequence of groups

$$\mathbb{I} \longrightarrow \mathrm{Aut}_V(\mathcal{B}) \xrightarrow{\ i\ } \mathrm{Aut}(\mathcal{B}) \xrightarrow{\ p\ } \mathrm{Diff}(M) \longrightarrow \mathbb{I} \qquad (4.1.7)$$

where $\mathbb{I} = \{e\}$ is the trivial group, the map $i : \mathrm{Aut}_V(\mathcal{B}) \longrightarrow \mathrm{Aut}(\mathcal{B})$ is the canonical inclusion while $p : \mathrm{Aut}(\mathcal{B}) \longrightarrow \mathrm{Diff}(M)$ is the canonical projection which associates to $\boldsymbol{\Phi} = (\Phi, \phi)$ the diffeomorphism $\phi : M \longrightarrow M$. If the bundle $\mathcal{B}$ is natural we can lift any diffeomorphism $\phi : M \longrightarrow M$ to a bundle automorphism $\hat{\boldsymbol{\phi}} = (\hat{\phi}, \phi)$, i.e. we can define a canonical splitting

$$\mathbb{I} \longrightarrow \mathrm{Aut}_V(\mathcal{B}) \xrightarrow{\ i\ } \mathrm{Aut}(\mathcal{B}) \underset{(\cdot)^\wedge}{\overset{p}{\rightleftarrows}} \mathrm{Diff}(M) \longrightarrow \mathbb{I} \tag{4.1.8}$$

Since one can use the functorial lift $\varphi \mapsto \hat{\varphi}$ to define a cocycle on $\mathcal{B}$ out of the cocycle of the transition functions of an atlas on M, the canonical splitting of the sequence (4.1.8) is equivalent to the naturality of $\mathcal{B}$. The splitting implies that any automorphism $\boldsymbol{\Phi} = (\Phi, \phi)$ of $\mathcal{B}$ can be canonically decomposed as follows:

$$\Phi = \Phi_V \circ \hat{\phi} \tag{4.1.9}$$

where $\hat{\phi}$ is the natural lift of the diffeomorphism ϕ and Φ_V which is called the *vertical part of* Φ, is defined by $\Phi_V = \Phi \circ \hat{\phi}^{-1} \in \mathrm{Aut}_V(\mathcal{B})$.

From the infinitesimal viewpoint, we have also a splitting of the following sequence of Lie algebras

$$0 \longrightarrow \mathfrak{X}_V(\mathcal{B}) \xrightarrow{\ i\ } \mathfrak{X}_{proj}(\mathcal{B}) \underset{(\cdot)^\wedge}{\overset{p}{\rightleftarrows}} \mathfrak{X}(M) \longrightarrow 0 \tag{4.1.10}$$

This splitting (which provides a natural lift from vector fields over M to projectable vector fields over $\mathcal{B}$) is canonical and it preserves the Lie algebra structure (i.e., if the lift is denoted by $\xi \in \mathfrak{X}(M) \mapsto \hat{\xi} \in \mathfrak{X}_{proj}(\mathcal{B})$, then $[\hat{\xi}, \hat{\zeta}] = [\xi, \zeta]^\wedge$ for all ξ, ζ; see proposition (4.2.11) below).

Notice that a formally analogous splitting $\xi \mapsto \hat{\xi} \equiv \omega(\xi)$ can be constructed for each connection in $\mathcal{B}$ by starting from the splitting (3.5.6) which defines the connection itself. However, using a connection ω we usually get a splitting of the sequence of vector spaces (3.5.6) which does not preserve the Lie algebra structure, due to proposition (3.8.36), unless the connection is flat, i.e. $F_\omega = 0$. Furthermore, even if the connection is flat the splitting is of course not canonical, as required by naturality, since it depends on the connection.

2. Natural Morphisms

As we remarked above, natural morphisms are the natural lifts of local diffeomorphisms given by equation (4.1.3). They are the only morphisms which preserve the natural structure. Of course one can consider bundle automorphisms $\boldsymbol{\Phi} = (\Phi, \phi)$ on the tangent bundle $T(M)$ which are not the lift of local diffeomorphisms, i.e. $\Phi \neq T(\phi)$. However, this corresponds to consider the tangent bundle $T(M)$ as a bundle *without* its natural structure; e.g., by regarding $T(M)$ as a gauge natural bundle associated to the structure bundle $L(M)$ (see Chapter 5 below) if the automorphism $\boldsymbol{\Phi}$ is induced by a principal automorphism of $L(M)$, or, more generally, to consider $T(M)$ as a *"bare"* fiber bundle if $\boldsymbol{\Phi}$ is a general (non-linear) automorphism.

For example, let consider the bundle $T(\mathbb{R}) = (\mathbb{R}^2, \mathbb{R}, \mathrm{pr}_1; \mathbb{R})$ with global fibered coordinates $(x; y)$; the morphism $\boldsymbol{\Phi}_1$ given by

$$\begin{cases} x' = x^3 + x \\ y' = (3x^2 + 1)\, y \end{cases} \tag{4.2.1}$$

is a natural morphism over $T(\mathbb{R})$ being the tangent lift of the base diffeomorphism $\phi : \mathbb{R} \longrightarrow \mathbb{R} : x \mapsto x^3 + x$. The morphism $\boldsymbol{\Phi}_2$ given by

$$\begin{cases} x' = x^3 + x \\ y' = \cos(x)\, y \end{cases} \tag{4.2.2}$$

is linear but not natural. The morphism $\boldsymbol{\Phi}_3$ given by

$$\begin{cases} x' = x^3 + x \\ y' = \cos(x)\, y + \sin(x) \end{cases} \tag{4.2.3}$$

is affine but not linear. Finally, the morphism $\boldsymbol{\Phi}_4$ given by

$$\begin{cases} x' = x^3 + x \\ y' = x\, y^2 \end{cases} \tag{4.2.4}$$

is a general morphism which is neither natural, nor linear, nor affine.

Locally let us consider fibered coordinates $(x^\mu; y^i)$ in a natural trivialization of a natural bundle $L^s(M) \times_\lambda F$. As an example, we already introduced the local explicit expression of the natural lift of a diffeomorphism $\phi : M \longrightarrow M$ up to the frame bundle $L^3(M)$ (see (3.7.22)). The general explicit expression for the induced automorphism on $L^s(M)$ for $s \geq 4$ can be easily worked out and it is left as an exercise to the reader. Anyway, the induced automorphism is a principal morphism, i.e. there exists a local map $j^s_0\gamma : U_\alpha \longrightarrow \mathrm{GL}^s(m) : x \mapsto j^s_0\gamma(x)$ such that we have

$$L^s(\phi) : L^s(M) \longrightarrow L^s(M) : j^s_0\epsilon \mapsto j^s_0\epsilon \cdot j^s_0\gamma \tag{4.2.5}$$

Here, according to the notation introduced in Chapter 2, we denoted by $j^s_0\epsilon$ a point in $L^s(M)$, namely the class selected by the representative $\epsilon : \mathbb{R}^m \longrightarrow M$ which is locally invertible around $0 \in \mathbb{R}^m$.

If we denote by $[j^s_0\epsilon, y]_\lambda$ a point in the natural bundle $L^s(M) \times_\lambda F$ and we fix a local trivialization $t^{(\alpha)}_s$ in $L^s(M)$ such that $j^s_0\epsilon = j^s_0\epsilon^{(\alpha)} \cdot j^s_0\alpha$ (α being a map $\alpha : \mathbb{R}^m \longrightarrow \mathbb{R}^m$ which is locally invertible at the origin and such that $\alpha(0) = 0$; see definition (2.7.3)), then the point $[j^s_0\epsilon, y]_\lambda$ corresponds to the fibered coordinates $(x; \lambda(j^s_0\alpha, y))$ where $x = \tau(j^s_0\epsilon)$. Thus the local expression of the induced morphism λ is implicitly given by:

$$\phi_\lambda : L^s(M) \times_\lambda F \longrightarrow L^s(M) \times_\lambda F : (x; y) \mapsto (\phi(x), \lambda(j^s_0\gamma, y)) \tag{4.2.6}$$

If F is a vector space and λ is a linear representation of the group $\mathrm{GL}^s(m)$ over F, i.e. $\lambda^i(j^s_0\gamma, y) = \lambda^i_{\ j}(j^s_0\gamma)y^j$, we obtain the expression

$$\begin{cases} x'^\mu = \phi^\mu(x) \\ y'^i = \lambda^i_{\ j}(j^s_0\gamma)\, y^j \end{cases} \tag{4.2.7}$$

If we consider a vector field $\xi = \xi^\mu(x)\partial_\mu$ on M and we denote by $\phi_s : M \longrightarrow M$ its flow, then $(\phi_s)_\lambda : L^s(M) \times_\lambda F \longrightarrow L^s(M) \times_\lambda F$ is a flow on $L^s(M) \times_\lambda F$ since $(\cdot)_\lambda$ is a functor. The infinitesimal generator of the flow $(\phi_s)_\lambda$ is a vector field $\hat{\xi}_\lambda = \xi^\mu\,\partial_\mu + \xi^i\,\partial_i$ given by:

$$\begin{cases} \xi^\mu = \dfrac{\mathrm{d}}{\mathrm{d}s}\phi^\mu_s\Big|_{s=0} \\ \xi^i = \partial^{\ k}_\mu \lambda^i_{\ j}(e)\, y^j \xi^\mu_k + \partial^{\ kh}_\mu \lambda^i_{\ j}(e)\, y^j \xi^\mu_{kh} + \ldots + \partial^{\ k_1\ldots k_s}_\mu \lambda^i_{\ j}(e)\, y^j \xi^\mu_{k_1\ldots k_s} \end{cases} \tag{4.2.8}$$

where e is the neutral element in $\mathrm{GL}^s(m)$, $(\gamma^\mu_k, \gamma^\mu_{kh}, \ldots, \gamma^\mu_{k_1\ldots k_s})$ are local coordinates on $\mathrm{GL}^s(m)$ and $\hat{\xi} = \xi^\mu\,\partial_\mu + \xi^\mu_k\,\partial^{\ k}_\mu + \xi^\mu_{kh}\,\partial^{\ kh}_\mu + \ldots + \xi^\mu_{k_1\ldots k_s}\,\partial^{\ k_1\ldots k_s}_\mu$ is the natural lift of ξ on $L^s(M)$ (locally given by expression (3.7.23) in the explicit case $s = 3$).

We warn the reader that in applications it is often easier to re-calculate the lifts on the particular natural bundle under consideration rather than to apply the general formulae obtained here. For example, on the tangent bundle $T(M)$ we can choose natural coordinates $(x^\mu; v^\mu)$, so that a base diffeomorphism $\phi : M \longrightarrow M$ is lifted according to the rule

$$\begin{cases} x'^\mu = \phi^\mu(x) \\ v'^\mu = (\partial_\nu\phi^\mu)\, v^\nu \end{cases} \tag{4.2.9}$$

(compare with equation (1.3.38)).

Then the lift of a vector field $\xi = \xi^\mu(x)\,\partial_\mu$ can be easily obtained by derivation from (1.3.38). The lifted vector field $\hat{\xi} = \xi^\mu\,\partial_\mu + \Xi^\mu\,\bar{\partial}_\mu$ is of the form

$$\Xi^\mu = (\partial_\nu\xi^\mu)\, v^\nu \tag{4.2.10}$$

where $\bar{\partial}_\mu = \frac{\partial}{\partial v^\mu}$ are the (local) basis for vertical vector fields.

Proposition (4.2.11): the natural lift is a Lie algebra morphism, i.e.:

$$[\xi, \zeta]^\wedge = [\hat{\xi}, \hat{\zeta}] \tag{4.2.12}$$

Proof: the general result is a direct consequence of the fact that natural functors are functors, i.e. they preserve compositions. ∎

We can check it directly on the tangent lifts (4.2.10). The lift of a commutator is given by:

$$\partial_\nu(\xi^\rho \partial_\rho \zeta^\mu - \zeta^\rho \partial_\rho \xi^\mu)\, v^\nu = (\partial_\nu \xi^\rho \partial_\rho \zeta^\mu + \xi^\rho \partial_{\nu\rho} \zeta^\mu - \partial_\nu \zeta^\rho \partial_\rho \xi^\mu - \zeta^\rho \partial_{\nu\rho} \xi^\mu)\, v^\nu \tag{4.2.13}$$

while the component along $\bar{\partial}_\mu$ of the commutator of lifts is given by:

$$(\xi^\rho \partial_{\rho\nu} \zeta^\mu + \partial_\nu \xi^\rho \partial_\rho \zeta^\mu - \zeta^\rho \partial_{\rho\nu} \xi^\mu - \partial_\nu \zeta^\rho \partial_\rho \xi^\mu)\, v^\nu \tag{4.2.14}$$

Thus we have $[\xi, \zeta]^\wedge = [\hat{\xi}, \hat{\zeta}]$.

3. Lie Derivatives of Sections

Let $\mathcal{B} = (B, M, \pi; F) = L^s(M) \times_\lambda F$ be a natural bundle of order s and $(x^\mu; y^i)$ be a system of local natural coordinates on B. For all base vector fields $\xi = \xi^\mu\, \partial_\mu$ we have a unique (projectable) vector field $\hat{\xi} \in \mathfrak{X}_{proj}(B)$ which is locally of the form $\hat{\xi} = \xi^\mu\, \partial_\mu + \xi^i\, \partial_i$, the explicit expression of which is given by equations (4.2.8). According to Section 2.6, we can consider the Lie derivative $\pounds_\Xi \sigma$ of a section $\sigma : M \longrightarrow B$ with respect to any projectable vector field Ξ on the bundle. This Lie derivative

$$\pounds_\Xi \sigma = T\sigma(\xi) - \Xi \circ \sigma \tag{4.3.1}$$

depends just on the section σ and on the vector field Ξ which projects onto the base vector field $\xi \in \mathfrak{X}(M)$.

Consequently, in natural bundles we can define the *Lie derivative of a section σ with respect to a base vector field* $\xi \in \mathfrak{X}(M)$ by simply setting:

$$\pounds_\xi \sigma \equiv \pounds_{\hat{\xi}} \sigma = T\sigma(\xi) - \hat{\xi} \circ \sigma \tag{4.3.2}$$

We stress that this is true thanks to the naturality of the bundle $\mathcal{B}$. In a general bundle $\mathcal{C} = (C, M, \pi; F)$ there is no natural way of associating a vector field $\Xi \in \mathfrak{X}_{proj}(C)$ to a base vector field $\xi \in \mathfrak{X}(M)$; on the contrary it is still true that we can associate a base vector field $\xi \in \mathfrak{X}(M)$ to a bundle vector field $\Xi \in \mathfrak{X}_{proj}(C)$ by setting ξ to be the vector field onto which Ξ is projected.

Proposition (4.3.3): if $\mathcal{B}$ is a natural bundle the following property holds for each k:

$$\forall \sigma \in \Gamma(\pi) \quad j^k(\pounds_\xi \sigma) = \pounds_\xi (j^k \sigma) \tag{4.3.4}$$

Proof: let $\hat{\xi}$ be the natural lift of ξ on the natural bundle $\mathcal{B}$ and $j^k\hat{\xi}$ be its prolongation to $J^k\mathcal{B}$. We remark that $J^k\mathcal{B}$ is still natural (see proposition (4.1.6)) and $j^k\hat{\xi}$ is the natural lift of ξ on $J^k\mathcal{B}$. Using proposition (2.6.7) we obtain:

$$j^k(\pounds_\xi \sigma) = j^k(\pounds_{\hat{\xi}} \sigma) = \pounds_{j^k\hat{\xi}}(j^k\sigma) = \pounds_\xi (j^k \sigma) \tag{4.3.5}$$

This proves our claim. ■

Proposition (4.3.6): let $(x^\mu; y^i, X^i)$ be natural fibered coordinates on the vertical bundle $\pi_V : VB \longrightarrow M$. Then we have:

$$[\pounds_{\xi_1}, \pounds_{\xi_2}]\sigma = (\pounds_{\xi_1} y^i - \pounds_{\xi_2} y^i)\partial_i + (\pounds_{[\xi_1,\xi_2]} y^i)\hat{\partial}_i \tag{4.3.7}$$

where we denoted by $\hat{\partial}_i$ the partial derivatives with respect to X^i.

Proof: it is a consequence of proposition (2.6.11) together with proposition (4.2.11). In fact we have:

$$\begin{aligned}[\pounds_{\xi_1}, \pounds_{\xi_2}]\sigma =&[\pounds_{\hat{\xi}_1}, \pounds_{\hat{\xi}_2}]\sigma = (\pounds_{\hat{\xi}_1} y^i - \pounds_{\hat{\xi}_2} y^i)\partial_i + (\pounds_{[\hat{\xi}_1,\hat{\xi}_2]} y^i)\hat{\partial}_i = \\ =&(\pounds_{\hat{\xi}_1} y^i - \pounds_{\hat{\xi}_2} y^i)\partial_i + (\pounds_{[\xi_1,\xi_2]^\wedge} y^i)\hat{\partial}_i = \\ =&(\pounds_{\xi_1} y^i - \pounds_{\xi_2} y^i)\partial_i + (\pounds_{[\xi_1,\xi_2]} y^i)\hat{\partial}_i\end{aligned} \tag{4.3.8}$$

which proves our claim. ■

We stress that on natural bundles we recover the standard situation encountered in Differential Geometry, where we can define the dragging of natural objects (e.g. vector fields or tensor fields) along base vector fields ξ and consequently define their Lie derivative along ξ. We remark that this is a characteristic feature of natural bundles which *does not* extend to general bundles, where there is no canonical way to define a natural action of base diffeomorphisms on the bundle and its sections. In this general situation only bundle automorphisms act on the bundle and its sections; consequently only Lie derivatives along (projectable) bundle vector fields can be defined.

4. Examples

As is well known, tensor fields on M, i.e. sections of tensor bundles $T^p_q(M)$, are natural objects. Natural coordinates on $T^p_q(M)$ are $(x^\mu; t^{\nu_1\dots\nu_p}_{\mu_1\dots\mu_q})$ and transition functions J^σ_ν act as follows:

$$t'^{\nu_1\dots\nu_p}_{\mu_1\dots\mu_q} = J^{\nu_1}_{\rho_1}\dots J^{\nu_p}_{\rho_p}\; t^{\rho_1\dots\rho_p}_{\sigma_1\dots\sigma_q}\; \bar{J}^{\sigma_1}_{\mu_1}\dots\bar{J}^{\sigma_q}_{\mu_q} \tag{4.4.1}$$

where, as usual, $\bar{J}^\sigma_\mu$ denotes the inverse matrix of the transition functions J^ν_σ. The equations (4.4.1) define a cocycle due to the chain rule iterated as many times as required.

This allows us to define the bundle $T^p_q(M)$ in an easy way so that it is evident that it is a natural bundle and that its sections are in a one-to-one correspondence with tensor fields over M. Let us in fact consider the representation of $\mathrm{GL}(m)$ on the vector space $T^p_q(\mathbb{R}^m) \simeq \mathbb{R}^k$ with $k = m^{p+q}$ given by:

$$\begin{aligned} \lambda :&\mathrm{GL}(m) \times T^p_q(\mathbb{R}^m) \longrightarrow T^p_q(\mathbb{R}^m) \\ :&(J^\sigma_\nu, t^{\nu_1\dots\nu_p}_{\mu_1\dots\mu_q}) \mapsto t'^{\nu_1\dots\nu_p}_{\mu_1\dots\mu_q} \end{aligned} \tag{4.4.2}$$

where $t'^{\nu_1\dots\nu_p}_{\mu_1\dots\mu_q}$ is given by equation (4.4.1).

Then we can consider the natural bundle $T^p_q(M) = L(M) \times_\lambda T^p_q(\mathbb{R}^m)$. Let $(x^\mu; t^{\nu_1\dots\nu_p}_{\mu_1\dots\mu_q})$ be natural coordinates on $T^p_q(M)$; a global section σ of $T^p_q(M)$ has local expression

$$\sigma : x^\mu \mapsto \left(x^\mu, t^{\nu_1\dots\nu_p}_{\mu_1\dots\mu_q}(x)\right) \tag{4.4.3}$$

The globality condition for the section σ tells us that, if we change the local chart in M and consequently the local natural trivialization on $T^p_q(M)$, the local expression $t^{\nu_1\dots\nu_p}_{\mu_1\dots\mu_q}(x)$ transforms according to the tensorial rule (4.4.1). Then there is a one-to-one correspondence between global sections of $T^p_q(M)$ and tensor fields of rank (p, q) over M.

This technique is very effective in physical applications since we often know the local expression of physical fields, we know how they transform with respect to diffeomorphisms and we want to regard them as natural objects, i.e. as global sections of some suitable natural bundle. Let us consider some relevant cases.

Some natural objects, as e.g. the volume form of a sub-manifold, are tensor densities. A *tensor density of rank (p, q) and weight k* is locally given by objects $t^{\nu_1\dots\nu_p}_{\mu_1\dots\mu_q}$ which transform according to the rule

$$t'^{\nu_1\dots\nu_p}_{\mu_1\dots\mu_q} = (\det J)^{-k} J^{\nu_1}_{\rho_1}\dots J^{\nu_p}_{\rho_p}\; t^{\rho_1\dots\rho_p}_{\sigma_1\dots\sigma_q}\; \bar{J}^{\sigma_1}_{\mu_1}\dots\bar{J}^{\sigma_q}_{\mu_q} \tag{4.4.4}$$

where $\det J$ is the determinant of the Jacobian matrix. The rule (4.4.1) is a particular case with $k = 0$.

Then let us consider the representation

$$\begin{aligned}\lambda :&\mathrm{GL}(m) \times T^p_q(\mathbb{R}^m) \longrightarrow T^p_q(\mathbb{R}^m)\\ :&(J^\sigma_\nu, t^{\nu_1\dots\nu_p}_{\mu_1\dots\mu_q}) \mapsto (\det J)^{-k} J^{\nu_1}_{\rho_1}\dots J^{\nu_p}_{\rho_p}\; t^{\rho_1\dots\rho_p}_{\sigma_1\dots\sigma_q}\; \bar J^{\sigma_1}_{\mu_1}\dots \bar J^{\sigma_q}_{\mu_q}\end{aligned} \tag{4.4.5}$$

and the natural bundle $\overset{k}{D}{}^p_q(M) = L(M) \times_\lambda T^p_q(\mathbb{R}^m)$ the sections of which are tensor densities of rank (p, q) and weight k. Again (4.4.4) defines a cocycle because of the chain rule together with the rule for the derivative of a determinant.

Analogously, we can consider linear connections on M as natural objects locally described by their coefficients $\Gamma^\alpha_{\beta\lambda}$ in a natural trivialization (see (3.7.6)). As we saw (see (3.7.11)) they transform according to the rule:

$$\Gamma'^\sigma_{\lambda\mu} = J^\sigma_\rho\Big(\Gamma^\rho_{\epsilon\nu}\bar J^\epsilon_\lambda \bar J^\nu_\mu + \bar J^\rho_{\lambda\mu}\Big) \tag{4.4.6}$$

Now $(J^\epsilon_\lambda, J^\epsilon_{\lambda\mu})$ are local coordinates on $\mathrm{GL}^2(m)$ and we can define a representation λ of the group $\mathrm{GL}^2(m)$ on the affine space $\mathbb{R}^{(m^3)}$ having coordinates $\Gamma^\rho_{\epsilon\nu}$ by setting

$$\begin{aligned}\lambda :&\mathrm{GL}^2(m) \times \mathbb{R}^{(m^3)} \longrightarrow \mathbb{R}^{(m^3)}\\ :&(J^\epsilon_\lambda, J^\epsilon_{\lambda\mu}, \Gamma^\rho_{\epsilon\nu}) \mapsto J^\sigma_\rho\Big(\Gamma^\rho_{\epsilon\nu}\bar J^\epsilon_\lambda \bar J^\nu_\mu + \bar J^\rho_{\lambda\mu}\Big)\end{aligned} \tag{4.4.7}$$

Consequently, we have a natural (affine) bundle

$$L^2(M) \times_\lambda \mathbb{R}^{(m^3)} \tag{4.4.8}$$

the sections of which are linear connections on M.

Finally, let us consider the metrics of signature (r, s) on a manifold M. These are locally expressed in a chart (U, x^λ) by $g = g_{\mu\nu}(x)\, \mathrm{d}x^\mu \otimes \mathrm{d}x^\nu$; here the coefficients $g_{\mu\nu}$ are symmetric and such that there exists in U a matrix e^a_μ which satisfies

$$g_{\mu\nu} = e^a_\mu\, \eta_{ab}\, e^b_\nu \qquad\qquad \eta = \|\eta_{ab}\| = \begin{pmatrix}\mathbb{I}_r & 0\\ 0 & -\mathbb{I}_s\end{pmatrix} \tag{4.4.9}$$

where $\mathbb{I}_r$ denotes the $r \times r$ identity matrix. Here $r + s = m = \dim(M)$.

The constraint (4.4.9) identifies a sub-manifold $M(r, s)$ of the vector space $\mathbb{R}^{\frac{m(m+1)}{2}}$ of symmetric coefficients. The sub-manifold $M(r, s) \subset \mathbb{R}^{\frac{m(m+1)}{2}}$ has codimension zero and actually $M(r, s)$ is an open set in $\mathbb{R}^{\frac{m(m+1)}{2}}$. Since metrics are tensors they transform as

$$g'_{\mu\nu} = \bar J^\rho_\mu\, g_{\rho\sigma}\, \bar J^\sigma_\nu \tag{4.4.10}$$

Thus we can define an action λ of $\mathrm{GL}(m)$ on $M(r,s)$ as follows

$$\begin{aligned}\lambda :&\mathrm{GL}(m) \times M(r,s) \longrightarrow M(r,s)\\ :&(J^{\epsilon}_{\lambda}, g_{\mu\nu}) \mapsto \bar{J}^{\rho}_{\mu}\, g_{\rho\sigma}\, \bar{J}^{\sigma}_{\nu}\end{aligned} \tag{4.4.11}$$

so that we can define the *bundle of all metrics on M of signature* (r,s). It is a natural bundle $\mathrm{Met}(M;r,s) = L(M) \times_{\lambda} M(r,s)$ but it is not a vector nor an affine bundle. As a consequence, we cannot be sure that $\mathrm{Met}(M;r,s)$ has global sections. If anyone of these exists, then global sections of $\mathrm{Met}(M;r,s)$ correspond to metrics of signature (r,s) on M. In the special case $r=1$, $s=m-1$ the metrics are called *Lorentzian* and the bundle is denoted by $\mathrm{Lor}(M) = \mathrm{Met}(M;1,m-1)$. It is known that global metrics always exist for the positive definite signature $(m,0)$ (as well as for the negative definite one $(0,m)$), because of the existence of a partition of unity (see Section 1.4). On the contrary, the existence of global metrics of other signatures imposes topological restrictions on the manifold M.

Exercises

1- Prove that the bundle $J^1T(M)$ is associated to $L^2(M)$. If $\hat{\xi}$ is the natural lift of ξ on $T(M)$ prove that $j^1\hat{\xi}$ is the natural lift of a vector field ξ on $J^1T(M)$. Try and generalize to the k-order jet prolongation $J^k\mathcal{B}$ of a natural bundle $\mathcal{B} = L^s(M) \times_{\lambda} F$.

Hint: write down transition functions of $J^1T(M)$; ξ can be regarded both as a vector field over M and as a section of $T(M)$.

2- Calculate the Lie derivative of a section of $J^1T(M)$.

Hint: a vector field ξ over M induces functorially a vector field over $J^1T(M)$.

3- Define the bundle of connections over $L^2(M)$ and show that it is a natural bundle. Calculate the expression of the Lie derivative of its sections.

Hint: find a pointwise local basis for vertical right invariant vector fields over $L^2(M)$; compute transformation rules of the principal connection coefficients w.r.t. a change of trivialization on $L^2(M)$.

4- Prove that the bundle $L^2(M) \times_{\lambda} \mathbb{R}^{(m^3)}$ defined in (4.4.8) is canonically isomorphic to the bundle $J^1L(M)/\mathrm{GL}(m)$.

Hint: compute transition functions of $J^1L(M)/\mathrm{GL}(m)$ and compare with the representation λ given by (4.4.7).

5- Prove that the representations (4.4.2), (4.4.5) and the action (4.4.11) are really left actions of $\mathrm{GL}(m)$. Prove also that the action (4.4.7) is a left action of $\mathrm{GL}^2(m)$.

Hint: check associativity.

Chapter 5

GAUGE NATURAL BUNDLES

Abstract We shall here define *gauge natural bundles* together with the structures defined on them, e.g. the *automorphisms and vector fields* induced by the structure bundle. Both the categorial and constructive definitions are given and compared. The *bundle of principal connections* is again considered as an example of gauge natural bundle. We shall also consider the *bundle of automorphisms* and *infinitesimal automorphisms* of a principal bundle $\mathcal{P}$ as further examples and applications.

References

The main reference about the category of gauge natural bundle is [KMS93]. The original material can be found in [K82], [E81]. For the use of gauge structures in Physics we refer to [A79], [B81], [N90], [F98].

1. Gauge Natural Bundles

Gauge natural bundles are a sub-category of the bundles with structure group. Natural bundles are particular gauge natural bundles. Gauge natural bundles are important for Physics since they provide the general framework for gauge theories, as we shall see all along the rest of this monograph. As natural bundles provide the modern perspective to define the objects of Differential Geometry (e.g. linear connections on manifolds), gauge natural bundles provide the framework to deal with objects defined on principal bundles (e.g. principal connections, principal automorphisms and infinitesimal generators of principal automorphisms). This analogy is very deeply founded. As natural bundles are functorially defined starting from a manifold, gauge natural bundles are functorially defined starting from a principal bundle $\mathcal{P}$, called the *structure bundle*. As natural bundles encode the properties of the manifold which they are defined on, gauge natural bundles encode the properties of their structure bundle.

A few simple examples of gauge natural bundles were already introduced in Chapter 3: the bundle $T\mathcal{P}/G$ of infinitesimal generators of principal automorphisms, the bundle $V\mathcal{P}/G \simeq \mathcal{P} \times_{\mathrm{Ad}} \mathfrak{g}$ of vertical infinitesimal automorphisms, the bundle $J^1\mathcal{P}/G$ of principal connections, as well as any associated bundle to $\mathcal{P}$ are gauge natural bundles associated to the structure bundle $\mathcal{P}$. Notice that, similarly to trivializations on natural bundles, once a trivialization on the structure bundle $\mathcal{P}$ is fixed, a trivialization on the gauge natural bundle is induced, and it is called a *canonical trivialization*.

Let us consider the category $\mathfrak{P}(G)$ of all principal bundles having G as structure group, with principal morphisms which project on local diffeomorphisms.

Definition (5.1.1): a covariant functor $\mathfrak{B}$ from the category $\mathfrak{P}(G)$ into the category of bundles with structure group G is a *gauge natural functor* if the following properties hold true for all $\mathcal{P} = (P, M, p; G) \in \mathfrak{P}(G)$:

(a) $\mathfrak{B}(\mathcal{P}) = (BP, M, \pi; F)$ is a bundle over the same base manifold M of $\mathcal{P}$;

(b) any principal morphism $\Phi : P \longrightarrow P'$ projecting onto $\phi : M \longrightarrow M'$ induces a fibered morphism $\mathfrak{B}(\Phi) : BP \longrightarrow BP'$ also projecting over ϕ;

(c) if $U \subseteq M$ is an open subset in the base manifold M of the structure bundle $\mathcal{P}$ and $i : p^{-1}(U) \longrightarrow P$ is the inclusion morphism then $\mathfrak{B}(i) : B(p^{-1}(U)) \longrightarrow BP$ is the inclusion of the sub-bundle $\mathfrak{B}(p^{-1}(U))$ into $\mathfrak{B}(\mathcal{P})$. ∎

Remark: we encourage the reader to notice the similarities and the differences between this definition and definition $(4.1.1)$. Natural bundles are in fact a very particular case of gauge natural bundles, since $(4.1.1)$ follows directly from $(5.1.1)$ in the appropriate sub-case.

We already encountered some examples of gauge natural functors. For example, we can define a covariant functor $T_G(\cdot)$ which to each principal bundle $\mathcal{P} = (P, M, p; G)$ associates the bundle $T_G(\mathcal{P})$ of infinitesimal generators of automorphisms, which is a bundle over M with total space TP/G, and to any principal morphisms $\Phi : P \longrightarrow P'$ the morphism $T_G(\Phi) : TP/G \longrightarrow TP'/G$. Being Φ a principal morphism, its tangent map $T\Phi : TP \longrightarrow TP'$ preserves the canonical right actions on P and P' so that it induces the quotient map $T_G(\Phi) : TP/G \longrightarrow TP'/G$. Of course: (a), $T_G(\cdot)$ preserves compositions (i.e. it is a covariant functor); (b), for all principal morphisms $\boldsymbol{\Phi} = (\Phi, \phi)$ the induced morphism $T_G(\Phi)$ also projects onto $\phi : M \longrightarrow M'$; (c), finally, inclusions are preserved since $T(\cdot)$ is a natural functor. Thence $T_G(\cdot)$ is a gauge natural functor.

Analogously, the functors $V_G(\cdot)$ and $J^1_G(\cdot)$ associate to the structure bundle $\mathcal{P}$ the bundles $V_G(\mathcal{P})$ with total space VP/G and $J^1_G(\mathcal{P})$ with total space J^1P/G, respectively. In both cases a principal morphism $\Phi : P \longrightarrow P'$ can be lifted to $V(\Phi) : VP \longrightarrow VP'$ and $J^1(\Phi) : J^1P \longrightarrow J^1P'$, respectively, and they both turn out to preserve the right actions, so that they induce quotient morphisms $V_G(\Phi) : VP/G \longrightarrow VP'/G$ and $J^1_G(\Phi) : J^1P/G \longrightarrow J^1P'/G$, respectively. See Exercise 1 below.

A further class of examples of gauge natural functors is the following. Let $\mathcal{P} = (P, M, p; G)$ be a principal bundle and $\lambda : G \times F \longrightarrow F$ be a left action of G on a manifold F. We can define a functor $(\cdot)_\lambda$ which associates to $\mathcal{P}$ the associated bundle $(\mathcal{P})_\lambda = \mathcal{P} \times_\lambda F$ and to a principal morphism $\boldsymbol{\Phi} = (\Phi, \phi)$ between $\mathcal{P}$ and $\mathcal{P}' = (P', M', p'; G')$ the bundle morphism defined by:

$$(\Phi)_\lambda : P \times_\lambda F \longrightarrow P' \times_\lambda F : [p, f]_\lambda \mapsto [\Phi(p), f]_\lambda \qquad (5.1.2)$$

Then $(\cdot)_\lambda$ is a gauge natural functor. As we shall see below, these *are not* the most general gauge natural functors we can define.

Definition (5.1.3): the *category of gauge natural bundles* is the sub-category of the category of fiber bundles with structure group in which objects are the bundles, called *gauge natural bundles*, which are in the image of some gauge natural functor $\mathfrak{B}$; morphisms, called *gauge natural morphisms*, are fibered morphisms obtained by the lift $\mathfrak{B}(\Phi)$ of some principal morphism $\boldsymbol{\Phi} = (\Phi, \phi)$. Then gauge natural functors restrict to functors from the category of principal bundles having structure group G to the category of gauge natural bundles. Sections of gauge natural bundles are called *gauge natural objects*. ∎

As for the natural bundles, we stress that the construction of a gauge natural bundle out of its structure group $\mathcal{P}$ must be a *canonical* construction. Consequently, a gauge natural bundle is never defined on its own. On the contrary we always define a whole family of gauge natural bundles, one for each structure bundle.

2. Gauge Natural Prolongations of a Principal Bundle

If G is a Lie group and r any integer, the set

$$J^r_m G = \{ j^r_0 a \mid a : \mathbb{R}^m \longrightarrow G \} \tag{5.2.1}$$

can be endowed with a group structure by defining the multiplication

$$j^r_0 a \cdot j^r_0 b = j^r_0 (a \cdot b) \tag{5.2.2}$$

so that it is a Lie group.

If g^a are local coordinates on G, then $(g^a, g^a_\alpha, \dots, g^a_{\alpha_1 \dots \alpha_r})$ are local coordinates on $J^r_m G$. We shall denote by lower Latin indices $a, b, \dots = 1 \dots n$ the quantities in the group G, by capital Latin indices $A, B, \dots = 1 \dots n$ the quantities in the Lie algebra $\mathfrak{g}$, by Greek indices such as $\mu, \nu, \dots = 1 \dots m$ the quantities on M and by Greek indices $\alpha, \beta, \dots = 1 \dots m$ in the first part of the alphabet the quantities on $\mathbb{R}^m$.

Let us also recall that the group $\mathrm{GL}^s(m)$ was defined in Section 2.7 (see expression (2.7.3)). A point $j^s_0 \alpha \in \mathrm{GL}^s(m)$ has local coordinates in the form $(\alpha^\alpha_\beta, \alpha^\alpha_{\beta\gamma}, \dots, \alpha^\alpha_{\beta_1 \dots \beta_s})$. The multiplication $j^s_0 \alpha \cdot j^s_0 \beta$ in $\mathrm{GL}^s(m)$ is locally given by

$$(\alpha^\alpha_\epsilon \, \beta^\epsilon_\beta, \; \alpha^\alpha_{\epsilon\gamma} \, \beta^\epsilon_\beta + \alpha^\alpha_\epsilon \, \beta^\epsilon_{\beta\gamma}, \; \dots) \tag{5.2.3}$$

Furthermore, if we fix an integer r with $r \le s$, we can then define a canonical right action of $j^s_0 \alpha \in \mathrm{GL}^s(m)$ on $J^r_m G$ by setting

$$j^r_0 a \cdot j^s_0 \alpha = j^r_0 (a \circ \alpha) \qquad\qquad \mathbb{R}^m \xrightarrow{\alpha} \mathbb{R}^m \xrightarrow{a} G \tag{5.2.4}$$

which is well defined since it does not depend on the representative of the point $j^s_0 \alpha$ because of $r \le s$.[1]

Thence, we have two Lie groups ($\mathrm{GL}^s(m)$ and $J^r_m G$) and a right action of the first group on the second one. We can define the semi-direct product[2]

$$W^{(r,s)}_m G = \mathrm{GL}^s(m) \rtimes J^r_m G \tag{5.2.6}$$

which will be called *prolongation of order* (r, s) *of the group* G for reasons which will be clear in a while. The multiplication on $W^{(r,s)}_m G$ is defined as

$$(j^s_0 \alpha, \; j^r_0 a) \cdot (j^s_0 \beta, \; j^r_0 b) = \big(j^s_0 (\alpha \circ \beta), \; j^r_0 ((a \circ \beta) \cdot b) \big) \tag{5.2.7}$$

---

[1] This pair of integers (r, s) has of course nothing to do with any signature notion. The notation, which is standard, is the same, but we feel obliged to warn the reader against any possible confusion.

[2] Let G and H be two Lie groups and let $\rho : G \times H \longrightarrow G$ be a *right action* of H on G such that $\rho_h \equiv \rho(h, \cdot) \in \mathrm{Hom}(G)$ is a group homomorphism. The semidirect product $H \rtimes G$ is the group having the pairs (h, g) ($g \in G$ and $h \in H$) as elements with (associative) multiplication given by:

$$(h_1, g_1) \cdot (h_2, g_2) = (h_1 \cdot h_2, \rho(g_1, h_2) \cdot g_2) \tag{5.2.5}$$

For each principal bundle $\mathcal{P} = (P, M, \pi; G)$ let us define the total space

$$W^{(r,s)}P = L^s(M) \times_M J^r P \qquad (5.2.8)$$

The total space $W^{(r,s)}P$ is made of pairs $(j^s_0\epsilon,\ j^r_x\sigma)$ such that $j^s_0\epsilon \in L^s(M)$ and $j^r_x\sigma \in J^rP$. We also define a canonical projection $p : W^{(r,s)}P \longrightarrow M : (j^s_0\epsilon,\ j^r_x\sigma) \mapsto x$.

A canonical right action of $W^{(r,s)}_m G$ on $W^{(r,s)}P$ can be defined by setting

$$(j^s_0\epsilon,\ j^r_x\sigma) \cdot (j^s_0\alpha,\ j^r_0 a) = \left(j^s_0(\epsilon \circ \alpha),\ j^r_x\left(\sigma \cdot (a \circ \alpha^{-1} \circ \epsilon^{-1})\right)\right) \qquad (5.2.9)$$

This right action turns out to be free so that $(W^{(r,s)}P, M, p; W^{(s,r)}_m G)$ is a principal bundle which will be denoted by $W^{(r,s)}\mathcal{P}$.

Definition (5.2.10): for each principal bundle $\mathcal{P} = (P, M, \pi; G)$ and for each integer pair (r, s) with $r \leq s$ we can define the principal bundle

$$W^{(r,s)}\mathcal{P} = L^s(M) \rtimes J^r\mathcal{P} \equiv (W^{(r,s)}P, M, p; W^{(s,r)}_m G) \qquad (5.2.11)$$

which is called the *gauge natural prolongation of order* (r, s) *of the principal bundle* $\mathcal{P}$.

■

Since gauge natural prolongations will play a prominent role in the sequel let us study them is some further detail. For notational simplicity let us specialize to a matrix group, e.g. $G = \mathrm{GL}(n)$, so that the matrix entries $\gamma^i_{\ j}$ are local coordinates on G. Doing that any Latin index (both lowercase and capital) becomes a pair of indices $i,\ j = 1 \dots n$. Let us then further specialize to $r = s = 1$. A point $j^1_x\sigma \in J^1P$ is locally given by $(x^\mu, g^i_{\ j}, g^i_{\ j\mu})$ (notice the world indices as derivatives; σ depends in fact on $x \in M$). Analogously, a point $j^1_0\epsilon \in L(M)$ is locally given by $(x^\mu, \epsilon^\mu_{\ \alpha})$ (remind that $\epsilon : \mathbb{R}^m \longrightarrow M$). Thus we can choose local coordinates $(x^\mu, \epsilon^\mu_{\ \alpha}, g^i_{\ j}, g^i_{\ j\mu})$ on $W^{(1,1)}P$. If $(\alpha^\epsilon_{\ \alpha}, a^i_{\ j}, a^i_{\ j\alpha})$ are local coordinates on $W^{(1,1)}_m G$, the right action (5.2.9) locally reads as:

$$(x^\mu,\ \epsilon^\mu_{\ \alpha},\ g^i_{\ j},\ g^i_{\ j\mu}) \cdot (\alpha^\alpha_{\ \beta},\ a^i_{\ j},\ a^i_{\ j\alpha}) = (x'^\mu,\ \epsilon'^\mu_{\ \alpha},\ g'^i_{\ j},\ g'^i_{\ j\mu}) \qquad (5.2.12)$$

Here we set

$$\begin{cases} x'^\mu = x^\mu \\ \epsilon'^\mu_{\ \beta} = \epsilon^\mu_{\ \alpha}\, \alpha^\alpha_{\ \beta} \\ g'^i_{\ j} = g^i_{\ k}\, a^k_j \\ g'^i_{\ j\mu} = g^i_{\ k\mu} a^k_j + g^i_{\ k}\, a^k_{j\alpha}\, \bar\alpha^\alpha_{\ \beta}\, \bar\epsilon^\beta_{\ \mu} \end{cases} \qquad (5.2.13)$$

where as usual $\bar\alpha^\alpha_{\ \beta}$ and $\bar\epsilon^\beta_{\ \mu}$ are the inverse matrices of $\alpha^\beta_{\ \alpha}$ and $\epsilon^\mu_{\ \beta}$, respectively.

Notice that we cannot recognize it as a multiplication in the group. This is due to the fact that natural coordinates on J^1P (in particular $g^i_{\ j\mu}$) do not correspond to natural coordinates on J^1G (in particular $a^i_{\ j\alpha}$) since the index μ does not match the index α as they describe objects in different (though locally diffeomorphic) spaces, i.e. M and $\mathbb{R}^m$. Anyway, we can define new local coordinates $(x^\mu, \epsilon^\mu_{\ \alpha}, g^i_{\ j}, g^i_{\ j\alpha})$ on the bundle $W^{(1,1)}\mathcal{P}$ by setting

$$g^i_{\ j\alpha} = g^i_{\ j\mu}\,\epsilon^\mu_{\ \alpha} \tag{5.2.14}$$

In these new coordinates the right action reads as

$$\begin{aligned} g'^i_{\ j\beta} =& g'^i_{\ j\mu}\,\epsilon'^\mu_{\ \beta} = (g^i_{\ k\mu}a^k_j + g^i_{\ k}\,a^k_{\ j\alpha}\,\bar\alpha^\alpha_{\ \beta}\,\bar\epsilon^\beta_{\ \mu})\epsilon^\mu_{\ \gamma}\,\alpha^\gamma_{\ \beta} = \\ =& g^i_{\ k\mu}a^k_{\ j}\epsilon^\mu_{\ \gamma}\,\alpha^\gamma_{\ \beta} + g^i_{\ k}\,a^k_{\ j\beta} = g^i_{\ k\gamma}a^k_{\ j}\,\alpha^\gamma_{\ \beta} + g^i_{\ k}\,a^k_{\ j\beta} \end{aligned} \tag{5.2.15}$$

It corresponds to the multiplication defined by (5.2.7) which under the present hypotheses locally reads as follows

$$(\alpha^\alpha_{\ \beta},\, a^i_{\ j},\, a^i_{\ j\alpha})\cdot(\beta^\alpha_{\ \beta}, b^i_{\ j}, b^i_{\ j\alpha}) = (\alpha^\alpha_{\ \gamma}\,\beta^\gamma_{\ \beta},\, a^i_{\ k}\,b^k_{\ j},\, a^i_{\ k\beta}\,b^k_j\,\beta^\beta_{\ \alpha} + a^i_{\ k}\,b^k_{\ j\alpha}) \tag{5.2.16}$$

Consequently, in a trivialization of $W^{(1,1)}\mathcal{P}$, the group element $(\epsilon^\mu_{\ \alpha}, g^i_{\ j}, g^i_{\ j\alpha})$ corresponds to the point $(j^1_0\epsilon,\ j^1_x\sigma)$, the right action of $(j^1_0\alpha,\ j^1_0 a) \in W^{(1,1)}_m G$ acts by right translation, as expected on principal bundles. The local coordinates $(x^\mu, \epsilon^\mu_{\ \alpha}, g^i_{\ j}, g^i_{\ j\alpha})$ on $W^{(1,1)}\mathcal{P}$ (with $g^i_{\ j\alpha}$ defined by (5.2.14)) should be used whenever we want to recognize it as a principal bundle.

Let $\boldsymbol{\Phi} = (\Phi, \phi)$ be a principal morphism between two principal bundles $\mathcal{P} = (P, M, p; G)$ and $\mathcal{P}' = (P', M', p'; G)$ having the same structure group G. It induces functorially a bundle morphism $W^{(r,s)}\boldsymbol{\Phi} = (W^{(r,s)}\Phi, \phi)$ by

$$\begin{aligned} W^{(r,s)}\Phi :\ & W^{(r,s)}P \longrightarrow W^{(r,s)}P' \\ :\ & (j^s_0\epsilon,\ j^r_x\sigma) \mapsto \big(L^s(\phi)(j^s_0\epsilon),\ j^r\Phi(j^r_x\sigma)\big) \\ :\ & (j^s_0\epsilon,\ j^r_x\sigma) \mapsto \big(j^s_0(\phi\circ\epsilon),\ j^r_{\phi(x)}(\Phi\circ\sigma\circ\phi^{-1})\big) \end{aligned} \tag{5.2.17}$$

Proposition (5.2.18): the bundle morphism $W^{(r,s)}\Phi$ preserves the right action, being thus principal.

Proof: we simply notice that:

$$\begin{aligned} &\Big[W^{(r,s)}\Phi(j^s_0\epsilon,\ j^r_x\sigma)\Big]\cdot(j^s_0\alpha,\ j^r_0 a) = \\ &\quad= \Big(j^s_0(\phi\circ\epsilon),\ j^r_{\phi(x)}(\Phi\circ\sigma\circ\phi^{-1})\Big)\cdot(j^s_0\alpha,\ j^r_0 a) = \\ &\quad= \Big(j^s_0(\phi\circ\epsilon\circ\alpha),\ j^r_{\phi(x)}\Big((\Phi\circ\sigma\circ\phi^{-1})\cdot(a\circ\bar\alpha\circ\bar\epsilon)\Big)\Big) = \\ &\quad= \Big(j^s_0(\phi\circ\epsilon\circ\alpha),\ j^r_{\phi(x)}\Big(\Phi\circ(R_g\sigma)\circ\phi^{-1}\Big)\Big) = \\ &\quad= W^{(r,s)}\Phi\Big[\Big(j^s_0(\epsilon\circ\alpha),\ j^r_x(R_g\sigma)\Big)\Big] \\ &= W^{(r,s)}\Phi\Big[(j^s_0\epsilon,\ j^r_x\sigma)\cdot(j^s_0\alpha,\ j^r_0 a)\Big] \end{aligned} \tag{5.2.19}$$

where we set $g = a \circ \bar{\alpha} \circ \bar{\epsilon}$. ∎

The functor $W^{(r,s)}(\cdot)$ from the category of principal bundles with structure group G into the category of principal bundles with structure group $W^{(r,s)}_m G$ has been thence defined. If we specialize to $G = \mathrm{GL}(n)$ and $r = s = 1$, we get the local expression of $W^{(1,1)}(\Phi)$ given by

$$\begin{cases} x'^{\mu} = \phi^{\mu}(x) \\ \epsilon'^{\mu}_{\ \alpha} = J^{\mu}_{\nu}\, \epsilon^{\nu}_{\ \alpha} \\ g'^{i}_{\ j} = \gamma^{i}_{\ k}(x)\, g^{k}_{\ j} \\ g'^{i}_{\ j\mu} = \bar{J}^{\nu}_{\mu}(\gamma^{i}_{\ k\nu}\, g^{k}_{\ j} + \gamma^{i}_{\ k}\, g^{k}_{\ j\nu}) \end{cases} \tag{5.2.20}$$

where we set J^{μ}_{ν} to be the Jacobian of ϕ and $\bar{J}^{\nu}_{\mu}$ for its inverse.

Notice that again we do not recognize it as acting by a left translation of the group element $(\gamma^{i}_{\ j},\ \gamma^{i}_{\ j\mu},\ J^{\mu}_{\nu})$. If we use again $g^{i}_{\ j\alpha} = g^{i}_{\ j\mu}\, \epsilon^{\mu}_{\ \alpha}$ instead of $g^{i}_{\ j\mu}$ the action of $W^{(1,1)}\Phi$ reads as

$$\begin{aligned} g'^{i}_{\ j\alpha} =& g'^{i}_{\ j\mu}\, \epsilon'^{\mu}_{\ \alpha} = \bar{J}^{\nu}_{\mu}(\gamma^{i}_{\ k\nu}\, g^{k}_{\ j} + \gamma^{i}_{\ k}\, g^{k}_{\ j\nu}) J^{\mu}_{\lambda}\, \epsilon^{\lambda}_{\ \alpha} = \\ =&(\gamma^{i}_{\ k\nu}\, g^{k}_{\ j}\, \epsilon^{\nu}_{\ \alpha} + \gamma^{i}_{\ k}\, g^{k}_{\ j\alpha}) \end{aligned} \tag{5.2.21}$$

which agrees with the left translation $(J^{\mu}_{\nu},\ \gamma^{i}_{\ j},\ \gamma^{i}_{\ j\mu}) \cdot (\epsilon^{\mu}_{\ \alpha},\ g^{i}_{\ j},\ g^{i}_{\ j\alpha})$. Thus principal automorphisms of $W^{(1,1)}\mathcal{P}$ act by left translation, as expected on principal bundles.

Prolongations of vector fields to $W^{(r,s)}\mathcal{P}$

Let $\mathcal{P} = (P, M, p; G)$ be a principal bundle, ρ_A be a pointwise local basis of vertical right invariant vector fields and $\Xi = \xi^{\mu}(x)\partial_{\mu} + \xi^{A}(x)\,\rho_A$ be the infinitesimal generator of the following 1-parameter family of automorphisms on $\mathcal{P}$:

$$\begin{cases} x' = \phi_s(x) \\ g' = \varphi_s(x) \cdot g \end{cases} \tag{5.2.22}$$

We recall (see (3.3.15)) that the components of Ξ are then given by

$$\begin{cases} \xi^{\mu}(x) = \left(\dfrac{\mathrm{d}\phi^{\mu}_s}{\mathrm{d}s}(x)\right)_{s=0} \\ \xi^{A}(x) = \bar{T}^{A}_{a}\left(\dfrac{\mathrm{d}\varphi^{a}_s}{\mathrm{d}s}(x)\right)_{s=0} \end{cases} \tag{5.2.23}$$

Being $W^{(r,s)}(\cdot)$ a functor, we can use the prolongation (5.2.17) to define a 1-parameter family of automorphisms on $W^{(r,s)}\mathcal{P}$. The infinitesimal generator of the prolonged 1-parameter family is denoted by $\hat{\Xi}$ and it is a right invariant vector field on $W^{(r,s)}\mathcal{P}$. Thence we have a correspondence $\Xi \mapsto \hat{\Xi}$ which is called the *prolongation of vector fields to* $W^{(r,s)}\mathcal{P}$.

Let us again specialize to the case $G = \mathrm{GL}(n)$ and $r = s = 1$. The prolonged 1-parameter family of automorphisms on $W^{(1,1)}\mathcal{P}$ is given by the expressions (5.2.20). The infinitesimal generator of the prolonged family is $\hat{\Xi} = \xi^\mu\,\partial_\mu + \xi^i{}_j\,\rho_i^j + \mathrm{d}_\mu\xi^i{}_j\,\rho_i^{j\mu} + \mathrm{d}_\mu\xi^\nu\,\rho_\nu^\mu$, where we set

$$\begin{cases} \rho_j^i = g^i{}_k\,\partial_j^k + g^i{}_{k\nu}\,\partial_j^{k\nu} \\ \rho_j^{i\mu} = g^i{}_k\,\partial_j^{k\mu} \\ \rho_\nu^\mu = \epsilon^\mu_\alpha\,\partial_\nu^\alpha - g^i_{j\nu}\partial_i^{j\mu} \end{cases} \tag{5.2.24}$$

Since $\hat{\Xi}$ is right invariant and the coefficients $(\xi^\mu,\ \mathrm{d}_\nu\xi^\mu,\ \xi^i_j,\ \mathrm{d}_\mu\xi^i_j)$ are pointwise independent, the local vector fields $(\rho_i^j,\ \rho_i^{j\mu},\ \rho_\nu^\mu)$ are a pointwise local basis of vertical vector fields on $W^{(1,1)}\mathcal{P}$ (see Theorem (3.4.1)).

Gauge Natural Bundles of Finite Order

We can now give another, more general, example of gauge natural functor. Let us consider a left action λ of the group $W_m^{(r,s)}G$ on a manifold F. We can define a correspondence $(\cdot)_\lambda$ which associates to each principal bundle $\mathcal{P} = (P, M, p; G)$, called the *structure bundle*, the associated bundle $(\mathcal{P})_\lambda = W^{(r,s)}\mathcal{P} \times_\lambda F$, and to each principal morphism $\boldsymbol{\Phi} = (\Phi, \phi)$ between $\mathcal{P}$ and $\mathcal{P}'$ the corresponding bundle morphism

$$\begin{aligned} (\Phi)_\lambda :&W^{(r,s)}P \times_\lambda F \longrightarrow W^{(r,s)}P' \times_\lambda F \\ :&[j_0^s\epsilon,\ j_x^r\sigma,\ f]_\lambda \mapsto [W^{(r,s)}\Phi(j_0^s\epsilon,\ j_x^r\sigma),\ f]_\lambda \end{aligned} \tag{5.2.25}$$

The correspondence $(\cdot)_\lambda$ preserves compositions so that it defines a covariant functor. It is a gauge natural functor since: (a), $W^{(r,s)}\mathcal{P} \times_\lambda F$ is still a bundle over M; (b), $W^{(r,s)}(\boldsymbol{\Phi}) = (W^{(r,s)}(\Phi), \phi)$ still projects over ϕ; (c), sub-bundles are preserved. Notice that, for example one can use (c) to prove that a trivialization on $\mathcal{P}$ induces a trivialization on $(\mathcal{P})_\lambda$.

The fiber bundles canonically associated to the gauge natural prolongation $W^{(r,s)}\mathcal{P}$ (of some finite order (r,s)) of a principal bundle $\mathcal{P}$ are not just an example of gauge natural bundles, since they constitute the whole category. The example above referring to $\mathcal{P} \times_\lambda F$ and equation (5.1.2) corresponds in fact to the particular sub-case $r = s = 0$. Let us set the following:

Definition (5.2.26): a *gauge natural bundle of finite order* (r, s) is a bundle which is canonically associated to $W^{(r,s)}\mathcal{P}$ so that (r, s) is minimal. ∎

One can then prove (see e.g. [KMS93] Chapter XII, or [E81] for a short proof) that *all* gauge natural bundles are associated to some $W^{(r,s)}\mathcal{P}$; the minimal integer pair (r, s) (with $r \leq s$) is called the *order of the gauge natural bundle*. This theorem is usually called *finite order theorem* for gauge natural bundles. Here we shall not prove the theorem because in the sequel we shall always define gauge natural bundles as bundles associated from the very beginning to some structure bundle $\mathcal{P}$. In this way they are always of finite order.

Let now $\mathcal{P}$ be a principal bundle and $W^{(r,s)}\mathcal{P}\times_\lambda F$ be a gauge natural bundle. Let us choose a local trivialization $\sigma^{(\alpha)} : x \mapsto \sigma^{(\alpha)}(x)$ of the structure bundle $\mathcal{P}$ and a local natural trivialization $\hat{\epsilon}_s^{(\alpha)} : x \mapsto j_0^s\epsilon^{(\alpha)}$ (defined on the same domain $U_\alpha \subseteq M$) of the frame bundle $L^s(M)$, both induced by a coordinate system on M. We can define a local trivialization induced on $W^{(r,s)}\mathcal{P}$ by setting

$$\sigma_{(r,s)}^{(\alpha)} : x \mapsto (\hat{\epsilon}_s^{(\alpha)}(x),\ j_x^r\sigma^{(\alpha)}) \tag{5.2.27}$$

Each point $[j_0^s\epsilon, j_x^r\sigma, \hat{y}]_\lambda$ in $W^{(r,s)}P \times_\lambda F$ allows a canonical representative of the form $[\sigma_{(r,s)}^{(\alpha)}(x), y]_\lambda$ (by using the canonical right action which is transitive on the fibers of $W^{(r,s)}P$). Then $(x^\mu; y^i)$ are fibered coordinates on the gauge natural bundle $W^{(r,s)}\mathcal{P}\times_\lambda F$. In these local coordinates we can locally express the morphism $(\Phi)_\lambda$ induced by a morphism Φ of the structure bundle:

$$\begin{cases} x' = \phi(x) \\ g' = \varphi(x) \cdot g \end{cases} \quad \rightsquigarrow \quad \begin{cases} x' = \phi(x) \\ y' = [\lambda(j_0^s\phi(x), j_x^r\varphi(x))](y) \end{cases} \tag{5.2.28}$$

As usual, let us specialize to the particular case $G = \mathrm{GL}(n)$ and $r = s = 1$. Let us further specialize to $F = V$ a vector space and λ a (linear) representation of the group $W_m^{(1,1)}G$ on V, so that $W^{(1,1)}P \times_\lambda V$ is a gauge natural vector bundle. The principal morphism on the structure bundle is of the form

$$\begin{cases} x'^\mu = \phi^\mu(x) \\ g'^i{}_j = \varphi^i{}_k\, g^k_j \end{cases} \tag{5.2.29}$$

Let $(\varphi^i_j,\ \varphi^i_{j\mu},\ J^\mu_\nu)$ be coordinates on the group $W_m^{(1,1)}G$. Thence the induced morphism is

$$\begin{cases} x'^\mu = \phi^\mu(x) \\ v'^i = \lambda^i{}_j(J^\mu_\nu,\ \varphi^i_j,\ \varphi^i_{j\mu})\, v^j \end{cases} \tag{5.2.30}$$

See the next subsection for a concrete example.

The Bundle of Connections on $\mathcal{P}$

Let us consider a structure bundle $\mathcal{P} = (P, M, p; G)$. We shall here introduce a gauge natural affine bundle $\mathcal{C}_{\mathcal{P}}$ the sections of which correspond to principal connections over $\mathcal{P}$. Let us choose the affine space $(\mathbb{R}^*)^m \otimes \mathfrak{g}$ with coordinates $\tilde{\omega}^A_\mu$ obtained by fixing the canonical dual basis d^μ on $\mathbb{R}^*$ and the basis T_A in $\mathfrak{g}$. A point in $(\mathbb{R}^*)^m \otimes \mathfrak{g}$ is thence of the form $\omega = \tilde{\omega}^A_\mu\, \mathrm{d}^\mu \otimes T_A$. Let us consider the left action

$$\begin{aligned} \lambda : W^{(1,1)}_m G \times [(\mathbb{R}^*)^m \otimes \mathfrak{g}] &\longrightarrow (\mathbb{R}^*)^m \otimes \mathfrak{g} \\ : (J^\mu_\nu,\, \varphi^a,\, \varphi^a_\mu; \tilde{\omega}^A_\mu) &\mapsto \tilde{\omega}'^A_\mu = \bar{J}^\nu_\mu (\mathrm{Ad}^A_B(\varphi)\, \tilde{\omega}^B_\nu - \bar{R}^A{}_a(\varphi)\, \varphi^a_\nu) \end{aligned} \tag{5.2.31}$$

Notice that $W^{(1,1)}_m G$ acts on $(\mathbb{R}^*)^m \otimes \mathfrak{g}$ by affine transformations.

Then the affine gauge natural bundle $\mathcal{C}_{\mathcal{P}} = W^{(1,1)}\mathcal{P} \times_\lambda [(\mathbb{R}^*)^m \otimes \mathfrak{g}]$ is defined. By comparing (5.2.31) with the transition functions of $J^1\mathcal{P}/G$ (see (3.5.21)) it is easy to prove that

Proposition (5.2.32): there is a canonical (global) isomorphism between the bundles $J^1\mathcal{P}/G$ and $\mathcal{C}_{\mathcal{P}}$.

Proof: in the fibered coordinates introduced on the two bundles, the correspondence is given by $(x^\mu, \omega^A_\mu) \mapsto (x^\mu, \tilde{\omega}^A_\mu)$. Show that these local morphisms patch together to define a global isomorphism. ■

Because of this, from now on we shall use ω^A_μ also for the gauge natural coordinates of $\mathcal{C}_{\mathcal{P}}$. Thence $J^1\mathcal{P}/G$ is a gauge natural bundle of order $(1,1)$. Notice that $\mathcal{C}_{\mathcal{P}}$ is automatically endowed with a structure of (affine) fiber bundle with structure group $W^{(1,1)}_m G$. We stress that this approach is quite general. Connections are global objects defined on a principal bundle $\mathcal{P}$, we know how to locally parametrize them by local functions ω^A_μ and we know how these local representations transform under principal automorphisms of the structure bundle $\mathcal{P}$. It is very easy to build a gauge natural bundle the sections of which correspond to connections. In fact, local representations fix the standard fibers while the transformation rules fix the transition functions.

The principal connections of the principal bundle $L(M)$ are nothing but the linear connections of the first example of Section 3.7 above.

3. The Infinitesimal Generators of Automorphisms (IGA)

In this Section we shall introduce the bundle of infinitesimal generators of principal automorphisms (IGA) on a structure bundle $\mathcal{P}$ and study a decomposition of that bundle induced by a principal connection ω. We shall also study the case of IGA on $W^{(1,1)}\mathcal{P}$, as usual in the particular case $G = \mathrm{GL}(n)$, and how a connection on $\mathcal{P}$ induces a connection on $W^{(1,1)}\mathcal{P}$.

This Section can be skipped in a first reading, although it provides an example of application and computation which is quite representative of what is needed in the sequel.

The Bundle of IGA

In Section 3.4 we introduced the bundle $T\mathcal{P}/G$ of infinitesimal generators of automorphisms (IGA). We did not investigate its structure bundle since we were not able to interpret some of the objects entering the transformation rules (3.4.4). We are now ready to do that by following what we already did for principal connections.

Let us consider the standard fiber $\mathbb{R}^m \oplus \mathfrak{g}$ having coordinates (ξ^μ, ξ^A) together with the representation

$$\begin{aligned}\lambda :&W^{(1,1)}_m G \times [\mathbb{R}^m \oplus \mathfrak{g}] \longrightarrow \mathbb{R}^m \oplus \mathfrak{g}\\ :&(J^\mu_\nu, \varphi^a, \varphi^a_\mu; \xi^\mu, \xi^A) \mapsto (\xi'^\mu, \xi'^A) = (J^\mu_\nu\, \xi^\nu, \mathrm{Ad}^A_B(\varphi)\, \xi^B + \bar{R}^A_{\ a}(\varphi)\, \varphi^a_\mu\, \xi^\mu)\end{aligned} \tag{5.3.1}$$

We can define a gauge natural vector bundle $W^{(1,1)}\mathcal{P} \times_\lambda (\mathbb{R}^m \oplus \mathfrak{g})$ which in virtue of (5.3.1) is canonically isomorphic to $T\mathcal{P}/G$ and it is then called the *bundle of IGA*. Consequently, $W^{(1,1)}\mathcal{P}$ is the structure bundle of $T\mathcal{P}/G$.

Decomposition of IGA

Let $\omega = \mathrm{d}x^\mu \otimes (\partial_\mu - \omega^A_\mu\, \rho_A)$ be a principal connection on $\mathcal{P}$. Then an IGA on $\mathcal{P}$ can be split into its horizontal and vertical parts

$$\begin{aligned}\Xi =&\xi^\mu(x)\, \partial_\mu + \xi^A(x)\, \rho_A = \xi^\mu(\,\partial_\mu - \omega^A_\mu\, \rho_A) \oplus (\xi^A + \omega^A_\mu \xi^\mu)\, \rho_A =\\ =&\xi^\mu(\,\partial_\mu - \omega^A_\mu\, \rho_A) \oplus \xi^A_{(V)}\, \rho_A = \Xi_{(H)} \oplus \Xi_{(V)}\end{aligned} \tag{5.3.2}$$

The coefficients of the horizontal and the vertical parts of Ξ transform as

$$\begin{cases}\xi'^\mu = J^\mu_\nu\, \xi^\nu\\ \xi'^A_{(V)} = \mathrm{Ad}^A_B(\varphi)\, \xi^B_{(V)}\end{cases} \tag{5.3.3}$$

which are to be interpreted as the transition functions of the bundle $T\mathcal{P}/G$ with respect to the local trivialization associated to the fibered coordinates $(x^\mu; \xi^\mu, \xi^A_{(V)})$.

Notice that transition functions (5.3.3) are the transition functions of the bundle $TM \oplus V\mathcal{P}/G$. Thence it immediately follows the next proposition.

Proposition (5.3.4): each connection ω on $\mathcal{P}$ induces a global isomorphism between $T\mathcal{P}/G$ and $TM \oplus V\mathcal{P}/G$. ■

Prolongation of a Connection

We know from Section 5.2 that vector fields Ξ on $\mathcal{P}$ can be prolonged to $\hat{\Xi}$ on $W^{(r,s)}\mathcal{P}$. By using a connection on $\mathcal{P}$ and one on the suitable $L^s(M)$, the prolongation $\hat{\Xi}$ can be split. As a consequence we shall be able to define an induced connection on $W^{(r,s)}\mathcal{P}$. It is called the *prolongation of the connection* (ω, Γ).

Let us once again specialize to the case $G = \mathrm{GL}(n)$ and $r = s = 1$. The prolongation of the IGA $\Xi = \xi^\mu\, \partial_\mu + \xi^A\, \rho_A$ is given by

$$\hat{\Xi} = \xi^\mu\, \partial_\mu + \xi^i_{\ j}\, \rho_i^{\ j} + \mathrm{d}_\mu \xi^i_{\ j}\, \rho_i^{\ j\mu} + \mathrm{d}_\mu \xi^\nu\, \rho_\nu^{\ \mu} \tag{5.3.5}$$

where the vertical right invariant vector fields $(\rho_i^{\ j},\ \rho_i^{\ j\mu},\ \rho_\nu^{\ \mu})$ are given by expressions (5.2.24).

We stress that these coefficients, and consequently the local vertical right invariant vector fields, transform in an awful way with respect to principal automorphisms of $W^{(1,1)}\mathcal{P}$. Then we can use the connection $(\omega^a_{\ b\mu}, \Gamma^\lambda_{\ \mu\nu})$ on $\mathcal{P}$ and $L(M)$ to split them in a better way:

$$\begin{aligned}\hat{\Xi} =&\xi^\mu\, (\partial_\mu - \omega^a_{\ b\mu}\, r_a^{\ b} - (\mathrm{d}_\nu \omega^a_{\ b\mu} + \omega^a_{\ c\nu}\omega^c_{\ b\mu} - \omega^a_{\ c\mu}\omega^c_{\ b\nu})\, r_a^{\ b\nu} - \Gamma^\lambda_{\ \mu\nu}\, r_\lambda^{\ \nu}) +\\ &+ (\xi_{(V)})^a_{\ b}\, r_a^{\ b} + \nabla_\mu (\xi_{(V)})^a_{\ b}\, r_a^{\ b\mu} + \nabla_\mu \xi^\nu\, r_\nu^{\ \mu}\end{aligned} \tag{5.3.6}$$

where we set for the basis vector fields

$$\begin{cases} r_a^{\ b} = \rho_a^{\ b} - \omega^c_{\ a\nu}\, \rho_c^{\ b\nu} + \omega^b_{\ c\nu}\, \rho_a^{\ c\nu} \\ r_a^{\ b\mu} = \rho_a^{\ b\mu} \\ r_\nu^{\ \mu} = \rho_\nu^{\ \mu} - \omega^a_{\ b\nu}\rho_a^{\ b\mu} \end{cases} \tag{5.3.7}$$

and for the coefficients

$$\begin{cases} (\xi_{(V)})^a_{\ b} = \xi^a_{\ b} + \omega^a_{\ b\mu}\xi^\mu \\ \nabla_\mu (\xi_{(V)})^a_{\ b} = \mathrm{d}_\mu (\xi_{(V)})^a_{\ b} + \omega^a_{\ c\mu}(\xi_{(V)})^c_{\ b} - \omega^c_{\ b\mu}(\xi_{(V)})^a_{\ c} \\ \nabla_\mu \xi^\nu = \mathrm{d}_\mu \xi^\nu + \Gamma^\nu_{\ \lambda\mu}\xi^\lambda \end{cases} \tag{5.3.8}$$

If we consider a principal automorphism $\Phi^{(1,1)}$ on $W^{(1,1)}\mathcal{P}$ of the usual form

$$\begin{cases} x'^\mu = \phi^\mu(x) \\ \epsilon'^\mu_{\ \alpha} = J^\mu_\nu\, \epsilon^\nu_{\ \alpha} \\ g'^i_{\ j} = \varphi^i_{\ k}\, g^k_j \\ g'^i_{\ j\mu} = \bar{J}^\nu_\mu(\varphi^i_{\ k\nu}\, g^k_j + \varphi^i_{\ k}\, g^k_{j\nu}) \end{cases} \tag{5.3.9}$$

we obtain the following transformation rules

$$\begin{cases} \Phi_*^{(1,1)}(\delta_\mu) = J^\nu_{\ \mu}\, \delta'_\nu \\ \Phi_*^{(1,1)}(r_a^{\ b}) = \bar\varphi^b_{\ d}\, r'^{\ d}_{c}\, \varphi^c_{\ a} \\ \Phi_*^{(1,1)}(r_a^{\ b\mu}) = \bar J^\mu_{\ \nu}\, \bar\varphi^b_{\ d}\, r'^{\ d\nu}_{a}\, \varphi^c_{\ a} \\ \Phi_*^{(1,1)}(r_\nu^{\ \mu}) = \bar J^\mu_{\ \rho}\, r'^{\ \rho}_{\sigma}\, J^\sigma_{\ \nu} \end{cases} \tag{5.3.10}$$

where primed quantities are evaluated at $x' = \phi(x)$, bars denote inverse matrices and we set

$$\delta_\mu = \partial_\mu - \omega^a_{\ b\mu}\, r_a^{\ b} - (\mathrm{d}_\nu \omega^a_{\ b\mu} + \omega^a_{\ c\nu}\omega^c_{\ b\mu} - \omega^a_{\ c\mu}\omega^c_{\ b\nu})\, r_a^{\ b\nu} - \Gamma^\lambda_{\ \mu\nu}\, r_\lambda^{\ \nu} \tag{5.3.11}$$

As a consequence, since $\hat\Xi$ is a global vector field on $W^{(1,1)}\mathcal{P}$, the coefficients transform according to the rule

$$\begin{cases} \xi'^\mu = J^\mu_{\ \nu}\, \xi^\nu \\ (\xi'_{(V)})^a_{\ b} = \varphi^b_{\ d}\, (\xi_{(V)})^d_{\ c}\, \bar\varphi^c_{\ a} \\ \nabla_\mu (\xi'_{(V)})^a_{\ b} = J^\nu_{\ \mu}\, \varphi^b_{\ d}\, \nabla_\nu (\xi_{(V)})^d_{\ c}\, \bar\varphi^c_{\ a} \\ \nabla_\mu \xi'^\nu = J^\nu_{\ \rho}\, \nabla_\sigma \xi^\rho\, \bar J^\sigma_{\ \mu} \end{cases} \tag{5.3.12}$$

These transformation rules can be as well regarded as the new transition functions of the fiber bundle $TW^{(1,1)}\mathcal{P}/W^{(1,1)}_m G$ of IGA on the principal bundle $W^{(1,1)}\mathcal{P}$ with respect to the trivialization associated to the fibered coordinates $(x^\mu; \xi^\mu, \nabla_\nu \xi^\mu, (\xi_{(V)})^a_{\ b}, \nabla_\nu (\xi_{(V)})^a_{\ b})$. Using these coordinates the following Propositions are clearly true:

Proposition (5.3.13): there is a global isomorphism between the gauge natural bundles $TW^{(1,1)}\mathcal{P}/W^{(1,1)}_m G$ and $J^1 V\mathcal{P}/G \times_M J^1 TM$. ■

An analogue of this Proposition holds in fact at any order (r, s).

Proposition (5.3.14): the induced connection on $W^{(1,1)}\mathcal{P}$ is of the form

$$\hat\omega = \mathrm{d}x^\mu \otimes (\partial_\mu - \omega^a_{\ b\mu}\, r_a^{\ b} - (\mathrm{d}_\nu \omega^a_{\ b\mu} + \omega^a_{\ c\nu}\, \omega^c_{\ b\mu} - \omega^a_{\ c\mu}\, \omega^c_{\ b\nu})\, r_a^{\ b\nu} - \Gamma^\lambda_{\ \mu\nu}\, r_\lambda^{\ \nu}) \tag{5.3.15}$$

which is a global principal connection. ■

4. Lie Derivatives of Sections

Let $\mathcal{P} = (P, M, p; G)$ be a principal bundle and $\mathcal{B} = (B, M, \pi; F)$ be a gauge natural bundle of order (s, r) associated to $\mathcal{P}$. Let $\Xi = \xi^\mu\, \partial_\mu + \xi^A\, \rho_A$ be an IGA on $\mathcal{P}$ and $\sigma : M \longrightarrow B$ a (local) section of $\mathcal{B}$. As we did on natural bundles we can define Lie derivatives of the section σ with respect to the vector field Ξ on $\mathcal{P}$ by setting

$$\pounds_\Xi \sigma = \pounds_{\hat{\Xi}} \sigma = T\sigma(\xi) - \hat{\Xi} \circ \sigma \tag{5.4.1}$$

where $\hat{\Xi}$ is the induced vector field on $\mathcal{B}$ and $\xi = \xi^\mu\, \partial_\mu$ is the vector field on the base manifold M on which both Ξ and $\hat{\Xi}$ project onto.

If $(x^\mu; y^i)$ are fibered coordinates on $\mathcal{B}$ then the vector field $\hat{\Xi}$ is of the form $\hat{\Xi} = \xi^\mu\, \partial_\mu + \xi^i\, \partial_i$ and ξ^i depends linearly on ξ^μ, ξ^A and their derivatives up to order (s, r), respectively. Since the gauge natural lift $\hat{\Xi}$ is obtained in a functorial way from Ξ we see immediately that the lift preserves the Lie algebra structure, i.e. $[\Xi_1, \Xi_2]^\wedge = [\hat{\Xi}_2, \hat{\Xi}_2]$. As a consequence, the following holds true:

Proposition (5.4.2): the Lie derivative preserve commutators, i.e.

$$[\pounds_{\Xi_1}, \pounds_{\Xi_2}]\sigma = (\pounds_{\Xi_1} y^i - \pounds_{\Xi_2} y^i)\partial_i + (\pounds_{[\Xi_1, \Xi_2]} y^i)\hat{\partial}_i \tag{5.4.3}$$

Proof: it is analogous to (4.3.6). ■

Notice that in general there is no lift to define Lie derivatives of sections of $\mathcal{B}$ with respect to base vector fields ξ. This is due to the fact that while Aut($\mathcal{P}$) canonically acts on $\mathcal{B}$, Diff(M) does not. This is the main difference between natural and gauge natural theories. We stress that from a physical viewpoint this is quite obvious. The automorphisms of $\mathcal{P}$ are naturally introduced in Physics and they are usually called *(generalized) gauge transformations.*

Actually, vertical automorphisms are called *pure gauge transformations*, while non-vertical transformations are not always introduced. This is due to a couple of reasons. First of all, locally an automorphism (Φ, ϕ) can be decomposed into a pure gauge transformation and an automorphism which is horizontal with respect to the current local trivialization. Such a splitting is of course local and it has no global meaning. In the second place, physical theories are often defined on a sub-category of manifolds, for example Minkowski flat spaces, in which less morphisms are available. Sometimes it is possible to lift a subclass of morphisms (e.g. Poincaré transformations in Minkowski space) and treat fields as a kind of *"natural objects"* in the restricted theory.

We shall learn how to cope with the general situations in which there is no naturality. Consequently, we shall be able to treat global theories from a general relativistic perspective.

5. Examples

We shall here analyze the kinematical aspects of Yang-Mills theories. This example is mainly devoted to the reader familiar with Physics phenomenology. The mathematician can instead start to envisage some aspects that will turn out to be important for physical applications.

Let us consider a semisimple group[3], e.g. a unitary group $G = \mathrm{SU}(n)$ with $n \geq 2$ or $G = U(1)$.[4]

Being it semisimple, there exists a non-degenerate symmetric metric on the group G called the *Cartan-Killing metric* on G, which induces an inner product on the semisimple algebra $\mathfrak{g}$, defined by $\kappa(\xi, \zeta) = \mathrm{Tr}(\mathrm{ad}\,\xi, \mathrm{ad}\,\zeta)$, where $\mathrm{ad}\,\xi : \mathfrak{g} \longrightarrow \mathfrak{g} : \zeta \mapsto [\xi, \zeta]$. The metric κ can be diagonalized by choosing a suitable basis T_A in the Lie algebra $\mathfrak{g}$ such that $\kappa(T_A, T_B) = \delta_{AB}$.

To introduce *Yang-Mills fields* (which are nothing but *Maxwell fields* if the group G is $U(1)$), let us consider a principal fiber bundle $\mathcal{P} = (P, M, p; G)$ as a structure bundle and the gauge natural bundle $\mathcal{B}_1 = J^1\mathcal{P}/G$ of principal connections on $\mathcal{P}$. Fibered coordinates on $\mathcal{B}_1$ are $(x^\mu; \omega^A_\mu)$ and a section $\sigma : M \longrightarrow J^1P/G$ represents a principal connection on $\mathcal{P}$. Such connections are called *Yang-Mills fields* (or *Maxwell fields* in the Abelian $U(1)$ case).

Let us also consider a n-tuple of scalar complex fields φ^A. They are also sections of an associated bundle to $\mathcal{P}$ with respect to a representation $\lambda : G \times \mathbb{C}^n \longrightarrow \mathbb{C}^n$. For example, for $G = \mathrm{SU}(n)$ we can consider the standard action

$$\lambda : \mathrm{SU}(n) \times \mathbb{C}^n \longrightarrow \mathbb{C}^n : (U^A_B, \varphi^B) \mapsto U^A_B\, \varphi^B \tag{5.5.1}$$

The associated bundle $\mathcal{B}_2 = \mathcal{P} \times_\lambda \mathbb{C}^n$ has local coordinates $(x^\mu; \varphi^A)$.

We stress that the expression *"scalar" complex fields* is not coherent with our framework. In Physics "scalar" is used to say that φ^A does not change with respect to $\mathrm{Diff}(M)$ (or more often with respect to the Poincaré group acting on M). This is undefined from a global viewpoint since $\mathrm{Diff}(M)$ does not act at all on field configurations. Locally, if we consider a principal automorphism on $\mathcal{P}$ ($G = \mathrm{SU}(n)$) given by

$$\begin{cases} x'^\mu = \phi^\mu(x) \\ g'^A = \gamma^A_B(x) g^B \end{cases} \tag{5.5.2}$$

it induces an automorphism on $\mathcal{B}_2$ locally given by $\varphi'^A(x') = \gamma^A_B(x)\varphi^B(x)$. Thus one can see that the fields φ^A locally transform with respect to pure gauge

---

[3] A group G is *semisimple* if it does not contain any normal subgroup. A *normal subgroup* is a subgroup which is invariant with respect to the adjoint action on G.

[4] Elements of $\mathrm{SU}(n)$ are $n \times n$ self adjoint complex matrices with determinant equal to 1. Elements of $U(1)$ are unimodular complex numbers, i.e. of the form $e^{i\theta}$.

transformations parametrized by γ^A_B and they are insensitive to the transformation ϕ on the base manifold, as if they were true scalar fields, i.e. sections of a trivial bundle on M.

When we want to describe both Yang-Mills fields and scalar fields we have to consider the fibered product $\mathcal{B} = \mathcal{B}_1 \times_M \mathcal{B}_2 = J^1\mathcal{P}/G \times_M (\mathcal{P} \times_\lambda \mathbb{C}^n)$ as configuration bundle. It has coordinates $(x^\mu; \omega^A_\mu, \varphi^A)$ so that a section σ on $\mathcal{B}$ is in fact a pair of sections (σ_1, σ_2) on $\mathcal{B}_1$ and $\mathcal{B}_2$, respectively. The bundle $\mathcal{B}$ is a gauge natural bundle (of order $(1,1)$) associated to the structure bundle $\mathcal{P}$. The action of a generalized gauge transformation $\boldsymbol{\Phi} \in \mathrm{Aut}(\mathcal{P})$ locally given by expression (5.5.2) is defined on $\mathcal{B}$ by the following (setting, as usual $\bar{J}^\nu_\mu = \partial_\mu(\phi^{-1})^\nu$):

$$\begin{cases} \omega'^A_\mu = \bar{J}^\nu_\mu(\mathrm{Ad}^A_B(\gamma)\,\omega^B_\nu - \bar{R}^A{}_a(\gamma)\,\gamma^a_\nu) \\ \varphi'^A = \gamma^A_B\,\varphi^B \end{cases} \tag{5.5.3}$$

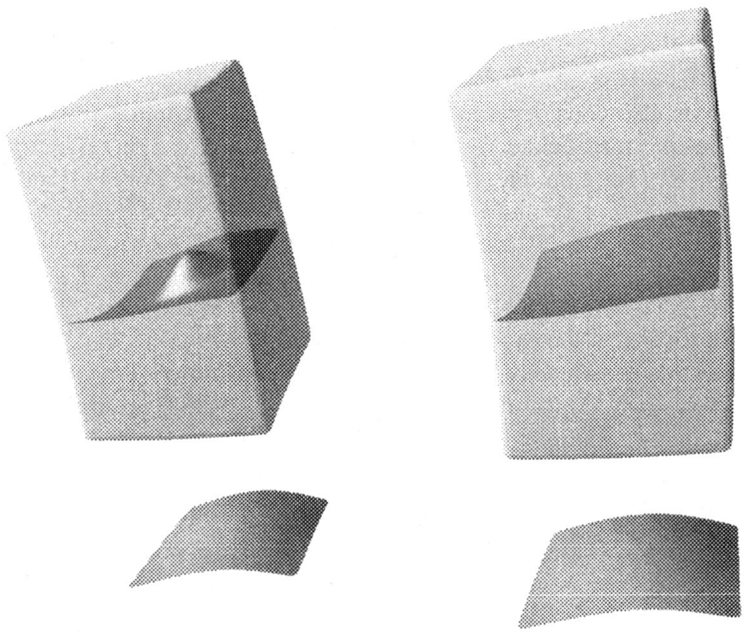

Fig. 17 – Two solutions differing by a gauge transformation.

One can argue what is the role played by the structure bundle $\mathcal{P}$. The answer is strikingly simple: it selects a class of global fields. In fact, it is often true that more that one possible structure bundle $\mathcal{P}$ can be chosen on the base manifold M once a group G has been fixed. [If M is contractible only trivial bundles can be defined on M, but if the topology of M is not trivial there are a lot of non-equivalent alternative possibilities amongst which to choose.]

This is very deeply encoded in some of the basic aspects of gauge theories. In fact, the classical configurations are selected by field equations. Field equations are required to *determine* (physicists are quite strict on this point) the configuration of the system once initial conditions are provided. But gauge transformations transform classical solutions into classical solutions (i.e. they are required to be symmetries of field equations). One can easily produce compact supported gauge transformations which thence produce a whole family of different classical solutions with the same initial conditions and consequently field equations are not truly able to associate a unique classical solution to a given initial or boundary condition. There is one only way out: *two configurations differing by a (compact supported) gauge transformation represent the same physical system configurations.* Classical gauge theories are thence underdetermined (i.e. they identify only a class of classical solutions, not a representative of that class).

Once this fact has been accepted, when we locally solve field equations on two patches and we want to glue them together there is no need to require the local expressions fully to agree on the overlap. We can simply require them to agree *up to a (pure) gauge transformation*; in that way they still in fact represent a unique physical system.

Let us give an example. Let us choose $G = U(1)$, $M = S^2$ and describe a Maxwell field $\omega^A_\mu (= \omega_\mu)$ (where the index $A = 1$ "runs" on a unique value and can be understood). An example of local field configuration in the open set $S^2_N = S^2 - \{S = (0,0,-1) \in S^2\}$, using spherical coordinates (θ, ϕ) on S^2, is $(\forall m \in \mathbb{Z})$:

$$\omega_\theta = 0 \qquad \omega_\phi = \frac{m}{2}(1 - \cos(\theta)) \tag{5.5.4}$$

It is well defined in $N = (0,0,1) \in S^2$ since, even if ϕ is there undetermined, the coefficient $1 - \cos(\theta)$ vanishes. On the contrary, it cannot be extended in $S = (0,0,-1) \in S^2$ since there the coefficient $1 - \cos(\theta) = 2$ does not vanish. Anyway, the curvature

$$F_{\theta\phi} = \frac{m}{2}\sin(\theta) \tag{5.5.5}$$

is well defined on the whole sphere S^2 since the coefficient $\sin(\theta)$ vanishes at both poles N and S.

This is a consequence of the fact that there exists a pure gauge transformation defined on $S^2 - \{N, S\}$ by the rule $\gamma = \exp(im\phi) \in U(1)$ which transforms, according to the rule $\omega'_\mu = \omega_\mu + \gamma^{-1}\mathrm{d}_\mu\gamma$, the local solution (5.5.4) into the following

$$\omega'_\theta = 0 \qquad \omega'_\phi = -\frac{m}{2}(1 + \cos(\theta)) \tag{5.5.6}$$

This fails to be well defined at $N \in S^2$ but it turns out to be well defined at $S \in S^2$.

Thence we have two local expressions ω'_μ and ω_μ which differ by a compact supported pure gauge transformation and cover the whole base manifold S^2. The pure gauge transformation γ forms trivially a cocycle over S^2 with values in $U(1)$ and it consequently defines a structure bundle $\mathcal{P}$ over M with group $U(1)$. It is easy to check that (for $m = 1$), the bundle $\mathcal{P} = \mathcal{H} = (S^3, S^2, p; U(1))$ is the Hopf bundle defined in Section 1.1 (see exercise 4 below). Thence the two local expressions (5.5.4) and (5.5.6) can be regarded as the local expressions of a global principal connection on $\mathcal{H}$ given by

$$\omega = \mathrm{d}x^\mu \otimes (\partial_\mu - \omega_\mu\, \rho) \otimes T \qquad \omega' = \mathrm{d}x^\mu \otimes (\partial_\mu - \omega'_\mu\, \rho) \otimes T \tag{5.5.7}$$

where $(\theta, \phi; f = e^{i\psi})$ are fibered coordinates on $\mathcal{H}$, $\rho = f\partial_f$ is a pointwise local basis of vertical (right) invariant vector fields on $\mathcal{P}$ and $T = i$ is a basis of the (one-dimensional) Lie algebra $\mathfrak{u}(1) \simeq i\mathbb{R}$.

We remark that despite the two local expressions (5.5.4) and (5.5.6) do not patch together on the overlap, the two local connections (5.5.7) do. This is due to the fact that when the local expressions are regarded as principal connections on $\mathcal{P}$ the pure gauge transformation needed to glue (5.5.4) with (5.5.6) is provided by transition functions of $\mathcal{P}$. This is a general feature of the bundle framework. Fiber bundles are needed whenever a family of local objects has to be regarded as a global object even if the local objects do not patch together. The globality of sections representing physical configurations will be a starting point for the applications of Variational Calculus which will be developed in the next Part of this book.

Exercises

1- Let $\boldsymbol{\Phi} = (\Phi, \phi)$ be a principal morphism between two principal bundles $\mathcal{P} = (P, M, p; G)$ and $\mathcal{P} = (P', M', p'; G')$. Prove that the lifts $V(\Phi) : VP \longrightarrow VP'$ and $J^1(\Phi) : J^1P \longrightarrow J^1P'$ preserve the right actions.
Hint: $\boldsymbol{\Phi}$ is principal and $J^1(\cdot)$ is a functor.

2- Prove that the multiplications (5.2.2) and (5.2.7) are associative. Prove also that the right actions (5.2.4) and (5.2.9) are really actions.
Hint: Use functorial properties of jet prolongations.

3- Prove that the vector fields (5.2.24) on $W^{(1,1)}\mathcal{P}$ are right invariant.
Hint: Apply directly the relevant right action.

4- Prove that the cocycle $\gamma = \exp(im\phi) \in U(1)$ ($m \in \mathbb{Z}$) defined on the sphere S^2 defines in fact the Hopf bundle $(S^3, S^2, p; U(1))$.
Hint: Compare with the transition functions of the Hopf bundle w.r.t. a suitable trivialization.

PART II

THE VARAIATIONAL STRUCTURE OF FIELD THEORIES

We are very much in the dark. We left the sunny grasses of Cartesian – Newtonian physics, and we are travelling through the woods, with everything we have learned and with our weak intuition always wishing we are smarter.
It would be a discouraging state of confusion and we would feel lost if it wasn't that the trip is wonderful and the landscape so breathtaking.

C. Rovelli, Halfway through the wood

Introduction

The expression *variational calculus* (or *Calculus of Variations*) usually identifies two different but related branches in Mathematics. The first aimed to produce theorems on the existence of solutions of (partial or ordinary) differential equations generated by a *variational principle* and it is a branch of local analysis (usually in $\mathbb{R}^n$); the second uses techniques of differential geometry to deal with the so-called *variational calculus on manifolds*. The common background is the existence of a *variational principle* (see later) which generates the equations.

The local-analytic paradigm is often aimed to deal with particular situations, when it is necessary to pay attention to the exact definition of the functional space which needs to be considered. That functional space is very sensitive to boundary conditions. Moreover, minimal requirements on data are investigated in order to allow the existence of (weak) solutions of the equations.

On the contrary, the global-geometric paradigm investigates the minimal structures which allow to pose the variational problems on manifolds, extending what is done in $\mathbb{R}^n$ but usually being quite *generous* about regularity hypotheses (e.g. hardly ever one considers less than C^∞-objects). Since, even on manifolds, the search for solutions starts with a local problem (for which one can use local analysis) the global-geometric paradigm hardly ever deals with exact solutions, unless the global geometric structure of the manifold strongly constrains the existence of solutions (e.g. a typical problem is the existence of solutions to Dirac equations; see for example [B97], [B81], [GP78]).

A further *a priori* different approach is the one of Physics. In Physics one usually has field equations which are locally given on a portion of an unknown manifold. One thence starts to solve field equations locally in order to find a local solution and only afterwards one tries to find the maximal analytical extension (if any) of that local solution. The maximal extension can be regarded as a global solution on a suitable manifold M, in the sense that the extension *defines* M as well. In fact, one first proceeds to solve field equations in a coordinate neighbourhood; afterwards, one changes coordinates and tries to extend the found solution out of the patches as long as it is possible. The coordinate changes are the (cocycle of) transition functions with respect to the atlas and they *define* the base manifold M. This approach is essential to physical applications when the base manifold is *a priori* unknown (for example in General Relativity) and it has to be determined by physical inputs.

Luckily enough, that approach does not disagree with the *standard* variational calculus approach in which the base manifold M is instead fixed from the very beginning. One can regard the variational problem as the search for a solution on that particular base manifold. Global solutions on other manifolds may be found using other variational principles on different base manifolds. Even for this reason, the variational principle should be *universal*, i.e. one defines a family of variational principles: one for each base manifold, or at least one for any base manifold in a *"reasonably"* wide class of manifolds. We stress that this is a strong requirement which is physically motivated by the belief that Physics should work more or less in the same way regardless of the particular spacetime which is actually realized in Nature. Of course, a scenario would be conceivable in which everything works because of the particular (topological, differentiable, etc.) structure of the spacetime. This position, however, is not desirable from a physical viewpoint since, in this case, one has to explain why *that* particular spacetime is realized (*a priori* or *a posteriori*).

In spite of the aforementioned strong regularity requirements, the spectrum of situations one can encounter is unexpectedly wide, covering, as we shall see, the whole fundamental physics (Mechanics, classical Field Theories – and in particular General Relativity – Gauge Theories and the Standard Model of particle physics; see [CB68], [N90], [F91], [MS00]). Moreover, it is surprising how the geometric formalism is effectual for what concerns identifications of basic structures of field theories. In fact, just requiring the theory to be globally well-defined and to depend on physical data only, it often constrains very strongly the choice of the local theories to be globalized. These constraints are one of the strongest motivations in choosing a variational approach in physical applications. Another motivation is, as we shall see in the sequel, a well formulated framework for conserved quantities. We stress here that a global-geometric framework is *a priori* necessary to deal with conserved quantities being them non-local.

We also have to mention that in the modern perspective of Quantum Field Theory (QFT) the basic object encoding the properties of any quantum system is the action functional, which also the classical field theory we are going to introduce is based on (see [CDF91], [S93]). From a quantum viewpoint the action functional is more fundamental than field equations which are obtained in the classical limit. In our perspective, of course, classical field theories are both necessary to those branches (e.g., General Relativity) for which we do not have a quantum version yet and serve as a first step towards quantization.

We believe, in fact, that a better understanding of the quantum scenario must pass through a better understanding of the Classical Field Theory.

Many examples of applications to Physics will be examined in the sequel: symmetries, conserved quantities, spinor theories, gauge theories are just some of them. As far as symmetries are concerned, it is clear that they play a fundamental role in describing the basic structure of physical systems. Nevertheless, there are many different ways of mathematically implementing this notion. They are usually interpreted as some kind of transformations which somehow preserve solutions. We shall hereafter characterize a sub-class of all symmetries, namely those which preserve the Poincaré-Cartan form. These are very special symmetries, called *Lagrangian symmetries*, since they preserve the Lagrangian structure of the field theory, though much more general classes could be investigated (see for example [AKO93], [FFML02]). Nevertheless, this class of symmetries provides a satisfactory framework for conserved quantities reproducing any expected result whenever a physical interpretation justifies an expectation value.

Furthermore, the geometric framework provides drastic simplifications of some key issues, such as the definition of the *variation* operator. The variation that we are going to define is deeply geometric though, in practice, it coincides with the definition given in the local-analytic paradigm. In the latter case, the functional derivative is usually the directional derivative of the action functional which is a function on the infinite-dimensional space of fields defined on a region D together with some boundary conditions on the boundary ∂D. To be able to define it one should first define the functional space, then define some notion of deformation which preserves the boundary conditions (or equivalently topologize the functional space), define a variation operator on the chosen space (i.e., loosely speaking, some differential structure on it), and, finally, prove the most commonly used properties of derivatives (see [CBWM82]). Once one has done it, one finds in principle the same results that would be found when using the geometric definition of variation (for which *no* infinite dimensional space is needed). In fact, in any case of interest for fundamental physics, the functional derivative is simply defined by means of the derivative of a real function of one real variable. We may say that the Lagrangian formalism we are going to define hereafter is a shortcut which translates the variation of (infinite dimensional) action functionals into the variation of the (finite dimensional) Lagrangian structure.

Another feature of the geometric framework is the possibility of dealing with non–local properties of field theories. There are, in fact, phenomena, such as monopoles or instantons, which are described by means of non-trivial bundles (see [GS87]). Their properties are tightly related to the non-triviality of the configuration bundle; and they are relatively obscure when regarded by any local paradigm. In some sense, a local paradigm *hides* global properties in the boundary conditions and in the symmetries of the field equations, which are in turn reflected in the functional space we choose and about which, being it infinite dimensional, we do not know almost anything *a priori*. We could say that the existence of these phenomena is a further hint that field theories *have* to be stated on bundles rather than on Cartesian products. Once again this statement is *phenomenologically* driven.

When a non–trivial bundle is involved in a field theory, from a physical viewpoint it has to be regarded as an unknown object. As for the base manifold, it has then to be constructed out of physical inputs. One can do that in (at least) two ways which are both actually used in applications. First of all, one can assume the bundle to be a natural bundle which is thence canonically constructed out of its base manifold. Since the base manifold is identified by the (maximal) extension of the local solutions, then the bundle itself is identified too. This approach is the one used in General Relativity. There are examples, however, in which the bundles involved are not natural, e.g. in Gauge Theories. In these applications, bundles are gauge natural and they are therefore constructed out of a structure bundle $\mathcal{P}$, which, usually, contains extra information which is not directly encoded into the spacetime manifolds. In physical applications the structure bundle $\mathcal{P}$ has also to be constructed out of physical observables. This can be achieved by using gauge invariance of field equations. In fact, two local solutions differing by a (pure) gauge transformation describe the same physical system. Then while extending from one patch to another we feel free both to change coordinates on M *and* to perform a (pure) gauge transformation before glueing two local solutions. Then coordinate changes define the base manifold M, while the (pure) gauge transformations form a cocycle (valued in the gauge group) which defines, in fact, the structure bundle $\mathcal{P}$. Once again solutions with different structure bundles can be found in different variational principles. Accordingly, the variational principle should be *universal* with respect to the structure bundle. This approach is the one used in gauge natural theories, which will be introduced in Chapter 8.

Of course, in spite of our preference for the geometric language, we stress that local results are by no means less important. They are often the foundations on which the geometric framework is based on. More explicitly, Variational Calculus is perhaps the branch of mathematics in which we had the strongest interaction between Analysis and Geometry. We believe that this is the actual reason which has made it one of the most fruitful disciplines both for Mathematics and Physics. Examples of this fruitfulness are the most fundamental understanding of Nature which is *based* on field theories; but also the recent history of exotic differential structures of $\mathbb{R}^4$, quantum-cohomology and knot theory is tightly related to variational calculus (see [BM94]).

In Chapter 6 we will give an essential account of the geometric formulation of classical field theories, using the geometric structures introduced in Part I. We shall also give basic definitions about field theories, i.e. the Lagrangian formalism and variational calculus on fiber bundles, and some preliminary and introductory version of the framework for symmetries and conserved quantities (in an as much unrestricted as possible way). In particular the framework shall be defined for a k-order field theory (with arbitrary k) even if all known applications to Physics are first order, except General Relativity which is second order (although it allows an equivalent first order formulation called the *metric-affine formulation*). The reasons to do that are twofold. First of all there is no evidence that any physical field theory has to be *in principle* first or second order. Secondly, dealing with the general case teaches us which is the essential structure for a field theory to have a sense.

In Chapter 7 we shall introduce the natural framework for field theories, which provides also a transcription in the field theory framework of all the geometric knowledge which was achieved until the first decades of the twentieth century. This knowledge culminated in the formulation of General Relativity, in 1916.

In Chapter 8 we shall introduce gauge natural field theories, which extend the framework of natural theories to encompass *all* the modern applications of fundamental physics, including Gauge Theories and their interactions with spinor fields.

Most of the notation introduced in Part I will be used hereafter. The reader is assumed to be familiar with jet bundles, bundle morphisms, contact forms and connections. In Chapter 7 natural bundles and natural lifts will be extensively used, while in Chapter 8 the reader will be assumed to be familiar with principal bundles, principal connections, gauge natural bundles and generalized gauge transformations too.

Chapter 6

THE LAGRANGIAN FORMALISM

Abstract We shall hereafter introduce the global Lagrangian formalism for field theories (which applies in particular to Mechanics). In Section 2, the most important algebraic Lemmas are presented in an abstract and general form and they are then applied to Variational Calculus to obtain the relevant quantities for a general Field Theory. The approach based on the Poincaré-Cartan form is also briefly discussed. Then symmetries and conserved quantities are defined and obtained by Nöther theorem. Finally some useful generalizations of symmetries and, consequently, of Nöther theorem are presented.

References

Many monographs are dedicated to the classical Lagrangian formalism. Wide sections, e.g. in [CBWM82], [S95], [F91], [CDF91], [MM92] are devoted to Lagrangian formalism and variational calculus. A further source of information are original papers, e.g. [T72], [FF91], [FFFR99] for general geometric structures; [K84], [F84] for the Poincaré-Cartan forms; [GS78] and [FF01] for definitions and details about Spencer cohomology and variational morphisms. For higher order theories see [F84], [FR], [KMS93], [K84] and references quoted therein. See also the monograph [FP99]. Legendre transformations and Hamiltonian formalism are presented in [FR], [GIMMPSY] and [CFT79] and references quoted therein, while some applications of generalized symmetries can be found in [FFML02]. Details and further references about Nöther theorem can be found in the papers [FF91] and [FFFR99].

Material about supermanifolds, superfields and supersymmetries can be found in [CDF91], [R81], [C90] and [M88]. For the standard facts about General Relativity we refer to [W72] or [HE73]. For Gauge Theories we refer to [GS87], [FFF97], [BD64] and [B81]. Finally, a brief account on Maple tensor package can be found in [CFFML97].

1. The Lagrangian and the Action Functional

As we mentioned in the Introduction of Part I, configurations of a field theory are, by definition, (differentiable) sections of a *configuration bundle* $\mathcal{C} = (C, M, \pi; F)$ over the base manifold M. A section of the bundle $\mathcal{C}$ describes the value that fields assume in the standard fiber F at any point of M and it thence describes the *evolution* of the system. Consequently, field equations will have to identify a section of $\mathcal{C}$ as a solution. In Chapters 7 and 8 below we shall introduce further structures.

We remark that, as a particular case, mechanical systems may be studied in this framework. Configurations of these systems are described by points in a manifold Q, called *configuration space*. In fact, we can choose $M = \mathbb{R}$ as the base manifold ($t \in \mathbb{R}$ is the independent variable) and $C = \mathbb{R} \times Q$ as total space of the configuration bundle, which is always trivial being the base manifold $\mathbb{R}$ contractible. A section of $\mathcal{C} = (\mathbb{R} \times Q, \mathbb{R}, \pi; Q)$ identifies a configuration $q \in Q$ at any time $t \in \mathbb{R}$. Such a section can be canonically identified with a curve in Q, i.e. it exactly describes the evolution of the mechanical system.

In the general case a *Lagrangian L of order k* is a bundle morphism:

$$\begin{array}{ccc} J^k\mathcal{C} & \xrightarrow{\ L\ } & A_m(M) \\ {\scriptstyle \pi^k}\downarrow & \swarrow{\scriptstyle \tau_M} & \\ M & & \end{array} \tag{6.1.1}$$

where $A_m(M)$ is the bundle of m-forms over M, $m = \dim(M)$. The bundle $J^k\mathcal{C}$ is called *the (Lagrangian) phase bundle.*

Even if this definition may seem strange, it directly expresses the formulation of *Hamilton stationary action principle.* In fact, if $D \subset M$ is any *m-region of M* (i.e. a compact submanifold of dimension m with a boundary ∂D which is a compact submanifold of dimension $m-1$) and $\rho : M \longrightarrow \mathcal{C}$ is a configuration, we may evaluate the Lagrangian L on the prolongation of order k of the section obtaining an m-form $L \circ j^k\rho$ over M. This form is a suitable object to be integrated on D. Then we define the *action of the section ρ over the m-region D* as follows:

$$A_D(\rho) = \int_D L \circ j^k\rho \tag{6.1.2}$$

In Mechanics, we have $k = 1$ and $J^1(\mathbb{R} \times Q) \simeq \mathbb{R} \times TQ$ (non-canonically). A first order Lagrangian is locally given by:

$$L : \mathbb{R} \times TQ \longrightarrow A_1(\mathbb{R}) : (t, q^i, u^i) \mapsto \mathcal{L}(t, q^i, u^i)\mathrm{d}t \tag{6.1.3}$$

thence coinciding with the standard notion given in time-dependent Mechanics.

In general, let us denote by $(x^\mu; y^i)$ fibered coordinates on the configuration bundle $\mathcal{C}$; then a k-order Lagrangian is locally expressed in the form

$$\begin{aligned} L :& J^k\mathcal{C} \longrightarrow A_m(M) \\ :& (x^\mu, y^i, y^i_\mu, \dots, y^i_{\mu_1\dots\mu_k}) \mapsto \mathcal{L}(x^\mu, y^i, y^i_\mu, \dots, y^i_{\mu_1\dots\mu_k})\, \mathrm{d}s \end{aligned} \tag{6.1.4}$$

where $\mathrm{d}s = \mathrm{d}x^1 \wedge \dots \wedge \mathrm{d}x^m$ is the standard local basis of m-forms over M.

The local map $\mathcal{L}(x^\mu, y^i, y^i_\mu, \dots, y^i_{\mu_1\dots\mu_k})$ is called the *Lagrangian density*. Of course the quantity $\mathcal{L}(x^\mu, y^i, y^i_\mu, \dots, y^i_{\mu_1\dots\mu_k})\, \mathrm{d}s$ can be also regarded as an m-form on $J^k\mathcal{C}$ which is horizontal with respect to the projection $\pi^k : J^k\mathcal{C} \longrightarrow M$. Such a form is called the *Lagrangian form*. As a consequence, all the results in Variational Calculus can be equivalently stated in terms of Lagrangian forms as well as in terms of bundle morphisms. However, we shall usually prefer the language of bundle morphisms.

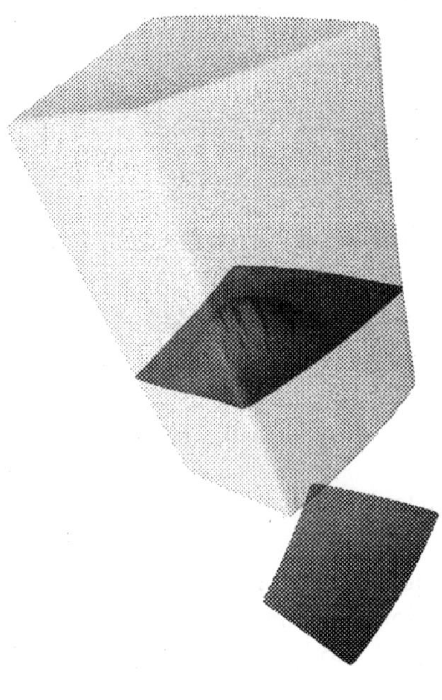

Fig. 18 – The deformation of a section.

Let us now consider a vertical vector field X over $\mathcal{C}$ the support of which is contained in an m-region $D \subset M$ which is in turn contained into an open set $U \subset M$, i.e. $\mathrm{supp}X \subseteq D \subset U$. Let us also require that $j^{k-1}X$ vanishes over the boundary ∂D. Notice that in particular the vector field X identically vanishes outside U (and in fact outside D). Such a vector field is called *deformation over* D.[1] We shall denote by $\Phi_s : \mathcal{C} \longrightarrow \mathcal{C}$ the flow of X which is a 1-parameter sub-group of vertical automorphisms of the configuration bundle $\mathcal{C}$. Since X vanishes outside D then $\Phi_s = \mathrm{id}$ out of D. Now we can *drag* a global section (though one can also use local ones) $\rho : M \longrightarrow \mathcal{C}$ defining in this way a 1-parameter family of sections:

$$\rho_s = \Phi_s \circ \rho \tag{6.1.5}$$

(all of them also coinciding with ρ out of the domain D).

Let us consider the action of these sections:

$$A^{\rho}_{D}(s) = A_D(\rho_s) = \int_D L \circ j^k \rho_s \tag{6.1.6}$$

Being ρ and X fixed, A^{ρ}_{D} is a function of $s \in \mathbb{R}$ giving the value of the action on the section ρ dragged along X in the region D. It is thence a (differentiable) function $A^{\rho}_{D} : \mathbb{R} \longrightarrow \mathbb{R}$ and it may be derived at $s = 0$. Let us thence define *the variation along X of the action (at ρ)* by:

$$\delta_X A_D(\rho) = \left[\frac{\mathrm{d}}{\mathrm{d}s} A^{\rho}_{D}(s)\right]_{s=0} \tag{6.1.7}$$

We can now state the *Hamilton principle of stationary action.*

Definition (6.1.8): we say that a section $\rho : M \longrightarrow \mathcal{C}$ is a *classical solution* (or a *critical section* or a *shell*) if the action is stationary, i.e.:

$$\delta_X A_D(\rho) = 0 \tag{6.1.9}$$

for any m-region D and for any deformation X over D. ∎

When a physical quantity is calculated over a critical section (i.e. along a solution of field equations) we shall often use the physical terminology saying that it is *evaluated on-shell.* Notice that the Hamilton principle involves only a 1-parameter family of configurations ρ_s. Most of the properties studied hereafter will hold regardless whether or not the family of sections ρ_s has been obtained by dragging ρ along a deformation X or not. In particular, especially when we shall deal with symmetries, the family will be sometimes obtained by dragging along some generalized vector field (which of course is not a deformation).

[1] In the classical notation X is called a *variation* (or a *deformation*) and it is locally expressed as $X = (\delta y^i)\,\partial_i$.

From a mathematical viewpoint, it is relatively easy to show that the Hamilton principle of stationary action implies that a section ρ is critical if and only if it satisfies Euler-Lagrange equations (as it will be shown below). However, from a classical viewpoint it may not be clear why some physical systems (and among them all systems relevant to *fundamental physics*) ought to satisfy this axiom. This is partially enlightened if one notices that the Lagrangian of a system is not at all an observable quantity. Actually, it is not even uniquely defined by its (local) field equations. The only thing that one can classically observe is the evolution of the system. Thus the only reason for the Lagrangian to exist is to provide (in a universal and canonical way) via variational calculus and Euler-Lagrange equations, the correct field equations.

In this way, the principle of stationary action is an implicit definition of what has to be regarded as the Lagrangian of the system. It is the one which provides the correct field equations and which determines the observed evolution of the system. However, from a quantum (and more fundamental) viewpoint, this is not true any longer. In the quantum approach the action functional provides the most fundamental description of the system. Field equations are obtained in the classical limit as an approximate description. In principle, quantum corrections can discriminate between classically equivalent Lagrangians (i.e. between Lagrangians which at least locally identify the same critical sections). In that perspective, the fact that fundamental physics can be presently described in the framework of variational calculus is to be regarded as an (unexpected and beautiful) coincidence. Of course, at our stage of understanding of Nature, we can accept this coincidence with pleasure; on the other hand we are completely exposed to future surprises. We can see no *a priori* insurance that future developments will still keep going that way.

2. The Variational Morphisms

In the present Section we shall cope with most of the technical difficulties that we shall encounter in Variational Calculus. In the sequel of this monograph we shall often refer to the definitions and results stated hereafter.

In the typical situation that will be encountered below we will be dealing with bundle morphisms of the following type:

Definition (6.2.1): let $\mathcal{E} = (E, M, \pi; \mathbb{R}^n)$ be a vector bundle. Let k, h and n be integers. A bundle morphism

$$\mathbb{V} : J^k\mathcal{C} \longrightarrow (J^h\mathcal{E})^* \otimes A_{m-n}(M) \tag{6.2.2}$$

is called a *variational $\mathcal{E}$-morphism on $\mathcal{C}$*. The (minimal) integer k is called the *order* of $\mathbb{V}$, h is called the *rank* and $(m-n)$ is called the *degree* of $\mathbb{V}$ (being n the *codegree*). ∎

We shall see below many examples of this general structure and we shall always point out how the various structures will be cast in this form. We already introduced a variational morphism: any Lagrangian of order k as defined by (6.1.1) is a variational morphism. In fact, one can consider the 0-rank vector bundle $\mathcal{E} = (M \times \{0\}, M, \pi; \{0\})$ so that $\mathcal{E}^* \otimes A_m(M) \simeq A_m(M)$. Thus the Lagrangian is a variational $\mathcal{E}$-morphism of order k, rank $h = 0$ and codegree $n = 0$.

We can define the divergence operator on variational morphisms as follows:

Definition (6.2.3): let $\mathbb{V} : J^k\mathcal{C} \longrightarrow A_{m-n}(M)$ be a morphism of rank $h = 0$. We define the *divergence* of V to be the variational morphism $\mathrm{Div}(\mathbb{V}) : J^{k+1}\mathcal{C} \longrightarrow A_{m-n+1}(M)$ such that

$$\mathrm{Div}(\mathbb{V}) \circ j^{k+1}\rho = \mathrm{d}(\mathbb{V} \circ j^k\rho) \tag{6.2.4}$$

for each section $\rho : M \longrightarrow \mathcal{C}$. ■

Notice that variational morphisms $\mathbb{V} : J^k\mathcal{C} \longrightarrow A_{m-n}(M)$ of rank $h = 0$ can be identified with horizontal $(m - n)$-forms over $J^k\mathcal{C}$. Under this identification the divergence operator corresponds to the formal differential defined by (2.2.13).

Lemma (6.2.5): let $\mathbb{V} : J^k\mathcal{C} \longrightarrow A_{m-n}(M)$ be a variational morphism. Then we have $\mathrm{Div}^2(\mathbb{V}) = 0$

Proof: in fact, for each section $\sigma : M \longrightarrow \mathcal{C}$ we have:

$$\begin{aligned}(\mathrm{Div}^2\mathbb{V}) \circ j^{k+2}\sigma =& \mathrm{Div} \circ \mathrm{Div}(\mathbb{V}) \circ j^{k+2}\sigma = \mathrm{d}((\mathrm{Div}\mathbb{V}) \circ j^{k+1}\sigma) = \\ =& \mathrm{d} \circ \mathrm{d}(\mathbb{V} \circ j^k\sigma) = 0\end{aligned} \tag{6.2.6}$$

Since it vanishes on all sections, we have $\mathrm{Div}^2\mathbb{V} = 0$. ■

In general, let $(x^\mu; v^A)$, $\mu = 1, \dots, m$ and $A = 1, \dots, n$ be fibered coordinates on $\mathcal{E}$ associated to a fiberwise basis e_A and $(x^\mu; y^i)$, $i = 1, \dots, \dim(F)$ be fibered coordinates on $\mathcal{C} = (C, M, \pi; F)$. Then $J^h\mathcal{E}$ is also a vector bundle and we can consider the natural trivialization induced by the fiberwise basis $(e_A, e_A^\lambda, \dots, e_A^{\lambda_1 \dots \lambda_h})$. The natural fibered coordinates on $J^h\mathcal{E}$ are thence $(x^\mu; v^A, v^A_\lambda, \dots, v^A_{\lambda_1 \dots \lambda_h})$. The fiberwise dual basis $(e^A, e^A_\lambda, \dots, e^A_{\lambda_1 \dots \lambda_h})$ induces on $(J^h\mathcal{E})^*$ the set of local fibered coordinates $(x^\mu; v_A, v_A^\lambda, \dots, v_A^{\lambda_1 \dots \lambda_h})$. Thence the local expression of the variational $\mathcal{E}$-morphism (6.2.2) is the following

$$\mathbb{V} = \frac{1}{n!}\left[v_A^{\mu_1 \dots \mu_n}\, e^A + v_A^{\mu_1 \dots \mu_n\, \lambda}\, e^A_\lambda + \dots v_A^{\mu_1 \dots \mu_n\, \lambda_1 \dots \lambda_h}\, e^A_{\lambda_1 \dots \lambda_h}\right] \otimes \mathrm{d}s_{\mu_1 \dots \mu_n} \tag{6.2.7}$$

where the coefficients $(v_A^{\mu_1\dots\mu_n}, v_A^{\mu_1\dots\mu_n\,\lambda}, \dots, v_A^{\mu_1\dots\mu_n\,\lambda_1\dots\lambda_h})$ are understood to be functions of the point $j^k_x\rho \in J^kC$. Here we recursively defined the standard local basis for $(m-n)$-forms, according to the following rules: $\mathrm{d}s_\mu = i_{\partial_\mu}\mathrm{d}s$, $\mathrm{d}s_{\mu\nu} = i_{\partial_\nu}\mathrm{d}s_\mu$, $\mathrm{d}s_{\mu\nu\lambda} = i_{\partial_\lambda}\mathrm{d}s_{\mu\nu}$, and so on.

We remark that the coefficients $v_A^{\mu_1\dots\mu_n\,\lambda_1\dots\lambda_l}$ in the local expression of $\mathbb{V}$ are skew-symmetric in the $[\mu_1\dots\mu_n]$ indices and symmetric in the $(\lambda_1\dots\lambda_l)$ indices (if any).

Let us now assume that we have a linear connection $\Gamma^\alpha_{\beta\mu}$ over M and a connection $\Gamma^A_{B\mu}$ over the vector bundle $\mathcal{E}$. The pair $(\Gamma^\alpha_{\beta\mu}, \Gamma^A_{B\mu})$ will be called a *fibered connection* on $\mathcal{E}$. Thanks to the fibered connection we can change coordinates in $J^h\mathcal{E}$ by setting

$$\begin{aligned}
&\hat{v}^A = v^A\\
&\hat{v}^A_\mu = \nabla_\mu v^A = d_\mu v^A + \Gamma^A_{B\mu}v^B\\
&\dots\\
&\hat{v}^A_{\mu_1\dots\mu_k} = \nabla_{(\mu_1}\dots\nabla_{\mu_k)}v^A =: \nabla_{\mu_1\dots\mu_k}v^A
\end{aligned} \tag{6.2.8}$$

which are associated to a new fiberwise basis which will be denoted by

$$(\hat{e}_A, \hat{e}^\lambda_A, \dots, \hat{e}^{\lambda_1\dots\lambda_h}_A) \tag{6.2.9}$$

For example, when $h=1$ one has

$$\begin{cases} \hat{e}_A = e_A + \Gamma^B_{A\lambda}\, e^\lambda_B\\ \hat{e}^\lambda_A = e^\lambda_A \end{cases} \tag{6.2.10}$$

The new basis on $J^h\mathcal{E}$ induces a new dual basis on $(J^h\mathcal{E})^*$ which will be denoted by $(\hat{e}^A, \hat{e}^A_\lambda, \dots, \hat{e}^A_{\lambda_1\dots\lambda_h})$. It is called the *basis of symmetrized covariant derivatives*. For example when $h=1$ one obtains

$$\begin{cases} \hat{e}^A = e^A\\ \hat{e}^A_\lambda = e^A_\lambda + \Gamma^A_{B\lambda}\, e^B \end{cases} \tag{6.2.11}$$

Accordingly, a new set of local (fibered) coordinates $(x^\mu;\, \hat{v}_A^{\mu_1\dots\mu_n}, \hat{v}_A^{\mu_1\dots\mu_n\,\lambda}, \dots, \hat{v}_A^{\mu_1\dots\mu_n\,\lambda_1\dots\lambda_h})$ is induced on $(J^h\mathcal{E})^* \otimes A_{m-n}(M)$. For example when $h=1$ and $n=0$ one has

$$\begin{cases} v_A = \hat{v}_A + \hat{v}^\lambda_B\Gamma^B_{A\lambda}\\ v^\lambda_A = \hat{v}^\lambda_A \end{cases} \tag{6.2.12}$$

The reason to introduce these new coordinates relies on the fact that when one changes trivializations in C by

$$\begin{cases} x'^\mu = \phi^\mu(x)\\ y'^i = Y^i(x,y) \end{cases} \tag{6.2.13}$$

and in $\mathcal{E}$ by

$$\begin{cases} x'^\mu = \phi^\mu(x) \\ v'^A = J^A_B\, v^B \end{cases} \tag{6.2.14}$$

the new coordinates $(\hat{v}_A, \hat{v}^\lambda_A, \ldots, \hat{v}^{\lambda_1\ldots\lambda_h}_A)$ change tensorially. For example, when $h = 1$ and $n = 1$ we have:

$$\begin{cases} \hat{v}_A = \bar{J}\, \hat{v}'_B\, J^B_A \\ \hat{v}^\mu_A = \bar{J}\, \bar{J}^\mu_\nu\, \hat{v}'^\nu_B\, J^B_A \\ \hat{v}^{\mu\,\lambda}_A = \bar{J}\, \bar{J}^\mu_\nu\, \bar{J}^\lambda_\epsilon \hat{v}'^{\nu\,\epsilon}_B J^B_A \end{cases} \tag{6.2.15}$$

where $\bar{J}^\mu_\nu$ is the inverse Jacobian of the transformation $\phi : x^\mu \mapsto x'^\mu$, $\bar{J} = \det(\|\bar{J}^\mu_\nu\|)$ is the determinant of the inverse Jacobian and J^B_A is the Jacobian of the transformation rule of v'^A.

Using these coordinates, the variational morphism $\mathbb{V}$ above can be recast in the following form:

$$\mathbb{V} = \frac{1}{n!}\left[\hat{v}^{\mu_1\ldots\mu_n}_A\, \hat{e}^A + \hat{v}^{\mu_1\ldots\mu_n\,\lambda}_A\, \hat{e}^A_\lambda + \ldots \hat{v}^{\mu_1\ldots\mu_n\,\lambda_1\ldots\lambda_h}_A\, \hat{e}^A_{\lambda_1\ldots\lambda_h}\right] \otimes \mathrm{d}s_{\mu_1\ldots\mu_n} \tag{6.2.16}$$

We remark that the coefficients $\hat{v}^{\mu_1\ldots\mu_n\,\lambda_1\ldots\lambda_l}_A$ of the term of rank $0 \le l \le k$ in the new local expression of $\mathbb{V}$ are skew-symmetric in the $[\mu_1 \ldots \mu_n]$ indices and symmetric in the $(\lambda_1 \ldots \lambda_l)$ indices (if any).

We stress that the morphism $\mathbb{V}$ can be *algorithmically* recast in the form (6.2.16). The recasting is nothing more that a change of basis though it becomes more and more complicated with the growth of the rank, since symmetrized covariant derivatives become more and more complicated. However, one gains simplicity with respect to the changes of trivializations since any object in (6.2.16) transforms tensorially. As a consequence it will be quite easy to keep globality properties under control. In particular, when we shall split a variational morphism into two variational morphisms it will be easy to check if the splitting is global. We also remark that because of the transformation rules (6.2.15) each term $\mathbb{V}_l = \frac{1}{n!}(\hat{v}^{\mu_1\ldots\mu_n\,\lambda_1\ldots\lambda_l}_A\, \hat{e}^A_{\lambda_1\ldots\lambda_l}) \otimes \mathrm{d}s_{\mu_1\ldots\mu_n}$ of each order l (with $0 \le l \le h$) in the local expression (6.2.16) of the variational morphism $\mathbb{V}$ is a global variational morphism on its own. It is called the *l-rank term of* $\mathbb{V}$. When $l = h$ it will be called the *highest rank term of* $\mathbb{V}$.

Sometimes the splitting may turn out not to depend on the fibered connection. In those cases we shall regard the fibered connection simply as a technical tool to define the splitting.

Canonical Splittings

The reason to introduce variational morphisms is that they admit canonical and algorithmic splitting into two variational morphisms. This property is basic for physical applications. It corresponds in fact (as we shall see below) to the possibility of defining *"first-variation-like-splittings"*, i.e. to the possibility of performing a global and covariant integration by parts which produces a volume plus a boundary term for $\mathbb{V}$.

Lemma (6.2.17): let $\mathbb{V} : J^k\mathcal{C} \longrightarrow (J^h\mathcal{E})^* \otimes A_m(M)$ be a global variational $\mathcal{E}$-morphism of codegree $n = 0$. Then we can define two global variational $\mathcal{E}$-morphisms

$$\begin{aligned} \mathbb{R} \equiv \mathbb{R}(\mathbb{V}) &: J^{k+h}\mathcal{C} \longrightarrow \mathcal{E}^* \otimes A_m(M) \\ \mathbb{T} \equiv \mathbb{T}(\mathbb{V}) &: J^{k+h-1}\mathcal{C} \longrightarrow (J^{h-1}\mathcal{E})^* \otimes A_{m-1}(M) \end{aligned} \tag{6.2.18}$$

such that the following splitting property holds true

$$< \mathbb{V} \mid j^h X > = < \mathbb{R} \mid X > + \operatorname{Div} < \mathbb{T} \mid j^{h-1} X > \tag{6.2.19}$$

for each section X of $\mathcal{E}$. The variational morphism $\mathbb{R}$ is called the *volume part* of $\mathbb{V}$ while $\mathbb{T}$ is called the *boundary part* of $\mathbb{V}$.

Proof: the proof is carried over by induction on the rank h of the variational morphism $\mathbb{V}$. For $h = 0$ one simply sets $\mathbb{R} = \mathbb{V}$ and $\mathbb{T} = 0$.

For $h = 1$, the morphism $\mathbb{V}$ can be locally expressed as follows

$$\begin{aligned} < \mathbb{V} \mid j^1 X > &= [\hat{v}_A\, \hat{X}^A + \hat{v}_A^\lambda\, \hat{X}_\lambda^A] \otimes \mathrm{d}s = \\ &= [(\hat{v}_A - \nabla_\lambda \hat{v}_A^\lambda)\, \hat{X}^A + \nabla_\lambda(\hat{v}_A^\lambda\, \hat{X}^A)] \otimes \mathrm{d}s \end{aligned} \tag{6.2.20}$$

Then the claim follows by setting explicitly

$$\begin{aligned} \mathbb{R} &: J^{k+1}\mathcal{C} \longrightarrow \mathcal{E}^* \otimes A_m(M) \\ \mathbb{T} &: J^k\mathcal{C} \longrightarrow \mathcal{E}^* \otimes A_{m-1}(M) \end{aligned} \qquad \begin{cases} \mathbb{R} = (\hat{v}_A - \nabla_\lambda \hat{v}_A^\lambda)\, \hat{e}^A \otimes \mathrm{d}s \\ \mathbb{T} = (\hat{v}_A^\lambda\, \hat{e}^A) \otimes \mathrm{d}s_\lambda \end{cases} \tag{6.2.21}$$

Let us now suppose that (6.2.19) holds for rank $(h - 1)$ and let us show that it holds for h. The coefficient of the highest rank term of $\mathbb{V}$ can be recast as follows:

$$\hat{v}_A^{\lambda_1 \ldots \lambda_h}\, \hat{X}^A_{\lambda_1 \ldots \lambda_h} = \nabla_{\lambda_1}(\hat{v}_A^{\lambda_1 \ldots \lambda_h}\, \hat{X}^A_{\lambda_2 \ldots \lambda_h}) - (\nabla_{\lambda_1} \hat{v}_A^{\lambda_1 \ldots \lambda_h})\, \hat{X}^A_{\lambda_2 \ldots \lambda_h} \tag{6.2.22}$$

The first term in the above expression is

$$\nabla_{\lambda_1}(\hat{v}_A^{\lambda_1 \ldots \lambda_h}\, \hat{X}^A_{\lambda_2 \ldots \lambda_h}) \otimes \mathrm{d}s = \operatorname{Div}\left(\hat{v}_A^{\lambda_1 \ldots \lambda_h}\, \hat{X}^A_{\lambda_2 \ldots \lambda_h} \otimes \mathrm{d}s_{\lambda_1}\right) \tag{6.2.23}$$

which thence contributes to $\mathbb{T}$ by the term $(\hat{v}_A^{\lambda_1 \lambda_2 \ldots \lambda_h}\, \hat{e}^A_{\lambda_2 \ldots \lambda_h}) \otimes \mathrm{d}s_{\lambda_1}$, which is a variational morphism of rank $(h - 1)$ and codegree $n = 1$.

The second term $(\nabla_{\lambda_1} \hat{v}_A^{\lambda_1\lambda_2\ldots\lambda_h}\, \hat{X}^A_{\lambda_2\ldots\lambda_h}) \otimes \mathrm{d}s$, together with the remaining part of $\mathbb{V}$, is a variational morphism of order $k+1$, of rank $(h-1)$ and codegree $n=0$, which splits by the inductive hypothesis.

The terms of the variational morphisms $\mathbb{R}$ and $\mathbb{T}$ all transform tensorially, and both $\mathbb{R}$ and $\mathbb{T}$ are thence global variational morphisms. ■

A similar result can be obtained for variational morphisms of higher codegree, though the case of codegree $n=0$ appears to be a degenerate case which has to be stated and proved separately. First of all, let us give the following:

Definition (6.2.24): let $\mathbb{V} : J^k\mathcal{C} \longrightarrow (J^h\mathcal{E})^* \otimes A_{m-n}(M)$ be a variational morphism and $\mathbb{V}_l = \frac{1}{n!}(\hat{v}_A^{\mu_1\ldots\mu_n\,\lambda_1\ldots\lambda_l}\, \hat{e}^A_{\lambda_1\ldots\lambda_l}) \otimes \mathrm{d}s_{\mu_1\ldots\mu_n}$ its term of rank $0 \le l \le h$. The term $\mathbb{V}_l$ is said to be *reduced with respect to the fibered connection* $(\Gamma^\alpha_{\beta\mu}, \Gamma^A_{B\mu})$ if $\hat{v}_A^{[\mu_1\ldots\mu_n\,\lambda_1]\lambda_2\ldots\lambda_l} = 0$. The variational morphism $\mathbb{V}$ is *reduced* if and only if all its terms are reduced. ■

Notice that when $m = \dim(M) = 1$, e.g. in the case of Mechanics, all variational morphisms are reduced.

We remark that the variational morphism $\mathbb{T}$ obtained in Lemma (6.2.17) is reduced because of the particular algorithm we have chosen to perform the splitting. We also stress that Definition (6.2.24) depends on the fibered connection we choose from the beginning. In other words, a variational morphism might be reduced with respect to a fibered connection $(\Gamma^\alpha_{\beta\mu}, \Gamma^A_{B\mu})$ but not with respect to another fibered connection $(\Gamma'^\alpha_{\beta\mu}, \Gamma'^A_{B\mu})$ whenever its rank h is at least two. We shall show it in the case $n=1$, $h=2$, just to explain why it works so. A general proof is cumbersome (as well as totally irrelevant for our present purposes).

In fact, by changing fibered connection we obtain two systems of coordinates on the vector bundle $J^h\mathcal{E}$ which (for $h=2$) are related by the following transformation laws:

$$\begin{cases} \hat{X}'^A = \hat{X}^A \\ \hat{X}'^A_\lambda = \hat{X}^A_\lambda + \Delta^A_{B\lambda}\hat{X}^B \\ \hat{X}'^A_{\lambda\sigma} = \hat{X}^A_{\lambda\sigma} + \Delta^{A\ \rho}_{\lambda\sigma B}\, \hat{X}^B_\rho + \Delta^A_{\lambda\sigma B}\, \hat{X}^B \end{cases} \qquad \begin{array}{l} \Delta^A_{B\lambda} = \Gamma'^A_{B\lambda} - \Gamma^A_{B\lambda} \\ \Delta^\mu_{\nu\lambda} = \Gamma'^\mu_{\nu\lambda} - \Gamma^\mu_{\nu\lambda} \end{array} \tag{6.2.25}$$

where the coefficients are defined by the following rules

$$\begin{cases} \Delta^{A\ \rho}_{\lambda\sigma B} = 2\,\Delta^A_{B(\lambda}\delta^\rho_{\sigma)} - \Delta^\rho_{\lambda\sigma}\delta^A_B \\ \Delta^A_{\lambda\sigma B} = \nabla_{(\sigma}\Delta^A_{B\lambda)} + \Delta^A_{C(\sigma}\Delta^C_{B\lambda)} - \Delta^\rho_{\lambda\sigma}\Delta^A_{B\rho} \end{cases} \tag{6.2.26}$$

A variational morphism $\mathbb{V}$ of codegree $n=1$ and of rank $h=2$ can be locally expressed as

$$\begin{aligned} < \mathbb{V} \,|\, j^2X > &= [\hat{v}^\mu_A \hat{X}^A + \hat{v}^{\mu\,\lambda}_A \hat{X}^A_\lambda + \hat{v}^{\mu\,\lambda\sigma}_A \hat{X}^A_{\lambda\sigma}] \otimes \mathrm{d}s_\mu = \\ &= [\hat{v}'^\mu_A \hat{X}'^A + \hat{v}'^{\mu\,\lambda}_A \hat{X}'^A_\lambda + \hat{v}'^{\mu\,\lambda\sigma}_A \hat{X}'^A_{\lambda\sigma}] \otimes \mathrm{d}s_\mu \end{aligned} \tag{6.2.27}$$

so that the coefficients are related by the following relations

$$\begin{cases} \hat{v}_A^{\mu\,\lambda\sigma} = \hat{v}'^{\mu\,\lambda\sigma}_A \\ \hat{v}_A^{\mu\,\lambda} = \hat{v}'^{\mu\,\lambda}_A + \hat{v}'^{\mu\,\rho\sigma}_B\, \Delta^{B\,\lambda}_{\rho\sigma A} \\ \hat{v}_A^{\mu} = \hat{v}'^{\mu}_A + \hat{v}'^{\mu\,\lambda}_B\, \Delta^B_{A\lambda} + \hat{v}'^{\mu\,\rho\sigma}_B\, \Delta^B_{\rho\sigma A} \end{cases} \tag{6.2.28}$$

Then, even if we suppose that $\mathbb{V}$ is reduced with respect to the fibered connection $(\Gamma'^{\alpha}_{\beta\mu}, \Gamma'^{A}_{B\mu})$ (i.e. we assume $\hat{v}'^{[\mu\,\lambda]}_A = 0$ and $\hat{v}'^{[\mu\,\lambda]\sigma}_A = 0$), this does not imply that $\mathbb{V}$ is reduced with respect to the other fibered connection $(\Gamma^{\alpha}_{\beta\mu}, \Gamma^{A}_{B\mu})$. In fact, we trivially get $\hat{v}_A^{[\mu\,\lambda]\sigma} = 0$, but the same does not happen for the first rank term, i.e. we get:

$$\hat{v}_A^{[\mu\,\rho]} = \hat{v}'^{\lambda\,\sigma[\rho}_A\, \Delta^{\mu]}_{\lambda\sigma} \tag{6.2.29}$$

which has no reason to be zero in general. In this case the variational morphism $\mathbb{V}$ is not reduced with respect to $(\Gamma^{\alpha}_{\beta\mu}, \Gamma^{A}_{B\mu})$ even if it is reduced with respect to $(\Gamma'^{\alpha}_{\beta\mu}, \Gamma'^{A}_{B\mu})$.

Notice, however, that if $\mathbb{V}$ is of rank $h = 1$, i.e. $\hat{v}'^{\lambda\,\sigma\mu}_A = 0$, then reduction does not depend on the fibered connection, as one immediately sees from (6.2.29). Notice also the fact that only the connection on the base manifold enters (6.2.29). This is not a coincidence: it can in fact be shown that at any rank h the reduction depends at most on the base connection (see [K84]).

We can now state the following splitting Lemma:

Lemma (6.2.30): let $\mathbb{V} : J^kC \longrightarrow (J^h\mathcal{E})^* \otimes A_{m-n}(M)$ be a global variational $\mathcal{E}$-morphism of codegree $n \geq 1$. Then we can define two global variational $\mathcal{E}$-morphisms

$$\begin{aligned} \mathbb{R} \equiv \mathbb{R}(\mathbb{V}) &: J^{k+h}C \longrightarrow (J^h\mathcal{E})^* \otimes A_{m-n}(M) \\ \mathbb{T} \equiv \mathbb{T}(\mathbb{V}) &: J^{k+h-1}C \longrightarrow (J^{h-1}\mathcal{E})^* \otimes A_{m-n-1}(M) \end{aligned} \tag{6.2.31}$$

where $\mathbb{R}$ is a reduced variational morphism and such that the following holds true

$$< \mathbb{V} \mid j^hX > = < \mathbb{R} \mid j^hX > + \mathrm{Div} < \mathbb{T} \mid j^{h-1}X > \tag{6.2.32}$$

for each section X of $\mathcal{E}$. Also the variational morphism $\mathbb{R}$ is called the *volume part* of $\mathbb{V}$ while $\mathbb{T}$ is called the *boundary part* of $\mathbb{V}$.

Proof: the proof is carried over by induction on the rank h of the variational morphism $\mathbb{V}$. For $h = 0$ one simply sets $\mathbb{R} = \mathbb{V}$ and $\mathbb{T} = 0$.

For $h = 1$, the morphism $\mathbb{V}$ can be locally expressed as follows

$$\begin{aligned} < \mathbb{V} \mid j^1X > &= \frac{1}{n!}\left[\hat{v}_A^{\mu_1\ldots\mu_n}\,\hat{X}^A + \hat{v}_A^{\mu_1\ldots\mu_n\,\lambda}\,\hat{X}^A_\lambda\right] \otimes \mathrm{d}s_{\mu_1\ldots\mu_n} = \\ &= \frac{1}{n!}\left[(\hat{v}_A^{\mu_1\ldots\mu_n} - \nabla_\lambda \hat{v}_A^{[\mu_1\ldots\mu_n\,\lambda]})\,\hat{X}^A + \hat{w}_A^{\mu_1\ldots\mu_n\,\lambda}\,\hat{X}^A_\lambda + \right. \\ &\quad \left. + \nabla_\lambda(\hat{v}_A^{[\mu_1\ldots\mu_n\,\lambda]}\,\hat{X}^A)\right] \otimes \mathrm{d}s_{\mu_1\ldots\mu_n} \end{aligned} \tag{6.2.33}$$

where we set $\hat{w}_A^{\mu_1\dots\mu_n\,\lambda} = \hat{v}_A^{\mu_1\dots\mu_n\,\lambda} - \hat{v}_A^{[\mu_1\dots\mu_n\,\lambda]}$.

Then our claim follows by setting

$$\begin{aligned} &\mathbb{R} : J^{k+1}\mathcal{C} \longrightarrow (J^1\mathcal{E})^* \otimes A_{m-n}(M) \\ &\mathbb{T} : J^k\mathcal{C} \longrightarrow \mathcal{E}^* \otimes A_{m-n-1}(M) \end{aligned} \tag{6.2.34}$$

with the following local expressions:

$$\begin{cases} \mathbb{R} = \dfrac{1}{n!} \left[(\hat{v}_A^{\mu_1\dots\mu_n} - \nabla_\lambda \hat{v}_A^{[\mu_1\dots\mu_n\,\lambda]})\,\hat{e}^A + \hat{w}_A^{\mu_1\dots\mu_n\,\lambda}\,\hat{e}^A_\lambda \right] \otimes \mathrm{d}s_{\mu_1\dots\mu_n} \\ \mathbb{T} = \dfrac{1}{n!} \left(\hat{v}_A^{[\mu_1\dots\mu_n\,\lambda]}\,\hat{e}^A \right) \otimes \mathrm{d}s_{\mu_1\dots\mu_n\lambda} \end{cases} \tag{6.2.35}$$

We remark that the volume part $\mathbb{R}$ is a variational morphism since it is skew in the indices $[\mu_1 \dots \mu_n]$ and (trivially) symmetric in the index λ. It is also reduced since

$$\hat{w}_A^{[\mu_1\dots\mu_n\,\lambda]} = \hat{v}_A^{[\mu_1\dots\mu_n\,\lambda]} - \hat{v}_A^{[\mu_1\dots\mu_n\,\lambda]} = 0 \tag{6.2.36}$$

Let us now suppose that the claim holds for rank $(h-1)$ and let us show that it holds for h. The coefficient of the highest rank term of $< \mathbb{V} \mid j^h X >$ can be recast as follows:

$$\hat{v}_A^{\mu_1\dots\mu_n\,\lambda_1\dots\lambda_h} = \hat{v}_A^{[\mu_1\dots\mu_n\,\lambda_1]\dots\lambda_h} + \hat{w}_A^{\mu_1\dots\mu_n\,[\lambda_1\lambda_2]\dots\lambda_h} + \hat{w}_A^{\mu_1\dots\mu_n\,(\lambda_1\lambda_2)\dots\lambda_h} \tag{6.2.37}$$

where we set $\hat{w}_A^{\mu_1\dots\mu_n\,\lambda_1\dots\lambda_h} = \hat{v}_A^{\mu_1\dots\mu_n\,\lambda_1\dots\lambda_h} - \hat{v}_A^{[\mu_1\dots\mu_n\,\lambda_1]\dots\lambda_h}$.

Then one can easily prove that:

$$\begin{aligned} \hat{w}_A^{\mu_1\dots\mu_n\,(\lambda_1\lambda_2)\dots\lambda_h} &= -\frac{h-1}{n+1}\hat{v}_A^{\mu_1\dots\mu_n\,\lambda_1\dots\lambda_h} + \\ &+ \frac{1}{2}\hat{v}_A^{[\mu_1\dots\mu_n\,\lambda_1]\dots\lambda_h} + \frac{1}{2}\hat{v}_A^{[\mu_1\dots\mu_n\,\lambda_2]\lambda_1\lambda_3\dots\lambda_h} + \\ &+ \hat{v}_A^{[\mu_1\dots\mu_n\,\lambda_3]\lambda_1\lambda_1\lambda_4\dots\lambda_h} + \dots + \hat{v}_A^{[\mu_1\dots\mu_n\,\lambda_h]\lambda_1\dots\lambda_{h-1}} + \\ &+ \frac{h+1}{n+1}\left[\hat{v}_A^{(\mu_1\underline{\mu_2}\dots\underline{\mu_n}\,\lambda_1\dots\lambda_h)} + \dots + \hat{v}_A^{(\underline{\mu_1}\dots\underline{\mu_{n-1}}\mu_n\,\lambda_1\dots\lambda_h)} \right] \end{aligned} \tag{6.2.38}$$

where we denoted by $(\mu_1\underline{\mu_2}\dots\underline{\mu_n}\,\lambda_1\dots\lambda_h)$ the complete symmetrization on non-underlined indices.

By substituting into equation (6.2.37) we obtain

$$\begin{aligned} \hat{v}_A^{\mu_1\dots\mu_n\,\lambda_1\dots\lambda_h} &= \frac{n+1}{n+h}\hat{w}_A^{\mu_1\dots\mu_n\,[\lambda_1\lambda_2]\dots\lambda_h} + \\ &+ \frac{h+1}{n+h}\left[\hat{v}_A^{(\mu_1\underline{\mu_2}\dots\underline{\mu_n}\,\lambda_1\dots\lambda_h)} + \dots + \hat{v}_A^{(\underline{\mu_1}\dots\underline{\mu_{n-1}}\mu_n\,\lambda_1\dots\lambda_h)} \right] + \\ &+ \frac{3(n+1)}{2(n+h)}\hat{v}_A^{[\mu_1\dots\mu_n\,\lambda_1]\dots\lambda_h} + \frac{n+1}{2(n+h)}\hat{v}_A^{[\mu_1\dots\mu_n\,\lambda_2]\lambda_1\lambda_3\dots\lambda_h} + \\ &+ \frac{n+1}{n+h}\hat{v}_A^{[\mu_1\dots\mu_n\,\lambda_3]\lambda_1\lambda_1\lambda_4\dots\lambda_h} + \dots + \frac{n+1}{n+h}\hat{v}_A^{[\mu_1\dots\mu_n\,\lambda_h]\lambda_1\dots\lambda_{h-1}} \end{aligned} \tag{6.2.39}$$

Thus we can expand the highest rank term $(\hat{v}_A^{\mu_1\dots\mu_n\ \lambda_1\dots\lambda_h}\nabla_{\lambda_1}\dots\nabla_{\lambda_h}\hat{X}^A)\otimes \mathrm{d}s_{\mu_1\dots\mu_n}$ of the variational morphism $\mathbb{V}$. Apart from their numerical coefficients, we obtain the following terms:

- $\hat{w}_A^{\mu_1\dots\mu_n\ [\lambda_1\lambda_2]\dots\lambda_h}\nabla_{[\lambda_1}\nabla_{\lambda_2]}\dots\nabla_{\lambda_h}\hat{X}^A\otimes \mathrm{d}s_{\mu_1\dots\mu_n}$ which, once one uses commutation properties of covariant derivatives, is reduced to a variational morphism of rank $h-2$;
- $\left[\hat{v}_A^{(\mu_1\underline{\mu_2}\dots\underline{\mu_n}\ \lambda_1\dots\lambda_h)}+\dots+\hat{v}_A^{(\underline{\mu_1}\dots\underline{\mu_{n-1}}\mu_n\ \lambda_1\dots\lambda_h)}\right]\hat{X}^A_{\lambda_1\dots\lambda_n}\ \otimes \mathrm{d}s_{\mu_1\dots\mu_n}$ which contributes to the reduced variational morphism $\mathbb{R}$ with a term of rank h, since the coefficient is reduced;
- $\hat{v}_A^{[\mu_1\dots\mu_n\ \lambda_1]\lambda_2\dots\lambda_h}\nabla_{\lambda_1}\nabla_{\lambda_2}\dots\nabla_{\lambda_h}\hat{X}^A\ \otimes \mathrm{d}s_{\mu_1\dots\mu_n}$ that can be integrated by parts obtaining

$$\begin{aligned}\mathrm{Div}\left(\frac{1}{(n+1)}\hat{v}_A^{[\mu_1\dots\mu_n\ \lambda_1]\lambda_2\dots\lambda_h}\ \hat{X}^A_{\lambda_2\dots\lambda_h}\ \otimes \mathrm{d}s_{\mu_1\dots\mu_n\lambda_1}\right)+\\ -\nabla_{\lambda_1}\hat{v}_A^{[\mu_1\dots\mu_n\ \lambda_1]\lambda_2\dots\lambda_h}\ \hat{X}^A_{\lambda_2\dots\lambda_h}\ \otimes \mathrm{d}s_{\mu_1\dots\mu_n}\end{aligned}\tag{6.2.40}$$

  the first of which contributes to $\mathbb{T}$ with a term of rank $(h-1)$, while the second is a variational morphism of rank $(h-1)$;
- in the other terms:
  $(\hat{v}_A^{[\mu_1\dots\mu_n\ \lambda_2]\lambda_1\lambda_3\dots\lambda_h}\nabla_{\lambda_1}\nabla_{\lambda_2}\dots\nabla_{\lambda_h}\hat{X}^A)\ \otimes \mathrm{d}s_{\mu_1\dots\mu_n}$,
  $(\hat{v}_A^{[\mu_1\dots\mu_n\ \lambda_3]\lambda_1\lambda_1\lambda_4\dots\lambda_h}\nabla_{\lambda_1}\nabla_{\lambda_2}\dots\nabla_{\lambda_h}\hat{X}^A)\ \otimes \mathrm{d}s_{\mu_1\dots\mu_n}$,
  $\dots$,
  $(\hat{v}_A^{[\mu_1\dots\mu_n\ \lambda_h]\lambda_1\dots\lambda_{h-1}}\nabla_{\lambda_1}\nabla_{\lambda_2}\dots\nabla_{\lambda_h}\hat{X}^A)\ \otimes \mathrm{d}s_{\mu_1\dots\mu_n}$
  one can use commutation properties of covariant derivatives followed by an integration by parts to get a pure divergence (that contributes to $\mathbb{T}$) and a variational morphism of rank $(h-1)$.

Then the remaining part is a variational morphism of rank $(h-1)$ which splits by the inductive hypothesis. Thus the variational morphism $\mathbb{V}$ splits as well.

The terms of the variational morphisms $\mathbb{R}$ and $\mathbb{T}$ all transform tensorially, and both $\mathbb{R}$ and $\mathbb{T}$ are thence global variational morphisms. ∎

For example let us specialize to the case $n=1$ and $h=2$ (which shall be often used in the sequel). If we consider a variational morphism $\mathbb{V}: J^k\mathcal{C}\longrightarrow (J^2\mathcal{E})^*\otimes A_{m-1}(M)$ locally expressed by

$$\mathbb{V}=[\hat{v}^\mu_A\hat{e}^A+\hat{v}_A^{\mu\ \lambda}\hat{e}^A_\lambda+\hat{v}_A^{\mu\ \lambda\nu}\hat{e}^A_{\lambda\nu}]\ \otimes \mathrm{d}s_\mu \tag{6.2.41}$$

it splits by means of the (reduced) volume part $\mathbb{R}: J^{k+2}\mathcal{C}\longrightarrow (J^2\mathcal{E})^*\otimes A_{m-1}(M)$ and the boundary part $\mathbb{T}: J^{k+1}\mathcal{C}\longrightarrow (J^1\mathcal{E})^*\otimes A_{m-2}(M)$ locally given by

$$\mathbb{R} = \left[\left(\hat{v}_A^{\mu} - \nabla_\sigma \hat{v}_A^{[\mu\,\sigma]} + \frac{1}{3}\hat{v}_B^{\rho\,\sigma\mu}\, R^B_{A\rho\sigma} - \frac{2}{3}\nabla_\rho\nabla_\sigma \hat{v}_A^{[\mu\,\rho]\sigma}\right)\hat{e}^A + \right.$$
$$\left. + \left(\hat{v}_A^{(\mu\,\lambda)} + \frac{2}{3}\nabla_\sigma \hat{v}^{\sigma\,\mu\lambda} - \frac{2}{3}\nabla_\sigma \hat{v}^{(\mu\,\lambda)\sigma}\right)\hat{e}^A_\lambda + \hat{v}_A^{(\mu\,\lambda\sigma)}\hat{e}^A_{\lambda\sigma}\right] \otimes \mathbf{d}s_\mu \tag{6.2.42}$$

$$\mathbb{T} = \frac{1}{2}\left[\left(\hat{v}^{[\mu\,\nu]} - \frac{2}{3}\nabla_\lambda \hat{v}^{[\mu\,\nu]\lambda}\right)\hat{e}^A + \frac{4}{3}\hat{v}^{[\mu\,\nu]\lambda}\hat{e}^A_\lambda\right] \otimes \mathbf{d}s_{\mu\nu}$$

Uniqueness Results

Let us first recall that any morphism $\mathbb{V} : J^kC \longrightarrow A_{m-n}(M)$ is a variational morphism of rank $h = 0$ relative to the vector bundle $\mathcal{E} = (M \times \{0\}, M, \pi; \{0\})$. Before to come to uniqueness of the splittings we shall prove the following two Lemmas:

Lemma (6.2.43): let $\hat{v}_A^{\mu_1\ldots\mu_n\,\lambda_1\ldots\lambda_l}$ denote the coefficients of a variational morphism; in particular it is skewsymmetric in the first n indices $[\mu_1\ldots\mu_n]$ and symmetric in the last l indices $(\lambda_1\ldots\lambda_l)$. If we have:

$$\begin{cases} v_A^{[\mu_1\ldots\mu_n\,\lambda_1]\lambda_2\ldots\lambda_l} = 0 \\ v_A^{\mu_1\ldots\mu_{n-1}(\mu_n\,\lambda_1\ldots\lambda_l)} = 0 \end{cases} \tag{6.2.44}$$

then $\hat{v}_A^{\mu_1\ldots\mu_n\,\lambda_1\ldots\lambda_h} = 0$.

Proof: We can sum over the permutations of indices $[\mu_1\ldots\mu_n]$ and $(\lambda_1\ldots\lambda_h)$ obtaining

$$\begin{cases} v_A^{[\mu_1\ldots\mu_{n-1}\mu_n\,\lambda_1\underline{\lambda}_2\ldots\underline{\lambda}_l]} = 0 \\ v_A^{[\mu_1\ldots\mu_n\,\underline{\lambda}_1\lambda_2\underline{\lambda}_3\ldots\underline{\lambda}_l]} = 0 \\ \ldots \\ v_A^{[\mu_1\ldots\mu_n\,\underline{\lambda}_1\ldots\underline{\lambda}_{l-1}\lambda_l]} = 0 \end{cases} \qquad \begin{cases} v_A^{(\mu_1\underline{\mu}_2\ldots\underline{\mu}_n\,\lambda_1\ldots\lambda_l)} = 0 \\ v_A^{(\underline{\mu}_1\mu_2\underline{\mu}_3\ldots\underline{\mu}_n\,\lambda_1\ldots\lambda_l)} = 0 \\ \ldots \\ v_A^{(\underline{\mu}_1\ldots\underline{\mu}_{n-1}\mu_n\,\lambda_1\ldots\lambda_l)} = 0 \end{cases} \tag{6.2.45}$$

where both symmetrizations and antisymmetrizations are performed on non–underlined indices only. By expanding and summing altogether these identities we immediately obtain that $v_A^{\mu_1\ldots\mu_n\,\lambda_1\ldots\lambda_l} = 0$. ■

The second Lemma is:

Lemma (6.2.46): let $\mathbb{V} : J^kC \longrightarrow (J^h\mathcal{E})^* \otimes A_{m-n}(M)$ be a reduced variational morphism with $n \geq 1$. Then Div$(< \mathbb{V} \mid j^hX >) = 0$ for all sections X of $\mathcal{E}$ implies $\mathbb{V} = 0$.

Proof: the claim is trivially true for rank $h = 1$.

Let us suppose that it is true for rank $h-1$ and we shall prove it for rank h. Let us consider a reduced variational morphism

$$\mathbb{V} = [\hat{v}_A^{\mu_1\dots\mu_n}\hat{e}^A + \hat{v}_A^{\mu_1\dots\mu_n\,\lambda}\hat{e}_\lambda^A + \dots + \hat{v}_A^{\mu_1\dots\mu_n\,\lambda_1\dots\lambda_h}\hat{e}_{\lambda_1\dots\lambda_h}^A] \otimes \mathrm{d}s_{\mu_1\dots\mu_n} \tag{6.2.47}$$

The only contribution to the highest rank term in Div$(< \mathbb{V} \mid j^hX >) = 0$ comes from the highest rank term in $\mathbb{V}$. In fact, since the variational morphism $\mathbb{V}$ is reduced the highest term satisfies the following properties

$$\begin{cases} \hat{v}_A^{[\mu_1\dots\mu_n\,\lambda_1]\dots\lambda_h} = 0 \\ \hat{v}_A^{\mu_1\dots(\mu_n\,\lambda_1\dots\lambda_h)} = 0 \end{cases} \tag{6.2.48}$$

Because of the Lemma (6.2.43) this implies $\hat{v}_A^{\mu_1\dots\mu_n\,\lambda_1\dots\lambda_l} = 0$. The variational morphism $\mathbb{V}$ is *a posteriori* of rank $h-1$ and it thence vanishes because of the inductive hypotheses. ∎

Let us now assume that the fibered connection has been fixed once for all. We can prove some uniqueness result on the canonical splitting defined above.

Proposition (6.2.49): in the canonical splitting given by Lemma (6.2.17) of the variational morphism $\mathbb{V} : J^kC \longrightarrow (J^h\mathcal{E})^* \otimes A_m(M)$ (i.e. for codegree $n = 0$)

$$< \mathbb{V} \mid j^hX > = < \mathbb{R} \mid X > + \mathrm{Div} < \mathbb{T} \mid j^{h-1}X > \tag{6.2.50}$$

the volume part is uniquely determined, while the boundary part is determined modulo a divergenceless term.

Proof: let us consider a section $X : M \longrightarrow \mathcal{E}$ in the vector bundle and a section $\sigma : M \longrightarrow C$; then we can build the quantity $< \mathbb{V} \mid j^hX >$: $J^kC \longrightarrow A_m(M)$ and evaluate it along (the suitable prolongation of) the section σ to obtain

$$< \mathbb{V} \mid j^hX > |_\sigma = < \mathbb{V} \mid j^hX > \circ j^k\sigma \tag{6.2.51}$$

Then we can integrate the above on an m-region $D \subset M$ and restricting to sections $X : M \longrightarrow \mathcal{E}$ such that $j^{h-1}X|_{\partial D} = 0$ we obtain:

$$\begin{aligned} \int_D < \mathbb{V} \mid j^hX > |_\sigma &= \int_D < \mathbb{R} \mid X > |_\sigma + \int_{\partial D} < \mathbb{T} \mid j^{h-1}X > |_\sigma = \\ &= \int_D < \mathbb{R} \mid X > |_\sigma \end{aligned} \tag{6.2.52}$$

The second integral does not contribute since $j^{k-1}X|_{\partial D} = 0$.

Let us now suppose that (for the connection chosen) there exist two different splittings satisfying Lemma $(6.2.17)$ exist, i.e. we have

$$\begin{aligned} < \mathbb{V} \mid j^h X > &= < \mathbb{R} \mid X > + \text{Div} < \mathbb{T} \mid j^{h-1} X > = \\ &= < \mathbb{R}' \mid X > + \text{Div} < \mathbb{T}' \mid j^{h-1} X > \end{aligned} \tag{6.2.53}$$

Then by integration we obtain

$$0 = \int_D < \mathbb{R} - \mathbb{R}' \mid X > |_\sigma \tag{6.2.54}$$

for each region D, for each section $X : M \longrightarrow \mathcal{E}$ such that $j^{h-1}X|_{\partial D} = 0$ and for each section $\sigma : M \longrightarrow \mathcal{C}$. This implies $\mathbb{R}' = \mathbb{R}$.

By substituting into the splittings $(6.2.53)$ we obtain

$$\text{Div} < \mathbb{T}' - \mathbb{T} \mid j^{h-1} X > = 0 \tag{6.2.55}$$

Thus we define a morphism $\omega : J^{k+h-1}\mathcal{C} \longrightarrow (J^{h-1}\mathcal{E})^* \otimes A_{m-1}(M)$ such that $\omega = \mathbb{T} - \mathbb{T}'$, i.e. $\mathbb{T}' = \mathbb{T} - \omega$, and such that $\text{Div}(< \omega \mid j^{h-1} X >) = 0$ for each section $X : M \longrightarrow \mathcal{E}$. ∎

A similar result can be stated and proved for variational morphisms of higher codegree.

Proposition (6.2.56): in the canonical splitting given by Lemma $(6.2.30)$ of the variational morphism $\mathbb{V} : J^k\mathcal{C} \longrightarrow (J^h\mathcal{E})^* \otimes A_{m-n}(M)$ $(n \geq 1)$

$$< \mathbb{V} \mid j^h X > = < \mathbb{R} \mid j^h X > + \text{Div} < \mathbb{T} \mid j^{h-1} X > \tag{6.2.57}$$

the volume part is uniquely determined, while the boundary part is determined modulo a divergenceless term.

Proof: let us suppose that (for the given connection) there exist two pairs of morphisms for which the splitting holds true, i.e:

$$\begin{aligned} < \mathbb{V} \mid j^h X > &= < \mathbb{R} \mid j^h X > + \text{Div} < \mathbb{T} \mid j^{h-1} X > = \\ &= < \mathbb{R}' \mid j^h X > + \text{Div} < \mathbb{T}' \mid j^{h-1} X > \end{aligned} \tag{6.2.58}$$

As a consequence we obtain

$$< \mathbb{R} - \mathbb{R}' \mid j^h X > + \text{Div} < \mathbb{T} - \mathbb{T}' \mid j^{h-1} X > = 0 \tag{6.2.59}$$

By Lemma $(6.2.5)$ we obtain

$$\text{Div}(< \mathbb{R} - \mathbb{R}' \mid j^h X >) = 0 \tag{6.2.60}$$

and by Lemma (6.2.46), being $\mathbb{R} - \mathbb{R}'$ reduced by construction, we obtain $\mathbb{R} = \mathbb{R}'$.

Now substituting into equation (6.2.59) we obtain $\text{Div}(< \mathbb{T} - \mathbb{T}' \mid j^{h-1}X >) = 0$. Thus we define a morphism $\omega : J^{k+h-1}\mathcal{C} \longrightarrow (J^{h-1}\mathcal{E})^* \otimes A_{m-n-1}(M)$ such that $\omega = \mathbb{T} - \mathbb{T}'$, i.e. $\mathbb{T}' = \mathbb{T} - \omega$, and such that $\text{Div}(< \omega \mid j^{h-1}X >) = 0$ for each section $X : M \longrightarrow \mathcal{E}$. ∎

We remark that when $h \geq 2$ one can proceed by further splitting. We thence obtain

$$< \mathbb{T} \mid j^{h-1}X > = < \mathbb{S} \mid j^{h-1}X > + \text{Div} < \mathbb{Q} \mid j^{h-2}X > \tag{6.2.61}$$

where the variational morphism $\mathbb{S} : J^{k+2h-2}\mathcal{C} \longrightarrow (J^{h-1}\mathcal{E})^* \otimes A_{m-n-1}(M)$ is reduced by construction. This allows to define a *canonical* boundary term which is uniquely determined by the variational morphism $\mathbb{V}$ we started from. In fact, if $\mathbb{T}' = \mathbb{T} - \omega$ is another boundary term in the splitting of $\mathbb{V}$, we obtain the splitting

$$< \mathbb{T}' \mid j^{h-1}X > = < \mathbb{S}' \mid j^{h-1}X > + \text{Div} < \mathbb{Q}' \mid j^{h-2}X > \tag{6.2.62}$$

By subtracting from the splitting (6.2.61) we obtain

$$\begin{aligned} < \mathbb{T} - \mathbb{T}' \mid j^{h-1}X > = \; & < \omega \mid j^{h-1}X > = \\ = \; & < \mathbb{S} - \mathbb{S}' \mid j^{h-1}X > + \text{Div} < \mathbb{Q} - \mathbb{Q}' \mid j^{h-2}X > \end{aligned} \tag{6.2.63}$$

By applying the appropriate Div and recalling that ω is divergenceless we get

$$\text{Div} < \mathbb{S} - \mathbb{S}' \mid j^{h-1}X > = 0 \tag{6.2.64}$$

which implies $\mathbb{S} = \mathbb{S}'$, being $\mathbb{S} - \mathbb{S}'$ reduced. Thus in two different splittings of the variational morphism $\mathbb{V}$ the volume part is uniquely determined and the boundary parts are not. However, the reduced part of the boundary parts are uniquely determined as well.

One can check that the splitting algorithm introduced in the proof of Lemma (6.2.30) already produced the reduced boundary part. In other words, if one considers the splitting of $\mathbb{V}$ obtained by the algorithm there introduced, the boundary part is already reduced, i.e. when one tries a further splitting one obtains

$$\begin{aligned} &< \mathbb{T} \mid j^{h-1}X > = < \mathbb{S} \mid j^{h-1}X > + \text{Div} < \mathbb{Q} \mid j^{h-2}X > \\ &\qquad\qquad \text{with } \mathbb{S} = \mathbb{T} \text{ and } \mathbb{Q} = 0 \end{aligned} \tag{6.2.65}$$

The proof of this is an easy check. We stress that the particular algorithm introduced by Lemma (6.2.30) is of course *one* of the many possible splitting algorithms. However, it is better than other choices, and in a sense *canonical*, since it directly produces reduced boundary parts.

We could now investigate what happens when the fibered connection is free to be changed. If the rank of $\mathbb{V}$ is $h \leq 1$ then we know that the splitting is unique regardless of the choice of the fibered connection, since the reduction of the morphism $\mathbb{R}$ is independent of the connection. If the rank is $h \geq 2$, however, the splitting generally depends explicitly on the connection on the base manifold.

Worked examples which are relevant to physical applications may be found, e.g., in [FF91] and [FFFR99]. We shall not consider here the general transformation rule of the uniqueness results above under a change of connection, leaving it as an exercise for the interested reader. However, in the applications that we shall consider in the sequel such a rule is never needed. We shall consider in fact natural or gauge natural field theories in which one has *a priori* a canonical choice of the connection.

Spencer Cohomology

This sub-Section is added just for completeness and it may be skipped by the uninterested reader since it is not essential for the sequel.

Some of the results of the previous sub-Sections can be easily expressed by using the so-called *(dual) Spencer cohomology*. In that cohomology the vector spaces are $A^n \otimes S^h$ the elements of which are objects with n antisymmetric indices and h symmetric indices. Those objects can be regarded as $(m-n)$-forms on $\mathbb{R}^m$ with coefficients which are homogeneous polynomials of degree h. We can define two operations $A : A^n \otimes S^h \longrightarrow A^{n+1} \otimes S^{h-1}$ and $S : A^n \otimes S^h \longrightarrow A^{n-1} \otimes S^{h+1}$ given by

$$\begin{cases} A(\hat{v}^{\mu_1 \dots \mu_n \, \lambda_1 \dots \lambda_h}) = (n+1) \, \hat{v}^{[\mu_1 \dots \mu_n \, \lambda_1] \dots \lambda_h} \\ S(\hat{v}^{\mu_1 \dots \mu_n \, \lambda_1 \dots \lambda_h}) = \dfrac{h+1}{h+n} \, \hat{v}^{\mu_1 \dots (\mu_n \, \lambda_1 \dots \lambda_h)} \end{cases} \tag{6.2.66}$$

respectively. We remark that the two operations S and A are the exterior differential and the integral along a segment, respectively, when the elements in $A^n \otimes S^h$ are regarded as $(m-n)$-forms on $\mathbb{R}^m$ with coefficients which are homogeneous polynomials of degree h. Thence, by Poincaré Lemma we easily obtain the identity

$$A \circ S + S \circ A = \mathbb{I} \tag{6.2.67}$$

These two operations are obviously involutive so that we obtain, e.g. for

$n + h = 3$, the following diagrams

$$
\begin{array}{ccccccccccc}
\dots & \longrightarrow & 0 & & & & & & & & \\
\downarrow & & \downarrow & & & & & & & & \\
\dots & \xrightarrow{A} & A^3 \otimes S^0 & \xrightarrow{A} & 0 & & & & & & \\
\downarrow & & \downarrow{\scriptstyle S} & & \downarrow & & & & & & \\
\dots & \xrightarrow{A} & A^2 \otimes S^1 & \xrightarrow{A} & A^3 \otimes S^0 & \xrightarrow{A} & 0 & & & & \\
\downarrow & & \downarrow{\scriptstyle S} & & \downarrow{\scriptstyle S} & & \downarrow & & & & \\
\dots & \xrightarrow{A} & A^1 \otimes S^2 & \xrightarrow{A} & A^2 \otimes S^1 & \xrightarrow{A} & A^3 \otimes S^0 & \xrightarrow{A} & 0 & & \\
\downarrow & & \downarrow{\scriptstyle S} & & \downarrow{\scriptstyle S} & & \downarrow{\scriptstyle S} & & \downarrow & & \\
0 \xrightarrow{A} & & A^0 \otimes S^3 & \xrightarrow{A} & A^1 \otimes S^2 & \xrightarrow{A} & A^2 \otimes S^1 & \xrightarrow{A} & A^3 \otimes S^0 & \xrightarrow{A} & 0 \\
 & & \downarrow & & \downarrow{\scriptstyle S} & & \downarrow{\scriptstyle S} & & \downarrow{\scriptstyle S} & & \downarrow \\
 & & 0 & \longrightarrow & \dots & \longrightarrow & \dots & \longrightarrow & \dots & \longrightarrow & \dots
\end{array}
\qquad (6.2.68)
$$

where all rows and columns are exact.

Reductions of a variational morphism is then equivalent to require $A(\hat{v}) = 0$. Accordingly, the canonical splitting of a homogeneous term $\hat{v} \in A^n \otimes S^h$ is simply given by $\hat{v} = A \circ S\,(\hat{v}) + S \circ A(\hat{v})$. The first term $A \circ S\,(\hat{v})$ is reduced by definition ($A^2 \circ S = 0$), and the second term $S \circ A(\hat{v})$ is ready to be integrated by parts.

3. Euler-Lagrange Equations

The variation of the action we introduced above (see (6.1.7)) can be recast as follows:

$$
\delta_X A_D(\rho) = \left[\frac{\mathrm{d}}{\mathrm{d}s} \int_D L \circ j^k\Phi_s \circ j^k\rho\right]_{s=0} = \int_D \left[\frac{\mathrm{d}}{\mathrm{d}s}(L \circ j^k\Phi_s \circ j^k\rho)\right]_{s=0} \qquad (6.3.1)
$$

where Φ_s denotes the (vertical) flow of the deformation $X = (\delta y^i)\partial_i$.

We can then define a variational morphism $\delta L : J^k\mathcal{C} \longrightarrow V^*(J^k\mathcal{C}) \otimes A_m(M)$ given by

$$
< \delta L \mid j^k X >= \left[\frac{\mathrm{d}}{\mathrm{d}s}(L \circ j^k\Phi_s \circ j^k\rho)\right]_{s=0} \qquad (6.3.2)
$$

Here we used the canonical bundle isomorphism $\left(J^k V(\mathcal{C})\right)^* \simeq V^*(J^k\mathcal{C})$. Locally we have

$$
\delta L = [p_i \bar{\mathrm{d}}y^i + p_i^\mu \bar{\mathrm{d}}y^i_\mu + \ldots + p_i^{\mu_1 \ldots \mu_k} \bar{\mathrm{d}}y^i_{\mu_1 \ldots \mu_k}] \otimes \mathrm{d}s \qquad (6.3.3)
$$

where $(p_i, p_i^\mu, \dots, p_i^{\mu_1\dots\mu_k})$ denote the partial derivatives of the Lagrangian density $\mathcal{L}(x^\mu, y^i, y^i_\mu, \dots, y^i_{\mu_1\dots\mu_k})$ with respect to $(y^i, y^i_\mu, \dots, y^i_{\mu_1\dots\mu_k})$, respectively, i.e.:

$$p_i^{\mu_1\dots\mu_s} = \frac{\partial \mathcal{L}}{\partial y^i_{\mu_1\dots\mu_s}} \tag{6.3.4}$$

and $\bar{\mathrm{d}}y^i_{\mu_1\dots\mu_s}$ are defined in (1.6.10), i.e. they form the dual basis of $\partial_i^{\mu_1\dots\mu_s}$.

The quantities $(p_i, p_i^\mu, \dots, p_i^{\mu_1\dots\mu_k})$ are called the *(naive) momenta of the Lagrangian L*. Conventionally, the set of all momenta is denoted by $p_i^{\mu_1\dots\mu_r}$ with $0 \le r \le k$, by assuming that $r = 0$ means no upper index μ.

We can now choose a fibered connection $(\Gamma^\alpha_{\beta\mu}, \Gamma^i_{j\mu})$ over C so that the variational morphism δL can be algorithmically recast as

$$< \delta L \mid j^k X >= [\hat{p}_i X^i + \hat{p}_i^\mu \nabla_\mu X^i + \dots + \hat{p}_i^{\mu_1\dots\mu_k} \nabla_{\mu_1\dots\mu_k} X^i]\, \mathrm{d}s \tag{6.3.5}$$

where $\nabla_{\lambda_1\dots\lambda_s} X^i$ denote the symmetrized covariant derivatives with respect to the fixed fibered connection $(\Gamma^\alpha_{\beta\mu}, \Gamma^i_{j\mu})$ and the new coefficients are algebraic functions of the old p's and the connection coefficients. The coefficients $(\hat{p}_i, \hat{p}_i^\mu, \dots, \hat{p}_i^{\mu_1\dots\mu_k})$ are also called the *covariant momenta*. We can now perform the canonical splitting of δL. This produces two variational morphisms $\mathbb{E}(L) \equiv \mathbb{R}(\delta L) : J^{2k}C \longrightarrow V^*(C) \otimes A_m(M)$ and $\mathbb{F}(L) \equiv \mathbb{T}(\delta L) : J^{2k-1}C \longrightarrow V^*(J^{k-1}C) \otimes A_{m-1}(M)$, respectively, called the *Euler-Lagrange morphism* and the *Poincaré-Cartan morphism* of the Lagrangian L. These variational morphisms will be also denoted simply by $\mathbb{E} = \mathbb{E}(L)$ and $\mathbb{F} = \mathbb{F}(L)$ when there is no danger of confusion.

The Euler-Lagrange morphism and the Poincaré-Cartan morphisms are hence defined so that, for each vertical vector field X over C, the so-called *first variation formula* holds, i.e.:

$$< \delta L \mid j^k X > = < \mathbb{E}(L) \mid X > + \mathrm{Div}\left[< \mathbb{F}(L) \mid j^{k-1} X >\right] \tag{6.3.6}$$

We stress that *a priori* the Euler-Lagrange morphism and the Poincaré-Cartan morphism depend on the fibered connection used to perform the splitting. However, by Lemma (6.2.49), the Euler-Lagrange morphism is unique and, since no reduction condition is required, it does not depend on the connection at all. One can easily check this by locally applying the splitting Lemma (6.2.17), using the particular connection the coefficients of which vanish in the fibered coordinates used (it always exists). In this case, in fact, the covariant derivative ∇_μ reduces to the total derivative d_μ and, being the Euler-Lagrange morphism unique and independent of the connection, locally we have the familiar expression

$$\mathbb{E} \equiv \mathbb{E}(L) = \left[p_i - \mathrm{d}_\mu p_i^\mu + \dots + (-1)^k\, \mathrm{d}_{\mu_1\dots\mu_k} p_i^{\mu_1\dots\mu_k}\right] \bar{\mathrm{d}}y^i \otimes \mathrm{d}s \tag{6.3.7}$$

When $m = \dim(M) = 1$ the Poincaré-Cartan morphism is also unique since the reduction condition (6.2.24) is trivially satisfied. On the contrary, in higher dimensions, if $k \geq 2$ the Poincaré-Cartan morphism is not unique. As a trivial application of the results of Section 6.2, for any fibered connection there exists in fact a unique reduced Poincaré-Cartan morphism. When $k = 2$ the reduced Poincaré-Cartan morphism is reduced for any other fibered connection, so that there exist infinitely many Poincaré-Cartan morphisms but only one is reduced; we select such a reduced Poincaré-Cartan morphism as a canonical representative. When $k \geq 3$, the Poincaré-Cartan morphism is not unique and there is a reduced representative which in principle varies with the fibered connection. Thus we have a whole family of Poincaré-Cartan morphisms, *a priori* one for each connection on the base manifold. Locally, each Poincaré-Cartan morphism is in the following form

$$\begin{aligned} &< \mathbb{F} \mid j^{k-1}X > = < \mathbb{F}(L) \mid j^{k-1}X > = \\ &=[f_i^\mu\, X^i + f_i^{\mu\,\lambda}\, X^i_\lambda + \ldots + f_i^{\mu\,\lambda_1\ldots\lambda_{k-1}}\, X^i_{\lambda_1\ldots\lambda_{k-1}}]\, \mathrm{d}s_\mu = \\ &=[\hat f_i^\mu\, X^i + \hat f_i^{\mu\,\lambda}\, \nabla_\lambda X^i + \ldots + \hat f_i^{\mu\,\lambda_1\ldots\lambda_{k-1}}\, \nabla_{\lambda_1\ldots\lambda_{k-1}} X^i]\, \mathrm{d}s_\mu = \end{aligned} \tag{6.3.8}$$

where $\nabla_{\lambda_1\ldots\lambda_s} X^i$ denote the symmetrized covariant derivatives with respect to the fibered connection. The coefficients $\hat f_i^{\mu\lambda_1\ldots\lambda_p}$ ($0 \leq p \leq k-1$ with the convention that $p = 0$ corresponds to the coefficient $\hat f_i^\mu$) are called the *covariant effective momenta* (with respect to the given connection). They are recursively defined as

$$\begin{cases} \hat f_i^{\mu_1\ldots\mu_k} = \hat p_i^{\mu_1\ldots\mu_k} \\ \hat f_i^{\mu_1\ldots\mu_{k-1}} = \hat p_i^{\mu_1\ldots\mu_{k-1}} - \nabla_\nu \hat f_i^{\nu\mu_1\ldots\mu_{k-1}} \\ \ldots \\ \hat f_i^\mu = \hat p_i^\mu - \nabla_\nu \hat f_i^{\nu\mu} \end{cases} \tag{6.3.9}$$

Notice the complete analogy with the quantities (6.4.6) defined in Section 6.4 below. The coefficients $(f_i^\mu, f_i^{\mu\,\lambda}, \ldots, f_i^{\mu\,\lambda_1\ldots\lambda_{k-1}})$ are non-covariant and they can depend on the fibered connection also.

In particular for $k = 2$ the canonical choice is

$$< \mathbb{F} \mid j^1 X > = [\hat f_i^\mu\, X^i + \hat f_i^{\mu\,\lambda}\, \nabla_\lambda X^i]\, \mathrm{d}s_\mu \tag{6.3.10}$$

with

$$\begin{cases} \hat f_i^{\lambda\mu} = \hat p_i^{\lambda\mu} \\ \hat f_i^\lambda = \hat p_i^\lambda - \mathrm{d}_\mu \hat p_i^{\lambda\mu} \end{cases} \tag{6.3.11}$$

One can now easily obtain the Euler-Lagrange equations from Hamilton's principle of stationary action as follows:

$$\begin{aligned}
\delta_X A_D(\rho) &= \int_D < \delta L \,|\, j^k X > \Big|_\rho = \\
&= \int_D < \mathbb{E} \,|\, X > \Big|_\rho + \int_D \mathrm{d} \left[< \mathbb{F} \,|\, j^{k-1} X > \Big|_\rho \right] = \\
&= \int_D < \mathbb{E} \,|\, X > \Big|_\rho + \int_{\partial D} < \mathbb{F} \,|\, j^{k-1} X > \Big|_\rho
\end{aligned} \tag{6.3.12}$$

having used Stoke's theorem $\int_D \mathrm{d}\theta = \int_{\partial D} \theta$.

The last integral vanishes since the integrand is calculated on ∂D where, by definition of deformation, we have $j^{k-1}X = 0$. Thence the following must hold

$$\int_D < \mathbb{E} \,|\, X > \Big|_\rho = 0 \tag{6.3.13}$$

in any m-region $D \subset M$ and for all deformations X. Since D is arbitrary, then the integrand $< \mathbb{E} \mid X > |_\rho$ itself must vanish. Since the deformation X is arbitrary we finally infer that the following equations must be satisfied:

$$\mathbb{E}(L) \circ j^{2k}\rho = 0 \tag{6.3.14}$$

In other words, a section $\rho : M \longrightarrow \mathcal{C}$ is a critical section if and only if $j^{2k}\rho(x)$ lies in the kernel of theEuler-Lagrange morphism. Consequently, the Euler-Lagrange morphism $\mathbb{E}$ defines a submanifold $\ker(\mathbb{E}) \subset J^{2k}\mathcal{C}$ which has to be interpreted as a system of partial differential equations of order $2k$ for the section ρ in the sense of definition (2.8.1). These equations are called *Euler-Lagrange equations*.

It is then easy to show that:

Theorem (6.3.15): if the Lagrangian is a formal m-divergence, i.e. there exists a morphism $\theta : J^{k-1}\mathcal{C} \longrightarrow A_{m-1}(M)$ such that $L = \mathrm{Div}\,(\theta)$, then any section $\rho : M \longrightarrow \mathcal{C}$ is critical.

Proof: it follows from the fact that the action is constant; in fact:

$$\begin{aligned}
A_D(\rho) &= \int_D L \circ j^k \rho = \\
&= \int_D \mathrm{Div}\,(\theta) \circ j^k \rho = \int_{\partial D} \theta \circ j^{k-1} \rho
\end{aligned} \tag{6.3.16}$$

Since any deformation X, by definition, keeps the section ρ unchanged on the boundary ∂D, then $\delta_X A_D(\rho) \equiv 0$ for any section ρ. ∎

This latter theorem shows that, as we already recalled above, the Lagrangian of a Lagrangian system of partial differential equations is not uniquely determined by the requirement that it has to produce the correct field equations. In fact, if L is a Lagrangian of order k then for any integer h and any morphism $\theta : J^{h-1}\mathcal{C} \longrightarrow A_{m-1}(M)$ one can define (by pull-back to the appropriate jet prolongation) another Lagrangian which, by an abuse of language, we shall denote by $L' = L + \text{Div}(\theta)$; this new Lagrangian is of order at most equal to $\max(k, h)$. The remark about the maximal order of $L + \text{Div}(\theta)$ is crucial; we say "at most", since, obviously, by choosing particular Lagrangians L and particular morphisms θ cancellations may occur which drastically reduce the order. [As a trivial example if $L = \text{Div}(\sigma)$, by choosing $\theta = -\sigma$ one gets $L' \equiv 0$ as new Lagrangian, which has of course order zero and trivial Euler-Lagrange equations.]

The two Lagrangians L and $L' = L + \text{Div}(\theta)$ have the same field equations, as it easily follows from the previous theorem (6.3.15). Thence L' has the same rights as the original Lagrangian L to be considered as *the* Lagrangian of the system of PDE's ensuing from L by (6.3.14). Whenever two Lagrangians L and L' on two (possibly different) jet prolongations of the same configuration bundle $\mathcal{C}$ produce the same Euler-Lagrange equations we say that the two given Lagrangians are *(strongly) equivalent* (see,e.g., [FR]). In the sequel we shall show that because of this ambiguity one can in fact provide a first order Lagrangian for General Relativity depending on the choice of a background.

4. Poincaré-Cartan Form

The variational formalism based on the Poincaré-Cartan form provides an alternative (equivalent) viewpoint for Mechanics and Field Theory. The Poincaré-Cartan form is particularly well suited to examine the relations between the Lagrangian and the Hamiltonian formalism and to deal with Lagrangian symmetries as well as with generalized symmetries. In particular the Lagrangian and the Poincaré-Cartan form differ by a contact form which is thence irrelevant once evaluation on a (not necessarily critical) section is performed. However, the additional information stored in the contact terms often improves the behavior of physical quantity at *bundle level* (i.e. *before* the evaluation on any section) still remaining completely equivalent along both critical and uncritical sections.

We remark that the evaluation along sections is not only peculiar of classical non-quantized models. One of the main differences between classical and quantum physics is that in a classical context evaluation is performed only along critical sections while in the quantum context the evaluation is performed along generic sections (think of path integral approach). In both cases the classical physical quantities are unaffected by contact terms. However, the formal improvement is already worth the effort of introducing the new formalism and moreover we stress that understanding the structure of these extra terms allows to better understand the various quantum counterparts of a given classical theory.

Axiomatics

We briefly recall here the main points of Poincaré-Cartan formalism. We also refer to [R00] for a more detailed exposition.

Definition (6.4.1): Let us consider a fiber bundle $\mathcal{C} = (C, M, \pi; F)$ over a manifold M of dimension $m = \dim(M)$. A *Poincaré-Cartan form* Θ *of order* k is an m-form over $J^{2k-1}\mathcal{C}$ which satisfies the following axioms:

(a) $\forall X, Y \in V(\pi^{2k-1})$, $i_X\, i_Y\, \Theta = 0$;[2]

(b) $\forall X \in V(\pi^{2k-1}_{k-1})$, $i_X\, \Theta = 0$;

(c) for any vertical vector field $X \in V(\pi^{2k-1}_0)$ the m-form $i_X\,(\mathrm{d}\Theta)$ is a contact form.

If the horizontal part of Θ is a Lagrangian form $L = \mathcal{L}\,\mathrm{d}s$, then we say that Θ is a *Poincaré-Cartan form associated to the Lagrangian* L and we write $\Theta = \Theta_L$. ■

Axiom (a) ensures that the form Θ is at most of first order in the *"vertical differentials"* $\mathrm{d}y^i_{\mu_1\ldots\mu_n}$. Axiom (b) ensures that no differentials $\mathrm{d}y^i_{\mu_1\ldots\mu_n}$ with $n \geq k$ enter the local expression of Θ. Consequently, any Poincaré-Cartan form Θ can be locally expressed as

$$\Theta = f\,\mathrm{d}s + [f^\mu_i\,\omega^i + \ldots + f^{\mu\alpha_1\ldots\alpha_{k-1}}_i\,\omega^i_{\alpha_1\ldots\alpha_{k-1}}] \wedge \mathrm{d}s_\mu \tag{6.4.2}$$

A priori the coefficients f, f^μ_i, ..., $f^{\mu\alpha_1\ldots\alpha_{k-1}}_i$ may depend on all the local coordinates in $J^{2k-1}\mathcal{C}$. Then the exterior differential $\mathrm{d}\Theta$ is expressed by

$$\begin{aligned}\mathrm{d}\Theta =& [\mathrm{d}f - f^\mu_i\,\omega^i_\mu - \ldots - f^{\mu\alpha_1\ldots\alpha_{k-1}}_i\,\omega^i_{\mu\alpha_1\ldots\alpha_{k-1}}] \wedge \mathrm{d}s +\\ &+ [\mathrm{d}f^\mu_i \wedge \omega^i + \ldots + \mathrm{d}f^{\mu\alpha_1\ldots\alpha_{k-1}}_i \wedge \omega^i_{\alpha_1\ldots\alpha_{k-1}}] \wedge \mathrm{d}s_\mu\end{aligned} \tag{6.4.3}$$

Let us now consider a vertical vector $X = X^i_\mu\,\partial^\mu_i + \ldots + X^i_{\mu_1\ldots\mu_{2k-1}}\,\partial^{\mu_1\ldots\mu_{2k-1}}_i \in V(\pi^{2k-1}_0)$. By the axiom (c) the horizontal part of $i_X\,(\mathrm{d}\Theta)$ vanishes. Thence we obtain

$$\begin{aligned}i_X\,\mathrm{d}\Theta =& [\partial^\mu_i f\,X^i_\mu + \ldots + \partial^{\mu_1\ldots\mu_{2k-1}}_i f\,X^i_{\mu_1\ldots\mu_{2k-1}}]\mathrm{d}s +\\ &- [\mathrm{d}_\mu f^{\mu\nu}_i\,X^i_\nu + \ldots + \mathrm{d}_\mu f^{\mu\mu_1\ldots\mu_{k-1}}_i\,X^i_{\mu_1\ldots\mu_{k-1}}]\mathrm{d}s +\\ &- [f^\mu_i\,X^i_\mu + \ldots + f^{\mu_1\ldots\mu_k}_i\,X^i_{\mu_1\ldots\mu_k}]\mathrm{d}s\\ &- [\partial^{\sigma_1\ldots\sigma_{2k-1}}_j f^{\mu\nu}_i\,X^i_\nu + \ldots +\\ &\quad + \partial^{\sigma_1\ldots\sigma_{2k-1}}_j f^{\mu\mu_1\ldots\mu_{k-1}}_i\,X^i_{\mu_1\ldots\mu_{k-1}}]\mathrm{d}y^j_{\sigma_1\ldots\sigma_{2k-1}} \wedge \mathrm{d}s_\mu +\\ &+ \text{contact terms}\end{aligned} \tag{6.4.4}$$

[2] Here i_X denotes as usual the contraction of a form along the vector field X.

Since the horizontal part has to vanish for all vertical vectors X, we obtain the conditions

$$\begin{cases} \partial_i^{\mu_1\dots\mu_{2k-1}} f = 0 \\ \dots \\ \partial_i^{\mu_1\dots\mu_{k+1}} f = 0 \end{cases} \tag{6.4.5}$$

These equations constrain the horizontal part of Θ, i.e. the coefficient f, to depend just on $(x^\mu, y^i, y^i_\mu, \dots, y^i_{\mu_1\dots\mu_k})$. It can be thence interpreted as a Lagrangian form of order k. Let us denote it by $L = \mathcal{L}\, \mathrm{d}s$ by setting $\mathcal{L} = f$. We stress that, as a consequence, the naive momenta of the Lagrangian are $p_i^{\mu_1\dots\mu_l} = \partial_i^{\mu_1\dots\mu_l}\mathcal{L}$ for all $0 \le l \le k$.

By the vanishing of the horizontal part of (6.4.4) we also obtain the following equations:

$$\begin{cases} f_i^{(\mu_1\dots\mu_k)} = p_i^{\mu_1\dots\mu_k} \\ f_i^{(\mu_1\dots\mu_{k-1})} = p_i^{\mu_1\dots\mu_k} - \mathrm{d}_\nu f_i^{\nu\mu_1\dots\mu_{k-1}} \\ \dots \\ f_i^\mu = p_i^\mu - \mathrm{d}_\nu f_i^{\nu\mu} \end{cases} \tag{6.4.6}$$

Notice the formal identity with equations (6.3.9) We can explicitly solve these equations by setting the ansatz

$$f_i^{\mu_1\dots\mu_k} = p_i^{\mu_1\dots\mu_k} + A_i^{\mu_1\dots\mu_k} \tag{6.4.7}$$

with $A_i^{(\mu_1\dots\mu_k)} = 0$. By substituting into the second equation of (6.4.6) we obtain

$$f_i^{\mu_1\dots\mu_{k-1}} = p_i^{\mu_1\dots\mu_k} - \mathrm{d}_\nu p_i^{\nu\mu_1\dots\mu_{k-1}} + A_i^{\mu_1\dots\mu_{k-1}} - \mathrm{d}_\nu A_i^{\nu\mu_1\dots\mu_{k-1}} \tag{6.4.8}$$

where again $A_i^{(\mu_1\dots\mu_{k-1})} = 0$. By iteration we obtain finally

$$\begin{cases} f_i^{\mu_1\dots\mu_k} = p_i^{\mu_1\dots\mu_k} + A_i^{\mu_1\dots\mu_k} \\ f_i^{\mu_1\dots\mu_{k-1}} = p_i^{\mu_1\dots\mu_{k-1}} - \mathrm{d}_\nu p_i^{\nu\mu_1\dots\mu_{k-1}} + A_i^{\mu_1\dots\mu_{k-1}} - \mathrm{d}_\nu A_i^{\nu\mu_1\dots\mu_{k-1}} \\ \dots \\ f_i^\mu = p_i^\mu - \mathrm{d}_\nu p_i^{\nu\mu} + \dots + (-1)^{k-1}\mathrm{d}_{\mu_1\dots\mu_{k-1}} p_i^{\mu_1\dots\mu_{k-1}\mu} + \\ \qquad - \mathrm{d}_\nu A_i^{\nu\mu} + \dots + (-1)^{k-1}\mathrm{d}_{\mu_1\dots\mu_{k-1}} A_i^{\mu_1\dots\mu_{k-1}\mu} \end{cases} \tag{6.4.9}$$

where all the coefficients $A_i^{\mu_1\dots\mu_h}$ have vanishing symmetric part.

In equation (6.4.4) one can recast the spurious terms involving $\mathrm{d}y^j_{\sigma_1\dots\sigma_{2k-1}} \wedge \mathrm{d}s_\mu$ by means of the identity

$$\mathrm{d}y^j_{\sigma_1\dots\sigma_{2k-1}} \wedge \mathrm{d}s_\mu = \omega^j_{\sigma_1\dots\sigma_{2k-1}} \wedge \mathrm{d}s_\mu + y^j_{\sigma_1\dots\sigma_{2k-1}\mu}\, \mathrm{d}s \tag{6.4.10}$$

Thence the coefficients of these terms have to vanish because of horizontality. This implies the following equations

$$\begin{cases} \partial_j^{\sigma_1\dots\sigma_{2k-1}} f_i^{\mu_1\dots\mu_k} = 0 \\ \partial_j^{\sigma_1\dots\sigma_{2k-1}} f_i^{\mu_1\dots\mu_{k-1}} = 0 \\ \dots \\ \partial_j^{\sigma_1\dots\sigma_{2k-1}} f_i^{\mu\nu} = 0 \end{cases} \tag{6.4.11}$$

Consequently, the coefficient $f_i^{\mu_1\dots\mu_k}$ (as well as $A_i^{\mu_1\dots\mu_k}$) may depend just on the first $2k-1$ derivatives. The same can be said for $f_i^{\mu_1\dots\mu_{k-1}}$. By comparing with the second equation in (6.4.9) we immediately obtain that $A_i^{\mu_1\dots\mu_k}$ (and consequently $f_i^{\mu_1\dots\mu_k}$) cannot depend on $y^i_{\mu_1\dots\mu_{2k-1}}$. By iteration we easily obtain at last that the coefficient $f_i^{\mu_1\dots\mu_k}$ depends on the derivatives of fields up to order k only, $f_i^{\mu_1\dots\mu_{k-1}}$ depends on the derivatives of fields up to order $k+1$ only, an so on up to f_i^μ which is the only coefficient to depend on the derivatives of fields up to the whole order $2k-1$. Thence we see that the axioms of Poincaré-Cartan forms very drastically constrain the local expressions allowed.

To summarize we have seen that (6.4.2) specializes to the following:

$$\Theta = \mathcal{L}(j^k y^i)\,\mathrm{d}s + [f_i^\mu(j^{2k-1}y^i)\,\omega^i + \dots + f_i^{\mu\alpha_1\dots\alpha_{k-1}}(j^k y^i)\,\omega^i_{\alpha_1\dots\alpha_{k-1}}] \wedge \mathrm{d}s_\mu \tag{6.4.12}$$

Another consequence of definition (6.4.1) is that the difference between the Poincaré-Cartan form and Lagrangian form is a contact form. Thence the same action $A_D(\rho)$ defined above by (6.1.2) can be recast as:

$$A_D(\rho) = \int_D (j^{2k-1}\rho)^*\Theta_L \tag{6.4.13}$$

since the contact terms do not affect the integral. The variation of the action along a vertical vector field X over $\mathcal{C}$ reads then as

$$\begin{aligned} \delta_X A_D(\rho) &= \int_D \frac{\mathrm{d}}{\mathrm{d}s}(j^{2k-1}\rho_s)^*\,\Theta_L = \int_D (j^{2k-1}\rho)^*\,\frac{\mathrm{d}}{\mathrm{d}s}(j^{2k-1}\Phi_s)^*\,\Theta_L = \\ &= \int_D (j^{2k-1}\rho)^*\,\pounds_{j^{2k-1}X}\Theta_L \end{aligned} \tag{6.4.14}$$

where for the last equality the definition of Lie derivative of forms has been used.

The variation of the action functional can be thence expressed by[3]

$$\delta_X A_D(\rho) = \int_D (j^{2k-1}\rho)^* i_{j^{2k-1}X}\,\mathrm{d}\Theta_L + \int_{\partial D} (j^{2k-1}\rho)^* i_{j^{2k-1}X}\,\Theta_L \tag{6.4.15}$$

[3] We use the standard identity $\pounds_Y = i_Y \circ \mathrm{d} + \mathrm{d} \circ i_Y$.

As a consequence a section ρ is critical if and only if one has

$$(j^{2k-1}\rho)^* \ [i_{J^{2k-1}X}\, \mathrm{d}\Theta_L] = 0 \tag{6.4.16}$$

for any vertical vector field X on $\mathcal{C}$. Equation (6.4.16) is thence totally equivalent to Euler-Lagrange equations (6.3.14). The Poincaré-Cartan form will be used again when dealing with symmetries and conserved quantities.

We stress that the same ambiguities in the definition of the Poincaré-Cartan morphism $\mathbb{F}$ reverberate in the definition of Poincaré-Cartan forms. In fact, in field theories of order $k = 1$ there exists a unique Poincaré-Cartan form, for $k = 2$ uniqueness is lost even if there exists a unique (reduced) canonical representative. For $k \geq 3$ also the possibility of choosing a canonical representative for the Poincaré-Cartan form depends explicitly on the fibered connection. As for Poincaré-Cartan morphisms, uniqueness is also achieved at any order in Mechanics or whenever $\dim(M) = 1$.

This is not a coincidence since we can build a Poincaré-Cartan form for each first variation formula. In fact, suppose we have a first variation formula, i.e. a global splitting

$$< \delta L \mid j^k X > = < \mathbb{E} \mid X > + \mathrm{Div} < \mathbb{F} \mid j^{k-1} X > \tag{6.4.17}$$

so that the corresponding Poincaré-Cartan morphism is locally expressed by equation (6.3.8). To such a variational morphism we can associate a contact m-form over $J^{2k-1}\mathcal{C}$ locally given by:

$$\mathbf{f} = \left[f^{\mu}_{a} \omega^a + f^{\mu\mu_2}_{a} \omega^a_{\mu_2} + \ldots + f^{\mu\mu_2\ldots\mu_k}_{a} \omega^a_{\mu_2\ldots\mu_k} \right] \wedge \mathrm{d}s_\mu \tag{6.4.18}$$

where $(\omega^a,\ \omega^a_\mu\ , \ldots , \omega^a_{\mu_2\ldots\mu_k})$ are contact 1-forms defined above by (2.3.3). We remark that the transformation rules of contact 1-forms together with the transformation rules of the coefficients of the Poincaré-Cartan morphisms automatically provide the globality of the form (6.4.18). We thus have the following:

Theorem (6.4.19): the form **f** as defined by (6.4.18) is a global form on $J^{2k-1}\mathcal{C}$

Proof: let us consider a fibered transformation on $\mathcal{C}$

$$\begin{cases} x'^\mu = \phi^\mu(x) \\ y'^i = \Phi^i(x, y) \end{cases} \tag{6.4.20}$$

and let us denote as usual the relevant Jacobians by $J^\mu_\nu = \partial_\nu \phi^\mu(x)$, $J^i_j = \partial_j y'^i$, $J^i_\mu = \partial_\mu y'^i$, and so on for higher orders.

The induced transformation rules of contact 1-forms are given by

$$\begin{cases} \omega'^i = J^i_j\, \omega^j \\ \omega'^i_{\mu_2} = \mathrm{d}_{\mu_2} J^i_j\, \omega^j + \partial^{\alpha_2}_j y'^i_{\mu_2}\, \omega^j_{\alpha_2} \\ \dots \\ \omega'^i_{\mu_2\dots\mu_k} = \partial_j y'^i_{\mu_2\dots\mu_k}\, \omega^j + \partial^{\alpha_2}_j y'^i_{\mu_2\dots\mu_k}\, \omega^j_{\alpha_2} + \dots + \\ \qquad + \dots + \partial^{\alpha_2\dots\alpha_k}_j y'^i_{\mu_2\dots\mu_k}\, \omega^j_{\alpha_2\dots\alpha_k} \end{cases} \tag{6.4.21}$$

On the other hand the coefficients of the Poincaré-Cartan morphism transform as follows:

$$\begin{cases} f^{\mu_1}_a = \mathcal{J}\bar{J}^{\mu_1}_{\alpha_1} \left(f'^{\alpha_1}_b \partial_a y'^b + \dots + f'^{\alpha_1\alpha_2\dots\alpha_k}_b \partial_a y'^b_{\alpha_2\dots\alpha_k} \right) \\ f^{\mu_1\mu_2}_a = \mathcal{J}\bar{J}^{\mu_1}_{\alpha_1} \left(f'^{\alpha_1\alpha_2}_b \partial^{\mu_2}_a y'^b_{\alpha_2} + \dots + f'^{\alpha_1\alpha_2\dots\alpha_k}_b \partial^{\mu_2}_a y'^b_{\alpha_2\dots\alpha_k} \right) \\ \dots \\ f^{\mu_1\mu_2\dots\mu_k}_a = \mathcal{J}\bar{J}^{\mu_1}_{\alpha_1} f'^{\alpha_1\alpha_2\dots\alpha_k}_b \partial^{\mu_2\dots\mu_k}_a y'^b_{\alpha_2\dots\alpha_k} \end{cases} \tag{6.4.22}$$

where $\mathcal{J}$ is the determinant of the Jacobian. Globality of the form $\mathbf{f}$ immediately follows. ∎

Corollary (6.4.23): $\Theta_L = (\pi^{2k-1}_k)^*(\mathcal{L}\,\mathrm{ds}) + \mathbf{f}$ is a global m-form on $J^{2k-1}\mathcal{C}$. ∎

Let us consider a projectable vector field $X = X^\mu\, \partial_\mu + X^i\, \partial_i$; locally, the quantity $i_{j^{2k-1}X}\,(\mathrm{d}\Theta)$ can be thence written as

$$\begin{aligned} i_{j^{2k-1}X}\,(\mathrm{d}\Theta) = \Big[& X^i\, (p_i - \mathrm{d}_\mu f^\mu_i) + X^i_\sigma\, (p^\sigma_i - \mathrm{d}_\mu f^{\mu\sigma}_i - f^\sigma_i) + \dots + \\ & + X^i_{\sigma_1\dots\sigma_{k-1}}\, (p^{\sigma_1\dots\sigma_{k-1}}_i - \mathrm{d}_\mu f^{\mu\sigma_1\dots\sigma_{k-1}}_i - f^{\sigma_1\dots\sigma_{k-1}}_i) + \\ & + X^i_{\sigma_1\dots\sigma_k}\, (p^{\sigma_1\dots\sigma_k}_i - f^{\sigma_1\dots\sigma_k}_i) \Big] \wedge \mathrm{ds} + (\text{contact terms}) = \\ = & X^i\, (p_i - \mathrm{d}_\mu f^\mu_i) \wedge \mathrm{ds} + (\text{contact terms}) \end{aligned} \tag{6.4.24}$$

where equations (6.4.6) have been used and where the coefficients of the Poincaré-Cartan form $\Theta = \mathcal{L}\,\mathrm{ds} + [f^\mu_i\, \omega^i + \dots + f^{\mu\alpha_1\dots\alpha_{k-1}}_i\, \omega^i_{\alpha_1\dots\alpha_{k-1}}] \wedge \mathrm{ds}_\mu$ are expressed by (6.4.9).

Thus using (6.4.9) the horizontal part of $i_{j^{2k-1}X}\,(\mathrm{d}\Theta)$ can be recast in the following form[4]

$$\begin{aligned} & X^i\, (p_i - \mathrm{d}_\mu f^\mu_i)\, \mathrm{ds} = \\ & = X^i\, (p_i - \mathrm{d}_\mu p^\mu_i + \mathrm{d}_{\mu\nu} p^{\mu\nu}_i + \dots + (-1)^k \mathrm{d}_{\mu_1\dots\mu_k} p^{\mu_1\dots\mu_k}_i)\, \mathrm{ds} \\ & \quad + X^i\, \left(d_{\mu\nu} A^{\mu\nu}_i + \dots + (-1)^k d_{\mu_1\dots\mu_k} A^{\mu_1\dots\mu_k}_i\right)\, \mathrm{ds} = \\ & = X^i\, (p_i - \mathrm{d}_\mu p^\mu_i + \mathrm{d}_{\mu\nu} p^{\mu\nu}_i + \dots + (-1)^k \mathrm{d}_{\mu_1\dots\mu_k} p^{\mu_1\dots\mu_k}_i)\, \mathrm{ds} \end{aligned} \tag{6.4.25}$$

[4] The terms involving $A^{\mu_1\dots\mu_s}_i$ vanish because of $A^{(\mu_1\dots\mu_s)}_i = 0$.

Notice that despite the Poincaré-Cartan form Θ_L associated to a Lagrangian L may be not uniquely defined, each one of the representatives for Θ_L identifies the Euler-Lagrange equations of the original Lagrangian L. We also remark that the same result can be obtained for an arbitrary vector field $\hat{X} \in \mathfrak{X}(J^{2k-1}\mathcal{C})$ which projects over $X \in \mathfrak{X}(\mathcal{C})$. This can be trivially obtained by considering the axiom (c). In fact, the vector $\left[j^{2k-1}X - \hat{X}\right] \in \mathfrak{X}(\pi_0^{2k-1})$ is vertical so that

$$i_{\hat{X}}\,\mathrm{d}\Theta = i_{j^{2k-1}X}\,\mathrm{d}\Theta - i_{(j^{2k-1}X-\hat{X})}\,\mathrm{d}\Theta = i_{j^{2k-1}X}\,(\mathrm{d}\Theta) + (\text{contact terms}) \quad (6.4.26)$$

An example: the Poincaré-Cartan form for second order theories

For example let us explicitly consider the case of a second order Lagrangian

$$L = \mathcal{L}(x^\mu, y^i, y^i_\mu, y^i_{\mu\nu})\,\mathrm{d}s \qquad (6.4.27)$$

which covers all the cases of physical interest. The first order case is obtained by making the obvious cancellations implied by $p_i^{\mu\nu} = 0$.

Let us fix a fibered connection $(\Gamma^\alpha_{\beta\mu}, \Gamma^a_{b\mu})$. We have:

$$\begin{aligned} < \delta L \mid j^k X > &= \left[p_a X^a + p_a^\mu X^a_\mu + p_a^{\mu\nu} X^a_{\mu\nu}\right] \otimes \mathrm{d}s = \\ &= \left[\hat{p}_a \hat{X}^a + \hat{p}_a^\mu \hat{X}^a_\mu + \hat{p}_a^{\mu\nu} \hat{X}^a_{\mu\nu}\right] \otimes \mathrm{d}s \end{aligned} \qquad (6.4.28)$$

where we defined:

$$\begin{cases} \hat{p}_a = p_a - p_b^\mu \Gamma^b_{\ a\mu} - p_c^{\mu\nu}(d_\nu \Gamma^c_{\ a\mu} - \Gamma^c_{\ b\mu}\Gamma^b_{\ a\nu}) \\ \hat{p}_a^\mu = p_a^\mu - 2p_b^{\mu\nu}\Gamma^b_{\ a\nu} + p_a^{\rho\sigma}\Gamma^\mu_{\ \rho\sigma} \\ \hat{p}_a^{\mu\nu} = p_a^{\mu\nu} \end{cases} \qquad (6.4.29)$$

One obtains the Euler-Lagrange morphism of L expressed as follows:

$$\begin{aligned} \mathbb{E} &= \left[p_i - d_\mu p_i^\mu + d_{\mu\nu} p_i^{\mu\nu}\right] \bar{\mathrm{d}}y^i \otimes \mathrm{d}s = \\ &= \left[\hat{p}_i - \nabla_\mu \hat{p}_i^\mu + \nabla_{\mu\nu}\hat{p}_i^{\mu\nu}\right] \bar{\mathrm{d}}y^i \otimes \mathrm{d}s \end{aligned} \qquad (6.4.30)$$

while the Poincaré-Cartan morphism of L is given by:

$$\begin{aligned} < \mathbb{F} \mid j^{k-1} X > &= \left[\hat{f}_a^\mu X^a + \hat{f}_a^{\mu\nu} \nabla_\nu X^a\right] \otimes \mathrm{d}s_\mu = \\ &= \left[f_a^\mu X^a + f_a^{\mu\nu} d_\nu X^a\right] \otimes \mathrm{d}s_\mu \end{aligned} \qquad (6.4.31)$$

where we set

$$\begin{aligned} \hat{f}_a^\mu &= \hat{p}_a^\mu - \nabla_\nu \hat{p}_a^{\mu\nu} = & f_a^\mu &= p_a^\mu - d_\nu p_a^{\mu\nu} \\ &= p_a^\mu - d_\nu p_a^{\mu\nu} - p_b^{\mu\nu}\Gamma^b_{\ a\nu} & f_a^{\mu\nu} &= p_a^{\mu\nu} \\ \hat{f}_a^{\mu\nu} &= \hat{p}_a^{\mu\nu} = p_a^{\mu\nu} \end{aligned} \qquad (6.4.32)$$

As we already remarked above, the Euler-Lagrange morphism is unique and its dependence in (6.4.30) on the fibered connection is just a computational tool which may turn to be useful.

As a consequence, the Poincaré-Cartan form of L is locally given by

$$\Theta_L = \mathcal{L}\, \mathrm{d}s + (p_a^\mu - d_\nu p_a^{\mu\nu})\omega^a \wedge \mathrm{d}s_\mu + p_a^{\mu\nu}\omega_\nu^a \wedge \mathrm{d}s_\mu \tag{6.4.33}$$

Of course also the Poincaré-Cartan form so obtained is independent of the fibered connection since we choose the canonical representative of the Poincaré-Cartan morphism, which, as discussed above, is independent of the fibered connection.

For first order theories one can simply set $p_a^{\mu\nu} = 0$ in the above expressions.

5. Symmetries and Nöther's Theorem

Definition (6.5.1): a *symmetry for the Poincaré-Cartan form* Θ is an automorphism $\boldsymbol{\Phi} = (\Phi, \phi)$ of the configuration bundle:

$$\begin{array}{ccc} C & \xrightarrow{\ \Phi\ } & C \\ {\scriptstyle\pi}\downarrow & & \downarrow{\scriptstyle\pi} \\ M & \xrightarrow{\ \phi\ } & M \end{array} \tag{6.5.2}$$

such that $(j^{2k-1}\Phi)^*\Theta = \Theta$, i.e. it leaves the Poincaré-Cartan form invariant. ■

Theorem (6.5.3): if (Φ, ϕ) is a symmetry for Θ then it brings solutions of (6.4.16) into solutions.

Proof: let $\rho : M \longrightarrow C$ be a solution of the Euler-Lagrange equations of Θ, i.e.:

$$\begin{array}{l} \forall D \text{ region in } M \text{ and } \forall X \text{ vertical vector field defined over } D \\ \text{such that } j^{k-1}X = 0 \text{ over } \partial D, \text{ one has } (j^{2k-1}\rho)^* i_{j^{2k-1}X}(\mathrm{d}\Theta) = 0 \end{array} \tag{6.5.4}$$

We may define a new section $\rho' = \Phi \circ \rho \circ \phi^{-1}$ by means of the symmetry (Φ, ϕ). We claim that ρ' is also a solution; in fact:

$$\begin{aligned} &(j^{2k-1}\rho')^* \left(i_{j^{2k-1}X}\mathrm{d}\Theta\right) = \\ &= (\phi^{-1})^*(j^{2k-1}\rho)^*(j^{2k-1}\Phi)^* \left(i_{j^{2k-1}X}\mathrm{d}\Theta\right) = \\ &= (\phi^{-1})^*(j^{2k-1}\rho)^* \left(i_{j^{2k-1}(\Phi_* X)}(j^{2k-1}\Phi)^*\mathrm{d}\Theta\right) = \\ &= (\phi^{-1})^*(j^{2k-1}\rho)^* \left(i_{j^{2k-1}(\Phi_* X)}\mathrm{d}\Theta\right) \end{aligned} \tag{6.5.5}$$

Thence:

$$\begin{aligned}&(j^{2k-1}\rho')^*\left(i_{j^{2k-1}X}\mathrm{d}\Theta\right) = (\phi^{-1})^*(j^{2k-1}\rho)^*\left(i_{j^{2k-1}(\Phi_* X)}\mathrm{d}\Theta\right) = 0 \Leftrightarrow\\ &\Leftrightarrow (j^{2k-1}\rho)^*\left(i_{j^{2k-1}X}\mathrm{d}\Theta\right) = 0\end{aligned} \tag{6.5.6}$$

■

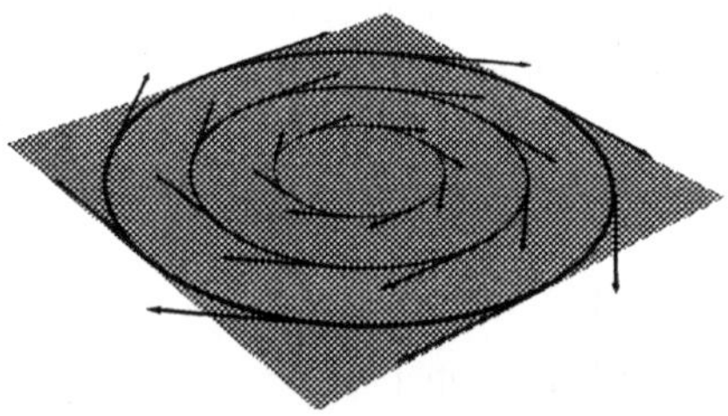

Fig. 19 – A symmetry for the plane harmonic oscillator.

Despite this is not the most general definition of symmetry that one can give for a system of PDEs related with a variational problem (see [AKO93]) it allows a general enough treatment of conserved quantities which covers most applications to fundamental physics.

Theorem (6.5.7): (Φ, ϕ) is a symmetry for Θ_L if and only if $\mathcal{J}\mathcal{L} \circ j^k\Phi = \mathcal{L}$ where $\mathcal{J} = \det(J)$ and J is the Jacobian of the diffeomorphism $\phi : M \longrightarrow M$ induced on the base manifold.

Proof: let (Φ, ϕ) be a symmetry and define $\mathcal{L} \circ j^k\Phi = \mathcal{L}'$. We have:

$$\begin{aligned}&(j^{2k-1}\Phi)^*\Theta_L =\\ &= \mathcal{L}'\mathrm{d}s' + \left[f'^{\mu_1}_i \omega'^i + f'^{\mu_1\mu_2}_i \omega'^i_{\mu_2} + \ldots + f'^{\mu_1\mu_2\ldots\mu_k}_i \omega'^i_{\mu_2\ldots\mu_k}\right] \wedge \mathrm{d}s'_{\mu_1} =\\ &= \mathcal{J}\mathcal{L}'\mathrm{d}s + \left[f^{\mu_1}_i \omega^i + f^{\mu_1\mu_2}_i \omega^i_{\mu_2} + \ldots + f^{\mu_1\mu_2\ldots\mu_k}_i \omega^i_{\mu_2\ldots\mu_k}\right] \wedge \mathrm{d}s_{\mu_1} =\\ &= \Theta_L\end{aligned} \tag{6.5.8}$$

Thence one has immediately $\mathcal{J}\mathcal{L}' = \mathcal{L}$.

Conversely, $\mathcal{J}\mathcal{L}' = \mathcal{L}$ implies:

$$\begin{aligned}&f^{\mu_1}_i = \mathcal{J}\bar{J}^{\mu_1}_{\alpha_1}\left(f'^{\alpha_1}_j \partial_i \varphi'^j + \ldots + f'^{\alpha_1\alpha_2\ldots\alpha_k}_j \partial_i \varphi'^j_{\alpha_2\ldots\alpha_k}\right)\\ &f^{\mu_1\mu_2}_i = \mathcal{J}\bar{J}^{\mu_1}_{\alpha_1}\left(f'^{\alpha_1\alpha_2}_j \partial^{\mu_2}_i y'^j_{\alpha_2} + \ldots + f'^{\alpha_1\alpha_2\ldots\alpha_k}_j \partial^{\mu_2}_i y'^j_{\alpha_2\ldots\alpha_k}\right)\\ &\ldots\\ &f^{\mu_1\mu_2\ldots\mu_k}_i = \mathcal{J}\bar{J}^{\mu_1}_{\alpha_1} f'^{\alpha_1\alpha_2\ldots\alpha_k}_j \partial^{\mu_2\ldots\mu_k}_i y'^j_{\alpha_2\ldots\alpha_k}\end{aligned} \tag{6.5.9}$$

which in turn imply $(j^{2k-1}\Phi)^*\Theta_L = \Theta_L$. ∎

Such a symmetry is also called a *Lagrangian symmetry*.

Let now $G \subset \mathrm{Aut}(\mathcal{C})$ be a subgroup of automorphisms of the configuration bundle.

Definition (6.5.10): a Lagrangian of order k is *G-covariant* if any automorphism $(\Phi, \phi) \in G$ is a Lagrangian symmetry. ∎

Consider then a 1-parameter subgroup (Φ_s, ϕ_s) of Lagrangian symmetries, locally given by:

$$\begin{cases} x'^\mu = \phi_s^\mu(x) \\ y'^i = \Phi_s^i(x, y) \end{cases} \qquad s \in \mathbb{R} \tag{6.5.11}$$

and denote by $\Xi = \xi^\mu \partial_\mu + \xi^a \partial_a$ its infinitesimal generator which is a projectable vector field $\Xi \in \mathfrak{X}_{proj}(\mathcal{C})$ projecting onto a vector field $\xi \in \mathfrak{X}(M)$.

The following holds:

Theorem (6.5.12): let Ξ be the infinitesimal generator of a group of Lagrangian symmetries. Then we have

$$< \delta L \mid j^k \pounds_\Xi y >= \mathrm{Div}\Big(i_\xi\, L\Big) \tag{6.5.13}$$

which is called the *covariance identity.*

Proof: since each (Φ_s, ϕ_s) is a symmetry, then we have $\mathcal{L} = \mathcal{JL} \circ j^k\phi_s$. Letting $s \longrightarrow 0$ we see that this implies the following infinitesimal condition:

$$0 = \Big(\partial_\mu \xi^\mu\Big)\mathcal{L} + \xi^\mu\,(\partial_\mu \mathcal{L}) + p_i\, \xi^i + p_i^\mu\, \xi^i_\mu + \ldots + p_i^{\mu_1 \ldots \mu_k}\, \xi^i_{\mu_1 \ldots \mu_k} \tag{6.5.14}$$

which can be recast as follows

$$\mathrm{d}_\mu(\mathcal{L}\xi^\mu) = p_i\, \Big(\pounds_\Xi y^i\Big) + p_i^\mu\, \Big(\pounds_\Xi y^i_\mu\Big) + \ldots + p_i^{\mu_1 \ldots \mu_k}\, \Big(\pounds_\Xi y^i_{\mu_1 \ldots \mu_k}\Big) \tag{6.5.15}$$

This is nothing but the local expression of (6.5.13) in local fibered natural coordinates. ∎

We stress that the intrinsic expression (6.5.13) can be much more useful than the local expression (6.5.15). Suppose in fact that the Lagrangian (let us say a first order one) depends explicitly on the derivatives of fields through some *a priori* given expression:

$$L = \mathcal{L}(x^\mu, y^i, R^A(y^i, y^i_\nu))\, \mathrm{d}s \tag{6.5.16}$$

as it often happens in Physics; e.g., in Gauge Theories, y^i are the components of the gauge potential and R^A are the components of the curvature.

Then the fundamental identity may be locally expressed as:

$$\mathrm{d}_\mu(\mathcal{L}\xi^\mu) = (\partial_i\mathcal{L})\left(\pounds_\Xi\varphi^i\right) + (\partial_A\mathcal{L})\left(\pounds_\Xi R^A\right) \tag{6.5.17}$$

where $\pounds_\Xi R^A$ is meant to be suitably expressed in terms of $\pounds_\Xi y^i$ and $\pounds_\Xi y^i_\mu$, while we set $\partial_A\mathcal{L} = \frac{\partial\mathcal{L}}{\partial R^A}$. Even if the functions R^A are complicated the expression (6.5.17) may be easier to treat than the expression (6.5.15) (see below for a number of examples). This is possibly due to the fact that the partial derivatives $\partial_A\mathcal{L}$ hide in themselves the dependence on first order derivatives y^i_μ and part of the dependence on the fields y^i. Such a possible simplification is paid in fact in the very definition of the Lie derivatives $\pounds_\Xi R^A$. On the other hand, such a Lie derivative is somehow insensitive to the expression of $R^A(y^i, y^i_\nu)$, being related only to the transformation rules of the object R^A itself. In Differential Geometry one may define a lot of objects which are complicated functions of the fields and their derivatives (e.g. the curvature Riemann tensor as a second order function of a given metric tensor) but which have beautiful and simple transformation rules (e.g. they are tensor fields). This is one of the reasons why Differential Geometry (as well as analogous techniques) may result to be very useful in Physics in general and in Field Theory in particular. They, in fact, provide guidelines on how to express physically relevant quantities.

The effective power of these naive remarks will be clear later, when we shall explicitly deal with physically relevant examples (see Section 7.5 for General Relativity, Section 6.7 for Gauge Theories and Section 10.1 for spinor fields). For a general treatment of this problem see [FR] and references quoted therein.

Using the covariance identity (6.5.13) and the first variation formula (6.3.6) we obtain:

$$< \mathbb{E} \mid \pounds_\Xi y > + \,\mathrm{Div} < \mathbb{F} \mid j^{k-1}\pounds_\Xi y > = \mathrm{Div}\,(i_\xi L) \tag{6.5.18}$$

which can be readily recast in the following more expressive form:

$$\mathrm{Div}\ [\mathcal{E}(L,\Xi)] = \mathcal{W}(L,\Xi) \tag{6.5.19}$$

where we set explicitly:

$$\begin{aligned} \mathcal{E}(L,\Xi) &= < \mathbb{F} \mid j^{k-1}\pounds_\Xi y > - i_\xi\, L \\ \mathcal{W}(L,\Xi) &= - < \mathbb{E} \mid \pounds_\Xi y > \end{aligned} \tag{6.5.20}$$

Equation (6.5.19) is called a *weak conservation law*. Equation (6.5.20) is one of the possible forms of the so-called *Nöther's theorem* which shows us how we can build, starting from the infinitesimal symmetry generator (Ξ, ξ), a horizontal $(m-1)$-form $\mathcal{E}(L,\Xi)$ over $j^{2k-1}\mathcal{C}$, called the *Nöther current* for the Lagrangian L. This $(m-1)$-form induces by *pull-back* along any section $\rho : M \longrightarrow \mathcal{C}$ an $(m-1)$-form over the base manifold M:

$$\mathcal{E}(L,\Xi,\rho) = (j^{2k-1}\rho)^*\mathcal{E}(L,\Xi) \tag{6.5.21}$$

which is closed whenever ρ is a solution of field equations. In fact, if ρ is a solution then $\mathcal{W}(L, \Xi, \rho) = (j^{2k}\rho)^*\mathcal{W}(L, \Xi) = 0$ because of the definition (6.5.20) of the horizontal m-form $\mathcal{W}(L, \Xi)$, which is thence called the *work form*, since it quantifies the lack in conservation of $\mathcal{E}(L, \Xi, \rho)$ when ρ is not critical.

Of course, whenever the Poincaré-Cartan morphism is not unique, then the Nöther current is not uniquely defined either. To encounter the first real mathematical ambiguities in the definition of Poincaré-Cartan morphism we already know that one should consider third order theories. However, also at the level of second order theories problems of ambiguity begin to occur. One can in fact choose a non-reduced representative

$$\bar{\mathbb{F}} = [(p_i^\mu - \mathrm{d}_\nu p_i^{\nu\mu} - \mathrm{d}_\nu A_i^{\nu\mu})\,\bar{\mathrm{d}}y^i + (p_i^{\mu\nu} + A_i^{\mu\nu})\,\bar{\mathrm{d}}\,y^i_\nu]\,\mathrm{d}s_\mu \tag{6.5.22}$$

for an arbitrary skew-symmetric term $A_i^{\mu\nu}$. As a consequence, in correspondence with $\bar{\mathbb{F}}$ one obtains another Nöther current $\bar{\mathcal{E}}(L, \Xi)$, which differs from the original one by a pure divergence

$$\begin{aligned}\bar{\mathcal{E}}(L,\Xi) - \mathcal{E}(L,\Xi) &= [-\mathrm{d}_\nu A_i^{\nu\mu}\,\pounds_\Xi y^i + A_i^{\mu\nu}\,\pounds_\Xi y^i_\nu]\,\mathrm{d}s_\mu = \\ &= [-\mathrm{d}_\nu A_i^{\nu\mu}\,\pounds_\Xi y^i + \mathrm{d}_\nu A_i^{\nu\mu}\,\pounds_\Xi y^i + \mathrm{d}_\nu(A_i^{\mu\nu}\,\pounds_\Xi y^i)]\,\mathrm{d}s_\mu = \\ &= \frac{1}{2}\mathrm{Div}\left[(A_i^{\mu\nu}\,\pounds_\Xi y^i)\,\mathrm{d}s_{\mu\nu}\right]\end{aligned} \tag{6.5.23}$$

We stress that this nice behavior is however peculiar of second order theories only. For example, at order 3 one has three new contributions

$$\begin{aligned}&\mathrm{d}_{\nu\lambda} A_i^{\nu\lambda\mu}\,\pounds_\Xi y^i - \mathrm{d}_\nu A_i^{\nu\mu\lambda}\,\pounds_\Xi y^i_\lambda + A_i^{\mu\nu\lambda}\,\pounds_\Xi y^i_{\nu\lambda} = \\ &= \mathrm{d}_\alpha[\mathrm{d}_\nu A_i^{\alpha\sigma\mu}\,\pounds_\Xi y^i + A_i^{\mu\alpha\nu}\,\pounds_\Xi y^i_\nu] = \mathrm{d}_\alpha\left[\mathrm{d}_\nu\left((A_i^{\alpha\nu\mu} + A_i^{\mu\alpha\nu})\,\pounds_\Xi y^i\right)\right] = \\ &= \mathrm{d}_\alpha\left[-\mathrm{d}_\nu\left(A_i^{\nu\mu\alpha}\,\pounds_\Xi y^i\right)\right] = \mathrm{d}_\alpha B^{\mu\alpha}\end{aligned} \tag{6.5.24}$$

where we set $B^{\mu\alpha} = -\mathrm{d}_\nu\left(A_i^{\nu\mu\alpha}\,\pounds_\Xi y^i\right)$. This last term is symmetric in the indices (μ, α) but, being it in general non vanishing, it cannot be skew-symmetric and thence cannot be the divergence of a variational morphism.

Thence in this last case the Nöther currents actually depend explicitly on the choice of the representative of the Poincaré-Cartan form. As a consequence one has to specify in any case a (possibly) canonical representative of the Poincaré-Cartan form to be chosen. The easiest way to do that is to fix *a priori* a fibered connection and to select the unique Poincaré-Cartan form which is reduced with respect to that fibered connection. As we said above, changing fibered connection amounts to change representative (remember that we are dealing with theories of order in principle $h \geq 3$) and thence Nöther currents. If we pretend the currents to be definitely fixed then we have to specify *canonically* the fibered connection. Despite this might seem to be a very strong and unphysical request, we shall see that in natural and gauge natural theories this can be done in fact in a reasonable way.

Another ambiguity in the definition of Nöther currents is originated by the two formalisms we have introduced. We have in fact two different prescriptions to obtain the Nöther currents. The first one is based on variational morphisms while the second one is based on Poincaré-Cartan forms. They are respectively:

$$\begin{aligned} \mathcal{E} &\equiv \mathcal{E}(L, \Xi) = < \mathbb{F} \mid j^{k-1} \pounds_{\Xi} y > - i_{\xi} L \\ \hat{\mathcal{E}} &\equiv \hat{\mathcal{E}}(L, \Xi) = i_{j^{2k-1}\Xi} \Theta_L \end{aligned} \tag{6.5.25}$$

However, we remark that these differ by a contact form so that they coincide once evaluated on any (arbitrary) section.

A further ambiguity is originated by the already discussed non-uniqueness of the Lagrangian. In fact, if L is a Lagrangian for the given system of PDEs and $\theta : J^h\mathcal{C} \longrightarrow A_{m-1}(M)$ is an arbitrary variational morphism, then, as we noticed above, the new Lagrangian $L' = L + \mathrm{Div}(\theta)$ is classically equivalent to L, i.e. they both produce the same field equations. *A priori*, however, the two Lagrangians produce two different sets of Nöther currents. Nevertheless, one can show in general that the two currents differ again for an exact form. For example, for $h = 1$ we have $\theta = \theta^\lambda(x^\mu, y^i, y^i_\mu)\, \mathrm{d}s_\lambda$ and $\mathrm{Div}(\theta) = (\mathrm{d}_\lambda \theta^\lambda)\, \mathrm{d}s$. The first variation formula for the Lagrangian $\mathrm{Div}(\theta)$ prescribes $\mathbb{E}[\mathrm{Div}(\theta)] \equiv 0$ and $\mathbb{F}[\mathrm{Div}(\theta)] = \delta\theta : J^1\mathcal{C} \longrightarrow V^*(J^1\mathcal{C}) \otimes A_{m-1}(M)$.

The covariance identity reads as

$$< \mathrm{Div}(\delta\theta) \mid j^1 \pounds_{\Xi} y > - \mathrm{Div}(i_\xi\, \delta\theta) = \mathrm{Div}(\alpha) \neq 0 \tag{6.5.26}$$

where we set $\alpha = \alpha^\mu\, \mathrm{d}s_\mu$ and

$$\alpha^\mu = \partial_i \theta^\mu\, \pounds_{\Xi} y^i + \ldots + \partial_i^{\sigma_1 \ldots \sigma_h} \theta^\mu\, \pounds_{\Xi} y^i_{\sigma_1 \ldots \sigma_h} - \xi^\mu \mathrm{d}_\sigma \theta^\sigma \tag{6.5.27}$$

Thus we see that in general a pure divergence Lagrangian $\mathrm{Div}(\theta)$ is not covariant with respect to arbitrary transformations, although they *are* symmetries in the sense they send critical sections into critical sections (each section is in fact critical for a pure divergence Lagrangian, which gives empty equations). This suggests the need of a wider definition of symmetries. First of all, let us notice that Nöther theorem relies on Lie derivatives, which in turn rely on the Lie dragging of objects. The possibility of dragging a section of a bundle is not necessarily related to vector fields of some prolongation $J^k\mathcal{C}$ (see Section 2.5 above). In fact, we can drag sections also along generalized vector fields (which are true vector fields only on $J^\infty\mathcal{C}$). In the second place, the covariance condition (6.5.13) is not the most general situation one can cope with. In fact, if one has (as above):

$$< \delta L \mid j^1 \pounds_{\Xi} y > - \mathrm{Div}(i_\xi\, L) = \mathrm{Div}(\alpha) + \omega\,, \qquad \omega \in \Omega_c \tag{6.5.28}$$

one obtains a weak conservation law by setting

$$\begin{cases} \mathcal{E}(L,\Xi) = <\mathbb{F}|\, j^{k-1}\pounds_\Xi y> -i_\xi\, L - \alpha \\ \mathcal{W}(L,\Xi) = -<\mathbb{E}|\, \pounds_\Xi y> \end{cases} \qquad \mathrm{Div}\, \mathcal{E}(L,\Xi) = \mathcal{W}(L,\Xi) \tag{6.5.29}$$

These remarks suggest the following:

Definition (6.5.30): an *(infinitesimal) generalized symmetry* for a Poincaré-Cartan form Θ is a generalized (projectable) vector field X such that

$$\pounds_{j\Xi}\, (\pi^\infty_{2k-1})^*\, \Theta = \mathrm{d}\alpha + \omega \tag{6.5.31}$$

where ω is a contact form and α is arbitrary.

Analogously, an *(infinitesimal) generalized symmetry* for a Lagrangian L is a generalized (projectable) vector field X such that there exists a variational morphism $\alpha : J^kC \longrightarrow V^*(J^{k-1}C) \otimes A_{m-1}(M)$ for which the following holds

$$<\delta L \mid j^k \pounds_\Xi y> -\mathrm{Div}(i_\xi\, L) = \mathrm{Div}\ <\alpha \mid j^{k-1} \pounds_\Xi y> + \omega \tag{6.5.32}$$

where ω is a variational morphism with values in contact forms. Of course these two notions of generalized symmetries are equivalent. ■

Nöther Theorem (6.5.33): if X is an infinitesimal generator of generalized symmetries for the Lagrangian L then there exists a *Nöther current* $\mathcal{E}(L,\Xi)$ and a *work form* $\mathcal{W}(L,\Xi)$ defined as in (6.5.29) such that a weak conservation law holds true. ■

By equation (6.5.26), if $L = \mathrm{Div}(\theta)$ is a pure divergence than any transformation turns out to be a generalized symmetry according to the definition (6.5.30).

In Mechanics (as more generally whenever $\dim(M) = m = 1$) this ends the matter. In this case, in fact, the weak conservation law $\mathrm{Div}\mathcal{E}(L,\Xi) = \mathcal{W}(L,\Xi)$ when evaluated along a critical section $\sigma : M \longrightarrow C$ gives:

$$\mathrm{d}\mathcal{E}(L,\Xi,\sigma) = 0 \qquad \begin{cases} \mathcal{E}(L,\Xi,\sigma) = (j^{2k-1}\sigma)^*\mathcal{E}(L,\Xi) \\ \mathcal{W}(L,\Xi,\sigma) = (j^{2k}\sigma)^*\mathcal{W}(L,\Xi) = 0 \end{cases} \tag{6.5.34}$$

Thus $\mathcal{E}(L,\Xi,\sigma)$ is a 0-form on the 1-dimensional manifold M which is closed on-shell, i.e. it is a function of one real variable which is constant along each motion curve. Thence it is a *first integral of motion* in the classical sense. In this case, thence, all the ambiguities in the definition of Nöther currents can be easily understood on physical grounds.

As well as we shall see below, in this 1-dimensional case the Nöther currents ensuing from different but equivalent Lagrangians all differ by an exact term, i.e. by a constant (the augmented de-Rham complex is used here[5]). This simply corresponds to a redefinition of the zero level of the corresponding first integral of motion. In fact, all first integrals produced by Nöther theorem (e.g. the energy, the linear momentum, the angular momentum, etc.; see Section 6.6 below) represent physical quantities which are *relative* in nature. As is well known, all of them have no absolute physical meaning, until their zero point is defined (i.e., the boundary conditions are kept fixed once for all).

6. Examples

Let us introduce here as an example all mechanical systems. Configurations are represented by points in a suitable *configuration space* Q. Local coordinates q^i on Q are the so-called *Lagrangian coordinates*. The evolution of the system is thence described by a curve $\gamma : \mathbb{R} \longrightarrow Q$ in the configuration space, or, almost equivalently, by a section of the trivial bundle $\mathcal{C} = (\mathbb{R}\times Q, \mathrm{pr}_1, \mathbb{R}; Q)$. Let us fix a global trivialization and denote by t the corresponding preferred coordinate in the base $\mathbb{R}$, which we conventionally assume to be the absolute time parameter. Changing the trivialization amounts to choose another time parameter; causality and logical coherence require the invertibility of the change of parameter $\tau = \tau(t)$. A Lagrangian L of order k is a bundle morphism $L : J^k\mathcal{C} \longrightarrow A_1(\mathbb{R})$; in view of the bundle isomorphism $J^k(\mathcal{C}) \simeq \mathbb{R} \times T^kQ$ induced by the given trivialization, the Lagrangian is locally expressed by

$$L = \mathcal{L}(t, q^i, u^i, a^i, \ldots)\, \mathrm{d}t \tag{6.6.1}$$

where the Lagrangian density depends on time t, on configuration variables q^i, on velocities u^i (i.e. first order derivatives of q^i), on accelerations a^i and on derivatives up to order k of the configuration variables q^i. As a particular case, we shall deal with first order Lagrangians

$$L = \mathcal{L}(t, q^i, u^i)\, \mathrm{d}t \tag{6.6.2}$$

which describe the classical case of time-dependent Mechanics. First order Lagrangians are particularly relevant to Physics since they produce second order equations of motion as it is expected as a consequence of Newton principles of dynamics.

---

[5] Augmented de-Rham complex is:

$$0 \longrightarrow \mathbb{R} \xrightarrow{\ i\ } \Lambda_0 \xrightarrow{\ \mathrm{d}\ } \Lambda_1 \xrightarrow{\ \mathrm{d}\ } \ldots \xrightarrow{\ \mathrm{d}\ } \Lambda_m \longrightarrow 0$$

Let us now consider an infinitesimal transformation $X = X^i\,\partial_i$ on the configuration space Q, i.e. a vertical vector field on $\mathcal{C} = \mathbb{R} \times Q$. According to the general theory developed above the field X is a Lagrangian symmetry if and only if the covariance identity (6.5.13) holds true. For a vertical vector field the covariance identity reduces in this case to the following condition:

$$\frac{\partial\mathcal{L}}{\partial q^i}\,\pounds_X q^i + \frac{\partial\mathcal{L}}{\partial u^i}\,\pounds_X u^i = 0 \qquad \Leftrightarrow \qquad \begin{cases} \pounds_X q^i = -X^i \\ \pounds_X u^i = \dfrac{\mathrm{d}}{\mathrm{d}t}\pounds_X q^i = -\dfrac{\mathrm{d}X^i}{\mathrm{d}t} \end{cases} \tag{6.6.3}$$

which can be recast into the more traditional form

$$0 = \frac{\partial\mathcal{L}}{\partial q^i}\,X^i + \frac{\partial\mathcal{L}}{\partial u^i}\,\frac{\mathrm{d}X^i}{\mathrm{d}t} = \hat{X}(\mathcal{L}) \tag{6.6.4}$$

where $\hat{X}$ denotes the natural lift of X to TQ or, equivalently, its lift to a vertical field over $J^1\mathcal{C} \simeq \mathbb{R} \times TQ$. Of course, the identity $\hat{X}(\mathcal{L}) = 0$ can be regarded under two viewpoints:

– *one can consider the Lagrangian L to be fixed and then the condition selects all the generators X of Lagrangian symmetries for that given L;*

– *one can instead fix the vector field X and look for all the Lagrangians which allow it as a symmetry generator.*

The two viewpoints are of course complementary. For example, specifying the condition (6.6.4) for the particular (local) vector field $X = \partial_i$ we obtain the identity

$$\frac{\partial\mathcal{L}}{\partial q^i} = 0 \tag{6.6.5}$$

This is identically satisfied if and only if the (local) coordinate q^i corresponding to the flow of $X = \partial_i$ is *cyclic*, i.e. it does not enter the Lagrangian density $\mathcal{L}$.

Consequently, whenever a coordinate q^i is cyclic, the corresponding (local) vector field $X = \partial_i$ is a generator of a Lagrangian symmetry. Then one has

$$0 = -\frac{\partial\mathcal{L}}{\partial q^i} = -\left(\frac{\partial\mathcal{L}}{\partial q^i} - \frac{\mathrm{d}}{\mathrm{d}t}\frac{\partial\mathcal{L}}{\partial u^i}\right) - \frac{\mathrm{d}}{\mathrm{d}t}\left(\frac{\partial\mathcal{L}}{\partial u^i}\right) \tag{6.6.6}$$

and the momentum

$$P \equiv \mathcal{E} = \frac{\partial\mathcal{L}}{\partial u^i} \tag{6.6.7}$$

is thence the Nöther current, which is conserved along motion curves.

Something more can be added by considering non-vertical transformations. To see this, let us in fact consider $\Xi = \partial_t$ (which is a good global vector field on $\mathcal{C}$ because of its triviality). Lie derivatives are

$$\begin{cases} \pounds_\Xi q^i = u^i \\ \pounds_\Xi u^i = a^i \end{cases} \tag{6.6.8}$$

Applying the method we see that a first order Lagrangian $L = \mathcal{L}(t, q^i, u^i)\, dt$ allows $\Xi \equiv \partial_t$ as a symmetry if and only if

$$\frac{d\mathcal{L}}{dt} = \frac{\partial \mathcal{L}}{\partial q^i}\, u^i + \frac{\partial \mathcal{L}}{\partial u^i}\frac{du^i}{dt}$$

i.e. whenever:

$$\frac{\partial \mathcal{L}}{\partial t} = 0 \tag{6.6.9}$$

as one should expect. In the sequel we shall consider a number of time- Lagrangians, so that ∂_t is a symmetry generator and the corresponding first integral of motion:

$$\mathcal{H} = \frac{\partial \mathcal{L}}{\partial u^i}\, u^i - \mathcal{L} \tag{6.6.10}$$

is called the *total energy.*

To such first order Lagrangians $L = \mathcal{L}(q^i, u^i)\, dt$ we can associate Poincaré-Cartan forms which are locally expressed by

$$\Theta_L = \mathcal{L}\, dt + p_i(dq^i - u^i\, dt) = -\mathcal{H}\, dt + p_i\, dq^i\,, \qquad p_i = \frac{\partial \mathcal{L}}{\partial u^i} \tag{6.6.11}$$

with $\partial_t \mathcal{H} = 0$. The simple examples in Mechanics and Field Theory which will be considered hereafter will be used as applications of the general theory above, as well as to introduce some standard techniques which will be later useful when dealing with more "interesting" systems.

The computations of the following Sections are often technically long to be carried over. We here limit ourselves to report just the main results which have been obtained by a direct application of the theoretical framework introduced above; we have to mention that the *Maple tensor package* was used to help in calculations in the more complicated cases.

The Free Particle System

Let us consider the Lagrangian for a free massive point

$$L = \tfrac{1}{2} g_{ij}\, u^i u^j \, \mathrm{d}t \tag{6.6.12}$$

where g_{ij} is a symmetric non-degenerate (usually positive definite) tensor, i.e. a metric over the configuration space Q. The Lagrangian density $\mathcal{L} = \frac{1}{2} g_{ij}\, u^i u^j$ is also called the *kinetic energy*. Here $i, j = 1, \dots, n = \dim(Q)$.

The Euler-Lagrange equations are thence

$$\ddot{q}^{\,i} + \Gamma^i{}_{jk}\, \dot{q}^j \dot{q}^k = 0 \tag{6.6.13}$$

i.e. they are the so-called *geodesic equations* of the (possibly pseudo-) Riemannian manifold (Q, g). Here $\Gamma^i{}_{jk} = \frac{1}{2} g^{il}(-\partial_l g_{jk} + \partial_j g_{kl} + \partial_k g_{lj})$ are the Christoffel symbols for the metric g. This reproduces the well-known fact that free particles move along geodesics on configuration spaces.

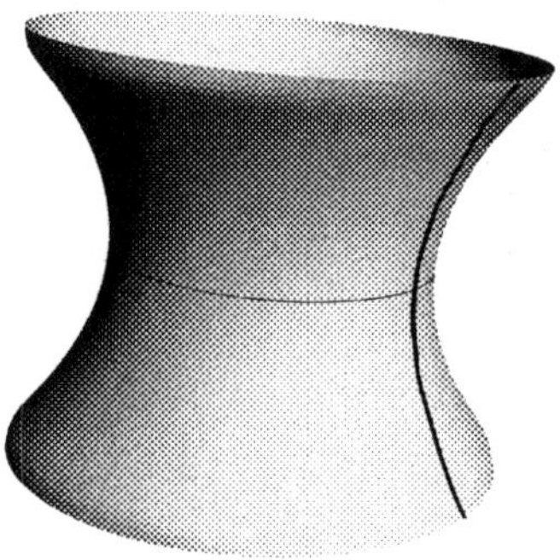

Fig. 20 – Geodesics.

Let us fix $Q = \mathbb{R}^3$ for the unconstrained particle and let us investigate which are the most general symmetries of the Lagrangian (6.6.12). Of course, part of the results mentioned hereafter generalize to an arbitrary dimension of Q. For simplicity let us consider Cartesian coordinates x^i on $\mathbb{R}^3$ in which we have $g_{ij} = m\, \delta_{ij}$, where m is the mass of the point. The velocities in Cartesian coordinates will be denoted by v^i.

The way to obtain generalized symmetries is to require that the Lie derivative of the Poincaré-Cartan form

$$\Theta_L = \tfrac{1}{2} m\, v_i v^i \, \mathrm{d}t - m\, v_i\, \omega^i\,, \qquad \omega^i = \mathrm{d}x^i - v^i \, \mathrm{d}t \tag{6.6.14}$$

is a closed form. This encompasses both the covariance conditions $\pounds_\Xi \Theta_L = 0$ and the more general condition $\pounds_\Xi \Theta_L = \mathrm{d}\alpha$.

For a candidate infinitesimal symmetry generator

$$\hat{\Xi} = \xi(t)\frac{\partial}{\partial t} + \xi^i(t,x)\frac{\partial}{\partial x^i} + \left(\frac{\mathrm{d}\xi^i}{\mathrm{d}t} - v^i\,\frac{\mathrm{d}\xi}{\mathrm{d}t}\right)\frac{\partial}{\mathrm{d}v^i} \qquad (6.6.15)$$

we have

$$\begin{aligned}\mathrm{d}\,(\pounds_{\hat{\Xi}}\Theta_L) =& \mathrm{d}\,i_{\hat{\Xi}}\,\mathrm{d}\Theta_L = m\Bigg[\left(\frac{\mathrm{d}^2\xi^i}{\mathrm{d}t^2} + \partial_i\xi_j\,a^j - \frac{\mathrm{d}}{\mathrm{d}t}(v_i\partial_t\xi)\right)\,\mathrm{d}t\wedge\omega^i + \\ &+\frac{\mathrm{d}}{\mathrm{d}t}(\partial_i\xi_j)\,\omega^i\wedge\omega^j + \left(\partial_j\xi_i - \partial_i\xi_j - \frac{\mathrm{d}\xi}{\mathrm{d}t}\delta_{ij}\right)\,\dot{\omega}^i\wedge\omega^j\Bigg] = 0\end{aligned} \qquad (6.6.16)$$

where we set $\dot{\omega}^i = \mathrm{d}v^i - a^i\,\mathrm{d}t$. Indices are raised and lowered by means of the metric $g_{ij} = \delta_{ij}$.

For this to be an identity it must be

$$\begin{cases}\dfrac{\mathrm{d}^2\xi^i}{\mathrm{d}t^2} + (\partial_i\xi_j)\,a^j - \dfrac{\mathrm{d}}{\mathrm{d}t}(v_i\partial_t\xi) = 0\\ \dfrac{\mathrm{d}}{\mathrm{d}t}(\partial_i\xi_j - \partial_j\xi_i) = 0\\ \partial_j\xi_i - \partial_i\xi_j - \dfrac{\mathrm{d}\xi}{\mathrm{d}t}\delta_{ij} = 0\end{cases} \qquad (6.6.17)$$

The first identity can be recast in the following form

$$\begin{aligned}\left(\partial_j\xi_i - \partial_i\xi_j - \frac{\mathrm{d}\xi}{\mathrm{d}t}\delta_{ij}\right)\,a^j + \partial_t^2\xi_i + \left(2\partial_t\partial_j\xi_i - \frac{\mathrm{d}^2\xi}{\mathrm{d}t^2x}\delta_{ij}\right)\,v^j + \\ + (\partial_j\partial_k\xi_i)\,v^jv^k = 0\end{aligned} \qquad (6.6.18)$$

The first term vanishes because of the third item of (6.6.17). For the other terms to vanish one has to have

$$\begin{cases}\xi^i = A^i + x^j\,A^i{}_j + t\,B^i + t\,x^j\,B^i{}_j\\ 2\partial_t\partial_j\xi_i = \dfrac{\mathrm{d}^2\xi}{\mathrm{d}t^2}\delta_{ij}\end{cases} \qquad \begin{aligned}&\Rightarrow\quad \partial_k\partial_t\,\xi_i = B_{ik}\\ &\Rightarrow\quad \xi = \alpha + \beta t + \tfrac{1}{2}\gamma t^2\\ &\qquad\ \ B_{ij} = \tfrac{1}{2}\gamma\delta_{ij}\end{aligned} \qquad (6.6.19)$$

where (α,β,γ) and $(A^i, A^i{}_j, B^i, B^i{}_j)$ are constants. With these expressions the second item of (6.6.17) is identically satisfied, while the third one can be recast as

$$A_{(ij)} = \tfrac{1}{2}\beta\delta_{ij} \qquad \Rightarrow \qquad A_{ij} = \tfrac{1}{2}\beta\delta_{ij} + \Omega_{ij} \qquad (6.6.20)$$

where Ω_{ij} is skew-symmetric and arbitrary. Thus the most general generalized symmetry for the free particle Lagrangian is

$$\begin{aligned}\Xi = \alpha\,\partial_t \oplus A^i\,\partial_i \oplus x_j\,\Omega^{ij}\,\partial_i \oplus t\,B^i\,\partial_i \oplus \\ \oplus\,\beta\,\left(t\,\partial_t + \tfrac{1}{2}x^i\partial_i\right) \oplus \tfrac{1}{2}\gamma\,\left(t^2\,\partial_t + tx^i\,\partial_i\right)\end{aligned} \qquad (6.6.21)$$

Let us analyze separately these generators, which in the case $\dim(Q) = 3$ are twelve.

– *Time translation* ∂_t: it is a standard Lagrangian symmetry since the Lagrangian is time independent. The Nöther current is the total energy

$$\mathcal{H} = \tfrac{1}{2}m\, v_i v^i \tag{6.6.22}$$

which is conserved along curves of motion.

– *Space translations* ∂_i: they are a standard Lagrangian symmetries since x^i are cyclic coordinates. The Nöther current is the *total momentum*

$$P_i = m v_i \tag{6.6.23}$$

which is a first integral of motion. In this case Nöther theorem reduces to the classical result which says that momentum is conserved in Mechanical systems which are invariant with respect to translations.

– *Space rotations* $\Xi_{ij} = (x_j \partial_i - x_i \partial_j)$: they are all standard symmetry generators. In fact, one has

$$\hat{\Xi}_{ij}(\mathcal{L}) = m\,(v_i v_j - v_j v_i) \equiv 0 \tag{6.6.24}$$

The corresponding conserved quantity is:

$$M_{ij} = m(x_i v_j - x_j v_i) \tag{6.6.25}$$

which is called the *angular momentum*. Notice that the generators Ξ_{ij} are generators of a representation of the algebra $\mathfrak{so}(3)$ of the rotation group SO(3) on $\mathbb{R}^3$. Angular momentum is thence related to spherical symmetry.

– *Galilei boosts* $(t\partial_i)$: they are infinitesimal generators of the transformations $x'^i = x^i + \epsilon^i\, t$ which are called *Galilei boosts*. They are *generalized symmetries* for the free mass point; in fact the generalized covariance identity reads as

$$m\, v_i = \frac{\mathrm{d}\alpha_i}{\mathrm{d}t}\,, \qquad\qquad \alpha_i = m\, x_i \tag{6.6.26}$$

Generalized Nöther theorem says that the quantity

$$M_{0i} = \frac{\partial \mathcal{L}}{\partial v^i} t - \alpha_i \equiv -m\,(x_i - v_i\, t) \tag{6.6.27}$$

is a first integral of motion. In fact, motion curves are $x^i = v_0^i t + x_0^i$ so that the quantity M_{0i} along a motion curve reads as

$$M_{0i} = -m\delta_{ij}(v_0^i t + x_0^i - v_0^i t) = -m\delta_{ij} x_0^i \tag{6.6.28}$$

which is clearly constant. The quantity M_{0i} is called the *motion of the center of mass* since its conservation implies that the center of mass moves with constant velocity. Of course in the case under consideration of one particle the center of mass coincides with the particle itself.

– *Time dilatation* $\Xi = (t\,\partial_t + \frac{1}{2}x^i\partial_i)$: it is the generator (at $\beta = 1$) of the transformations

$$\begin{cases} t' = \beta\, t \\ x'^i = \sqrt{\beta}\, x^i \end{cases} \tag{6.6.29}$$

It is a generalized symmetry since the covariance identity reduces to

$$mv_i(ta^i + \tfrac{1}{2}v^i) = \frac{\mathrm{d}\alpha}{\mathrm{d}t}\,, \qquad \alpha = \tfrac{m}{2}v_iv^it \tag{6.6.30}$$

The corresponding conserved quantity is

$$\begin{aligned} \mathcal{E} =& mv_i(tv^i + \tfrac{1}{2}x^i) - t\tfrac{m}{2}v_iv^i - \tfrac{m}{2}v_iv^it = \\ =& \tfrac{m}{2}v^i(x_i - v_it) = \tfrac{1}{2m}P^iM_{oi} \end{aligned} \tag{6.6.31}$$

Notice that the conserved quantity associated to this symmetry is however functionally dependent on the previously known first integrals of motion.

– *Time inversion* $\Xi = (t^2\,\partial_t + tx^i\,\partial_i)$: it is the generator of the transformations

$$\begin{cases} t' = -\dfrac{t}{\gamma t + 1} \\ x'^i = -\dfrac{x^i}{\gamma t + 1} \end{cases} \tag{6.6.32}$$

It is a generalized symmetry since the covariance identity reduces to

$$mv_i(t^2a^i - x^i + tv^i) = mv_i(tv^i + t^2a^i) + \frac{\mathrm{d}\alpha}{\mathrm{d}t}\,, \qquad \alpha = -\tfrac{m}{2}x_ix^i \tag{6.6.33}$$

The conserved quantity is

$$\begin{aligned} \mathcal{E} =& mv_i(t^2v^i - tx^i) - t^2\tfrac{m}{2}v_iv^i + \tfrac{m}{2}x_ix^i = \\ =& t^2\tfrac{m}{2}v_iv^i - mtv_ix^i + \tfrac{m}{2}x_ix^i = \tfrac{1}{2m}M_{0i}M_0{}^i \end{aligned} \tag{6.6.34}$$

which, also in this case, is functionally dependent on the previously known first integrals of motion.

In conclusion, we have shown that the Lagrangian of the free particle in $\mathbb{R}^3$ is covariant with respect to the Galilei group, i.e. with respect to transformations on $\mathbb{R} \times \mathbb{R}^3$ preserving both the absolute time and the Euclidean structure on $\mathbb{R}^3$. The Galilei group is composed by time translations, space translations, rotations and Galilei boosts defined by $x' = x + vt$.

Furthermore, time dilatation and time inversion are generalized symmetries, which do not introduce further constants. Thence, for $\dim(Q) = 3$, one recovers the correct numbers $10 = 12 - 2$ of constants of motion.

Classical Holonomic Systems

Mechanical systems are often subjected to external forces. Let us consider a Lagrangian of the form

$$L = \left[\tfrac{1}{2} g_{ij}\, u^i u^j + V(q, u)\right] \mathrm{d}t \tag{6.6.35}$$

where $V(q, u)$ is called a *generalized potential*. Let us also denote by

$$F_i = \frac{\partial V}{\partial q^i} - \frac{\mathrm{d}}{\mathrm{d}t}\frac{\partial V}{\partial u^i} \tag{6.6.36}$$

the so-called *Lagrangian forces*. The Euler-Lagrange equations read now as

$$\ddot{q}^i + \Gamma^i{}_{jk}\dot{q}^j\dot{q}^k = F^i_{\cdot} \tag{6.6.37}$$

where, as usual, we set $F^i_{\cdot} = g^{ij}\, F_j$ for the contravariant components of the Lagrangian forces.

One can easily prove that these equations are the same equations obtained by d'Alembert principle, i.e. by imposing the Newton principles of dynamics for (constrained) mechanical systems. Thence they represent the most genuine mechanical system one can conceive. One has also to impose that Lagrangian forces are independent of the accelerations, so that the potential $V(q, u)$, and as a consequence also the Lagrangian forces F_i, are at most linear in the velocities u^i. If the potential V is independent of the velocities then forces $F_i = \partial_i V$ are *potential* in the usual sense. For vanishing forces we see that equation (6.6.37) reduces to the geodesic equations which describe the evolution of the free mass particle.

As discussed above in the general case, if a Lagrangian coordinate q^i is cyclic one obtains conservation of the corresponding momenta

$$p_i = \frac{\partial \mathcal{L}}{\partial u^i} = g_{ij}\, u^i + \frac{\partial V}{\partial u^i} \tag{6.6.38}$$

For example, let us take $Q = \mathbb{R}^3$ and consider a point of mass m moving in a central force field described in polar coordinates (r, θ) in the plane of motion. The Lagrangian is in the form

$$L = \left[\tfrac{m}{2}(\dot{r}^2 + r^2\,\dot{\theta}^2) + V(r)\right] \mathrm{d}t \tag{6.6.39}$$

and the vector field $X = \partial_\theta$ is a symmetry since θ is cyclic. The conserved quantity so obtained is

$$\mathcal{E} = \frac{\partial \mathcal{L}}{\partial \dot{\theta}^i} = mr^2 \dot{\theta} \tag{6.6.40}$$

which is a well-known first integral of motion corresponding to the norm of the *angular momentum* (if, more generally, one starts with $\mathbb{Q} = \mathbb{R}^3$ and considers central forces, then the direction is conserved and it can be used to reduce the system on the plane of motion).

We can look for more general symmetries. Let us still take $Q = \mathbb{R}^2$ and specialize to the Kepler potential $V(r) = \kappa\, r^{-1}$. Consider the lift $\hat{\Xi}$ to $J^1 C$, with $C = (\mathbb{R} \times \mathbb{R}^2, \mathbb{R}, \mathrm{pr}_1; \mathbb{R}^2)$, of a generalized projectable vector field $\Xi = \xi(t)\partial_t + \xi^i(t, x, v)\partial_i$ (i=1,2); it is a generalized symmetry if

$$\pounds_{\hat{\Xi}} \Theta_L = \mathrm{d}\alpha + \omega\,, \qquad\qquad \omega \in \Omega_c \tag{6.6.41}$$

As a consequence, the horizontal part of $\pounds_{\hat{\Xi}} \Theta_L$ has to be a pure divergence, i.e. a total time derivative. One could easily write down explicitly the horizontal part of the Lie derivative of the Poincaré-Cartan form $\pounds_{\hat{\Xi}} \Theta_L$ for the Lagrangian (6.6.39) with $V(r) = \kappa\, r^{-1}$. How can one check if the (fairly complicated) result is a pure divergence? Of course, for a specific generator Ξ one can try to build an explicit primitive. This technique, of course, does not apply when Ξ is totally unknown. The answer relies on the fact that Euler-Lagrange equations of a pure divergence Lagrangian identically vanish. One can then consider $L' = \mathrm{hor}(\pounds_{\hat{\Xi}} \Theta_L)$ as an auxiliary Lagrangian and require its Euler-Lagrange equation to vanish identically.

If one deals with a generalized vector field of order h then one can compute $\pounds_{\hat{\Xi}} \Theta_L$ on $J^\infty C$ and realize that it projects on $J^h C$; consequently L' is regarded as a Lagrangian of order h on C. Let us thence consider a first order generalized vector field Ξ on $\mathbb{R} \times \mathbb{R}^2$; let us assume that the components ξ^i $(i, j = 1, 2)$ are polynomials of degree 1 in the velocities, i.e.:

$$\xi^i(t, x, v) = A^i(t, x) + A^i_j(t, x)\, v^j \tag{6.6.42}$$

The conditions for Ξ to be a generalized symmetry in polar coordinates (r, θ) imply that Ξ is in the following form

$$\begin{aligned}\Xi =& \mu\, \partial_t \oplus \lambda\, \partial_\theta \oplus \alpha\, [r^2 \sin(\theta)\, \dot{\theta}\, \partial_r + (2r \cos(\theta) + \dot{r} \sin(\theta))\, \partial_\theta] \oplus \\ &\oplus \beta[r^2 \cos(\theta)\, \dot{\theta}\, \partial_r + (-2r \sin(\theta) + \dot{r} \cos(\theta))\, \partial_\theta] \oplus \epsilon\, \dot{\theta}\, r^2\, \partial_\theta\end{aligned} \tag{6.6.43}$$

Let us analyze these symmetry generators one at a time.

– *Time translation* $\Xi = \partial_t$: it is an ordinary symmetry since the Lagrangian (6.6.39) is time . The corresponding conserved quantity is the total energy

$$\mathcal{H} = \tfrac{1}{2} m(\dot{r}^2 + r^2\, \dot{\theta}^2) - \tfrac{\kappa}{r} \tag{6.6.44}$$

– *Rotation* $\Xi = \partial_\theta$: it is an ordinary symmetry since the coordinate θ is cyclic. The corresponding conserved quantity is the norm of the angular momentum

$$p_\theta = mr^2\,\dot{\theta} \tag{6.6.45}$$

– *Runge-Lenz first component:*

$$\Xi = r^2\ \sin(\theta)\,\dot{\theta}\,\partial_r + (2r\ \cos(\theta) + \dot{r}\ \sin(\theta))\,\partial_\theta \tag{6.6.46}$$

it is a generalized symmetry with

$$\alpha = mr^3\cos(\theta)\dot{\theta}^2 + mr^2\sin(\theta)\dot{r}\dot{\theta} + \kappa\cos(\theta) \tag{6.6.47}$$

The corresponding conserved quantity is

$$L_x = m\dot{\theta}^2\cos(\theta)r^3 + m\dot{r}\dot{\theta}\sin(\theta)r^2 - \kappa\cos(\theta) \tag{6.6.48}$$

– *Runge-Lenz second component:*

$$\Xi = r^2\ \cos(\theta)\,\dot{\theta}\,\partial_r + (\dot{r}\ \cos(\theta) - 2r\ \sin(\theta))\,\partial_\theta \tag{6.6.49}$$

it is a generalized symmetry with

$$\alpha = -mr^3\sin(\theta)\dot{\theta}^2 + mr^2\cos(\theta)\dot{r}\dot{\theta} - \kappa\sin(\theta) \tag{6.6.50}$$

The corresponding conserved quantity is

$$L_y = m\dot{\theta}^2\sin(\theta)r^3 - m\dot{r}\dot{\theta}\cos(\theta)r^2 - \kappa\sin(\theta) \tag{6.6.51}$$

Their combination

$$\begin{aligned}\vec{L} =& L_x\,\vec{\imath} + L_y\,\vec{\jmath} = (mr^3\dot{\theta}^2 - k)\vec{u}_r - mr^2\dot{r}\dot{\theta}\vec{u}_\theta = \\ =& m\vec{v}\times(\vec{r}\times\vec{v}) - k\vec{u}_r\end{aligned} \tag{6.6.52}$$

is a vector which remains conserved for the Kepler system. Here $\vec{u}_r$ and $\vec{u}_\theta$ are the unit vectors in polar coordinates. It is called the *Runge-Lenz vector* or *Laplace vector*. It is related to the precession of perihelia and it is useful for the theoretical analysis of the famous test of General Relativity.

– $\Xi = \dot{\theta}\,r^2\,\partial_\theta$: it is a generalized symmetry with

$$\alpha = \tfrac{m}{2}r^4\dot{\theta}^2 \tag{6.6.53}$$

The corresponding conserved quantity is

$$-\tfrac{1}{2m}\,p_\theta^2 \tag{6.6.54}$$

which is a function of the previously known first integrals of motion.

Computations are not trivial but they can be carried over by computer algebra systems, as we already mentioned above.

Relativistic Particle

The principle of Relativity tells us that there is no preferred *time* to describe the evolution of a system. Each observer has its own different *time* and physical laws can be formulated under a form which does not depend on the observer. This is the celebrated *principle of covariance*. A first step towards the covariant formulation of Mechanics is based on the so-called *homogeneous formalism* in which time and space are treated on an equal footing as coordinates on *space-time*. In the homogeneous formalism the configuration space $\mathbb{R} \times \mathbb{R}^3$ for the free particle is replaced by $\mathbb{R}^4$, meaning that there is no preferred fibration onto $\mathbb{R}$ which in classical Mechanics is given by the choice of *absolute time*.

The history of a particle is described by a *trajectory* (not a parametrized curve, since no preferred parameter can be chosen *a priori*!) into spacetime. Of course, the homogeneous formalism *can* be used also in classical Newtonian Mechanics while in Relativistic Mechanics one *has to* use homogeneous formalism. In the classical case, however, being the absolute time observer-, histories in $\mathbb{R}^4$ canonically project onto curves in $\mathbb{R}^3$ while in Relativistic Mechanics each different observer uses his own *time* to regard the trajectory in $\mathbb{R}^4$ as the evolution of the system in $\mathbb{R}^3$.

Since we are interested in *trajectories in* $\mathbb{R}^4$ and variational calculus is best suited to provide critical (parametrized) curves, if we want to prescribe a variational principle for the relativistic particle one has to pretend that the Lagrangian is symmetric under arbitrary reparametrizations. In other words, one considers the space $\mathbb{R}$ where the parameter s lives. Notice however that s has lost its characterization of *time* but it is simply an arbitrary parameter along the trajectory. Curves in $\mathbb{R}^4$ are identified as section of the bundle $\mathcal{C} = (\mathbb{R} \times \mathbb{R}^4, \mathrm{pr}_1, \mathbb{R}; \mathbb{R}^4)$ and histories are maps $\gamma : \mathbb{R} \longrightarrow \mathbb{R}^4$. The Lagrangian $L = \mathcal{L}(s, q^\mu, u^\mu)\, \mathrm{d}s$ is a horizontal 1-form over $T(\mathcal{C})$, where (q^μ, u^μ) are natural coordinates on $T\mathbb{R}^4$ and $\mu = 0, 1, 2, 3$. Any arbitrary reparametrization

$$\begin{cases} s' = T(s) \\ q'^i = q^i \\ u'^i = \bar{\tau}\, u^i \end{cases} \tag{6.6.55}$$

with $\tau = \mathrm{d}_s T(s)$ and $\bar{\tau}$ its inverse, is thence required to be a Lagrangian symmetry. As a consequence if a curve γ is critical then any reparametrization of the same trajectory is critical as well. Euler-Lagrange equations identify classes of curves which uniquely determine trajectories.

Of course the classical Lagrangian for the free point of mass m, i.e.

$$L = \frac{1}{2} m(\delta_{ij}\, v^i v^j)\, \mathrm{d}s \tag{6.6.56}$$

is not covariant with respect to arbitrary reparametrizations (6.6.55) since

$$L' = \frac{m}{2}\delta_{ij}\, v'^i v'^j\, \mathrm{d}s' = \bar{\tau}\left(\frac{m}{2}\delta_{ij}\, v^i v^j\, \mathrm{d}s\right) = \bar{\tau} L \tag{6.6.57}$$

which differs from L unless one restricts to transformations which do not change the time scale factor (which is not enough for our present purposes).

According to Special Relativity, the spacetime $\mathbb{R}^4$ is equipped with the pseudo-Riemannian (flat) metric η of signature $(1,3)$ which in Cartesian coordinates reads as

$$\eta = \mathrm{d}t^2 - (\mathrm{d}x^2 + \mathrm{d}y^2 + \mathrm{d}z^2) = \eta_{\mu\nu}\, \mathrm{d}x^\mu \otimes \mathrm{d}x^\nu \tag{6.6.58}$$

The pseudo-metric space $(\mathbb{R}^4, \eta)$ is called *Minkowski space.* In arbitrary coordinates q^μ over $\mathbb{R}^4$ the Minkowski metric will read as $\eta = g_{\mu\nu}\, \mathrm{d}q^\mu \otimes \mathrm{d}q^\nu$, with $g_{\mu\nu} = \bar{J}^\rho_\mu\, \eta_{\rho\sigma}\, \bar{J}^\sigma_\nu$ where $\bar{J}^\rho_\mu$ is the (inverse) Jacobian connecting the arbitrary coordinates (q^μ) to a system of Cartesian coordinates (x^μ).

One can thence consider the new Lagrangian for the relativistic free particle

$$L = \sqrt{|\, g_{\alpha\beta}\, u^\alpha u^\beta\, |}\, \mathrm{d}s \tag{6.6.59}$$

Notice that the pseudo-metric η is not definite positive. As a consequence a curve may have positive, null or negative (scalar) velocity. Such curves are called *timelike*, *lightlike* or *spacelike*, respectively.

The Lagrangian (6.6.59) is covariant with respect to arbitrary reparametrizations (6.6.55) since the action functional

$$A_{[a,b]}\sigma = \int_a^b \sqrt{|\, g_{\alpha\beta}\, u^\alpha u^\beta\, |}\, \mathrm{d}s \tag{6.6.60}$$

is the arc-length in Minkowski space.

Euler-Lagrange equations read as

$$\frac{\mathrm{d}}{\mathrm{d}s}\left(\frac{g_{\mu\nu}\, u^\nu}{\sqrt{|\, g_{\alpha\beta}\, u^\alpha u^\beta\, |}}\right) = \frac{\partial_\mu g_{\nu\sigma}\, u^\nu u^\sigma}{2\sqrt{|\, g_{\alpha\beta}\, u^\alpha u^\beta\, |}} \tag{6.6.61}$$

which can be recast as

$$\frac{a^\lambda + \Gamma^\lambda_{\nu\sigma}\, u^\nu u^\sigma}{\sqrt{|\, g_{\alpha\beta}\, u^\alpha u^\beta\, |}} = \frac{\mathrm{d}}{\mathrm{d}s}\left(\frac{-1}{\sqrt{|\, g_{\alpha\beta}\, u^\alpha u^\beta\, |}}\right) u^\lambda \tag{6.6.62}$$

This is of course different from the geodesic equation on Minkowski space, due to the non-vanishing right-hand side. However, this is exactly the place where symmetries enter. For any trajectory one can use the arc-length parametrization in which $|g_{\alpha\beta}\, u^\alpha u^\beta| = 1$. We stress in fact that the geodesic equations (6.6.13) do not select geodesic trajectories but *uniform motions along geodesics*. By choosing the arc-length parametrization we are checking if such uniform motions are indeed critical. Because of the condition $|g_{\alpha\beta}\, u^\alpha u^\beta| = 1$, equation (6.6.62) simplifies to

$$a^\lambda + \Gamma^\lambda_{\nu\sigma}\, u^\nu u^\sigma = 0 \tag{6.6.63}$$

which is satisfied if the trajectory is geodesic. Then *any* curve supported on a geodesic trajectory is critical.

We stress that all these considerations hold for an arbitrary parameter. In particular any observer can use adapted coordinates in which the first coordinate $s = x^0 \equiv t$ is its own time coordinate. In this way $u^0 = 1$ and u^i can be identified with the velocities observed. In Cartesian coordinates $x^i = q^i$ and $v^i = u^i$, the equation of motion (6.6.61) reads now as

$$\begin{cases} \dfrac{\mathrm{d}}{\mathrm{d}t}\left(\dfrac{1}{\sqrt{|\,1 - |\vec{v}|^2\,|}}\right) = 0 \\ \dfrac{\mathrm{d}}{\mathrm{d}t}\left(\dfrac{v^i}{\sqrt{|\,1 - |\vec{v}|^2\,|}}\right) = 0 \qquad \Rightarrow \qquad a^i = 0 \end{cases} \tag{6.6.64}$$

i.e. the motion of the particle is *"linear and uniform"*. In these special coordinates the Lagrangian (6.6.59) reads as

$$L = \sqrt{|\,1 - |\vec{v}|^2\,|}\,\mathrm{d}t \tag{6.6.65}$$

which can be considered as a Lagrangian with 3 degrees of freedom. One can try to calculate the Hamiltonian of such a Lagrangian. Momenta are

$$p_i = -\frac{v_i}{\sqrt{|\,1 - |\vec{v}|^2\,|}}\,, \qquad \sqrt{1 + |\vec{p}|^2} = \frac{1}{\sqrt{|\,1 - |\vec{v}|^2\,|}}\,, \qquad v^i = -\frac{p^i}{\sqrt{1 + |\vec{p}|^2}} \tag{6.6.66}$$

so that the Hamiltonian is

$$H = -\sqrt{1 + |\vec{p}|^2} \tag{6.6.67}$$

However, doing that one looses manifest covariance. Here space and time are not treated on equal footing any longer. The parameter t has been frozen to be the proper time of a fixed observer and cannot be changed. Of course, the result does not depend on the observer in the end but to prove it one should transform everything (time included!) into quantities measured by another arbitrary observer.

This is why one wished to perform a Legendre transformation directly on the Lagrangian (6.6.59), which is manifestly covariant and in which all objects are well-defined from a four-dimensional viewpoint. However, this is not a trivial task. In fact, in the homogeneous formalism the momenta are

$$p_\mu = \frac{g_{\mu\nu}\, u^\nu}{\sqrt{|\, g_{\alpha\beta}\, u^\alpha u^\beta\, |}} \tag{6.6.68}$$

From this one obtains

$$|p|^2 = \frac{|u|^2}{|\, g_{\alpha\beta}\, u^\alpha u^\beta\, |} = 1 \tag{6.6.69}$$

which obviously cannot be inverted to give u^μ as a function of the momenta p_σ. In other words, the Lagrangian (6.6.59) is degenerate and one cannot rely on the standard theory of Legendre transformations.

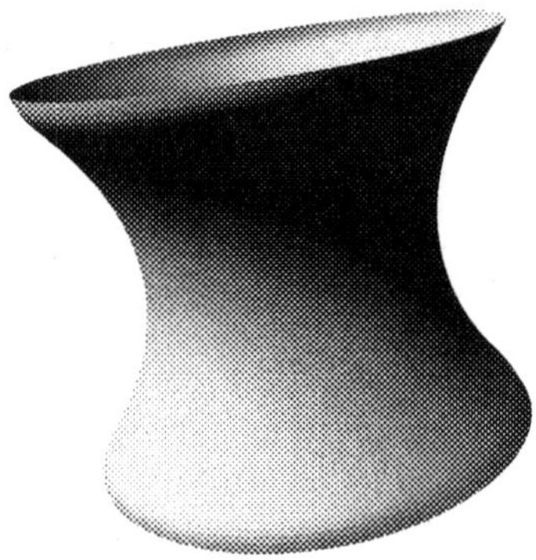

Fig. 21 – The mass shell $|p|^2 = 1$.

However, there is a way out: one can extend the theory of Legendre transformations to degenerate Lagrangians and in this way one preserves covariance. As it happens also for regular Lagrangians, a prominent role in the appropriate setting for this generalization is played by Poincaré-Cartan forms. One can in fact define a degenerate Legendre map $\Phi_L : \mathbb{R} \times TM \longrightarrow \mathbb{R} \times T^*M$ which is not surjective because of the constraints. For example, in the case under investigation, Φ_L restricts to a map $\Phi_L : \mathbb{R} \times TM \longrightarrow \mathcal{C}$, where we set $\mathcal{C} = \{(q, p) : |p|^2 = 1\} \subset T^*M$ for the constraint (6.6.69).

The Poincaré-Cartan form Θ_L on $\mathbb{R} \times TM$ is the pull-back of a Poincaré-Cartan form Θ on $\mathcal{C}$. The differential $\Omega = \mathrm{d}\Theta$ is a presymplectic 2-form. On that basis one starts a Dirac reduction algorithm and then one defines a symplectic submanifold on which Hamilton equations are defined. Alternatively (as well as more conveniently in our perspective), one can use the (unique) Poincaré-Cartan form Θ on $\mathcal{C}$ to define equation of motions on $\mathcal{C}$. This last viewpoint can be easily generalized to field theories and it is called *covariant Hamiltonian formalism*. We shall not further treat here the Hamiltonian formalism; we refer the reader to [FR].

In any case, here we are interested in the example of relativistic particle for another reason. As we shall see this situation is quite typical also in field theory and in particular in General Relativity as well as in Gauge Theories. Here we considered a system with quite a big symmetry group (for now arbitrary reparametrizations). The Lagrangian (6.6.59) of this system is covariant with respect to this group transformation but it is degenerate. One can thence break the symmetry by choosing a suitable reference, so loosing covariance but gaining regularity. This procedure is called, in a general sense, a *gauge fixing*. Regularity is important to define a Hamiltonian formalism as well as for a number of other (important and fundamental) matters (such as, for example, defining propagators and quantization of the system or computing conserved quantities and, in general, observables). On the other hand, once one obtains such quantities after fixing the gauge one has to prove *a posteriori* that they are independent of the gauge fixing. In fact, the gauge fixing is a trick *we* introduce and physical quantities cannot depend on which particular gauge we have fixed (unless there is a canonical gauge fixing, which is hardly ever the case).

The situation is completely analogous to what happens in Differential Geometry when one chooses coordinates (which can be regarded as a gauge fixing) to define intrinsic quantities and thence one has to prove that nothing depends in the end on the choice of the coordinates themselves (or, better to say, apart from their natural transformation rules in the sense of Chapter 4). If the quantity depends instead non-naturally on the choice of coordinates, then what we defined is not a *good* quantity for Differential Geometry since it is directly related, e.g., to the specific manifold we are studying. These considerations hide most of what we mean by *Differential Geometry* (as well as, *mutatis mutanda*, what we mean by Physics) and which is usually understood in the first claim of virtually any Introduction to the subject: Differential Geometry studies *curves, surfaces* and so on.

Let us thence consider a projectable vector field $\Xi = \xi(s)\partial_s + \xi^\mu(s,x)\,\partial_\mu$ on $\mathbb{R} \times \mathbb{R}^4$. The covariance identity for Ξ can be recast as

$$0 = (\nabla_\nu \xi_\mu)\, u^\mu u^\nu + u_\lambda \frac{\partial \xi^\lambda}{\partial s} \tag{6.6.70}$$

which is an identity precisely for an arbitrary $\xi(s)$ (which corresponds thence to general reparametrization invariance) if ξ^λ does not depend on s, thus inducing a vector field X on Minkowski space such that $\pounds_X g = 0$ (i.e. if X is a Killing vector). These latter symmetries correspond to Poincaré invariance of the theory.[6]

Let us consider invariance with respect to reparametrizations (6.6.55). Infinitesimally they correspond to vector fields $\Xi = \xi(s)\,\partial_s$. The covariance identity is in this case the following

$$p_\mu\,\pounds_\Xi u^\mu = \frac{g_{\mu\nu}\,u^\nu}{\sqrt{|\,g_{\alpha\beta}\,u^\alpha u^\beta\,|}}\,\frac{\mathrm{d}}{\mathrm{d}s}(\xi u^\mu) = \frac{\mathrm{d}}{\mathrm{d}s}\left(\xi\sqrt{|\,g_{\alpha\beta}\,u^\alpha u^\beta\,|}\right) = \pounds_\xi L \tag{6.6.71}$$

One can prove that this is actually an identity; in fact it can be recast as:

$$\frac{g_{\mu\nu}\,u^\nu}{\sqrt{|\,g_{\alpha\beta}\,u^\alpha u^\beta\,|}}\left(\frac{\mathrm{d}\xi}{\mathrm{d}s}u^\mu + \xi a^\mu\right) = \frac{\mathrm{d}\xi}{\mathrm{d}s}\sqrt{|\,g_{\alpha\beta}\,u^\alpha u^\beta\,|} + \xi\frac{g_{\mu\nu}\,a^\mu u^\nu}{\sqrt{|\,g_{\alpha\beta}\,u^\alpha u^\beta\,|}} \tag{6.6.72}$$

Thence $\Xi = \xi(s)\,\partial_s$ is an infinitesimal generator of Lagrangian symmetries. The corresponding Nöther current is

$$\mathcal{E} = \frac{g_{\mu\nu}\,u^\nu}{\sqrt{|\,g_{\alpha\beta}\,u^\alpha u^\beta\,|}}\xi u^\mu - \xi\sqrt{|\,g_{\alpha\beta}\,u^\alpha u^\beta\,|} \equiv 0 \tag{6.6.73}$$

This is clearly *"conserved"* along critical curves, being off-shell constant. Thus Nöther theorem is not condemned to produce physically relevant first integrals of motion.

We remark that the triviality of this latter Nöther current is deeply related to the degeneration of the Lagrangian discussed above; in fact, if the Lagrangian were non-degenerate the quantity (6.6.73) would have resulted to be the Lagrangian counterpart of the Hamiltonian.

The Lagrangian (6.6.59) is also covariant with respect to Poincaré transformations, i.e. with respect to isometries of Minkowski space. If x^μ are Cartesian coordinates on Minkowski space then a Lorentz transformation is

$$x'^\mu = \Lambda^\mu_\nu\, x^\nu\,, \qquad\qquad \Lambda^\mu_\rho\,\eta_{\mu\nu}\,\Lambda^\nu_\sigma = \eta_{\rho\sigma} \tag{6.6.74}$$

to which one has to add translations to recover the whole Poincaré group

$$x'^\mu = x^\mu + x^\mu_0 \tag{6.6.75}$$

[6] An *isometry* of a pseudo-Riemannian manifold (M, g) is a transformation $\phi : M \longrightarrow M$ which preserves the metric, i.e. $\phi^*(g) = g$. A *Killing vector* is an infinitesimal generator of isometries. A vector field $X = X^\mu\,\partial_\mu$ is Killing if it satisfies the following equation

$$\nabla^\mu X^\nu + \nabla^\nu X^\mu = 0$$

which is equivalent to $\pounds_\xi g = 0$. Here the covariant derivative is induced by the Levi-Civita connection of g.

Infinitesimal generators are thence vector fields on Minkowski space which read as follows

$$\begin{cases} \Xi = \xi^\mu_\nu \, x^\nu \, \dfrac{\partial}{\partial x^\mu} \,, & \text{where } \eta_{\mu(\sigma} \, \xi^\mu_{\rho)} = 0 \\ \Xi = \dfrac{\partial}{\partial x^\mu} \end{cases} \tag{6.6.76}$$

For Lorentz transformations the covariance identity is now

$$-\frac{1}{\sqrt{|\, \eta_{\alpha\beta} \, v^\alpha v^\beta \,|}} \, \eta_{\mu\sigma} \, \xi^\mu_\lambda v^\lambda v^\sigma = 0 \tag{6.6.77}$$

which holds because of the antisymmetric quantity $\eta_{\mu\sigma} \, \xi^\mu_\lambda$ which appears to be saturated with the symmetric object $v^\lambda v^\sigma$. The covariance identity for translations reduces to a trivial $0 = 0$. The Nöther current for Lorentz transformations corresponding to a generator $(\xi_\alpha\beta)^\mu_\lambda = \eta_{\alpha\lambda}\delta^\mu_\beta - \eta_{\beta\lambda}\delta^\mu_\alpha$ is instead

$$\mathcal{E} = -\frac{v_\beta \, x_\alpha - v_\alpha \, x^\beta}{\sqrt{|\, \eta_{\alpha\beta} \, v^\alpha v^\beta \,|}} \equiv M_{\alpha\beta} \tag{6.6.78}$$

which is called the *covariant angular momentum* (even if it is a four dimensional object and thence contains also the conserved quantity associated to Lorentz boosts which generalize to Special Relativity Galilei boosts).

The Nöther current associated to the generator ∂_λ of translations is

$$\mathcal{E} = -\frac{\eta_{\lambda\sigma} v^\sigma}{\sqrt{|\, \eta_{\alpha\beta} \, v^\alpha v^\beta \,|}} \equiv P_\lambda \tag{6.6.79}$$

which is called the *4-momentum*. In the coordinates adapted to an observer one has explicitly

$$\begin{cases} P_0 = -\dfrac{1}{\sqrt{|\, 1 - |\vec{v}|^2 \,|}} \\ P_i = \dfrac{v_i}{\sqrt{|\, 1 - |\vec{v}|^2 \,|}} \end{cases} \tag{6.6.80}$$

One can expand the 0-th component of the 4-momentum getting

$$P_0 \sim -\left(1 + \tfrac{1}{2}|\vec{v}|^2\right) \tag{6.6.81}$$

(here $\sim$ means *"proportional to"*) which (not setting $m = 1$ and $c = 1$ as we instead did above) corresponds to the celebrated Einstein's formula

$$E = mc^2 + \tfrac{m}{2}|\vec{v}|^2 \tag{6.6.82}$$

Even if we have introduced reparametrization and Poincaré covariance for the free relativistic particle, these symmetries are fundamental for any special relativistic system. Most of Special Relativity may be in fact encoded into two axioms which require that *any* relativistic system is covariant with respect to reparametrizations and Poincaré transformations. Also under this viewpoint the free relativistic particle is absolutely representative of the modern perspective about Physics. Special Relativity axioms here (as well as axioms of General Relativity, Gauge Theories, supersymmetries and many more below) express the universality of some symmetry group. Then on one hand Nöther theorem provides conserved quantities for *isolated* systems; on the other hand, the request that *any* system is symmetric with respect to those transformations quite severely constrains the potential or the interaction term in the Lagrangian. This is a very positive feature of symmetries: they can be used as a guideline to look for the possible Lagrangians of some given system. This is more effective the more we can trust on the symmetry we impose. Fortunately enough physicists presently rely completely on the symmetries imposed to fundamental systems since often they represent or correspond to axioms of very well tested and general theories. One could say that almost anything we know in fundamental physics can be expressed in terms of the symmetry groups coming from General Relativity and Gauge Theories. The modern attempts to quantize gravity (supersymmetric theories and superstrings) also rely on this approach (even if they do not appear to be empirically tested yet).

7. Conserved Quantities in Field Theory

As we saw, Nöther theorem in Mechanics produces 0-forms $\mathcal{E}(L,\xi)$ which are conserved along solutions of Euler-Lagrange equations. These conserved quantities are first integrals of motion and that is the end of the story. On the contrary, in Field Theory (i.e. whence $\dim(M) = m > 1$) Nöther currents are $(m-1)$-forms which are thence ready to be integrated on spacetime hypersurfaces.

Weak conservation laws imply that the Nöther current computed on-shell, i.e. $\mathcal{E}(L,\xi,\sigma) = (j^{2k-1}\sigma)^*\mathcal{E}(L,\xi)$, is a closed spacetime form. In local coordinates $(x^\mu) \equiv (x^0, x^i)$ any weak conservation law reads as

$$\mathrm{d}_0\mathcal{E}^0(L,\xi,\sigma) + \mathrm{d}_i\mathcal{E}^i(L,\xi,\sigma) = 0 \tag{6.7.1}$$

i.e. the Nöther currents obey a *continuity equation*[7] and are thence also called *conserved currents*.

In Field Theory, one can investigate whether the Nöther currents are also exact on-shell forms. Let us suppose as an ansatz that one can write

$$\mathcal{E}(L,\xi) = \tilde{\mathcal{E}}(L,\xi) + \operatorname{Div}\mathcal{U}(L,\xi) \tag{6.7.2}$$

[7] A continuity equation is an equation of the form $\partial_t A + \operatorname{Div}\vec{B} = 0$.

with a suitable $\tilde{\mathcal{E}}(L,\xi)$ vanishing on-shell (i.e., $(j^{2k-1}\sigma)^*\tilde{\mathcal{E}}=0$ if σ is critical) and let us set $\mathcal{U}(L,\xi,\sigma)=(j^{2k-2}\sigma)^*\mathcal{U}(L,\xi)$. Then because of the very definition of the operator Div$(\cdot)$ we have on-shell $\mathcal{E}(L,\xi,\sigma)=\mathrm{d}\,\mathcal{U}(L,\xi,\sigma)$, i.e. $\mathcal{E}(L,\xi,\sigma)$ is exact. In this case $\tilde{\mathcal{E}}(L,\xi)$ is called a *reduced current* for L and $\mathcal{U}(L,\xi,\sigma)$ is called a *superpotential* for L.

Under the weak hypotheses presently required, nothing can be said about existence or uniqueness of superpotentials, nor a general constructive algorithm can be provided. In the next Chapters we shall introduce stricter hypotheses which will allow a satisfactory analysis of the matter and show that the ansatz holds for practically all cases of physical interest. Fortunately enough, in fact, the further hypotheses needed are deeply related to very general physical requirements which any fundamental field theory should in principle obey: they are essentially the General Principle of Relativity and various forms of the Principle of Gauge Covariance.

Yang-Mills Fields

Let us first consider Yang-Mills fields as an instructive example in field theory. This Section corresponds to the standard *ad hoc* approach to gauge theories. In Chapters 8 we shall restate these results within the gauge natural framework so that the achievements and formal improvements will become evident. Let G be a semisimple Lie group, $\mathcal{P}=(P,M,\pi;G)$ be a principal bundle, g a metric over spacetime M, δ the Cartan-Killing metric over the Lie group G which is by definition Ad-invariant. Let $\mathcal{C}_{\mathcal{P}}$ be the bundle of principal connections over $\mathcal{P}$ as defined in (5.2.31), ω a section of $\mathcal{C}_{\mathcal{P}}$ and F its curvature 2-form valued into the Lie algebra $\mathfrak{g}$ of G. In this context F is also called the *field strength.*

We can define an inner product between k-forms valued into $\mathfrak{g}$:

Definition (6.7.3): let T_A be any δ-orthonormal basis of the Lie algebra $\mathfrak{g}$, so that any two Ad-invariant k-forms valued into the Lie algebra $\mathfrak{g}$ are locally expressed by $\zeta=\frac{1}{k!}\mathrm{Ad}^A_B(g^{-1})\zeta^B{}_{\mu_1\dots\mu_k}(x)T_A\otimes \mathrm{d}x^{\mu_1}\wedge\dots\wedge \mathrm{d}x^{\mu_k}$ and $\theta=\frac{1}{k!}\mathrm{Ad}^A_B(g^{-1})\theta^B{}_{\mu_1\dots\mu_k}(x)T_A\otimes \mathrm{d}x^{\mu_1}\wedge\dots\wedge \mathrm{d}x^{\mu_k}$; we set

$$\ll\zeta,\theta\gg=\delta_{AB}\,\zeta^A{}_{\mu_1\dots\mu_k}\,g^{\mu_1\nu_1}\dots g^{\mu_k\nu_k}\,\theta^B{}_{\nu_1\dots\nu_k} \tag{6.7.4}$$

where $g^{\mu\nu}$ are the components of the spacetime metric g and δ_{AB} are the components of the Cartan-Killing metric δ. Let us define also:

$$\theta_A{}^{\mu_1\dots\mu_k}=\delta_{AB}\,\theta^B{}_{\nu_1\dots\nu_k}\,g^{\mu_1\nu_1}\dots g^{\mu_k\nu_k} \tag{6.7.5}$$

We shall always rise and lower Greek indices by g and capital Latin indices by δ. ∎

We can now define a first order Lagrangian on the configuration bundle $\mathcal{C}_\mathcal{P}$ by setting:

$$L_{YM} = \mathcal{L}_{YM}\mathrm{d}s = -\frac{1}{4} \ll F, F \gg \sqrt{g}\,\mathrm{d}s = -\frac{1}{4}F^A_{\mu\nu}\, g^{\mu\rho}g^{\nu\sigma}\, F^B_{\rho\sigma}\,\delta_{AB}\sqrt{g}\,\mathrm{d}s \quad (6.7.6)$$

which is called *Yang-Mills Lagrangian*; here $\mathcal{L}_{YM} = \mathcal{L}_{YM}(j^1\omega)$. The Yang-Mills Lagrangian (6.7.6) is known to be *gauge invariant*, i.e. it is kept invariant under any automorphism $\Phi \in \mathrm{Aut}(\mathcal{P})$ of the principal bundle $\mathcal{P}$. We have then the following:

Proposition (6.7.7): let Ξ be the infinitesimal generator of a flow of automorphisms of $\mathcal{P}$. Then the covariance identity holds:

$$\mathrm{d}_\mu(\mathcal{L}_{YM}\xi^\mu) = p_A{}^{\mu\nu}\,\pounds_\Xi F^A{}_{\mu\nu} + p_{\mu\nu}\,\pounds_\Xi g^{\mu\nu} \qquad (6.7.8)$$

where we set

$$\begin{aligned} p_A{}^{\mu\nu} &= \frac{\partial \mathcal{L}_{YM}}{\partial F^A{}_{\mu\nu}} = -\frac{1}{2}F_A{}^{\mu\nu}\sqrt{g} \\ p_{\mu\nu} &= \frac{\partial \mathcal{L}_{YM}}{\partial g^{\mu\nu}} = -\frac{1}{2}\left(F^A{}_{\rho\mu}F_A{}^{\rho}{}_{\cdot\nu} - \frac{1}{4}g_{\mu\nu}\, F^A{}_{\rho\sigma}F_A{}^{\rho\sigma}\right)\sqrt{g} \equiv -\frac{1}{2}\hat{H}_{\mu\nu} \end{aligned} \qquad (6.7.9)$$

Proof: the right-hand side of (6.7.8) can be recast as follows:

$$\begin{aligned} &p_A{}^{\rho\nu}\,\pounds_\Xi F^A{}_{\rho\nu} + p_{\mu\nu}\,\pounds_\Xi g^{\mu\nu} = \\ &= p_A{}^{\rho\nu}(\xi^\mu\nabla_\mu F^A{}_{\rho\nu} + 2\nabla_\rho\xi^\mu F^A{}_{\mu\nu} - c^A{}_{BC}\xi^B_{(V)}F^C_{\rho\nu}) - 2\,p_{\mu\nu}\,g^{\mu\rho}\,\nabla_\rho\xi^\nu = \quad (6.7.10) \\ &= \nabla_\mu\mathcal{L}_{YM}\xi^\mu + \mathcal{L}_{YM}\nabla_\mu\xi^\mu = \nabla_\mu(\mathcal{L}_{YM}\xi^\mu) = \mathrm{d}_\mu(\mathcal{L}_{YM}\xi^\mu) \end{aligned}$$

The covariance identity thence holds true. ∎

Notice that in (6.7.8) symmetries act also on the metric $g^{\mu\nu}$ giving the term proportional to $\pounds_\Xi g^{\mu\nu}$. In the present perspective, however, the metric of spacetime in principle is *not* a dynamical field (if so, the appropriate configuration bundle should rather be $\mathcal{C}_\mathcal{P} \times_M \mathrm{Lor}(M)$; the whole matter becomes more clear, of course, in the context of gauge natural theories, which will be considered later in Section 8.2 to which we refer for details).

The fact that $\pounds_\Xi g^{\mu\nu}$ plays a role in (6.7.8) even if g is non-dynamical depends on two reason:

- first, trivially, without that term the covariance identity does not hold for general automorphisms on $\mathcal{P}$;
- second, the *gauge group* $\mathrm{Aut}(\mathcal{P})$ of all automorphisms of the structure bundle $\mathcal{P}$ is not *a priori* reduced to the subgroup of the vertical ones (i.e. those which

project onto the identity of M). As a consequence, also the group Diff(M), which naturally acts on $g^{\mu\nu}$, is embedded (although non-canonically) into the invariance group, so that, roughly speaking, Diff(M)-invariance is reflected into the term $\pounds_\Xi g^{\mu\nu}$.

From a puristic viewpoint, the Yang-Mills theory should be formulated as a theory of a (pure) gauge field over a (possibly) curved *background.* As such, one should consider the metric g to be fixed (e.g., the Minkowski metric $g = \eta$ on $M = \mathbb{R}^m$). In this case the covariance identity (6.7.8) could seem to be ambiguous because of the following sensitive question: *should the metric tensor g either be varied in the Lagrangian or not?* Actually one is free to treat g as a dynamical field or as a constant parameter which the Lagrangian depends on.

In fact, one can decide to consider it as a genuine dynamical field and rely on (6.7.8) as covariance identity; of course in this case one should vary the Lagrangian (6.7.6) with respect to g too, so to obtain some kind of field equations for g, namely $p_{\mu\nu} = 0$. These equations are not reasonable from a physical viewpoint but, once interpreted in the correct way, they are the base for the gauge natural formulation we shall consider in details in Chapter 8.

Conversely, if one wants then to keep g fixed *a posteriori*, the additional term $p_{\mu\nu}\pounds_\Xi g^{\mu\nu}$ in (6.7.8) has to be killed because of the variation field Ξ chosen (i.e. if we choose not to change g, then Ξ has to be restricted to be a Killing vector of the metric). Equivalently, one can consider g as a constant parameter in the Lagrangian from the very beginning. In this case a different covariance identity is obtained, namely

$$\mathrm{d}_\mu(\mathcal{L}_{YM}\xi^\mu) = p_A{}^{\mu\nu}\pounds_\Xi F^A{}_{\mu\nu} \tag{6.7.11}$$

which coincides with (6.7.8) provided that the symmetry generator Ξ projects onto an isometry of the metric g, i.e. $\pounds_\Xi g^{\mu\nu} = 0$. This sounds physically and mathematically reasonable: *if the metric g is a mere parameter in the Lagrangian we are allowed to consider only transformations which preserve such a structure.*

Let us thence turn our attention to the Euler-Lagrange equations of the Lagrangian (6.7.6). The most general variation of the Lagrangian is in the following form:

$$\begin{aligned}
\delta\mathcal{L}_{YM} &= p_A{}^{\mu\nu}\delta F^A{}_{\mu\nu} + p_{\mu\nu}\delta g^{\mu\nu} = \\
&=2p_A{}^{\mu\nu}\mathrm{d}_\mu\delta\omega^A{}_\nu + 2p_A{}^{\mu\nu}c^A_{\cdot BC}\delta\omega^B{}_\mu\omega^C{}_\nu + p_{\mu\nu}\delta g^{\mu\nu} = \\
&=\mathrm{d}_\mu\Big(2p_A{}^{\mu\nu}\delta\omega^A{}_\nu\Big) - 2\Big(p_A{}^{\mu\nu}c^A_{\cdot BC}\omega^C{}_\mu + \mathrm{d}_\mu p_B{}^{\mu\nu}\Big)\delta\omega^B{}_\nu + p_{\mu\nu}\delta g^{\mu\nu} = \\
&=\mathrm{d}_\mu \ll \mathbb{F}^\mu \mid X \gg + \mathbb{E}_A{}^\mu\delta\omega^A{}_\mu - \tfrac{1}{2}\hat{H}_{\mu\nu}\delta g^{\mu\nu}
\end{aligned} \tag{6.7.12}$$

where the vertical vector field $X = (\delta F^A{}_{\mu\nu})\partial_A{}^{\mu\nu} + (\delta g_{\mu\nu})\partial^{\mu\nu}$ is the field along which our deformation is considered and where we set

$$\begin{aligned}
&\mathbb{F}(\mathcal{L}_{YM}) = \mathbb{F}^\mu \, \mathrm{d}s_\mu = 2p_A{}^{\mu\nu}\bar{\mathrm{d}}\omega^A{}_\nu \otimes \mathrm{d}s_\mu \\
&\mathbb{E}(\mathcal{L}_{YM}) = \mathbb{E}^\mu_A \, \bar{\mathrm{d}}\omega^B{}_\mu \otimes \mathrm{d}s = -2\Big(p_A{}^{\mu\nu}c^{A}_{\cdot BC}\omega^C{}_\mu + \mathrm{d}_\mu p_B{}^{\mu\nu}\Big)\bar{\mathrm{d}}\omega^B{}_\nu \otimes \mathrm{d}s \\
&\hat{H}_{\mu\nu}(\mathcal{L}_{YM}) = \hat{H}_{\mu\nu} = -2p_{\mu\nu}
\end{aligned} \tag{6.7.13}$$

The tensor density $\hat{H}_{\mu\nu}$ is improperly called the *Hilbert stress tensor*. According to (6.7.12), if g were considered as a dynamical field then $\hat{H}_{\mu\nu} = 0$ should be interpreted as field equations for $g^{\mu\nu}$. Since we want instead to regard g as a parameter, as we discussed above, we consider deformations with $\delta g^{\mu\nu} = 0$ only. Yang-Mills equations have then the following local expression:

$$\mathrm{d}_\mu(\sqrt{g}\, F_B{}^{\mu\nu}) + \sqrt{g}\, F_A{}^{\mu\nu} c^{A}_{\cdot BC}\omega^C{}_\mu =: D_\mu(\sqrt{g}\, F_B{}^{\mu\nu}) = 0 \tag{6.7.14}$$

where the operator D_ν is called the *exterior covariant differential* (see equation (3.5.33)). Recall that here the metric g has been fixed. Accordingly, a solution of Yang-Mills equations is not in general a critical section for the Lagrangian L_{YM}, due to the usually non vanishing term $\hat{H}_{\mu\nu}\delta g^{\mu\nu}$ in the variation (6.7.12). It is critical only in the *"directions"* characterized by $\delta g^{\mu\nu} = 0$. We stress that all these considerations are due to the fact that we are trying to *artificially* regard g as a parameter to obtain a *"special relativistic"* field theory. As we did already mention, all these considerations about the role and the interpretation of g will be in fact overcome in the next Chapters, where we shall introduce the viewpoint of General Relativity in which *all* fields have to be considered as dynamical fields.

Let us now consider an infinitesimal generator of automorphisms of $\mathcal{P}$ to be explicitly decomposed in the form $\Xi = \xi^\mu(\partial_\mu + \omega^A_\mu \rho_A) + (\xi^A - \omega^A_\mu \xi^\mu)\rho_A$; here $\Xi_{(H)} = \xi^\mu(\partial_\mu + \omega^A_\mu \rho_A)$ and $\Xi_{(V)} = (\xi^A - \omega^A_\mu \xi^\mu)\rho_A$ denote the horizontal and vertical components of Ξ, respectively. The Nöther current with respect to the *"horizontal part"* $\Xi_{(H)}$ is:

$$\begin{aligned}
\mathcal{E}(L_{YM}, \Xi_{(H)}) &= \Big(2p_A{}^{\nu\mu}\pounds_{\Xi_{(H)}}\omega^A{}_\mu - \mathcal{L}_{YM}\xi^\nu\Big)\mathrm{d}s_\nu = \\
&= -\sqrt{g}\,\Big(F_A{}^{\mu\nu}F^A{}_{\mu\sigma} - \tfrac{1}{4}F_A{}^{\mu\rho}F^A{}_{\mu\rho}\delta^\nu_\sigma\Big)\xi^\sigma \mathrm{d}s_\nu
\end{aligned} \tag{6.7.15}$$

which reproduces the well known expression of the energy-momentum tensor of the Yang-Mills field.

We stress that (6.7.15) is not conserved due to the Hilbert stress tensor term in (6.7.12). Once again one can understand the particular case by restricting to transformations which project onto isometries. In this case the covariance identity reduces to (6.7.11) from which conservation of the current (6.7.15) follows easily. Thus the "horizontal" Nöther current (6.7.15) is conserved only if one restricts to gauge transformations which project onto isometries.

As far as vertical symmetries are concerned, the "vertical" Nöther current is:

$$\begin{aligned}\mathcal{E}(L_{YM}, \Xi_{(V)}) &= -\, 2p_A{}^{\mu\nu}\nabla_\mu \xi^A_{(V)}\,\mathrm{d}s_\nu = \\ &= \Big[-\nabla_\mu\Big(2p_A{}^{\mu\nu}\xi^A_{(V)}\Big) + 2\nabla_\mu p_A{}^{\mu\nu}\xi^A_{(V)}\Big]\mathrm{d}s_\nu\end{aligned} \tag{6.7.16}$$

One could now investigate whether the superpotentials are defined for both currents. The decomposition (6.7.16) allows one to define the superpotential and the reduced current for $\Xi_{(V)}$

$$\begin{aligned}\mathcal{U}(L_{YM}, \Xi_{(V)}) &= p_A{}^{\mu\nu}\,\xi^A_{(V)}\,\mathrm{d}s_{\mu\nu} = -\frac{1}{2}F_A{}^{\mu\nu}\xi^A_{(V)}\sqrt{g}\,\mathrm{d}s_{\mu\nu} \\ \tilde{\mathcal{E}}(L_{YM}, \Xi_{(V)}) &= 2\nabla_\mu p_A{}^{\mu\nu}\xi^A_{(V)}\,\mathrm{d}s_\nu = \nabla_\mu F_A{}^{\nu\mu}\xi^A_{(V)}\sqrt{g}\,\mathrm{d}s_\nu\end{aligned} \tag{6.7.17}$$

This is a good definition since the reduced current $\tilde{\mathcal{E}}(L_{YM}, \Xi_{(V)})$ does in fact vanish on shell. Thus the Nöther current is exact on-shell and the superpotential is anyone of its primitives.

The same procedure, however, cannot be applied to the horizontal part. In fact, the Nöther current $\mathcal{E}(L_{YM}, \Xi_{(H)})$ does not depend on the derivatives of the symmetry generator Ξ; thence, according to the general theory, it should be interpreted as the already reduced current with an identically vanishing superpotential. Unfortunately, this reduced current does not vanish as a consequence of field equations. This is due to the fact that we used the trick of restricting variations by setting $\delta g^{\mu\nu} = 0$ rather than accepting $\hat{H}_{\mu\nu} = 0$ as field equations for the metric g. Thence we would be tempted to say at a first glance that the horizontal part of the Nöther current does not allow a superpotential. We stress that those choices were however motivated by the axioms of Special Relativity. This quite unsatisfactory situation will be fortunately overcome in the next Chapters when we shall quit Special Relativity in favor of General Relativity.

Let us now consider as a very special example the Yang-Mills theory for the group $U(1)$. Being its Lie algebra $\mathfrak{u}(1) \simeq \mathbb{R}$ of dimension 1, we shall have the right of omitting the only group index; furthermore, being $U(1)$ a commutative group, its only structure constant trivially vanishes. The theory so obtained will be called *Maxwell theory*, since it describes electromagnetism in vacuum.

Accordingly, the Maxwell field is

$$\omega = \mathrm{d}x^\mu \otimes (\partial_\mu - A_\mu(x)\rho) \tag{6.7.18}$$

where $A = A_\mu \mathrm{d}x^\mu$ is a (local) 1-form over space-time. The curvature is given by the following 2-form:

$$F_{\mu\nu} = \partial_\mu A_\nu - \partial_\nu A_\mu \tag{6.7.19}$$

and vacuum Maxwell equations, together with Bianchi identities, are:

$$\begin{cases} \mathrm{d}_\nu F^{\mu\nu} = 0 \\ \mathrm{d}_{[\sigma} F_{\mu\nu]} = 0 \end{cases} \qquad \Leftrightarrow \qquad \begin{cases} * \, \mathrm{d} * F = 0 \\ \mathrm{d}F = 0 \end{cases} \tag{6.7.20}$$

where $*$ denotes the Hodge duality.[8] Bianchi identities $\mathrm{d}F = 0$ are identically satisfied because of the definition $F = \mathrm{d}A$. Notice that $F = \mathrm{d}A$ does not uniquely determine the field A. This is the starting point for locally defining gauge transformations $A' = A + \mathrm{d}\alpha$. However, we stress that for us A is a principal connection on a suitable principal bundle $\mathcal{P}$, not just a 1-form; gauge transformations then arise naturally as the canonical representations of the automorphisms of $\mathcal{P}$.

Let us choose for simplicity $M = \mathbb{R}^4$ with $g = \eta$ and consider a reference frame inducing a spacetime slicing:

$$\begin{array}{ccc} \mathbb{R}^4 \simeq \mathbb{R} \times \mathbb{R}^3 & \xrightarrow{\;p_2\;} & \mathbb{R}^3 \\ \Big\downarrow{\scriptstyle p_1} & & \\ \mathbb{R} & & \end{array} \tag{6.7.21}$$

Let us suppose that an observer, represented by the frame, chooses two (time-dependent) 1-forms $\mathbf{E}$ and $\mathbf{B}$ over $\mathbb{R}^3$, which is assumed to inherit from the slicing the Euclidean metric. Being $*_3$ Hodge's duality in $\mathbb{R}^3$, one can build a two form:

$$F = p_2^*\mathbf{E} \wedge \mathrm{d}t + p_2^*\big(*_3 \, \mathbf{B}\big) \tag{6.7.22}$$

Now, writing Maxwell equations and Bianchi identities (6.7.20) for the 2-form F, one obtains the well known classical equations:

$$\begin{cases} \nabla \cdot E = 0 \\ \nabla \times B - \partial_t E = 0 \\ \nabla \cdot B = 0 \\ \nabla \times E + \partial_t B = 0 \end{cases} \tag{6.7.23}$$

where E and B denote the dual vector fields to the 1-forms $\mathbf{E}$ and $\mathbf{B}$.

As we said, equations (6.7.23) are nothing but Maxwell equations in vacuum for the electromagnetic field described by the electric field E and the magnetic

[8] The Hodge duality $* : \Omega_n(M) \longrightarrow \Omega_{m-n}(M)$ is defined (in an orientable manifold) by the property

$$\alpha \wedge *\beta = g(\alpha, \beta)\eta_g \qquad \alpha,\ \beta \in \Omega_n(M)$$

where η_g is the volume form induced by the metric g.

field B in the chosen frame. As is well known, a generic Yang-Mills theory may be regarded as a non-commutative generalization of electromagnetism.

The conserved quantity associated to vertical symmetries of Maxwell theory, choosing a $t = constant$ surface D, is:

$$Q = \int_{\partial D} E \cdot n \, \mathrm{dS} \tag{6.7.24}$$

which, by Gauss theorem, coincides with the electric charge in D.

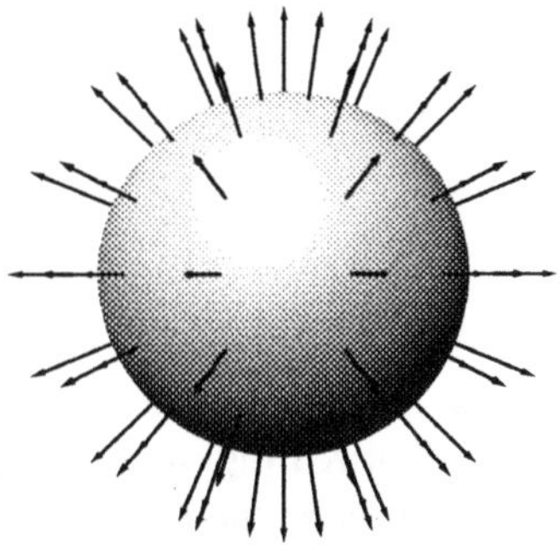

Fig. 22 – The flow of a Coulomb field.

Yang-Mills theories for $G = \mathrm{SU}(2)$ describe weak nuclear interactions according to the electroweak unified model which is now tested to a good degree of accuracy and the fundamental characteristics of which may be considered to be empirically tested.

The particular case for $G = \mathrm{SU}(3)$ is certainly related to chromodynamics and strong nuclear interactions even if the interpretation of empirical results is somehow trickier. One could thence say that Yang-Mills theories describe *all* interactions in fundamental particle physics being the standard model itself a Yang-Mills theory.

Thus Yang-Mills theories (and gauge theories in general) should be included in any framework for field theories. Next Chapters are devoted to the definition of natural and gauge natural theories, which encompass General Relativity, Gauge Theories and Yang-Mills theories as particular cases. This will also bring to a better setting for conserved quantities and points towards overcoming the problems related with the presence of a background non-dynamical field g which is ingrained to Special Relativity.

The Charged Particle

Let us now consider the most general interactions for the relativistic particle. If we require the potential to be invariant with respect to reparametrizations and under the action of the Poincaré group this strongly constrains the form of the generalized potentials we can conceive. In general a potential has the form $V(s,q,u) = U(s,q) + A_\mu(s,q)u^\mu$. Any vector field $\Xi = \xi(s)\,\partial_s + \xi^\mu(s,q)\,\partial_\mu$ lifts to $\hat{\Xi} = \xi(s)\,\partial_s + \xi^\mu(s,q)\,\partial_\mu + \hat{\xi}^\mu(s,q,u)\,\partial_\mu$ with $\hat{\xi}^\mu(s,q,u) = \mathrm{d}_s\xi^\mu - u^\mu \mathrm{d}_s\xi$. This lift $\hat{\Xi}$ is a generalized symmetry in the sense of (6.5.30) if the following identities hold true:

$$\frac{\mathrm{d}\alpha}{\mathrm{d}s} + \frac{\mathrm{d}}{\mathrm{d}s}(\xi V) = \frac{\partial V}{\partial q^\mu}\pounds_\Xi q^\mu + \frac{\partial V}{\partial u^\mu}\pounds_\Xi u^\mu \tag{6.7.25}$$

which gives the following

$$\frac{\mathrm{d}\alpha}{\mathrm{d}s} = -\frac{\partial U}{\partial q^\mu}\xi^\mu - \frac{\partial A_\nu}{\partial q^\mu}u^\mu\xi^\mu - A_\mu \mathrm{d}_s\xi^\mu \tag{6.7.26}$$

As for the free particle case, let us assume that $\xi = \xi^\mu\partial_\mu$ does not depend on the parameter s and that it is a Killing vector of the metric $g_{\mu\nu}$ appearing in the kinetic term (i.e. the Minkowski metric). Let us also specify to the case in which $\pounds_\Xi\Theta_L = \mathrm{d}\alpha + \omega$ with α independent of s. The covariance identity can be then recast as

$$0 = -\partial_\mu U\xi^\mu + [(\partial_\nu A_\mu - \partial_\mu A_\nu)\xi^\mu - \partial_\nu(\alpha + A_\mu\xi^\mu)]u^\nu \tag{6.7.27}$$

An isolated system has to be Poincaré invariant. In particular the (local) vector fields ∂_μ are symmetry generators in Cartesian coordinates x^μ. As a consequence U is a constant irrelevant in the Lagrangian. We can thence assume $U = 0$ without any loss of generality. The rest of the fundamental identity can be easily expressed: if we assume that $A = \mathrm{d}x^\mu \otimes (\partial_\mu - A_\mu(q)\,\rho)$ is a principal connection of a principal $U(1)$-bundle, with ρ a right invariant vertical vector field on $\mathcal{P}$ and $\Xi' = \xi^i(q)\,\partial_\mu + \alpha(q)\rho$ is a right invariant vector field on $\mathcal{P}$, the Lie derivative reads as

$$\pounds_{\Xi'}A = (\xi^\rho F_{\rho\mu} + \nabla_\mu\xi_{(V)})\partial^\mu = [(\partial_\mu A_\nu - \partial_\nu A_\mu)\xi^\mu + \partial_\nu(\alpha + A_\mu\xi^\mu)]\partial^\nu = \pounds_{\Xi'}A_\mu\partial^\mu \tag{6.7.28}$$

where ∂_μ is now the natural basis of the vertical vectors on the bundle of connections. The covariance identity is thence

$$0 = -(\pounds_{\Xi'}A_\mu)u^\mu \tag{6.7.29}$$

It says that Ξ is a symmetry if and only if Ξ' is a Killing vector for the connection A_μ, i.e $\pounds_{\Xi'}A_\mu = 0$. As a consequence we are not very free to fix a Lagrangian for a charged particle moving in a fixed electromagnetic field described by a potential A_μ. Let us consider, e.g., the *"good"* Lagrangian

$$L = [\sqrt{g_{\mu\nu}u^\mu u^\nu} + \epsilon A_\mu u^\mu]\,\mathrm{d}s \tag{6.7.30}$$

If we fix Cartesian coordinates adapted to an observer ($t = x^0$) and we set $A_\mu = (\phi, A_i)$ we get the Lagrangian

$$L = [\sqrt{1-|\vec{v}|^2} + \epsilon\phi + \epsilon A_i v^i]\mathrm{d}t \tag{6.7.31}$$

As a consequence the force experienced by the particle is

$$F_i = \epsilon\partial_i\phi + \epsilon\partial_i A_j v^j - \epsilon \mathrm{d}_t A_i = \epsilon\partial_i\phi + \epsilon(\partial_i A_j - \partial_j A_i)v^j - \epsilon\partial_t A_i \tag{6.7.32}$$

which, under the identifications of the previous Section and by a suitable fixing of the coupling constant ϵ, reduces to the Lorentz force $\vec{F} = e(\vec{E} + \vec{v}\times\vec{B})$.

8. BRST Transformations

As an example of generalized symmetries in field theories, we shall hereafter consider BRST transformations which are very relevant to the contemporary approach to Quantum Field Theories. Supersymmetries are a further example and they will be considered in Chapter 10 below, after we will have introduced spinor fields which are essential to supersymmetric theories. BRST transformations appear as a tool for the quantization of Yang-Mills theories. The Yang-Mills Lagrangian (6.7.6) introduced above has very attractive symmetry properties related to gauge transformations. This properties will become even more attractive when Yang-Mills theories will be introduced as gauge natural theories.

From a quantum viewpoint, however, the situation is not so cheerful. The Yang-Mills Lagrangian is degenerate; as a very direct consequence one cannot define the propagator of the Yang-Mills field. This is ultimately due to the fact that the gauge field A^A_μ contain both physical and unphysical degrees of freedom. One thence usually fixes a gauge, for example the Lorentz gauge $g^{\mu\nu}\mathrm{d}_\mu A^A_\nu = 0$, by modifying the Lagrangian to

$$L^{Lor}_{YM} = \left[-\tfrac{1}{4}F^A_{\mu\nu}F^{\mu\nu}_A + \tfrac{\alpha}{2}\delta_{AB}(g^{\mu\nu}\mathrm{d}_\mu A^A_\nu)(g^{\rho\sigma}\mathrm{d}_\rho A^B_\sigma)\right]\sqrt{g}\,\mathrm{d}s \tag{6.8.1}$$

This solves the problem of propagators, as the Lagrangian L^{Lor}_{YM} is no longer degenerate. Unfortunately, the new Lagrangian is no longer gauge invariant (and thus the propagator defined depends explicitly on the gauge fixing); furthermore, the theory is no longer unitary, i.e. its dynamics does not conserve probability normalization. Unitarity is the first requirement to be satisfied by a quantum field theory to be consistent. Fortunately enough unitarity can be restored by introducing a *ghost field* c^A together with its antighost $\bar{c}^A$. By ghost field it is meant a field which violates the spin-statistics connection, i.e. anticommuting fields with integer spin or half-integer spin fields which commute. In our case $(c^A, \bar{c}^A)$ are scalar (spin 0) anticommuting fields. In canonical quantization (anti)-commutation is realized by replacing canonical conjugate fields in (some) Hamiltonian formalism with (anti)-commuting operators on a suitable Hilbert space of states. This is quite cumbersome to be done in field theory away from some canonical simple cases. In general one prefers to try a Feynman path integral approach, in which commutation properties have to be inserted directly into the classical model to be quantized. There is an easy way to do that; the configuration bundle is assumed to have as standard fiber F which is a *"manifold"* modelled on the exterior algebra $\Lambda_n(V)$ of an auxiliary vector space V, rather than the usual manifolds modelled on $\mathbb{R}^n$. Depending on the case, fields can take their value in the even part $\Lambda_+(V)$ of the exterior algebra for commutative fields or in the odd part $\Lambda_-(V)$ for anticommutative fields.

We stress that both $\Lambda_+(V)$ and $\Lambda_-(V)$ are ordinary vector spaces on the field $\mathbb{R}$ (or $\mathbb{C}$) so that F is in the end a real manifold (possibly infinite dimensional if V is infinite dimensional). It is useful to introduce commuting (anticommuting, respectively) *"coordinates"* $(y^i; \tilde{y}^i)$ on $\Lambda(V) \simeq \Lambda_+(V) \oplus \Lambda_-(V)$. Each of these coordinates $(y^i; \tilde{y}^i)$ takes its value in $\Lambda_+(V)$ (or $\Lambda_-(V)$, respectively) and thence they represent a whole collection of real valued *"true"* coordinates. When these (anti)-commuting coordinates are used the wedge product in the exterior algebra $\Lambda(V)$ is understood so that if $\tilde{y}^i \in \Lambda_-(V)$ are odd fields their product is anticommutative. The same trick is required also for the quantization of Dirac fields in a Feynman path integral approach, which have to be anticommuting fields as a consequence of the spin-statistics connection. See also Section 10.4 for some further detail.

Here, the ghost fields are scalar anticommuting fields, i.e. both $(c^A, \bar{c}^A)$ are odd. Of course, as the name suggests, ghost fields have to be impossible to be observed. They are introduced to restore unitarity of the theory. The use of ghost fields is nowadays quite spread in quantum field theory techniques. For ghost fields and anticommuting fields we refer the reader to [R81], [C90], [CDF91] and [M88]. Here these considerations are used just as a motivation to obtain the correct BRST Lagrangian.

Let us consider a Lorentzian spacetime (M, g) on which we choose local coordinates x^μ and we define the dynamical fields $(A^C_\mu, B^A, c^A, \bar{c}^A)$. The field A^C_μ is a Yang-Mills field as introduced in Section 6.7 above and $F^A{}_{\mu\nu}$ denotes its curvature. We recall that A^C_μ is a principal connection on a suitable principal bundle $\mathcal{P}$ which will be called the *structure bundle* and which is *a priori* fixed.

We remark that usually the gauge fixing needed is just a local one; although one can do something on a more general background (and/or more general coordinates) the case under consideration is often considered as a somehow privileged case. We shall abandon in this Section global and geometrical considerations. Notice for example that the Lagrangian L_{BRST} we are going to define is not gauge covariant, which eventually implies that it is a *local* Lagrangian, at least when the structure bundle $\mathcal{P}$ on which Yang-Mills fields are defined is not a trivial principal bundle. Of course if we assume that (M, g) is Minkowski spacetime $(\mathbb{R}^4, \eta)$, as we shall do hereafter, i.e. no singularities are considered, $M \simeq \mathbb{R}^4$ is contractible and any $\mathcal{P}$ is trivial; in that case the Lagrangian may be seen as a global gauge-dependent Lagrangian.

The field B^A is a *"scalar"* real field, i.e. a section of the associated bundle $\mathcal{P} \times_{\mathrm{id}} \mathbb{R}$, where id is the trivial representation of G on $\mathbb{R}$ which maps the whole G onto $\mathbb{I} \in \mathrm{GL}(\mathbb{R})$. The ghost fields $(c^A, \bar{c}^A)$ are scalar anticommuting fields.

Let us consider the Lagrangian

$$L_{BRST} = [-\tfrac{1}{4} F^A{}_{\mu\nu} F_A{}^{\mu\nu} + \mathrm{d}^\mu \bar{c}_A \mathrm{D}_\mu c^A - \mathrm{d}^\mu B_A A^A_\mu + \tfrac{\alpha}{2} B_A B^A]\, \mathrm{d}s \tag{6.8.2}$$

where α is a coupling constant and $\mathrm{D}_\mu c^A = \mathrm{d}_\nu c^A + c^A{}_{BC} A^B_\mu c^A$ is the covariant derivative of the ghost field. Modulo a pure divergence term $-\mathrm{d}^\mu (B_A A^A_\mu)$ added to L^{Lor}_{YM}, one can identify the Lagrange multiplier term $B_A \mathrm{d}^\mu A^A_\mu$ to introduce the Lorentz gauge and the ghost kinetic term $\mathrm{d}^\mu \bar{c}_A \mathrm{D}_\mu c^A$. As we noticed many times, this Lagrangian is no longer gauge covariant but it defines good propagators and it yields a unitary quantum field theory. As we noticed above propagators are gauge dependent, but in any case propagators are not observable physical quantities. Only Green functions will be observable and one can prove that they are not affected by the gauge choice; gauge covariance is thence restored on an observational level.

Field equations are

$$\begin{cases} \mathrm{D}_\nu F_C{}^{\nu\lambda} = \mathrm{d}^\lambda \bar{c}_A\, c^A{}_{BC}\, c^B + \mathrm{d}^\lambda B_C \\ \Box \bar{c}_A + \bar{c}_C\, c^C{}_{AB}\, \mathrm{d}^\mu A^B_\mu = \mathrm{d}^\mu (\bar{c}_C\, c^C{}_{AB}\, A^B_\mu) \\ \Box c^A = \mathrm{d}^\mu (c^A{}_{BC}\, A^B_\mu\, c^C) \\ \alpha B^A + \mathrm{d}^\mu A^A_\mu = 0 \end{cases} \tag{6.8.3}$$

where as usual we restrict to deformations which keep the background η fixed and we denote by $\Box$ the box operator $\Box = \mathrm{d}^\mu \mathrm{d}_\mu$.

In the quantization of such a theory, which is of course out of the scope of this monograph, it turn useful to notice that the Lagrangian L_{BRST} is covariant with respect to the following generalized transformations;

$$\begin{cases} \delta A^A{}_\mu = \epsilon \mathrm{D}_\mu c^A = \epsilon(\mathrm{d}_\mu c^A + c^A{}_{BC}\, A^B_\mu\, c^C) \\ \delta c^A = -\tfrac{\epsilon}{2} \{c, c\}^A \\ \delta \bar{c}^A = \epsilon B^A \\ \delta B^A = 0 \end{cases} \tag{6.8.4}$$

where $\{c,c\}^A = c^A{}_{BC}\, c^B\, c^C$ is the relevant (anti)-commutator. We remark that $\delta c^A \neq 0$ since c^A is an anticommuting field. Notice also the asymmetry between c^A and $\bar{c}^A$. These asymmetry does not imply a violation of CPT invariance since c^A and $\bar{c}^A$ are ghost fields while CPT theorem assumes non-ghost fields.

These are generalized symmetries both because they depend on the derivatives of fields (thus they are generalized vector fields Ξ)

$$\Xi = (\mathrm{D}_\mu c^A)\frac{\partial}{\partial A^A{}_\mu} - \tfrac{1}{2}\{c,c\}^A \frac{\partial}{\partial c^A} + B^A\frac{\partial}{\partial \bar{c}^A} \tag{6.8.5}$$

and because they do not leave the Poincaré-Cartan form invariant. In fact one can easily check that the Poincaré-Cartan form is

$$\begin{aligned}\Theta_L =& \mathcal{L}_{BRST}\,\mathrm{d}s + [F_A{}^{\mu\nu}(\mathrm{d}A^A_\nu - \mathrm{d}_\lambda A^A_\nu\,\mathrm{d}x^\lambda) + \mathrm{d}^\mu\bar{c}_A(\mathrm{d}c^A - \mathrm{d}_\lambda c^A\,\mathrm{d}x^\lambda) + \\ &+ (\mathrm{d}\bar{c}_A - \mathrm{d}_\lambda\bar{c}_A\,\mathrm{d}x^\lambda)(\mathrm{d}_\nu c^A + c^A{}_{BC}\,A^B_\nu\,c^C)\eta^{\mu\nu} + \\ &- A^A_\nu\eta^{\mu\nu}(\mathrm{d}B^A - \mathrm{d}_\lambda B^A\,\mathrm{d}x^\lambda)] \wedge \mathrm{d}s_\mu\end{aligned} \tag{6.8.6}$$

Its Lie derivative along Ξ is

$$\pounds_\Xi\Theta = -F_A{}^{\mu\nu}(\mathrm{d}c^A_\mu - c^A_{\mu\lambda}\,\mathrm{d}x^\lambda)\wedge\mathrm{d}s_\nu + \mathrm{d}^\mu B_A(\mathrm{d}c^A - c^A_\lambda\,\mathrm{d}x^\lambda)\wedge\mathrm{d}s_\mu \tag{6.8.7}$$

Thence we are in the hypotheses of the generalized Nöther theorem. The conserved currents are given by:

$$\mathcal{E}(L_{BRST},\Xi) = [F_A{}^{\mu\nu}\mathrm{D}_\nu c^A - B_A\mathrm{D}^\mu c^A + \tfrac{1}{2}\mathrm{d}^\mu\bar{c}_A\{c,c\}^A]\,\mathrm{d}s_\mu \tag{6.8.8}$$

One can now consider two (different) infinitesimal transformations, Ξ_1 and Ξ_2, respectively. The commutator of these transformations is $[\Xi_1,\Xi_2]=0$; this implies that the BRST algebra is closed. This opens many interesting possibilities to be investigated. In principle one could exponentiate the infinitesimal generators to define the BRST group G. If this turns out to be a Lie group then one has an action of G on the infinite jet prolongation of the configuration bundle $\mathcal{C}$ and could hope to define a structure bundle $\mathcal{P}$ to which the infinite jet prolongation $J\mathcal{C}$ is associated. If this could be possible a way to a gauge natural interpretation of BRST transformations will be opened.

Exercises

1- Prove that the BRST transformations close to form an algebra.

Hint: the infinitesimal generators are closed with respect to commutators.

2- Define BRST transformations on a pseudo-Riemannian manifold (M, g) with respect to an arbitrary background (M, g).

Hint: generalize first to non-Cartesian coordinates.

3- Check that the Dirac monopole

$$A = -\frac{m}{2}\cos(\theta)\mathrm{d}\varphi \tag{6.8.9}$$

is a solution of Maxwell equations on the sphere S^2.

Hint: check locally Maxwell equations and then prove that the field strength is global on S^2. Can A be extended to S^2?

4- Compute the conserved quantities of the Dirac monopole.

Hint: Pull-back along the solution the off-shell expressions of conserved quantities. Stress the geometric meaning of the corresponding integrals.

5- Write the equation of motions for a relativistic particle with charge e in the Dirac monopole field.

Hint: use Lagrangian (6.7.30).

Chapter 7

NATURAL THEORIES

Abstract We shall introduce *Natural Field Theories* and investigate the consequences of naturality in term of conserved quantities. As applications we shall consider geodesic motions on a surface, General Relativity in its metric formulation and metric-affine formulation.

The reader should master Natural Bundles together with their lift properties and Lie derivatives (see Chapter 4) as well as the Lagrangian formalism introduced in Chapter 6.

References

We refer to [F91], [FF91], [FFFR99], [KMS93] for the natural structure and in particular for naturality of General Relativity. Cauchy problem in General Relativity, causal structure and global hyperbolicity are analyzed in detail in [CB79], [HE73], [GIMMPSY], [ON83], [CBCF78]. The original paper about ADM techniques is [ADM62]; further details can be found in [RT74], [FFS92]. The original paper about Komar potential is [K59]; anomalous factor is treated in [K85], [FF88] and [FF91]. For a good review on classical Utiyama theorem for natural operators see [J04] and references quoted therein. A precious catalogue of exact solutions of Einstein equations can be found in [HE73] and [KSHMC80]. The original papers about Taub-Bolt solutions are [M63], [T51]; further details can be found in [HHP98] and [FFFR00].

Naturality of Maxwell theory can be found in [FFRxx]. The standard reference about matter fields is [BD64]. Various formulations of General Relativity are discussed in detail in [FR].

1. Definition of Natural Theories

Natural Field Theories are field theories which are invariant with respect to spacetime diffeomorphisms. Since in the passive viewpoint spacetime diffeomorphisms are locally identified with changes of coordinates, natural field theories implement the general covariance principle of General Relativity. One could say that Natural Theories are the realization of the first axiom of General Relativity, i.e. the *general covariance* which does not talk about gravitation, yet. This analyzes the property that *any* physical theory should have; then, in a second place, one is entitled to define a field theory which meets this criteria to describe the gravitational field.

We remark that the general covariance axiom of Einstein's program appeared so tightly related to the second part and it is considered so self-evident that all standard empirical tests on General Relativity actually have been aimed to prove the second part. Actually this is also due to the fact that the physical contents of the symmetry principles, back since Galilei relativity and up to gauge covariance and supersymmetric theories, are so deeply related to formal mathematical issues that they are considered as tested by testing their "more or less direct" consequences.

We shall hereafter present a number of very satisfactory applications to basically any fundamental Field Theory previous to the advent of gauge field theories, from the theory of General Relativity to the theories with tensor matter fields. In our perspective also gauge natural theories will rely on a suitable extension of the same principles which for natural theories have been proven to be necessary in order to meet the contemporary phenomenology of fundamental physics. In this wider meaning, also spinor fields, Gauge Theories and supersymmetries will be later treated in a proper way and, what is essential, under a unifying perspective.

Natural theories are field theories that implement the general covariance principle of General Relativity. We shall not enter here in a discussion about the various equivalent formulations of the general covariance principle; we shall just state the form of the principle we shall use below.

Definition (7.1.1): a *natural field theory* is a Lagrangian field theory such that:

(a) the configuration bundle $\mathcal{C}$ is a natural bundle;

(b) all spacetime diffeomorphisms, represented on $\mathcal{C}$ by means of the natural action, are Lagrangian symmetries;

(c) one can choose a bundle morphism $\Gamma : J^h\mathcal{C} \longrightarrow L(M)/\mathrm{GL}(m)$ which allows to define a linear connection on M out of fields and their derivatives up to a suitable (finite) order h.

In this case the Lagrangian $L = \mathcal{L}\,\mathrm{d}s$ is called a *natural Lagrangian*. ∎

Axiom (c) needs some extra comment. It postulates the existence of a mechanism for choosing a connection, called the *dynamical connection*, on spacetime M; this dynamical connection is to be used whenever a connection is needed to uniquely determine otherwise ambiguous constructions, such as the Poincaré-Cartan morphism, the Poincaré-Cartan form, the splitting of variational morphisms and, consequently, the superpotentials. It may often happen that in a field theory one has more than one way to choose a dynamical connection, e.g. if one has (at least) two different metrics among the dynamical fields. In this case, different choices correspond in principle to different natural theories. In other words the dynamical connection is considered to be fixed once for all from the beginning.

We stress that a dynamical connection is a map which associates a connection to each configuration; when we say *fix a dynamical connection* this by no means implies fixing a single connection. We just fix the dependence of the connection on fields and their derivatives up to some finite order, e.g. by saying that we fix the Levi-Civita connection induced by the metric field. Every time we consider a configuration then a connection is produced.

This point is essential, e.g., in understanding the (possibly different) natural structures associated with relativistic theories of gravitation, whereby one or more than one linear connections appear from the very beginning among dynamical variables. In purely metric theories, i.e. when $\mathcal{C} = \mathrm{Lor}(M)$, one has the Levi-Civita connection of the dynamical metric; in purely affine theories, i.e. when $\mathcal{C} = L(M)/\mathrm{GL}(m)$, the dynamical connection itself is the dynamical field of the theory; in metric-affine theories, i.e. when the metric and the connection on M are considered *a priori* independent by setting $\mathcal{C} = \mathrm{Lor}(M) \times_M (L(M)/\mathrm{GL}(m))$, one has both the Levi-Civita connection of the dynamical metric and the (independent) linear connection on M. Whenever gravity is considered in interaction with matter, if one can suitably define a tensor $t^\alpha_{\beta\mu}$ using the matter fields, then the connection $\Gamma'^\alpha_{\beta\mu} = \Gamma^\alpha_{\beta\mu} + t^\alpha_{\beta\mu}$ is another option (e.g., Einstein-Cartan theory).

As a local consequence of axiom (b), if we consider two patches U_α and U_β of spacetime, naturality allows to define a well-defined theory on the union $U_\alpha \bigcup U_\beta$ of the two patches. In fact, the fields are first of all *global* sections of the configuration bundle and they are thence well-defined on $U_\alpha \bigcup U_\beta$ by construction. In second place, since diffeomorphisms (i.e. *changes of coordinates* passively speaking) are Lagrangian symmetries, also field equations are preserved by coordinate changes. As a consequence, also field equations are well-defined global objects.

We can summarize the principle of general covariance by saying that all the objects involved in the theory must be *well-defined geometrical quantities*. Some research groups use this fact to drastically postulate that everything in a natural field theory should be written without using coordinates. Of course this is a formal postulate, not really speaking about contents but just matters of language. We believe instead that local expressions using coordinates can often enhance comprehension, provided the use of coordinates is known (or can be proven) to be unessential. The *intrinsic language* is just one of the possible ways of proving that coordinates are unessential. We already saw many examples in Part I and many more will be presented hereafter.

We also remark that in natural theories one can overcome the problem of the definition of the topological and differential structure of spacetime. In Variational Calculus, in fact, the spacetime manifold M should be fixed from the beginning. On the other hand, in physical applications, e.g. cosmology, one does not really know anything *a priori* about the global structure of spacetime. One simply knows the local structure of spacetime (supposedly $\mathbb{R}^4$) and, because of naturality, this is enough to recover the global structure of spacetime out of the dynamics of the theory together with a countable number of physical observations. According to this approach one should not fix the global structure of the theory from the beginning. On the contrary only the local theory on a (small) open patch $U_\alpha \subset M$ is considered. This local version of the theory together with the observation of a set of *initial conditions* allows one to determine the local expression of fields on the patch U_α. Since one can cover any manifold with a countable atlas one can obtain a countable number of local solutions on each patch of spacetime. Because of the general covariance principle two such local expressions differ by a spacetime local diffeomorphism defined on the intersection of the two patches. Thus we obtained a countable number of patches, each one with a local solution on it, and a local diffeomorphism $\phi_{(\alpha\beta)}$ on each patch intersection $U_{\alpha\beta}$. One can easily prove that the local diffeomorphisms $\phi_{(\alpha\beta)}$ satisfy cocycle identities on $U_{\alpha\beta\gamma}$, i.e.

$$\phi_{(\alpha\beta)} \circ \phi_{(\beta\gamma)} \circ \phi_{(\gamma\alpha)} = \mathrm{id} \qquad\qquad \phi_{(\alpha\beta)} = [\phi_{(\beta\alpha)}]^{-1} \tag{7.1.2}$$

By a standard result, this is enough to determine a unique (up to global diffeomorphisms) manifold M which allows $(U_{\alpha\beta}, \phi_{(\alpha\beta)})$ as transition functions. Thence we can *a posteriori* define a natural theory on that particular M together with one of its global solutions which agree with the countable number of observations made on a local ground.

These considerations, modulo some technical considerations about what has to be intended by *initial conditions* and Cauchy problems in field theory, basically prove the equivalence between the general covariance principle and naturality. A natural theory collects and represents all the local solutions which bring to a definite global structure of spacetime. Of course other global structures may be obtained by other local solutions and they are obtained as global solutions of *another* natural theory on *another* spacetime manifold M'. This is the price one has to pay for it to be geometrically well-posed.

In return, we can understand very easily some quite far-reaches of the contemporary developments of field theories. For example, we can build out of observations only the global structure of spacetime *modulo* global diffeomorphisms. There are a number of classical arguments in different contexts, called *hole arguments*, which in fact prove that the actual manifold structure of spacetime is not observable. The hole argument in General Relativity hindered Einstein for a period while he was developing his theory. Hole arguments are always based on the existence of compactly supported symmetries (in natural theories compactly supported spacetime diffeomorphisms). These symmetries are easily proven to exist using a partition of unity. Unless one postulates that spacetime has necessarily an analytic structure and all fields have to be globally analytic too –which frankly seems to be both theoretically and empirically unmotivated if not wrong– then one can easily build two solutions of field equations which differ only in a compact region D. These two solutions (which exactly agree in the *"far past"*, differ in the *"present"* and again agree in the *far future*) clearly violate determinism. Since physicists are quite fond of determinism, the only possibility is that the two solutions of field equations actually represent the same physical situation. Thence the mathematical representation we choose is physically underdetermined and we have a group of transformations, which we may call the *gauge group*, which maps a representative into another representative of the same physical situation. In this way field equations have to be intended to identify the gauge orbit of a section. This can be done only if gauge transformations are symmetries of the theory.

In the simplest version of General Relativity the dynamical field is a metric tensor g (see below for details). In the above sense, $\mathrm{Diff}(M)$ is the gauge group of the theory and Einstein field equations identify an equivalence class of metrics containing all spacetimes (M, g) which differ by a global diffeomorphism.

Another example is provided by quantum field theory in its path integral formulation. It is generally accepted that in any version of quantum gravity the sum over histories has to run over all (equivalence classes of) topological and differential structures of spacetime. This is again due to the fact that Variational Calculus does not describe all physical situations (on which path integrals should sum over) but only the situations corresponding to a particular manifold structure M fixed from the beginning. In summing along all histories then one has to integrate somehow on all equivalence classes of sections of the configuration bundles, then summing up over all topological and differential structures of spacetime. See, e.g., [CDF91], [R81]for hints and details.

2. Bianchi Identities and Superpotentials

In a Natural Theory a set of symmetries of the theory has been singled out. Being any diffeomorphism a Lagrangian symmetry, the natural lift $\hat{\xi}$ of any spacetime vector field ξ is an infinitesimal generator of symmetry. As a consequence Nöther theorem produces a weak conservation law associated to any spacetime vector field ξ

$$\mathrm{Div}\, \mathcal{E}(L, \xi) = \mathcal{W}(L, \xi) \tag{7.2.1}$$

where we set

$$\begin{cases} \mathcal{E}(L, \xi) =< \mathbb{F} \,|\, j^{k-1} \pounds_\xi > -i_\xi L \\ \mathcal{W}(L, \xi) = - < \mathbb{E} \,|\, \pounds_\xi > \end{cases} \tag{7.2.2}$$

for the Nöther current and the work form, respectively.

We recall that, if $(x^\mu; y^i)$ are fibered coordinates on the (natural) configuration bundle $\mathcal{C}$, $\xi = \xi^\mu \partial_\mu$ is a spacetime vector field and $\hat{\xi} = \xi^\mu(x)\, \partial_\mu + \hat{\xi}^i(x, y)\, \partial_i$ is its (natural) lift to $\mathcal{C}$, then $\pounds_\xi = (\xi^\mu(x) y^i_\mu - \hat{\xi}^i(x, y))\partial_i$ is the formal Lie derivative and it is a generalized vertical vector field on $\mathcal{C}$ in the sense of definition (2.5.2).

We also recall that each natural bundle $\mathcal{C}$ is associated to some frame bundle $L^s(M)$, where the (minimal) integer $s \geq 1$ is called the *geometric order* of the natural bundle $\mathcal{C}$. Sections of $\mathcal{C}$ are called *geometric objects of order s*. If one recalls the definition of natural lift of a vector field ξ, the lifted vector field $\hat{\xi}$ is readily seen to depend linearly on ξ^μ together with its derivatives up to the (finite) order s (see for example equation (4.2.8)). The possibility of regarding Lie derivatives as linear operators on JTM encapsulates all the issues at the core of Natural Theories. As a direct consequence, in fact, one can define two variational morphisms

$$\begin{cases} \mathcal{E}(L) : J^{2k-1}\mathcal{C} \longrightarrow J^{k+s-1}TM \otimes A_{m-1}(M) \\ \mathcal{W}(L) : J^{2k}\mathcal{C} \longrightarrow J^s TM \otimes A_m(M) \end{cases} \tag{7.2.3}$$

such that the Nöther current and the work form may be finally written as

$$\begin{cases} \mathcal{E}(L, \xi) = < \mathcal{E}(L) \,|\, j^{k+s-1}\xi > \\ \mathcal{W}(L, \xi) = < \mathcal{W}(L) \,|\, j^s \xi > \end{cases} \tag{7.2.4}$$

We can thence apply the spitting Lemmas of the previous Chapter to decompose the Nöther current and the work form. First of all we can naturally decompose $\mathcal{W}(L)$ as follows

$$< \mathcal{W}(L) \,|\, j^s\xi > = < \mathcal{B}(L) \,|\, \xi > + \mathrm{Div}\, < \tilde{\mathcal{E}}(L) \,|\, j^{s-1}\xi > \tag{7.2.5}$$

The variational morphism $\mathcal{B}(L)$ is called the *Bianchi morphism* while $\tilde{\mathcal{E}}(L)$ is called the *reduced current morphism.* Of course, $\mathcal{B}(L)$ is unique while $\tilde{\mathcal{E}}(L)$ is defined up to a pure divergence; however, $\tilde{\mathcal{E}}(L)$ is uniquely determined if one requires it to be the reduced boundary term in the splitting.

We stress that the weak conservation law (7.2.1) provides a further splitting of the work morphism $\mathcal{W}(L)$ with an identically vanishing volume part and with $\mathcal{E}(L)$ as a boundary part. Because of the uniqueness result (6.2.49), by comparing the two splittings we obtain the *strong identities*

$$\mathcal{B}(L) = 0 \tag{7.2.6}$$

which are called *(generalized) Bianchi identities.* By *"strong"* we mean that Bianchi identities hold off-shell, i.e. they hold also out of solutions to field equations. They are *true* identities, with both a geometrical and a physical meaning prior to any dynamics. We stress that they are a direct consequence of the general covariance of the Lagrangian. We shall see that in General Relativity they are nothing but the standard Bianchi identities for the Levi-Civita connection of g, while in the Chapter 8 we shall see that in the gauge natural formulation of Yang-Mills theories they reproduce instead the Bianchi identities (3.8.21) for the field strength.

The uniqueness result given by Proposition (6.2.49) also implies the following strong identity:

$$\Delta_\xi \equiv \mathcal{E}(L,\xi) - \tilde{\mathcal{E}}(L,\xi) = \mathrm{Div}\, \mathcal{U}(L,\xi) \tag{7.2.7}$$

so that the l.h.s. Δ_ξ is a quantity which turns out to be covariantly conserved off-shell, i.e.

$$\mathrm{Div}\left(\mathcal{E}(L,\xi) - \tilde{\mathcal{E}}(L,\xi)\right) = 0 \tag{7.2.8}$$

We thence again say that the quantity Δ_ξ is *strongly conserved,* meaning that it is conserved also off-shell. We recall that the reduced current $\tilde{\mathcal{E}}(L,\xi)$ vanishes instead on-shell, so that Δ_ξ coincides on-shell with the Nöther current. However, we stress that in view of a Quantum Field Theory also off-shell quantities are endowed with a physical meaning. In this context Δ_ξ and the Nöther current actually differ and strong conservation laws may turn out to be physically important.

Equation (7.2.7) may be interpreted as a splitting of the Nöther current $\mathcal{E}(L,\xi)$. By using the splitting Lemma (6.2.56), recalling that the variational morphism $\tilde{\mathcal{E}}(L)$ is reduced, we obtain a superpotential morphism given by $\mathcal{U}(L) : J^{2k-2}\mathcal{C} \longrightarrow J^{k+s-2}TM \otimes A_{m-2}(M)$ such that the superpotential is given by $\mathcal{U}(L,\xi) =< \mathcal{U}(L) \mid j^{k+s-2}\xi >$. We remark that although the superpotential morphism is defined modulo a divergenceless term, for each fibered connection there is just one reduced superpotential, which is the only one produced by the algorithm introduced in Lemma (6.2.30). Unless otherwise stated the superpotential $\mathcal{U}(L,\xi)$ will be the reduced one.

Let us summarize: Nöther theorem associates to infinitesimal symmetries the so-called Nöther currents $\mathcal{E}(L,\xi)$ which are (weakly) conserved. This holds in any field theory even if the theory is not natural. In natural theories we have a lot of symmetries and thence a lot of Nöther currents, one for each vector field over spacetime. But naturality has also other side effects: first of all, thanks to linearity of the Lie derivative $\pounds_\xi\sigma$ with respect to the spacetime vector field ξ, the Nöther currents can be regarded as variational morphisms themselves. Secondly, the variational morphism representing Nöther currents can be split (canonically thanks to the choice of the dynamical connection) and this defines the superpotentials $\mathcal{U}(L,\xi)$ and proves (generalized) Bianchi identities $\mathcal{B}(L,\xi)=0$. Existence of superpotentials can be interpreted in two ways: first of all Nöther currents are not only closed on-shell but they are always exact on-shell. A second by-product of naturality is the definition of a strongly conserved current Δ_ξ, as in (7.2.7), so that following holds:

$$\mathcal{E}(L,\xi)=\tilde{\mathcal{E}}(L,\xi)+\operatorname{Div}\mathcal{U}(L,\xi) \tag{7.2.9}$$

3. Conserved Quantities

The first attempt to define *conserved quantities* is to use Nöther currents integrated along an $(m-1)$-region Σ, i.e.

$$\mathfrak{E}_\Sigma(\xi,\sigma)=\int_\Sigma \mathcal{E}(L,\xi,\sigma) \tag{7.3.1}$$

We shall call these quantities *regular conserved quantities*, since of course all objects have to be well-defined (i.e. *regular*) over the whole Σ. In particular the on-shell Nöther current $\mathcal{E}(L,\xi,\sigma)$ has to be non-singular on Σ. In this context the Nöther current is regarded as the *density* of the corresponding regular conserved quantity.

The limits of this definition appear when the solution diverges on Σ, as it happens, for example in electromagnetism, when an electric point charge is present on a 3-surface Σ. In that case one can try to impose constraints on how fast the solution grows in the neighbourhood of the singularity so that the integral can be given a generalized meaning.

On the other hand, in natural theories the Nöther current admits superpotentials $\mathcal{U}(L,\xi,\sigma)$ which can be integrated on a closed $(m-2)$-region Ω to define the so-called *covariantly conserved currents*, i.e.:

Definition (7.3.2): let us consider a natural theory; let $\Omega \subset M$ be a closed spacetime $(m-2)$-region, $\mathcal{U}(L,\xi)$ the superpotential for a spacetime vector field ξ and σ a solution of Euler-Lagrange equations. The *conserved quantity* Q_Ω associated to ξ is:

$$Q_\Omega(\xi,\sigma) = \int_\Omega \mathcal{U}(L,\xi,\sigma) = \int_\Omega (j^{2k-2}\sigma)^*\mathcal{U}(L,\xi) \tag{7.3.3}$$

∎

If Ω is a "small ball" around a singularity (see *Fig.23*), Q_Ω is a characteristic quantity of the singularity which is invariant with respect to diffeomorphisms. If $\phi : M \longrightarrow M$ is a (local) spacetime diffeomorphism and $\hat{\phi} : \mathcal{C} \longrightarrow \mathcal{C}$ is its lift to the configuration bundle, we can drag anything along ϕ. We can in fact define a new solution $\sigma' = \hat{\phi}\circ\sigma\circ\phi^{-1}$, a new trapping surface $\Omega' = \phi\circ\Omega$ and a new vector field $\xi' = (\phi)_*\xi$.

Then we have

$$\mathcal{U}(L,\xi) = (j^{2k-2}\hat{\phi})^*\mathcal{U}(L,\xi') \qquad \Rightarrow \qquad Q_\Omega(\xi,\sigma) = Q_{\Omega'}(\xi',\sigma') \tag{7.3.4}$$

In view of this latter proposition we can say that Q_Ω is an intrinsic quantity of the singularity.

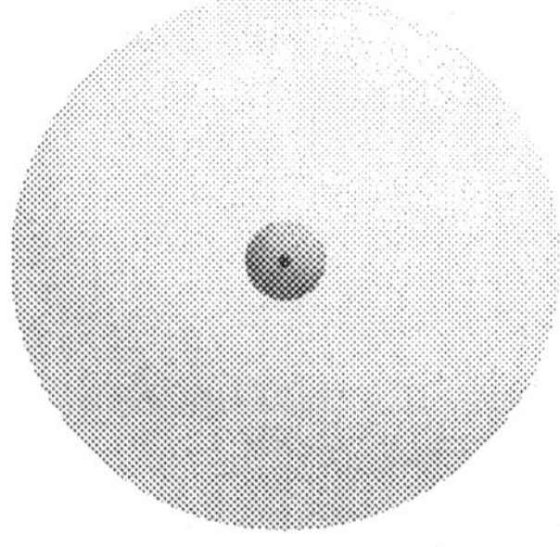

Fig. 23 – A singularity.

Calling that quantity a *conserved quantity* is, from a purely formal viewpoint, a matter of definition. From a physical viewpoint, it is motivated by the complete analogy with the definition of the *electric charge* by means of the Gauss theorem (see (6.7.24)) as the flow of the electric field on a trapping sphere. For example, one can consider the electrostatic field $\vec{E} = \frac{\alpha}{r^2}\vec{u}_r$ in $\mathbb{R}^3 - \{0\}$, $\vec{u}_r$ being the radial unit vector in spherical coordinates. Then let us consider the following flow integral on any sphere (actually, because of the homological invariance of flow integrals, any surface) surrounding the origin:

$$Q = \int_{S^2} \vec{E} \cdot \vec{n} = \frac{\alpha}{r^2} 4\pi r^2 = 4\pi\alpha \tag{7.3.5}$$

where $\vec{n} = \vec{u}_r$ denotes the normal unit vector to the surface S^2 we are integrating on. Such a quantity is an homological invariant associated to the field $\vec{E}$ above and it is called the *electric charge* generating $\vec{E}$.

Accordingly, let us now consider two encapsulated trapping surfaces $\Omega_{(1)}$ and $\Omega_{(2)}$ (see *Fig.* 8) so that $\Omega_{(2)} - \Omega_{(1)}$ is the homological boundary ∂D of an $(m-1)$-region D such that there are no singularities in between the two trapping surfaces.

In general, the two conserved quantities $Q_{\Omega_{(2)}}$ and $Q_{\Omega_{(1)}}$ actually differ by:

$$\begin{aligned} Q_{\Omega_{(2)}} - Q_{\Omega_{(1)}} &= \int_{\Omega_{(2)}-\Omega_{(1)}} \mathcal{U}(L,\xi,\sigma) = \int_{\partial D} \mathcal{U}(L,\xi,\sigma) = \\ &= \int_D \mathrm{d}\,\mathcal{U}(L,\xi,\sigma) = \int_D \mathcal{E}(L,\xi,\sigma) = \mathfrak{E}_D(\xi,\sigma) \end{aligned} \tag{7.3.6}$$

Since $\mathcal{E}$ obeys a continuity equation, the quantity Q_Ω is a *conserved quantity* and singularities are seen as sources.

We remark that in the case of electromagnetism the electric charge does not depend in fact on the trapping surface. This is a consequence of a particular situation, that can be traced back to the fact that the electromagnetic field does not carry electric charges, so that the region of space within D does not contribute. It is well known that this is not a general feature in field theory. More general Yang-Mills fields (e.g. SU(2) or SU(3)) as well as the gravitational field carry their own sources.

Back to Physics

It is now time to discuss the relation between this geometrical framework and real physical world. First of all, we have to remind the reader that all our framework deals with homogeneous formalism. In our field theories there is no time as required by General Relativity. Secondly, a class of quite general considerations (the *hole arguments*) proved that if the theory has to retain some sort of determinism, than our fields are not completely determined by observations. Finally, we defined *conserved quantities* providing them with a geometrical meaning, that *a priori* seems to have almost nothing to do with Physics.

The absence of *time* is the main source of problems because all observers (and consequently any observations) experience and participate to time passing by. Moreover, if there is no time then some very basic physical notions (some of which have been already mentioned) are in danger of loosing their meaning. *Determinism* means that the present (possibly together with the past) uniquely determines the future of the system. *What does determinism mean if there is no time?*

Conserved quantities means that when one measures them, their values tend to be maintained. Or at least if their value changes (e.g. an electric charge escaping the apparatus or two annihilating charges) one can control this change by controlling the flows on the border of the apparatus or enlarging the system (e.g. to include the gamma ray emitted by annihilation). Feynman gave a very detailed and deep description of this situation (see Section $4-1$ in [FLS66]). The fact is that *conserved quantities* ultimately mean *not transforming*. But without a *time* any transformation looses its meaning. In our framework there is nothing transforming. Field equations are partial differential equations; in classical Physics PDEs are often used to described stationary phenomena (i.e. without evolution). Whenever PDEs have to describe evolution (e.g. transport equations) a time parameter ought to be selected. In our picture the solution to field equations are there to describe the system in its four-dimensional extension, not in its evolution.

On the other hand, in real laboratories there exist clocks and rules and physical quantities are numbers, not points on a manifold. Thus sooner or later we have to break down the general covariance in order to compare with the actual observations. Our choice is to do that as late as possible maintaining manifest covariance to the end. Of course other options are viable: some researches choose to break manifest covariance very soon, before writing field equations, in order to stress the evolution of the system in an arbitrary time. We believe that the main reason to use this approach is to provide a framework for Hamiltonian formalism for field theory in view of a possible canonical quantization. We stress however that canonical quantization for General Relativity has not proved of being able to provide reasonable results yet. Also in view of this situation we preferred to keep a more classical attitude and to keep manifest covariance as long as possible.

When we finally break down covariance, the *time* may be identified with a timelike vector field ζ over spacetime (M, g). The flow ϕ_t of the vector field ζ drags a point $x_0 \in M$ into a worldline that describes a point staying steady with respect to the observer. The observer then chooses a (possibly local) spacelike hypersurface Σ representing points of M at time $t = 0$. Consequently, $\Sigma_t = \phi_t(\Sigma)$ gives a (local) foliation of M called an *ADM splitting*.

An ADM splitting defines a time since one restricts to coordinates adapted to the foliation in which the first coordinate $x^0 = t$ is the affine parameter of the flow ϕ_t. The hypersurface Σ is called a *Cauchy surface* since field equations determine the evolution of initial Cauchy data on Σ along the foliation. In other words, the leaves Σ_t of the foliation represent the (spatial) world of the observer at the various times. Roughly speaking, one can give the initial condition on Σ_0 and the field equations determine the situation at later (or earlier) times Σ_t.

In general (M, g) has no global Cauchy surface; if one wants to keep thinking in terms of time evolution then one has to restrict *a priori* to Lagrangians and *a posteriori* to solutions such that for any timelike vector field ζ there exists at least one global Cauchy surface. Field equations with respect to the induced ADM foliation becomes evolution equations, i.e. the Cauchy problem is well posed. One of the most important theorems in this area is the one relating existence of Cauchy surfaces to *global hyperbolicity*. In other words, global Cauchy surfaces exist if and only if the Laplacian operator (in the relevant signature) is globally hyperbolic.

Clearly, global hyperbolic spacetimes allow global foliations $M \simeq \mathbb{R} \times \Sigma$ and problems about causality are prevented on them.

We also have to mention that the situation is in many cases fairly more problematic than the one presented here, even if this picture is basically correct. The further problems are due to the fact that most of physically relevant field equations are degenerate, i.e. when they are written with respect to an ADM foliation they determine the evolution of some degrees of freedom leaving the others completely undetermined. This is the ADM consequence of the hole arguments. On the other hand, among field equations some determine the evolution while others do not depend on time derivatives at all and they have to be interpreted as constraints on the possible Cauchy data allowed. Thus from an ADM viewpoint field equations are usually underdetermined since the evolution of some fields is left undetermined and at the same time they are overdetermined since even when a Cauchy surface exists one is not free to set initial conditions arbitrarily.

This strange behavior can be thoroughly understood by investigating the simple example of Maxwell equations for the electromagnetic 2-form F on Minkowski space. In general these equations read as

$$\mathrm{d}F = 0 \qquad\qquad \mathrm{d} * F = J \tag{7.3.7}$$

where $*$ denotes Hodge duality and J denotes the source current, a vector density.

Let us now choose Cartesian coordinates (t, x, y, z) in Minkowski space so that the metric takes the form $\eta = \mathrm{d}t^2 - \mathrm{d}x^2 - \mathrm{d}y^2 - \mathrm{d}z^2$ and let us consider the following 2-forms depending on an arbitrary function $f : \mathbb{R} \longrightarrow \mathbb{R}$:

$$\begin{aligned} F_+ &= f(t+z)\mathrm{d}t \wedge \mathrm{d}x + f(t+z)\mathrm{d}x \wedge \mathrm{d}z \\ F_- &= f(t-z)\mathrm{d}t \wedge \mathrm{d}x + f(t-z)\mathrm{d}x \wedge \mathrm{d}z \end{aligned} \tag{7.3.8}$$

One can easily check that both these 2-forms satisfy Maxwell equations (7.3.7) with $J = 0$. It is also clear that once one has a solution $\bar{F}$ of Maxwell equations (7.3.7) (possibly for a non-zero source J) then $\bar{F}' = \bar{F} + \alpha F_+ + \beta F_-$ are solutions too. In this way, field equations alone are not able to determine the evolution of the combinations $F \pm i * F$ in the direction $t = z$ ($t = -z$, respectively).

On the other hand the electromagnetic field F decomposes as

$$F = \left(-E_i \, \mathrm{d}t \wedge \mathrm{d}x^i + B_k \, \epsilon^k_{\cdot ij} \, \mathrm{d}x^i \wedge \mathrm{d}x^j\right) \tag{7.3.9}$$

(where $\epsilon^k_{\cdot ij}$ is the Ricci skew-symmetric symbol) and Maxwell equations in vacuum in these coordinates read as

$$\begin{cases} \nabla \cdot B = 0 \\ \mathrm{d}_t B + \nabla \times E = 0 \end{cases} \qquad \begin{cases} \nabla \cdot E = 0 \\ -\mathrm{d}_t E + \nabla \times B = 0 \end{cases} \tag{7.3.10}$$

Now the equations $\nabla \cdot B = 0$ and $\nabla \cdot E = 0$ do not depend on time derivatives and they have to be interpreted as constraints on initial coordinates. When one specifies Cauchy data on the Cauchy surface $\Sigma = \{t = 0\}$ the electric field E_0 and the magnetic field B_0 on Σ_0, one is forced to choose them so that the following holds

$$\nabla \cdot E_0 = 0 \qquad \nabla \cdot B_0 = 0 \tag{7.3.11}$$

which shows the Maxwell equations to be overdetermined too.

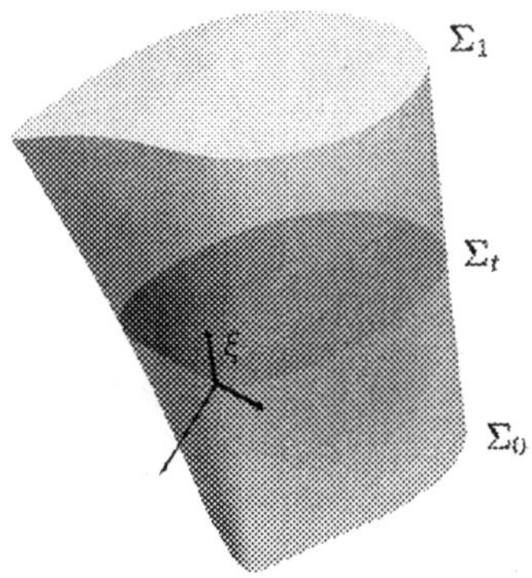

Fig. 24 – ADM local splitting.

As long as conserved currents are concerned we have to specify that *conserved* means here *covariantly conserved*, i.e. such that its integral on a homological boundary vanishes. They are not to be confused with *time conserved currents* which means that the value of the integral over a leaf of an ADM foliation does not depend on the leaf.

Despite the fact that quantities conserved along the *time* of a fixed ADM foliation are not manifestly covariant in nature they may be interesting to be investigated. In our perspective, in fact, they can be obtained from covariantly conserved quantities. To be precise, we can consider a variational principle, a spacetime vector field ξ which is an infinitesimal generator of Lagrangian symmetries and a solution σ of field equations. Then we can compute covariantly conserved currents $\mathcal{E}(L, \xi, \sigma)$ by Nöther theorem.

Let us then fix a spacelike $(n-1)$–region Σ_0 (with an exterior boundary B_0) and integrate the superpotential to define a conserved quantity $Q_{B_0}(\xi, \sigma)$. Any timelike vector field ζ allows then to evolve the region Σ_0 along its flow, parametrized by its affine parameter t. We consider than a *world tube* D (as in *Fig.24*) having topology $\Sigma_0 \times [t_0, t_1]$, with lateral surface $\mathcal{B}$ and *lids* $\Sigma_0 = \Sigma_{t_0}$ and $\Sigma_1 = \Sigma_{t_1}$.

Under this viewpoint, the question arises whether there exists a vector field ζ (possibly depending on the region Σ_0) such that the covariantly conserved quantity generated by ξ is also conserved in the local *time* induced by ζ. At a first glance, if we have $\partial D = \Sigma_{t_1} - \Sigma_{t_0} + \mathcal{B}$ (see *Fig.24*), conservation in time is equivalent to require that the integral of the Nöther current $\mathcal{E}(L, \xi, \sigma)$ on $\mathcal{B}$ vanishes (for any time interval $[t_0, t_1]$). In fact, we have the covariant conservation law $d\mathcal{E}(L, \xi, \sigma) = 0$, so that

$$\begin{aligned} 0 &= \int_D d\mathcal{E}(L,\xi,\sigma) = \int_{\partial D} \mathcal{E}(L,\xi,\sigma) = \\ &= \int_{\Sigma_{t_1}} \mathcal{E}(L,\xi,\sigma) - \int_{\Sigma_{t_0}} \mathcal{E}(L,\xi,\sigma) + \int_{\mathcal{B}} \mathcal{E}(L,\xi,\sigma) \Rightarrow \\ &\Rightarrow \int_{\Sigma_{t_1}} \mathcal{E}(L,\xi,\sigma) - \int_{\Sigma_{t_0}} \mathcal{E}(L,\xi,\sigma) = -\int_{\mathcal{B}} \mathcal{E}(L,\xi,\sigma) \end{aligned} \tag{7.3.12}$$

Thence the conserved quantity $\int_{\Sigma_t} \mathcal{E}(L, \xi, \sigma)$ computed on a leaf does not depend on the particular leaf if and only if $\int_{\mathcal{B}} \mathcal{E}(L, \xi, \sigma) = 0$ (for any time interval $[t_0, t_1]$). Physically speaking, this amount to require that the flow of the current $\mathcal{E}(L, \xi, \sigma)$ through $\mathcal{B}$ is vanishing. Clearly, different ADM foliations may evolve Σ_0 in different ways. In general, just few of them will lead to time–conserved quantities. The vanishing of $\int_{\mathcal{B}} \mathcal{E}(L, \xi, \sigma)$ has then to be guaranteed by additional hypotheses, possibly in many different ways. Under stronger hypotheses on ξ (or on ζ, or on the boundary conditions which g has to satisfy) the set of ADM foliations leading to time–conserved quantities with respect to different times may be possibly enlarged. Anyway this problem, which should be analyzed in detail and strongly depends on the model, is out of the scope of this monograph. We refer to $[FR]$ for a complete analysis of the situation in General Relativity.

The rest of this Chapter will be devoted to applications and examples. We shall present hereafter a number of worked examples illustrating various situations and cases of natural theories together with their conserved quantities. These examples, here appearing as simple applications of the framework presented above, are in fact the motivation of this framework. Thanks to these applications, the physically oriented reader can recognize that our framework reproduces and generalizes the standard models relevant to Physics. On the other hand, the mathematically oriented reader can regard the following examples as guidelines to trace a direct connection between the mathematical framework introduced above back down to the physical world which motivated it.

4. Geodesics

Let us start with the problem of geodesics which is relevant to Differential Geometry and relativistic Mechanics. On a (pseudo)-Riemannian manifold (M, g) a geodesic is a curve $\gamma : \mathbb{R} \longrightarrow M$ which extremizes the arc length functional on any compact interval $I \in \mathbb{R}$

$$\ell(\gamma) = \int_I \sqrt{|g_{\alpha\beta}\dot{\gamma}^\alpha \dot{\gamma}^\beta|}\mathrm{d}s \tag{7.4.1}$$

where s denotes an arbitrary parameter along the curve γ. Being the arc length a quantity depending on the trajectory $\mathrm{Im}(\gamma)$ but not on the parametrization, the functional ℓ is invariant with respect to arbitrary reparametrizations.

We already studied the functional in a particular case, describing the relativistic point particle. We are here interested in showing to what extent it defines a *natural* variational principle.

The configuration bundle $\mathbb{R} \times M$ is natural of order 0. It is associated to $L^0(\mathbb{R}) \simeq \mathbb{R}$ which is trivially fibered onto $\mathbb{R}$ by means of the obvious projection $\mathrm{id} : \mathbb{R} \longrightarrow \mathbb{R}$. It is also a *"principal bundle"* with the trivial group $\mathbb{I} = \{e\}$. Let us choose the standard fiber M on which the group $\mathbb{I}$ is represented by means of the trivial representation $0 : \mathbb{I} \longrightarrow \mathrm{Diff}(M) : e \mapsto \mathrm{id}_M$; then the associated bundle is $\mathcal{C} \simeq L^0(\mathbb{R}) \times_0 M \simeq \mathbb{R} \times M$.

Diffeomorphisms $\phi : \mathbb{R} \longrightarrow \mathbb{R}$ lift to $\mathcal{C}$ to give the automorphisms

$$\hat{\phi} : \mathcal{C} \longrightarrow \mathcal{C} : (s; q^\mu) \mapsto (\phi(s); q^\mu) \tag{7.4.2}$$

The Lagrangian $L = \sqrt{|g_{\alpha\beta}\dot{\gamma}^\alpha \dot{\gamma}^\beta|}\,\mathrm{d}s$ is covariant with respect to such transformations, which correspond to curve reparametrizations. Finally there is a canonical connection on (M, g) because of the metric background we have fixed. Strictly speaking that is not a dynamical connection, since it is not built out of dynamical fields but it is a background. Accordingly, the theory of geodesics is not a natural theory in the sense of Definition (7.1.1).

We have to mention, however, that many authors define natural theories as field theories just obeying axioms (a) and (b) of Definition (7.1.1) without any request on the existence of a dynamical connection (axiom (c)). In this weaker sense the variational theory of geodesics is a *"natural theory"*.

We preferred here our stronger definition as dynamical connections enhance our understanding of conserved quantities. Furthermore, nearly all natural theories used in fundamental physics allow a dynamical connection, as we did already mention.

5. General Relativity

As an application we now introduce here the standard (second order) framework for General Relativity. We shall show how to define superpotentials and how to define also a convenient first order Lagrangian for Einstein equations by choosing a reference background metric. As a specific application we shall analyze the Schwarzschild, the Kerr-Newman, the BTZ and the Taub-Bolt solutions. The material presented here is an essential part of the standard physical background.

Gravitational field is described by a pseudo-Riemannian metric g (with Lorentzian signature $(1, m-1)$) over the spacetime M of dimension $\dim(M) = m$; in standard General Relativity, $m = 4$. The configuration bundle is thence the bundle of Lorentzian metrics over M, denoted by $\mathrm{Lor}(M)$. The Lagrangian is second order and it is usually chosen to be the so-called *Hilbert Lagrangian*:

$$\begin{aligned} L_H &: J^2\mathrm{Lor}(M) \longrightarrow \Lambda^0_m(M) \\ L_H &= \mathcal{L}_H(g^{\alpha\beta}, R_{\alpha\beta})\, \mathrm{ds} = \frac{1}{2\kappa}(R - 2\Lambda)\sqrt{g}\, \mathrm{ds} \end{aligned} \tag{7.5.1}$$

where $R = g^{\alpha\beta} R_{\alpha\beta}$ denotes the scalar curvature, $\sqrt{g}$ the square root of the absolute value of the metric determinant and Λ is a real constant (called the *cosmological constant*). The coupling constant $(2\kappa)^{-1}$, which is completely irrelevant until the gravitational field is not coupled to some other field, depends on conventions; in natural units, i.e. $c = 1$, $h = 1$, $G = 1$, dimension 4 and signature $(+, -, -, -)$ one has $\kappa = -8\pi$.

Field equations are the well known Einstein equations with cosmological constant

$$R_{\alpha\beta} - \frac{1}{2} R\, g_{\alpha\beta} = -\Lambda\, g_{\alpha\beta} \tag{7.5.2}$$

We define the Lagrangian momenta by:

$$
\begin{aligned}
p_{\alpha\beta} &= \frac{\partial \mathcal{L}_H}{\partial g^{\alpha\beta}} = \frac{1}{2\kappa}\Big(R_{\alpha\beta} - \frac{1}{2}(R-2\Lambda)\, g_{\alpha\beta}\Big)\sqrt{g} \\
P^{\alpha\beta} &= \frac{\partial \mathcal{L}_H}{\partial R_{\alpha\beta}} = \frac{1}{2\kappa}\, g^{\alpha\beta}\sqrt{g}
\end{aligned}
\tag{7.5.3}
$$

Thus the covariance identity is the following:

$$
\mathrm{d}_\alpha(\mathcal{L}_H\, \xi^\alpha) = p_{\alpha\beta}\, \pounds_\xi g^{\alpha\beta} + P^{\alpha\beta}\, \pounds_\xi R_{\alpha\beta} \tag{7.5.4}
$$

or, equivalently:

$$
\nabla_\alpha(\mathcal{L}_H\, \xi^\alpha) = p_{\alpha\beta}\, \pounds_\xi g^{\alpha\beta} + P^{\alpha\beta}\, \nabla_\epsilon \left(\pounds_\xi \Gamma^\epsilon_{\alpha\beta} - \delta^\epsilon_\beta \pounds_\xi \Gamma^\lambda_{\alpha\lambda}\right) \tag{7.5.5}
$$

where ∇_ϵ denotes the covariant derivative with respect to the Levi-Civita connection of g. Thence we have a weak conservation law for the Hilbert Lagrangian

$$
\mathrm{Div}\, \mathcal{E}(L_H, \xi) = \mathcal{W}(L_H, \xi) \tag{7.5.6}
$$

Conserved currents and work forms have respectively the following expressions:

$$
\begin{aligned}
\mathcal{E}(L_H, \xi) &= \left[P^{\alpha\beta}\, \pounds_\xi \Gamma^\epsilon_{\alpha\beta} - P^{\alpha\epsilon}\, \pounds_\xi \Gamma^\lambda_{\alpha\lambda} - \mathcal{L}_H\, \xi^\epsilon\right] \mathrm{d}s_\epsilon = \\
&= \frac{\sqrt{g}}{2\kappa}\left(g^{\alpha\beta} g^{\epsilon\sigma} - g^{\sigma\beta} g^{\epsilon\alpha}\right) \nabla_\alpha \pounds_\xi g_{\beta\sigma}\, \mathrm{d}s_\epsilon - \frac{\sqrt{g}}{2\kappa} \xi^\epsilon R\, \mathrm{d}s_\epsilon = \\
&= \frac{\sqrt{g}}{2\kappa}\Bigg[\left(\tfrac{3}{2} R^\alpha_{\cdot\lambda} - (R-2\Lambda)\,\delta^\alpha_\lambda\right)\xi^\lambda + \\
&\qquad + \left(g^{\beta\gamma}\delta^\alpha_\lambda - g^{\alpha(\gamma}\delta^{\beta)}_\lambda\right)\nabla_{\beta\gamma}\xi^\lambda\Bigg]\, \mathrm{d}s_\alpha
\end{aligned}
\tag{7.5.7}
$$

$$
\mathcal{W}(L_H, \xi) = \frac{\sqrt{g}}{\kappa}\Big(R_{\alpha\beta} - \frac{1}{2}(R-2\Lambda)\, g_{\alpha\beta}\Big)\nabla^{(\alpha}\xi^{\beta)}\, \mathrm{d}s \tag{7.5.8}
$$

As any other natural theory, General Relativity allows superpotentials. In fact, the current can be recast into the form:

$$
\mathcal{E}(L_H, \xi) = \tilde{\mathcal{E}}(L_H, \xi) + \mathrm{Div}\, \mathcal{U}(L_H, \xi) \tag{7.5.9}
$$

where we set

$$
\begin{aligned}
\tilde{\mathcal{E}}(L_H, \xi) &= \frac{\sqrt{g}}{\kappa}\left(R^\alpha_{\cdot\beta} - \frac{1}{2}(R-2\Lambda)\,\delta^\alpha_\beta\right)\xi^\beta \mathrm{d}s_\alpha \\
\mathcal{U}(L_H, \xi) &= \frac{1}{2\kappa}\, \nabla^{[\beta}\xi^{\alpha]}\sqrt{g}\, \mathrm{d}s_{\alpha\beta}
\end{aligned}
\tag{7.5.10}
$$

We remark that the decomposition of the conserved current is here achieved by the standard splitting algorithm introduced in Lemma (6.2.30).

The superpotential (7.5.10) generalizes to an arbitrary vector field ξ the well known Komar superpotential which was originally derived for timelike Killing vectors. Whenever spacetime is assumed to be asymptotically flat, then the superpotential of Komar is known to produce upon integration at spatial infinity ∞ the correct value for angular momentum (e.g. for Kerr-Newman solutions) but just one half of the expected value of the mass. The *classical* prescriptions are in fact:

$$\begin{aligned} m &= 2\int_\infty \mathcal{U}(L_H, \partial_t, g) \\ J &= \int_\infty \mathcal{U}(L_H, \partial_\phi, g) \end{aligned} \tag{7.5.11}$$

For an asymptotically flat solution (e.g. the Kerr-Newman black hole solution) m coincides with the so-called *ADM mass* and J is the so-called *(ADM) angular momentum*. For the Kerr-Newman solution in polar coordinates (t, r, θ, ϕ) the vector fields ∂_t and ∂_ϕ are the Killing vectors which generate stationarity and axial symmetry, respectively. Thence, according to this prescription, $\mathcal{U}(L_H, \partial_\phi)$ is the superpotential for J while $2\mathcal{U}(L_H, \partial_t)$ is the superpotential for m. This is known as the *anomalous factor problem* for the Komar potential. To obtain the expected values for all conserved quantities from the same superpotential, one has to *correct* the superpotential (7.5.10) by some *ad hoc* additional boundary term, as e.g. done in [K85], [FF88]. Equivalently and alternatively, one can deduce a corrected superpotential as the canonical superpotential for a corrected Lagrangian, which is in fact the first order Lagrangian for standard General Relativity. This can be done covariantly, provided that one introduces an *extra* connection $\bar{\Gamma}^\alpha_{\beta\mu}$. The need of a reference connection $\bar{\Gamma}$ should be also motivated by physical considerations, according to which the conserved quantities have no absolute meaning but they are *intrinsically relative* to an arbitrarily fixed vacuum level. The simplest choice consists, in fact, in fixing a background metric $\bar{g}$ (not necessarily of the correct Lorentzian signature) and assuming $\bar{\Gamma}$ to be the Levi-Civita connection of $\bar{g}$. This is rather similar to the gauge fixing *à la Hawking* which allows to show that Einstein equations form in fact an *essentially hyperbolic* PDE system. Nothing prevents, however, from taking $\bar{\Gamma}$ to be any (in principle torsionless) connection on spacetime; also this corresponds to a gauge fixing towards hyperbolicity.

We stress that we shall use hereafter the term *background* for a field which enters a field theory in the same way as the metric enters Yang-Mills theory as presented in the previous Section. Thus, a background has to be fixed once for all and thence preserved, e.g. by symmetries and deformations. A background has no field equations since deformations fix it; it eventually destroys the naturality of a theory, since fixing the background results in allowing a smaller group of symmetries $G \subset \mathrm{Diff}(M)$. Accordingly, in *truly* natural field theories one should not consider background fields either if they are endowed with a physical meaning (as the metric in Yang-Mills theory does) or if they are not.

On the contrary we shall use the expression *reference* or *reference background* to denote an extra dynamical field which is not endowed with a *direct* physical meaning. As long as variational calculus is concerned, reference backgrounds behave in exactly the same way as other dynamical fields do. They obey field equations and they can be dragged along deformations and symmetries. It is important to stress that such a behavior has nothing to do with a *direct* physical meaning: even if a reference background obeys field equations this does not mean that it is *observable*, i.e. it can be measured in a laboratory. Of course, not any dynamical field can be treated as a reference background in the above sense. The Lagrangian has in fact to depend on reference backgrounds in a quite peculiar way, so that a reference background cannot interact with any other physical field, otherwise its effect would be observable in a laboratory. We shall hereafter consider the prototype of reference background fields. We remark however that, because of the discussion above about the need of choosing a vacuum state, in our perspective *any* field theory should have its reference, though in many occasions particular choices of the reference, e.g. the zero section in a vector bundle, may hide its presence.

Let then $\bar\Gamma$ be any (torsionless) reference connection. We shall systematically denote by a bar all quantities referred to the reference background, i.e. here and hereafter we shall use the following notation:

$g_{\mu\nu}$	$\bar g_{\mu\nu}$	covariant metric
$g^{\mu\nu}$	$\bar g^{\mu\nu}$	contravariant metric
$\Gamma^\alpha_{\beta\nu}$	$\bar\Gamma^\alpha_{\beta\nu}$	Levi–Civita connection
$R^\alpha_{\beta\mu\nu}$	$\bar R^\alpha_{\beta\mu\nu}$	Riemann tensor
$R_{\mu\nu} = R^\alpha_{\mu\alpha\nu}$	$\bar R_{\mu\nu} = \bar R^\alpha_{\mu\alpha\nu}$	Ricci tensor
$R = g^{\mu\nu} R_{\mu\nu}$	$\bar R = \bar g^{\mu\nu} \bar R_{\mu\nu}$	scalar curvature
$u^\mu_{\alpha\beta} = \Gamma^\mu_{\alpha\beta} - \delta^\mu_{(\alpha}\Gamma^\epsilon_{\beta)\epsilon}$	$\bar u^\mu_{\alpha\beta} = \bar\Gamma^\mu_{\alpha\beta} - \delta^\mu_{(\alpha}\bar\Gamma^\epsilon_{\beta)\epsilon}$	

(7.5.12)

We also introduce the following *relative quantities*, which are both tensors:

$$\begin{aligned} q^\mu_{\alpha\beta} &= \Gamma^\mu_{\alpha\beta} - \bar\Gamma^\mu_{\alpha\beta} \\ w^\mu_{\alpha\beta} &= u^\mu_{\alpha\beta} - \bar u^\mu_{\alpha\beta} \end{aligned} \tag{7.5.13}$$

For any linear torsionless connection $\bar\Gamma$, the Hilbert-Einstein Lagrangian (7.5.1) can be covariantly recast as:

$$\begin{aligned} L_H &= \mathrm{d}_\alpha(P^{\beta\mu} u^\alpha_{\beta\mu})\,\mathrm{d}s + \tfrac{1}{2\kappa}\left[g^{\beta\mu}(\Gamma^\rho_{\beta\sigma}\Gamma^\sigma_{\rho\mu} - \Gamma^\alpha_{\alpha\sigma}\Gamma^\sigma_{\beta\mu}) - 2\Lambda\right]\sqrt{g}\,\mathrm{d}s = \\ &= \mathrm{d}_\alpha(P^{\beta\mu} w^\alpha_{\beta\mu})\,\mathrm{d}s + \tfrac{1}{2\kappa}\left[g^{\beta\mu}(\bar R_{\beta\mu} + q^\rho_{\beta\sigma}q^\sigma_{\rho\mu} - q^\alpha_{\alpha\sigma}q^\sigma_{\beta\mu}) - 2\Lambda\right]\sqrt{g}\,\mathrm{d}s \end{aligned} \tag{7.5.14}$$

The first expression for L_H shows that $\bar{\Gamma}$ (or $\bar{g}$, if $\bar{\Gamma}$ are assumed *a priori* to be Christoffel symbols of the reference metric $\bar{g}$) has no dynamics, i.e. field equations for the reference connection are identically satisfied (since any dependence on it is hidden under a divergence). The second expression shows instead that the same Einstein equations for g can be obtained as the Euler-Lagrange equation for the Lagrangian:

$$L_1 = \frac{1}{2\kappa}\left[g^{\beta\mu}(\bar{R}_{\beta\mu} + q^\rho_{\beta\sigma}q^\sigma_{\rho\mu} - q^\alpha_{\alpha\sigma}q^\sigma_{\beta\mu}) - 2\Lambda\right]\sqrt{g}\,\mathrm{d}s \tag{7.5.15}$$

which is first order in the dynamical field g and it is covariant since q is a tensor. The two Lagrangians L_H and L_1 are thence said to be equivalent, since they provide the same field equations (see Proposition (6.3.15) above).

We stress that we did not yet define a natural theory; in fact we still have to declare our attitude towards the reference field $\bar{\Gamma}$. One possibility is to mimic the procedure used in Yang-Mills theories, i.e. restrict ourselves to variations which keep the reference background fixed. In the worked example of Yang-Mills theory over a fixed non-dynamical background we already noticed that the result is affected by a number of problems, e.g. in dealing with conserved quantities. Alternatively we can consider $\bar{\Gamma}$ (or $\bar{g}$) as a dynamical field exactly as g is, even though the reference is not endowed with a physical meaning. In other words, we consider arbitrary variations and arbitrary transformations even if we declare that g is *"observable"* and genuinely related to the gravitational field, while $\bar{\Gamma}$ is not observable and it just sets the reference level of conserved quantities. A further important role played by $\bar{\Gamma}$ is that it allows covariance of the first order Lagrangian L_1. No first order Lagrangian for Einstein equations exists, in fact, if one does not allow the existence of a reference background field (a connection or something else, e.g. a metric or a tetrad field).

We stress that at this point of our development it is still not definitely clear whether and why natural theories should be considered as a paradigm of (al least a class of) fundamental field theories. Natural theories are just a special class of field theories with a number of good mathematical and physical features, but there is no *a priori* need to believe that all theories should be natural. In any case, if we discover that a theory is natural we get a number of relevant advantages, e.g. about conserved quantities.

To obtain a good and physically sound theory out of the Lagrangian L_1 we still have to improve its dependence on the reference background $\bar{\Gamma}$. For the sake of simplicity we shall assume from now on that $\bar{\Gamma}$ is the Levi-Civita connection of a metric $\bar{g}$ which thence becomes the reference background. Let us also assume (even if this is not at all necessary) that the reference background $\bar{g}$ is Lorentzian. We shall introduce a dynamics for the reference background

$\bar{g}$, (thus transforming its Levi-Civita connection into a truly dynamical connection), by considering a new Lagrangian

$$\begin{aligned} L_{1B} =& \frac{1}{2\kappa}\left[\sqrt{g}\,(R-2\Lambda) - \mathrm{d}_\alpha\big(\sqrt{g}\,g^{\mu\nu}w^\alpha_{\mu\nu}\big) - \sqrt{\bar{g}}\,(\bar{R}-2\Lambda)\right]\,\mathrm{d}s = \\ =& \frac{1}{2\kappa}\left[(\bar{R}-2\Lambda)\,(\sqrt{g}-\sqrt{\bar{g}}) + \sqrt{g}\,g^{\beta\mu}(q^\rho_{\beta\sigma}q^\sigma_{\rho\mu} - q^\alpha_{\alpha\sigma}q^\sigma_{\beta\mu})\right]\,\mathrm{d}s \end{aligned} \tag{7.5.16}$$

which is obtained from L_1 by subtracting the kinetic term $(\bar{R}-2\Lambda)\,\sqrt{\bar{g}}$.[9] Of course, the field $\bar{g}$ is no longer undetermined by field equations (as it was for the simpler Lagrangian (7.5.14)) but it has to be a solution of the variational equations for L_{1B} with respect to $\bar{g}$, which are easily seen to coincide with Einstein field equations. Why should a reference field, which we pretend not to be observable, obey some field equation? Field equations are here functional to the role that $\bar{g}$ plays in our framework. If $\bar{g}$ has to fix the zero value of conserved quantities of g which are *relative to the reference configuration* $\bar{g}$ it is thence reasonable to require that $\bar{g}$ is a solution of Einstein equations as well. Under this assumption, in fact, both g and $\bar{g}$ represent a physical situation and relative conserved quantities represent, for example, the energy *"spent to go"* from the configuration $\bar{g}$ to the configuration g.

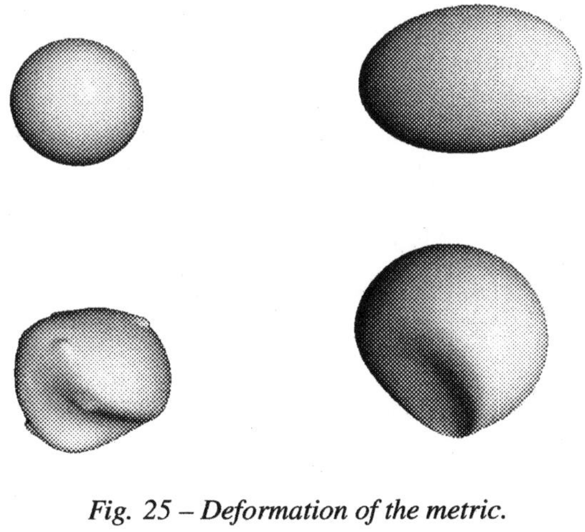

Fig. 25 – Deformation of the metric.

To be strictly precise, further hypotheses should be made to make the whole matter physically meaningful in concrete situations. In a suitable sense we

[9] In other words, the Lagrangian (7.5.16) is obtained from the second order Hilbert–Einstein Lagrangian $\left[\sqrt{g}\,(R-2\Lambda) - \sqrt{\bar{g}}\,(\bar{R}-2\Lambda)\right]\,\mathrm{d}s$ by performing the reduction to first order only in g.

have to ensure that g and $\bar{g}$ belong to the same equivalence class under some (yet undetermined equivalence relation), e.g. that g can be homotopically deformed onto $\bar{g}$ or that they satisfy some common set of boundary (or asymptotic) conditions.

From now on the Lagrangian $L_{1\mathrm{B}}$ of (7.5.16) will be thence considered as a function of the two dynamical fields g and $\bar{g}$, first order in g and second order in $\bar{g}$.[10] The field g is endowed with a physical meaning ultimately related to the gravitational field, while $\bar{g}$ is not observable and it provides at once covariance and the zero level of conserved quantities. We stress that, from now on, deformations will be ordinary (unrestricted) deformations both on g and $\bar{g}$, and symmetries will drag both g and $\bar{g}$. Of course, a natural framework has to be absolute to have a sense; any further trick or limitation does eventually destroy the naturality. The Lagrangian $L_{1\mathrm{B}}$ is thence a Lagrangian $L_{1\mathrm{B}} : J^2\mathrm{Lor}(M) \times_M J^1\mathrm{Lor}(M) \longrightarrow A_m(M)$.

Let ξ be a spacetime vector field, which lifts to the configuration bundles giving

$$\hat{\xi} = \xi^\mu\,\partial_\mu - (g_{\alpha\nu}\partial_\mu\xi^\alpha + g_{\alpha\mu}\partial_\nu\xi^\alpha)\,\partial^{\mu\nu} - (\bar{g}_{\alpha\nu}\partial_\mu\xi^\alpha + \bar{g}_{\alpha\mu}\partial_\nu\xi^\alpha)\,\bar{\partial}^{\mu\nu} \qquad (7.5.17)$$

which is a Lagrangian symmetry. We can choose the dynamical connection of axiom (c) to be the Levi-Civita connection $\Gamma^\alpha_{\beta\mu}$ of the dynamical metric g. We have then a natural theory; a number of mathematical consequences are automatically drawn and they have to be checked against the phenomenology of gravitational physics. In these consequences the opportunity of the natural framework for General Relativity relies.

We can now calculate the contributions to the Nöther current and the superpotential from the divergence term in the first order Lagrangian $L_{1\mathrm{B}}$, namely $L_D = -\frac{1}{2\kappa}\mathrm{d}_\alpha\big(\sqrt{g}\,g^{\mu\nu}w^\alpha_{\mu\nu}\big)\,\mathrm{d}s$, obtaining explicitly:

$$\mathcal{E}(L_D,\xi) = -\tfrac{1}{2\kappa}\big[\pounds_\xi(\sqrt{g}\,w^\alpha_{\mu\nu}g^{\mu\nu}) - \xi^\alpha\nabla_\rho(\sqrt{g}\,w^\rho_{\mu\nu}g^{\mu\nu})\big]\,\mathrm{d}s_\alpha \qquad (7.5.18)$$

which can be recast in the form $\mathcal{E}(L_D,\xi) = \tilde{\mathcal{E}}(L_D,\xi) + \mathrm{Div}\,\mathcal{U}(L_D,\xi)$ by setting:

$$\begin{aligned} \tilde{\mathcal{E}}(L_D,\xi) &= 0 \\ \mathcal{U}(L_D,\xi) &= \frac{\sqrt{g}}{2\kappa}w^{[\sigma}_{\mu\nu}\xi^{\alpha]}\,g^{\mu\nu}\,\mathrm{d}s_{\alpha\sigma} \end{aligned} \qquad (7.5.19)$$

These have to be understood as the corrections to the quantities computed out of the Hilbert Lagrangian L_H to obtain the corresponding quantities for the first order Lagrangian $L_{1\mathrm{B}}$. For example, for the superpotential we have:

$$\mathcal{U}(L_{1\mathrm{B}},\xi) = \frac{1}{2\kappa}\left[\nabla^{[\sigma}\xi^{\alpha]}\sqrt{g} + w^{[\sigma}_{\mu\nu}\xi^{\alpha]}\,g^{\mu\nu}\sqrt{g} - \bar{\nabla}^{[\sigma}\xi^{\alpha]}\sqrt{\bar{g}}\right]\mathrm{d}s_{\alpha\sigma} \qquad (7.5.20)$$

[10] The name $L_{1\mathrm{B}}$ has in fact been chosen to recall that $L_{1\mathrm{B}}$ is obtained from L_1 by adding a dynamical background.

where $\bar{\nabla}$ denotes the covariant derivative with respect to the reference background metric $\bar{g}$. Notice that the reduced current is not changed with respect to (7.5.10), while the Komar superpotential suffers the addition of two extra terms.

The Nöther current $\mathcal{E}(L_{1\mathrm{B}}, \xi)$ is conserved *on-shell*, being $\tilde{\mathcal{E}}(L_{1\mathrm{B}}, \xi) = 0$ in virtue of field equations, while $\mathcal{U}(L_{1\mathrm{B}}, \xi)$ is the (reduced) superpotential. The corresponding conserved quantity has to be interpreted as the *relative conserved quantity with respect to* $\bar{g}$.

We stress that no *a priori* hypothesis has been done yet on the asymptotic behavior of solutions. In particular the solution is not assumed to be *asymptotically flat* (for this we refer to [RT74], [FFS92] and [FFFR99]).

Before going on considering various interactions between gravity and matter fields let us consider some of the standard solutions to vacuum Einstein field equations.

Schwarzschild Solution

As an example we can consider the Schwarzschild solution in $\dim(M) = 4$. Let us fix the signature $(+,-,-,-)$ so that $\kappa = -8\pi$. The metric is given in the following form:

$$\begin{aligned} g_{\mathrm{Sch}} =& \mathrm{d}t^2 - \delta_{ij}\left(\mathrm{d}x^i - f(r)\, n^i\, \mathrm{d}t\right)\left(\mathrm{d}x^j - f(r)\, n^j\, \mathrm{d}t\right) \\ & r^2 = x_i\, x^i, \quad n^i = \frac{x^i}{r}, \quad f(r) = \pm\sqrt{\frac{2m}{r}}, \qquad i = 1,2,3 \end{aligned} \tag{7.5.21}$$

under which the singularity lies at $r = 0$. The metric (7.5.21) is in fact defined in the topology $\mathbb{R} \times \mathbb{R} \times S^2$. It is a solution of vacuum Einstein field equations without cosmological constant, i.e. $\Lambda = 0$. It is an easy task to see that (7.5.21) is in fact an equivalent form of the classical Schwarzschild solution in spherical polar coordinates (τ, r, θ, ϕ)

$$g_{\mathrm{Sch}} = \left(1 - \frac{2m}{r}\right)\mathrm{d}\tau^2 - \left(1 - \frac{2m}{r}\right)^{-1}\mathrm{d}r^2 - r^2\mathrm{d}\theta^2 - r^2\sin^2\theta\,\mathrm{d}\phi^2 \tag{7.5.22}$$

which (for $r > 2m$) describes the exterior space out of a (non-rotating) black hole of mass m. We are not interested here in its maximal analytic extension (*à la Kruskal*; see [HE73] pages $153 - 155$) nor to its interpretation as the gravitational field of a pointlike massive source located at $r = 0$. The relation between the two sets of coordinates is simply

$$t = -\tau + 2\sqrt{2mr} - 4m\,\mathrm{arctanh}\left(\frac{r}{2m}\right) \tag{7.5.23}$$

while x^i are simply the spatial Cartesian coordinates which correspond to the spherical coordinates (r, θ, ϕ).

When one chooses local coordinates adapted to an ADM foliation the metric can be written in the form

$$g = g_{\mu\nu} \mathrm{d}x^\mu \otimes \mathrm{d}x^\nu = -N^2\, \mathrm{d}t^2 + h_{ij}(\mathrm{d}x^i + N^i\, \mathrm{d}t) \otimes (\mathrm{d}x^j + N^j\, \mathrm{d}t) \tag{7.5.24}$$

The (local) function N is called the *lapse* and the (local) 3-vector N^i is called the *shift*.

Notice that the shift vector $N^i = f(r)\, n^i$ of (7.5.21) goes to zero as r goes to infinity but it does not decay fast enough for the solution to be asymptotically flat in a reasonable sense; accordingly, the mass of the black hole cannot be calculated in these coordinates using standard ADM methods.

Let Σ be a spatial 2-surface including the singularity at $r = 0$ (e.g. a sphere r=constant); in the coordinates (t, r, θ, ϕ) we have the superpotential:

$$\mathcal{U}(L_{1\mathrm{B}}, \xi) = \frac{m}{2\pi} \sin(\phi)\, \mathrm{d}\theta \wedge \mathrm{d}\phi \tag{7.5.25}$$

Let us choose the vector $\xi = \partial_t$ and integrate over Σ; the result is $Q_\Sigma(\partial_t) = m$, which is the expected value for the total mass in this case.

Let us now consider (7.5.21) as a 1-parameter family of metrics $g_{(m)}$ indexed by the real (positive) parameter m. Letting $m \longrightarrow 0$ one obtains of course Minkowski spacetime as a limiting metric (in this case no singularity appears and the metric is in fact globally defined on $\mathbb{R}^4$). Accordingly, we may envisage (7.5.21) as a *finite deformation* of the Minkowski reference background.

Of course, the Minkowski background we chose is just one among infinitely many possibilities each leading to a different result for the relative energy. For example, we can choose a different Minkowski space which is matched to the Schwarzschild solution on a spatial sphere of radius $r = R$. One can consider the Minkowski solution inside the sphere (i.e. for $r \leq R$) prolonged with the Schwarzschild solution outside the sphere (i.e. for $r > R$). This metric $\bar{g}$ agrees with g everywhere but in the sphere $r \leq R$; the relative conserved quantity can be interpreted as the energy of the gravitational field within the sphere.

Let us now consider the Schwarzschild metric g in standard isotropic coordinates (t, ρ, θ, ϕ):

$$g = \frac{(2\rho - M)^2}{(2\rho + M)^2}\, \mathrm{d}t^2 - \left(1 + \frac{M}{2\rho}\right)^4 \left(\mathrm{d}\rho^2 + \rho^2\, (\mathrm{d}\theta^2 + \sin^2\theta\, \mathrm{d}\phi^2)\right) \tag{7.5.26}$$

where we set $2\rho = -M + r + \sqrt{r(r - 2M)}$, i.e. $r = (4\rho)^{-1}(M + 2\rho)^2$. Let us choose as a background the Minkowski spacetime, written in the unusual form

$$\bar{g} = \frac{(2R - M)^2}{(2R + M)^2}\, \mathrm{d}t^2 - \left(1 + \frac{M}{2R}\right)^4 \left(\mathrm{d}\rho^2 + \rho^2\, (\mathrm{d}\theta^2 + \sin^2\theta\, \mathrm{d}\phi^2)\right) \tag{7.5.27}$$

The background $\bar{g}$ is in fact a flat metric as one can easily check by direct computation of its Riemann tensor. Furthermore, the two metrics g and $\bar{g}$ are matched on the hypersurface $\mathcal{B}$ defined by the equation $\rho = R$.

Let us then choose the vector field

$$\xi = \alpha\, \partial_t + \beta \partial_\phi \tag{7.5.28}$$

(where α and β are two real constants) which is a generator of symmetries for the first order Lagrangian. We remark that ξ is a well-defined vector field on $\mathcal{B}$; in particular it extends to $\theta = 0$ and $\theta = \pi$. For $\alpha = 1$ and $\beta = 0$, ξ reduces to the ordinary time translation $\xi = \partial_t$ so that we expect the corresponding conserved quantity $Q_\Sigma(\partial_t)$ to be interpreted as the *mass* of g *relative* to $\bar{g}$ on the leaf $\Sigma = \{t \text{ constant}\}$ in the region $\rho \leq R$.

If we calculate the Nöther conserved quantity for (7.5.28) we get the following result

$$Q_\Sigma(\xi) = \alpha \left(m - \frac{m^2}{2R} \right) \tag{7.5.29}$$

which shows no dependence on β.

Thus, when ξ is the Killing vector ∂_t ($\alpha = 1, \beta = 0$) we get the *relative energy* while of course no contribution to angular momentum arises. We also stress that the *conserved quantity* (i.e. letting R tend to infinity) is always $Q_\infty[\partial_t] = m$, i.e. it reduces to the expected value for the total mass. Classically speaking $\frac{m^2}{2R}$ corresponds to the energy gained in pushing a mass m from infinity onto a shell of radius $\rho = R$ so that $Q_\Sigma[\partial_t]$ may be interpreted as the true energy of the gravitational field inside the shell.

Kerr-Newman Solution

Of course, the anomalous factor problem is not very relevant when dealing with the Schwarzschild solution which has no angular momentum. Let us thence discuss the Kerr-Newman solution (again with $\Lambda = 0$) which has both a mass and an angular momentum. In coordinates (t, r, θ, ϕ) we have the well known expression

$$g_{\text{KN}} = \bar{g} - 2mr\rho^{-2} \left[\mathrm{d}t + \mathrm{d}r - a \sin^2 \theta \mathrm{d}\phi \right]^2 \tag{7.5.30}$$

where $\rho^2 = r^2 + a^2 \cos^2 \theta$, $m^2 \geq a^2$ and the reference background $\bar{g}$ is

$$\bar{g} = \mathrm{d}t^2 - \left[\mathrm{d}r - a \sin^2 \theta \, \mathrm{d}\phi \right]^2 - \rho^2 \left[\mathrm{d}\theta^2 + \sin^2 \theta \, \mathrm{d}\phi^2 \right] \tag{7.5.31}$$

This background is chosen so that *a posteriori* the correct conserved quantities are produced.

Let us again choose the vector

$$\xi = \alpha\,\partial_t + \beta\,\partial_\phi \tag{7.5.32}$$

as symmetry generator, which is a Killing vector for both g and $\bar{g}$ (α and β are two real constants).

The Nöther conserved quantities one obtains are now

$$Q_\Sigma(\xi) = \alpha\, m - \beta\, ma \tag{7.5.33}$$

which reproduce the expected values of the *relative mass* and *angular momentum* in the region $\Sigma = \{t \text{ constant}\}$ and $r \leq R$. Notice that the result is independent on R meaning that all the energy and angular momentum is *"buried in the singularity"*. We remark that setting everywhere $a = 0$ the Schwarzschild solution is recovered. The relative mass m we obtain in this case does not agree with the value found above in (7.5.29) because of the two different matches selected. A more complete discussion can be found in [FFS92], [FFFR99] and [FFFR01].

BTZ Solution

Both the Schwarzschild and the Kerr-Newman solutions are asymptotically flat in the sense that they match a Minkowski metric at infinity. Let us now consider $\dim(M) = 3$ and the so-called *BTZ metric*

$$g_{\text{BTZ}} = -N^2 \text{d}t^2 + N^{-2}\text{d}r^2 + r^2(N_\phi \text{d}t + d\phi)^2 \tag{7.5.34}$$

where we set

$$N^2 = -m + \frac{r^2}{l^2} + \frac{J^2}{4r^2}\,, \qquad\qquad N_\phi = -\frac{J}{2r^2} \tag{7.5.35}$$

Here m, l, J are real parameters interpreted as the mass, the characteristic constant related to the cosmological constant Λ (with $l \neq 0$), and the angular momentum, respectively.

This 3-metric is a solution of Einstein field equations with a (negative) cosmological constant $\Lambda = -l^{-2}$. We remark that in dimension 3 the suitable factor in front of the Lagrangian is obtained for $\kappa = \pi$. Being $\Lambda < 0$ the metric g_{BTZ} is asymptotically anti-de-Sitter, i.e. it matches at infinity the (3-dimensional) anti-de-Sitter metric g_{adS} which is obtain in the limit $m \to -1$ and $J \to 0$:

$$g_{\text{adS}} = -(1 - \Lambda r^2)\text{d}t^2 + (1 - \Lambda r^2)^{-1}\text{d}r^2 + r^2 d\phi^2 \tag{7.5.36}$$

To be general enough we shall choose as a reference background another BTZ metric $\bar{g}_{\text{BTZ}}$ of the form (7.5.34) for other parameters (m_0, J_0) but with the same l. The anti-de-Sitter background is of course a particular case. Another background which is used in the literature on the subject is the limit solution obtained for $m \to 0$ and $J \to 0$, when the black hole disappears.

The conserved quantities obtained are[11]

$$\begin{aligned} Q(L_{1B}, \partial_t; g_{BTZ}, \bar{g}_{BTZ}) &= \int_{S^1_r} \mathcal{U}(L_{1B}, \partial_t; g_{BTZ}, \bar{g}_{BTZ}) = \\ &= \left(2\frac{r^2}{l^2} \oplus (m - m_0) \ominus 2\frac{r^2}{l^2}\right) = m - m_0 \\ Q(L_{1B}, \partial_\phi; g_{BTZ}, \bar{g}_{BTZ}) &= \int_{S^1_r} \mathcal{U}(L_{1B}, \partial_\phi; g_{BTZ}, \bar{g}_{BTZ}) = \\ &= -\left(J \oplus 0 \ominus J_0\right) = -(J - J_0) \end{aligned} \tag{7.5.37}$$

We remark that the conserved quantities (7.5.37) are integrated on a spatial sphere S^2 of radius $R \to \infty$. On a finite sphere the conserved quantities (7.5.37) are just the first terms of two power series in R^{-1}. In the specific case the contributions due to Komar superpotentials are the exact result, while the divergence term is due to a series

$$\begin{aligned} &(m - m_0) - l^2 \frac{(m - m_0)^2}{r^2} + \\ &+ l^2 (m - m_0) \frac{4(m^2 - m_0^2)l^2 - 3(J^2 - J_0^2)}{4r^4} + O(r^{-6}) \end{aligned} \tag{7.5.38}$$

Here we keep separate the contribution of the Komar potential for the metric g, the contribution of the divergence $\mathrm{Div}(P^{\alpha\beta}\, w^\lambda_{\alpha\beta}\, \mathrm{d}s_\lambda)$ and the contribution of the Komar potential of the reference background $\bar{g}$, respectively.

Notice that the Komar superpotential alone produces an infinite result for the energy. That is typical of solutions which are not asymptotically flat. Also the first order Lagrangian (7.5.15) produces an infinite prediction for energy. In this case one can see how essential is the Komar potential of the background $\bar{g}$ to obtain a meaningful definition of the energy. We stress that in all cases the correct value is obtained by cancellations or sum. Accordingly, one should talk about *anomalous term* rather than *anomalous factor*. In the Schwarzschild case, in fact, the correct value is not obtained by naively doubling the integral $m/2$ resulting from Komar. On the contrary the correct value m is obtained by considering the contribution of the reference background which contributes by an extra term in the superpotential which, in the end, gives $m = m/2 \oplus m/2$. This remark is important since it shows once more that the correct energy

[11] We use here and hereafter $\oplus$ and $\ominus$ to better enucleate the separate contributions.

cannot be obtained as the conserved quantity associated to $\xi = 2\partial_t$. The factor in front of the symmetry generator, in fact, depends on the particular solution under consideration and in some cases, as for BTZ solution, there is no "factor" which can be used to regularize the result.

Taub-Bolt Solution

Finally let us discuss an example in the Euclidean sector (i.e. for positive definite signature); the field theory considered is completely analogous to Hilbert-Einstein General Relativity but in the Euclidean signature $(4,0)$ in $\dim(M) = 4$. The Taub-Bolt metric

$$g_{\text{Bolt}} = V_{\text{Bolt}}\,(\mathrm{d}\tau + 2N\,\cos\theta\,\mathrm{d}\phi)^2 + \frac{\mathrm{d}r^2}{V_{\text{Bolt}}} + (r^2 - N^2)(\mathrm{d}\theta^2 + \sin^2\theta\,\mathrm{d}\phi^2) \quad (7.5.39)$$

where we set

$$V_{\text{Bolt}} = \frac{(r-2N)(2r-N)}{2(r^2-N^2)} \quad (7.5.40)$$

is a solution of vacuum field equations. It is a *locally asymptotically flat* solution in the sense that at infinity it locally matches a Minkowski metric but not in a global way (see [M63]). As a reference background we shall choose the Taub-NUT metric

$$g_{\text{NUT}} = V_{\text{NUT}}\,(\mathrm{d}\tau + 2N\,\cos\theta\,\mathrm{d}\phi)^2 + \frac{\mathrm{d}r^2}{V_{\text{NUT}}} + (r^2 - N^2)(\mathrm{d}\theta^2 + \sin^2\theta\,\mathrm{d}\phi^2) \quad (7.5.41)$$

where we set

$$V_{\text{NUT}} = \frac{r-N}{r+N} \quad (7.5.42)$$

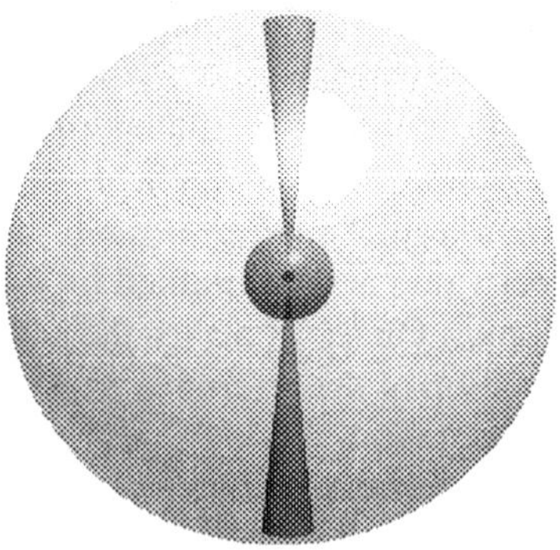

Fig. 26 – Taub Bolt solution.

Here N is a constant parameter related to the global topology of the maximal analytical extension of the Riemannian manifold (M, g_{NUT}); we refer the reader to the exhaustive analysis given in $[M63]$.

Let us choose the vector

$$\xi = \alpha\, \partial_\tau + \beta\, \partial_\phi \tag{7.5.43}$$

as symmetry generator.

The Komar potential of the Taub-Bolt metric (7.5.39) contributes to conserved quantities by a series

$$\alpha\, \frac{5}{8} N - \alpha\, \frac{N^2}{R} + \alpha\, \frac{5N^3}{4R^2} - \alpha\, \frac{N^4}{R^3} + O(R^{-4}) \tag{7.5.44}$$

The divergence in the Lagrangian contributes by

$$\alpha\, \frac{1}{8} N + \alpha\, \frac{N^3}{16R^2} + \alpha\, \frac{5N^4}{32R^3} + O(R^{-4}) \tag{7.5.45}$$

while the Komar potential of the Taub-NUT solution contributes by

$$\alpha\, \frac{1}{2} N - \alpha\, \frac{N^2}{R} + \alpha\, \frac{N^3}{R^2} - \alpha\, \frac{N^4}{R^3} + O(R^{-4}) \tag{7.5.46}$$

In the limit for $R \to \infty$ only the leading terms survive and the total conserved quantity is

$$Q(L_{1\mathrm{B}}, \xi; g_{\mathrm{Bolt}}, g_{\mathrm{NUT}}) = \alpha \frac{N}{4} \tag{7.5.47}$$

which is the expected value for the energy of the Taub-Bolt solution.

6. On Maxwell and Yang-Mills Naturality

In Section 6.7 we discussed Yang-Mills and Maxwell field theories as an example of gauge theory. One could investigate if and when those theories can be also endowed with a natural structure. In particular, in Maxwell theory the field strength $F = \frac{1}{2} F_{\mu\nu}\, dx^\mu \wedge dx^\nu$ is a real 2-form on spacetime, i.e. a section of $A_2(M)$ and thence a natural object at the same time. The homogeneous Maxwell equation $dF = 0$ is an equation involving forms and it has a well-known local solution $F = d\tilde{A}$, i.e. there exists a local spacetime 1-form $\tilde{A}$ which is a potential for the field strength F. Of course, if spacetime is contractible, as e.g. for Minkowski space, the solution is also a global one.

As is well-known, in the non-commutative Yang-Mills theory case the field strength $F = \frac{1}{2} F^A_{\mu\nu}\, T_A \otimes dx^\mu \wedge dx^\nu$ is no longer a spacetime form (see equation (3.8.15)). This is a somewhat trivial remark since the transformation laws of such field strength are obtained as the transformation laws of the curvature of a principal connection with values in the Lie algebra of some (semisimple) non-Abelian Lie group G (e.g. $G = \mathrm{SU}(n)$, $n \geq 2$). However, the common belief that electromagnetism is to be intended as the particular case (for $G = U(1)$) of a non-commutative theory is not really physically evident. Even if we subscribe this common belief, which is motivated also by the tremendous success of the quantized theory, let us for a while discuss electromagnetism as a standalone theory.

From a mathematical viewpoint this is a (different) approach to electromagnetism and the choice between the two can be dealt with on a physical ground only. Of course the 1-form $\tilde{A}$ is defined modulo a closed form, i.e. locally $\tilde{A}' = \tilde{A} + \mathrm{d}\alpha$ is another solution.

How can one decide whether the potential of electromagnetism should be considered as a 1-form or rather as a principal connection on a $U(1)$-bundle? First of all we notice that by a standard *hole argument* (one can easily define compact supported closed 1-forms, e.g. by choosing the differential of compact supported functions which always exist on a paracompact manifold) the potentials $\tilde{A}$ and $\tilde{A}'$ represent the same physical situation. On the other hand, from a mathematical viewpoint we would like the dynamical field, i.e. the potential $\tilde{A}$, to be a global section of some suitable configuration bundle. We stress that this requirement is a mathematical one, motivated on the wish of a well-defined geometrical perspective based on global Variational Calculus. We shall hereafter analyze some ways to reconcile these characteristics with naturality. In the following Chapter we shall finally resort to a gauge natural formalism, which encompasses both viewpoints.

The first mathematical way out is to restrict our attention to contractible spacetimes, where $\tilde{A}$ may be always chosen to be global. Then one can require the gauge transformations $\tilde{A}' = \tilde{A} + \mathrm{d}\alpha$ to be Lagrangian symmetries. In this way, in as much the same way as we did for the relativistic particle, field equations select a whole equivalence class of gauge-equivalent potentials, a procedure which solves the hole argument problem. In this picture the potential $\tilde{A}$ is really a 1-form, which can be dragged along spacetime diffeomorphism and which admits the ordinary Lie derivatives of 1-forms. Unfortunately, the restriction to contractible spacetimes is physically unmotivated and probably wrong.

Alternatively, one can restrict electromagnetic fields F, deciding that only exact 2-forms F are allowed. That actually restricts the observable physical situations, by changing the homogeneous Maxwell equations (i.e. Bianchi identities) by requiring that F is not only closed but exact. One should in principle be able to empirically reject this option.

On non-contractible spacetimes, one is necessarily forced to resort to a more *"democratic"* attitude. The spacetime is covered by a number of patches U_α.

On each patch U_α one defines a potential $A^{(\alpha)}$. In the intersection of two patches the two potentials $A^{(\alpha)}$ and $A^{(\beta)}$ may not agree. In each patch, in fact, the observer chooses his own conventions and he finds a different representative of the electromagnetic potential, which is related by a gauge transformation to the representatives chosen in the neighbour patch(es). Thence we have a family of gauge transformations, one in each intersection $U_{\alpha\beta}$, which obey cocycle identities. If one recognizes in them the action of $U(1)$ then one can build a principal bundle $\mathcal{P} = (P, M, \pi; U(1))$ and interpret the ensuing potential as a connection on $\mathcal{P}$. As we shall see this is the leading way to the *gauge natural formalism*.

Anyway this does not close the matter. One can investigate if and when the principal bundle $\mathcal{P}$, in addition to the obvious principal structure, can be also endowed with a natural structure. If that were possible then the bundle of connections $\mathcal{C}_\mathcal{P}$ (which is *associated* to $\mathcal{P}$) would also be natural. The problem of deciding whether a given gauge natural bundle can be endowed with a natural structure is quite difficult in general and no full theory is yet completely developed in mathematical terms. That is to say, there is no complete classification of the topological and differential geometric conditions which a principal bundle $\mathcal{P}$ has to satisfy in order to ensure that, among the principal trivializations which determine its gauge natural structure, one can choose a sub-class of trivializations which induce a purely natural bundle structure. Nor it is clear how many inequivalent natural structures a *good* principal bundle may support. First of all let us remark that there are important examples of bundles which support at the same time a natural and a gauge natural structure. Actually any natural bundle is associated to some frame bundle $L^s(M)$, which is principal; thence each natural bundle is also gauge natural in a trivial way. Nevertheless there are also less trivial examples. Since on any paracompact manifold one can choose a global Riemannian metric g, the corresponding tangent bundle $T(M)$ can be associated to the orthonormal frame bundle $\mathrm{O}(M, g)$ besides being obviously associated to $L(M)$. Thence the natural bundle $T(M)$ may be also endowed with a gauge natural bundle structure with structure group $\mathrm{O}(m)$. And if M is orientable the structure can be further reduced to a gauge natural bundle with structure group $\mathrm{SO}(m)$.

Roughly speaking, the task is achieved by imposing restrictions to cocycles which generate $T(M)$ according to the prescription (1.4.13) (i.e. by imposing a privileged class of changes of local laboratories and sets of measures). In our case, one imposes the cocycle $\varphi_{(\alpha\beta)}$ to take its values in $\mathrm{O}(m)$ rather than in the larger group $\mathrm{GL}(m)$. Inequivalent gauge natural structures are in one-to-one correspondence with (non isometric) Riemannian metrics on M.

Actually whenever we have a Lie group homomorphism $\rho : \mathrm{GL}^s(m) \longrightarrow G$ for some s onto some given Lie group G we can build a natural G-principal bundle on M. In fact, let $(U_\alpha, \varphi_{(\alpha)})$ be an atlas of the given manifold M,

$\varphi_{(\alpha\beta)}$ be its transition functions and $j^s\varphi_{(\alpha\beta)}$ be the induced transition functions of $L^s(M)$; then we can define a G-valued cocycle on M by setting $\rho(j^s\varphi_{(\alpha\beta)})$ and thence a (unique up to fibered isomorphisms) G-principal bundle $\mathcal{P}(M) = (P(M), M, \pi; G)$. The bundle $\mathcal{P}(M)$, as well as any gauge natural bundle associated to it, is natural by construction.

With this in mind, we can define a whole family of natural $U(1)$-bundles $\mathcal{P}_q(M)$ by using the bundle homomorphisms

$$\rho_q : \mathrm{GL}(m) \longrightarrow U(1) : J \mapsto \exp(iq \ln \det |J|) \tag{7.6.1}$$

where q is any real number and $\ln$ denotes the natural logarithm. In the case $q = 0$ the image of ρ_0 is the trivial group $\{\mathbb{I}\}$; as we shall see below, all the induced bundles are trivial, i.e. $P = M \times U(1)$.

The natural lift $\hat{\phi}$ of a diffeomorphism $\phi : M \longrightarrow M$ is given by

$$\hat{\phi}[x, e^{i\theta}]_\alpha = [\phi(x), e^{iq \ln \det |J|} \cdot e^{i\theta}]_\alpha \tag{7.6.2}$$

where J is the Jacobian of the morphism ϕ.

The bundles $\mathcal{P}_q(M)$ are all trivial since they allow a global section. In fact, on any manifold M one can define a (global) Riemannian metric g; the local sections

$$\sigma_\alpha : x \mapsto [x, \exp(iq \ln \sqrt{g})]_\alpha \tag{7.6.3}$$

glue together, since changing chart one has

$$\begin{aligned}\sigma_\beta =&[x, \exp(iq \ln \det \sqrt{g})]_\beta = \\ =&[x, \exp(iq \ln \det |J|) \cdot \exp(iq \ln \sqrt{g})]_\beta = [x, \exp(iq \ln \sqrt{g})]_\alpha = \sigma_\alpha\end{aligned} \tag{7.6.4}$$

Since the bundles $\mathcal{P}_q(M)$ are all trivial, they are all isomorphic to $M \times U(1)$ as principal $U(1)$-bundles, though in a non-canonical way unless $q = 0$. Any two of the bundles $\mathcal{P}_{q_1}(M)$ and $\mathcal{P}_{q_2}(M)$ for two different values of q are isomorphic as principal bundles but the isomorphism obtained is not the lift of a spacetime diffeomorphism because of the two different values of q. Thence they are *not* isomorphic as natural bundles. We are thence facing a very interesting situation: a gauge natural bundle $\mathcal{C}$ associated to the trivial principal bundle $\mathcal{P}$ can be endowed with an infinite family of natural structures, one for each $q \in \mathbb{R}$; each of these natural structures can be used to regard principal connections on $\mathcal{P}$ as natural objects on M and thence one can regard electromagnetism as a natural theory.

Now that the mathematical situation has been a little bit clarified, it is again a matter of physical interpretation. One can in fact restrict to electromagnetic potentials which are *a priori* connections on a *trivial* structure bundle $\mathcal{P} \simeq M \times U(1)$ or to accept that more complicated situations may occur in Nature. Once again we have to stress that these non-trivial situations are still empirically unsupported, at least at a fundamental level.

We shall see in the next Chapter how to overcome all difficulties arising from the fact that in general connections of general structure bundles are not natural. In particular we shall see how to define the energy-momentum tensor of the electromagnetic field even if dynamical fields cannot be dragged along spacetime diffeomorphisms. Of course this does not prove that electromagnetism on non-trivial structure bundles is physically sound. In the same way, the fact that the trivial situation can be geometrically unified with gravity does not prove the contrary. Which one is the true physical description can be decided only on physical grounds. It depends on whether one can accept a situation in which the field strength observed is the curvature of a non-trivial connection. One example would be the Dirac monopole, which, however, has never been observed yet.

7. Gravity and Bosonic Matter

Let us now draw our attention back to General Relativity and its interaction with matter fields. We shall present a general framework for describing the interactions and then specify to a number of applications considering the most common situations.

Let us consider a field theory aimed to describe gravity coupled to some arbitrary Bosonic matter. Let us assume as dynamical fields a (Lorentzian) metric g, a reference background metric $\bar{g}$ and a matter field φ associated to a representation $\lambda : \mathrm{GL}(m) \times V \longrightarrow V$, $m = \dim(M)$ (see Section 7.5 for a discussion about the meaning of the reference background field). The configuration bundle is thence the natural bundle

$$\mathcal{C} = \mathrm{Lor}(M) \times_M \mathrm{Lor}(M) \times_M \mathcal{B} \tag{7.7.1}$$

where $\mathcal{B} = L(M) \times_\lambda V$ is the *matter vector bundle*. For example we can fix the representation

$$\begin{aligned} \lambda :&\mathrm{GL}(m) \times T^q_p(\mathbb{R}) \longrightarrow T^q_p(\mathbb{R}) \\ :&(J^\mu_\nu, \varphi^{\mu_1\mu_2\dots\mu_q}_{\nu_1\nu_2\dots\nu_p}) \mapsto (\det J)^k\, J^{\mu_1}_{\rho_1} J^{\mu_2}_{\rho_2} \dots J^{\mu_q}_{\rho_q} \,(\varphi^{\rho_1\rho_2\dots\rho_q}_{\sigma_1\sigma_2\dots\sigma_p})\, \bar{J}^{\sigma_1}_{\nu_1} \bar{J}^{\sigma_2}_{\nu_2} \dots \bar{J}^{\sigma_p}_{\nu_p} \end{aligned} \tag{7.7.2}$$

to describe a tensorial density of rank (p, q) and weight k. In general we shall denote by v^a the coordinates on the vector space V so that the representation λ reads as

$$\lambda : \mathrm{GL}(m) \times V \longrightarrow V : (J^a_b, v^b) \mapsto \lambda^a_b(J)\, v^b \tag{7.7.3}$$

Here and in the sequel Latin indices $a, b, \dots$ will denote components with respect to a basis in V. Whenever V is a tensor space over $\mathbb{R}^m$ the index represents a whole collection of tensorial indices, i.e. $a = \binom{\mu_1\mu_2\dots\mu_q}{\nu_1\nu_2\dots\nu_p}$.

The Lagrangian $L = L_{1\mathrm{B}} + L_{\mathrm{mat}}$ on $\mathcal{C}$ is the sum of the gravitational Lagrangian (7.5.16) and of a *matter Lagrangian* on $J^1 Lor(M) \times_M J^1\mathcal{B}$, i.e.

$$L_{\mathrm{mat}} = \mathcal{L}_{\mathrm{mat}}(x^\mu, g_{\mu\nu}, \partial_\lambda g_{\mu\nu}, v^a, v^a_\lambda)\, \mathrm{ds} \tag{7.7.4}$$

The *total Lagrangian* L, and consequently the matter Lagrangian L_{mat} separately (since $L_{1\mathrm{B}}$ is such), must to be covariant with respect to spacetime diffeomorphisms, i.e. they have to be natural. As we already saw in some examples above, the symmetry requirements strongly constrain the form of the matter Lagrangian. Whenever we impose a symmetry group and we look for the general form of the Lagrangians which allow those symmetries, the requirement results in a proposition which will be called *Utiyama-like theorem.*

Proposition (7.7.5): the matter Lagrangian L_{mat} on $\mathcal{C}$ is natural if and only if it has the following form

$$L_{\mathrm{mat}} = \mathcal{L}^*_{\mathrm{mat}}(g_{\mu\nu}, v^a, \hat{v}^a_\lambda)\, \mathrm{d}s \tag{7.7.6}$$

(where $\hat{v}^a_\lambda = \nabla_\lambda v^a$ is the covariant derivative with respect to the Levi-Civita connection of the dynamical metric g) and the Lagrangian density $\mathcal{L}^*_{\mathrm{mat}}$ is locally Lorentz covariant, i.e.

$$\mathcal{L}^*_{\mathrm{mat}}(g_{\mu\nu}, v^a, \hat{v}^a_\lambda) = \det\bar{\theta}\, \mathcal{L}^*_{\mathrm{mat}}(g'_{\mu\nu}, v'^a, \hat{v}'^a_\lambda) \tag{7.7.7}$$

where we set

$$\begin{cases} g'_{\mu\nu} = \bar{\theta}^\rho_\mu\, g_{\rho\sigma}\, \bar{\theta}^\sigma_\nu \\ v'^a = \lambda^a_b(\theta)\, v^b \\ \hat{v}'^a_\lambda = \lambda^a_b(\theta)\, \hat{v}^b_\epsilon\, \bar{\theta}^\epsilon_\lambda \end{cases} \qquad \theta^\alpha_\beta \in \mathrm{O}(1,3) \tag{7.7.8}$$

and $\bar{\theta}$ denotes the inverse of θ.

Proof: we can choose new coordinates $(x^\mu, g_{\mu\nu}, \Gamma^\alpha_{\beta\mu}, v^a, \hat{v}^a_\lambda)$ on $J^1\mathcal{C}$ by setting

$$\begin{cases} \Gamma^\alpha_{\beta\mu} = \frac{1}{2} g^{\alpha\lambda}(-\mathrm{d}_\lambda g_{\beta\mu} + \mathrm{d}_\beta g_{\mu\lambda} + \mathrm{d}_\mu g_{\lambda\beta}) \\ \hat{v}^a_\lambda = \mathrm{d}_\lambda v^a + \Gamma^a_{b\lambda} v^b \end{cases} \tag{7.7.9}$$

where $\Gamma^a_{b\lambda}$ is the connection induced on $\mathcal{B} = L(M) \times_\lambda V$ by the Levi-Civita connection $\Gamma^\alpha_{\beta\lambda}$, i.e.

$$\Gamma^a_{b\lambda} = \partial^\beta_\alpha \lambda^a_b(\mathbb{I})\, \Gamma^\alpha_{\beta\lambda} \tag{7.7.10}$$

In this way the Lagrangian (7.7.4) can be recast in the following form

$$L_{\mathrm{mat}} = \mathcal{L}^{**}_{\mathrm{mat}}(x^\mu, g_{\mu\nu}, \Gamma^\alpha_{\beta\mu}, v^a, \hat{v}^a_\lambda)\, \mathrm{ds} \tag{7.7.11}$$

without any loss of generality. Now the covariance identity for the Lagrangian L_{mat} reads as

$$\nabla_\epsilon(\xi^\epsilon \mathcal{L}^{**}_{\mathrm{mat}}) = p^{\mu\nu}\pounds_\xi g_{\mu\nu} + p^{\beta\mu}_\alpha \pounds_\xi \Gamma^\alpha_{\beta\mu} + p_a \pounds_\xi v^a + \hat{p}^\lambda_a \pounds_\xi \hat{v}^a_\lambda \tag{7.7.12}$$

where we set for the momenta

$$\begin{cases} p^{\mu\nu} = \dfrac{\partial \mathcal{L}^{**}_{\mathrm{mat}}}{\partial g_{\mu\nu}} \\ p_\alpha^{\beta\mu} = \dfrac{\partial \mathcal{L}^{**}_{\mathrm{mat}}}{\partial \Gamma^\alpha_{\beta\mu}} \end{cases} \qquad \begin{cases} p_a = \dfrac{\partial \mathcal{L}^{**}_{\mathrm{mat}}}{\partial v^a} \\ \hat{p}_a^\lambda = \dfrac{\partial \mathcal{L}^{**}_{\mathrm{mat}}}{\partial \hat{v}^a_\lambda} \end{cases} \tag{7.7.13}$$

Notice that for the covariance identity to be meaningful we have to know how to define the Lie derivative of non-tensorial natural objects, namely of the linear connection $\Gamma^\alpha_{\beta\mu}$. In view of the general theory about Lie derivatives we know that

$$\pounds_\xi \Gamma^\alpha_{\beta\mu} = R^\alpha_{\beta\epsilon\mu}\xi^\epsilon - \nabla_{\beta\mu}\xi^\alpha \tag{7.7.14}$$

where $\nabla_{\beta\mu} \equiv \nabla_{(\beta}\nabla_{\mu)}$ denotes the symmetrized covariant derivative.

The covariance identity is thence equivalent to the following identities

$$\begin{cases} \nabla_\epsilon \mathcal{L}^{**}_{\mathrm{mat}} = p_a\, \nabla_\epsilon v^a + p_a^\lambda\, \nabla_\epsilon \hat{v}^a_\lambda + p_\alpha^{\beta\mu}\, R^\alpha_{\beta\epsilon\mu} \\ \mathcal{L}^{**}_{\mathrm{mat}}\, \delta^\mu_\epsilon = 2p^{\mu\nu}\, g_{\nu\epsilon} - p_a\, \partial^\mu_\epsilon \lambda^a_b(\mathbb{I})\, v^b - \hat{p}_a^\lambda\, \partial^\mu_\epsilon \lambda^a_b(\mathbb{I})\, \hat{v}^b_\lambda + \hat{p}^\mu_a\, \hat{v}^a_\epsilon \\ 0 = p_\alpha^{\beta\mu} \end{cases} \tag{7.7.15}$$

As a consequence the Lagrangian does not depend on the connection $\Gamma^\alpha_{\beta\mu}$. The first identity of (7.7.15) can be rewritten as

$$\nabla_\epsilon \mathcal{L}^{**}_{\mathrm{mat}} = p_a\, \nabla_\epsilon v^a + p_a^\lambda\, \nabla_\epsilon \hat{v}^a_\lambda \tag{7.7.16}$$

On the other hand, inserting the second of (7.7.15) into the left hand side of the first identity we have

$$\begin{aligned} \nabla_\alpha \mathcal{L}^{**}_{\mathrm{mat}} &= \mathrm{d}_\alpha \mathcal{L}^{**}_{\mathrm{mat}} - \Gamma^\epsilon_{\mu\alpha} \mathcal{L}^{**}_{\mathrm{mat}} \delta^\mu_\epsilon = \partial_\alpha \mathcal{L}^{**}_{\mathrm{mat}} + p^{\mu\nu} \nabla_\alpha g_{\mu\nu} + \\ &\quad + p_a\, \nabla_\epsilon v^a + p_a^\lambda\, \nabla_\epsilon \hat{v}^a_\lambda \end{aligned} \tag{7.7.17}$$

Since $\Gamma^\alpha_{\beta\mu}$ is the Levi-Civita connection of g (so that $\nabla_\alpha g_{\mu\nu}$ holds) the first identity of (7.7.15) becomes

$$\partial_\alpha \mathcal{L}^{**}_{\mathrm{mat}} = 0 \tag{7.7.18}$$

i.e. the Lagrangian density $\mathcal{L}^{**}_{\mathrm{mat}}$ cannot depend explicitly on spacetime coordinates.

Accordingly, the matter Lagrangian should be taken in the form $L = \mathcal{L}^*_{\mathrm{mat}}(g_{\mu\nu}, v^a, \hat{v}^a_\lambda)\, \mathrm{d}s$. With this expression the first and the third covariance identities (7.7.15) are identically satisfied, while $\mathcal{L}^*_{\mathrm{mat}}$ is still subjected to the second identity of (7.7.15), i.e.

$$\mathcal{L}^*_{\mathrm{mat}}\, \delta^\mu_\epsilon = 2p^{\mu\nu}\, g_{\nu\epsilon} - p_a\, \partial^\mu_\epsilon \lambda^a_b(\mathbb{I})\, v^b - \hat{p}_a^\lambda\, \partial^\mu_\epsilon \lambda^a_b(\mathbb{I})\, \hat{v}^b_\lambda + \hat{p}^\mu_a\, \hat{v}^a_\epsilon \tag{7.7.19}$$

which is exactly the infinitesimal form of the (7.7.7). ∎

This Utiyama-like proposition completely justifies the traditional physical prescription to build covariant Lagrangians: one first takes a Lorentz covariant Lagrangian and then replaces ordinary derivatives by means of covariant derivatives. We stress however that the notion of "local Lorentz covariance" is really a local one (unless $M = \mathbb{R}^m$ and g is Minkowski metric), so that it is just a shortcut in the formulation of the Utiyama-like theorem above, which is instead geometrically well-posed.

We shall now proceed further into our analysis and we shall present some more specific examples below. We have the Lagrangian $L = L_{1\mathrm{B}} + L_{\mathrm{mat}}$ on $\mathcal{C}$ which depends on the dynamical fields $(g_{\mu\nu}, \bar{g}_{\mu\nu}, v^a)$ and it is second order in the background metric $\bar{g}_{\mu\nu}$ and first order in the other fields $(g_{\mu\nu}, v^a)$. Here we stress that the bundle $\mathcal{B}$ of matter fields is a vector bundle, so that it has a canonical zero section which is assumed as a reference background for the matter. If more general fields were present, one would have to add an interaction term between the matter reference background and the metric reference background to ensure that the metric $\bar{g}$ is subjected to the same field equations of g. In our case, instead, the matter simplifies a bit. Let us first of all introduce a short notation for the first variation formula, namely:

$$< \delta L_{\mathrm{mat}} \,|\, jX >= \frac{\delta \mathcal{L}^*_{\mathrm{mat}}}{\delta g^{\mu\nu}}\, \delta g^{\mu\nu} + \frac{\delta \mathcal{L}^*_{\mathrm{mat}}}{\delta \bar{g}^{\mu\nu}}\, \delta \bar{g}^{\mu\nu} + \frac{\delta \mathcal{L}^*_{\mathrm{mat}}}{\delta v^a}\, \delta v^a + \text{Div terms} \quad (7.7.20)$$

where the coefficients

$$\frac{\delta \mathcal{L}^*_{\mathrm{mat}}}{\delta g^{\mu\nu}}, \qquad \frac{\delta \mathcal{L}^*_{\mathrm{mat}}}{\delta \bar{g}^{\mu\nu}}, \qquad \frac{\delta \mathcal{L}^*_{\mathrm{mat}}}{\delta v^a} \quad (7.7.21)$$

are called the *variational derivatives* with respect to g, $\bar{g}$ and v, respectively.

The dynamical metric g is subjected to field equations of the form

$$R_{\mu\nu} - \frac{1}{2}(R - 2\Lambda)\, g_{\mu\nu} = \kappa\, H_{\mu\nu}\,, \qquad \sqrt{g}\, H_{\mu\nu} = -2\frac{\delta \mathcal{L}^*_{\mathrm{mat}}}{\delta g^{\mu\nu}} \quad (7.7.22)$$

where the tensor $H_{\mu\nu}$ is called the *Hilbert stress tensor*. The Hilbert stress tensor obtained is linear in the matter fields v and in particular it is zero when calculated along the zero section $\bar{v} = 0$. Accordingly, the reference background metric is subjected to the field equations

$$\bar{R}_{\mu\nu} - \frac{1}{2}(\bar{R} - 2\Lambda)\, \bar{g}_{\mu\nu} = \kappa\, \bar{H}_{\mu\nu} \equiv 0 \quad (7.7.23)$$

i.e. Einstein vacuum field equations. Moreover, the matter field is subjected to matter field equations

$$\mathbb{E}_a \equiv \mathbb{E}_a(\mathcal{L}_{\mathrm{mat}}) = \frac{\delta \mathcal{L}^*_{\mathrm{mat}}}{\delta v^a} = 0 \quad (7.7.24)$$

which are assumed to be satisfied also by the matter reference background $\bar{v} = 0$. In the following worked examples matter field equations will be linear so that the zero section is automatically a solution.

The theory is natural and the required dynamical connection is fixed to be the Levi-Civita connection of g. All spacetime vector fields ξ are Lagrangian symmetries. After easy calculations, one sees that the superpotential receives a contribution from the gravitational Lagrangian $L_{1\mathrm{B}}$ which is given in the form

$$\mathcal{U}(L_{1\mathrm{B}},\xi) = \frac{1}{2\kappa}\left[\nabla^{[\sigma}\xi^{\alpha]}\,\sqrt{g} + w^{[\sigma}_{\mu\nu}\xi^{\alpha]}\,g^{\mu\nu}\sqrt{g} - \bar{\nabla}^{[\sigma}\xi^{\alpha]}\,\sqrt{\bar{g}}\right]\mathrm{d}s_{\alpha\sigma} \qquad (7.7.25)$$

The other contribution is from the matter Lagrangian L_{mat}. The covariance identity for the matter Lagrangian reads as

$$\nabla_\sigma(\xi^\sigma\mathcal{L}^*_{\mathrm{mat}}) = p_{\mu\nu}\pounds_\xi g^{\mu\nu} + p_a\pounds_\xi v^a + p^\lambda_a\pounds_\xi\hat{v}^a_\lambda \qquad (7.7.26)$$

In the last term we can exchange the Lie derivative with the covariant derivative by means of the following result

$$\pounds_\xi\hat{v}^a_\lambda = \nabla_\lambda\pounds_\xi v^a + \partial^\beta_\alpha\lambda^a_b(\mathbb{I})\,v^b\,\pounds_\xi\Gamma^\alpha_{\beta\lambda} \qquad (7.7.27)$$

which is a direct consequence of the definition (7.7.9) of the covariant derivative. We stress that the Lie derivative of the connection is meant to be expressed in terms of the Lie derivative of the metric by the so-called *Palatini formula*

$$\pounds_\xi\Gamma^\alpha_{\beta\mu} = \frac{1}{2}g^{\alpha\epsilon}\left(-\nabla_\epsilon\pounds_\xi g_{\beta\mu} + \nabla_\mu\pounds_\xi g_{\epsilon\beta} + \nabla_\beta\pounds_\xi g_{\mu\epsilon}\right) \qquad (7.7.28)$$

After integrating covariantly by parts we obtain the first variation formula

$$\nabla_\sigma(\xi^\sigma\mathcal{L}^*_{\mathrm{mat}}) = -\tfrac{1}{2}\hat{H}_{\mu\nu}\pounds_\xi g^{\mu\nu} + \mathbb{E}_a\pounds_\xi v^a + \nabla_\epsilon\left(\hat{p}^\epsilon_a\pounds_\xi v^a + \tfrac{1}{2}H^{\mu\nu\epsilon}\pounds_\xi g_{\mu\nu}\right) \qquad (7.7.29)$$

where we set

$$\begin{cases} \hat{H}_{\mu\nu} = \sqrt{g}H_{\mu\nu} = -2p_{\mu\nu} - \nabla_\epsilon H_{\mu\nu}{}^{\cdot\,\cdot\,\epsilon} \\ H^{\mu\nu\epsilon} = -A^{(\mu\nu)\epsilon} + A^{\epsilon(\mu\nu)} + A^{(\mu\underline{\epsilon}\nu)} \\ A^{\mu\nu\epsilon} = \hat{p}^\mu_a\,\partial^\nu_\alpha\lambda^a_b(\mathbb{I})\,v^b\,g^{\epsilon\alpha} \end{cases} \qquad (7.7.30)$$

the underbar $\underline{\epsilon}$ meaning that the index ϵ is not subjected to symmetrization.

Thence we have the matter contribution to the Nöther current

$$\begin{aligned}\mathcal{E}(L_{\mathrm{mat}},\xi) &= \left[\hat{p}^\epsilon_a\pounds_\xi v^a + \tfrac{1}{2}H^{\mu\nu\epsilon}\pounds_\xi g_{\mu\nu} - \xi^\epsilon\mathcal{L}^*_{\mathrm{mat}}\right]\mathrm{d}s_\epsilon = \\ &= \left[T^\epsilon_\mu\,\xi^\mu + T^{\epsilon\nu}_\mu\,\nabla_\nu\xi^\mu\right]\mathrm{d}s_\epsilon\end{aligned} \qquad (7.7.31)$$

where the tensor densities

$$\begin{cases} T^\epsilon_\mu = \hat{p}^\epsilon_a\nabla_\mu v^a - \mathcal{L}^*_{\mathrm{mat}}\delta^\epsilon_\mu \\ T^{\epsilon\nu}_\mu = H_\mu{}^{\cdot\,\nu\epsilon} - A^{\epsilon\nu}{}_\mu{}^{\cdot} \end{cases} \qquad (7.7.32)$$

are called the *canonical tensor densities*.

We stress that such a Nöther current is not conserved separately due to the Hilbert stress tensor term. However, the total Nöther current is conserved when (7.7.31) is considered together with the purely gravitational contribution (7.7.25).

If the reader wishes to compare our results with the standard results usually accepted in Special Relativity, i.e. on Minkowski space, he will find some discrepancies. They are due to the different definition given here of conserved current and conserved quantity. In Special Relativity in fact, as we noticed above, the Minkowski background is fixed and just Poincaré transformations (i.e. isometries) are considered as symmetry generators. In the Lagrangian the metric $g = \eta$ is thence considered as a constant and its variations are never taken into account. As a consequence one is led to consider the restricted covariance identity (compare with (7.7.29))

$$\nabla_\sigma(\xi^\sigma \mathcal{L}^*_{\text{mat}}) = \mathbb{E}_a \pounds_\xi v^a + \nabla_\epsilon \left(\hat{p}^\epsilon_a \pounds_\xi v^a\right) \tag{7.7.33}$$

from which the restricted Nöther current is obtained (compare with (7.7.31))

$$\begin{aligned}\mathcal{E}(L_{\text{mat}}, \xi) &= [\hat{p}^\epsilon_a \pounds_\xi v^a - \xi^\epsilon \mathcal{L}^*_{\text{mat}}]\ \mathrm{d}s_\epsilon = \\ &= \left[(T_{\text{mat}})^\epsilon_\mu\, \xi^\mu + (T_{\text{mat}})^{\epsilon\nu}_\mu\ \nabla_\nu \xi^\mu\right]\ \mathrm{d}s_\epsilon\end{aligned} \tag{7.7.34}$$

where the restricted canonical densities are defined as

$$\begin{cases} (T_{\text{mat}})^\epsilon_\mu = \hat{p}^\epsilon_a \nabla_\mu v^a - \mathcal{L}^*_{\text{mat}} \delta^\epsilon_\mu \equiv T^\epsilon_\mu \\ (T_{\text{mat}})^{\epsilon\nu}_\mu = -A^{\epsilon\nu}{}_{\cdot\,\mu} \equiv T^{\epsilon\nu}_\mu - H_\mu{}^{\cdot\nu\epsilon} \end{cases} \tag{7.7.35}$$

These restricted canonical tensor densities reproduce all the standard results on Minkowski background. The reader could argue that these restricted densities are those which should be always used to represent the physically relevant quantities. Unfortunately, this viewpoint cannot be extended to the general case since a general background metric $\bar{g}$ does not allow any Killing vector (i.e. infinitesimal symmetry). This has a simple physical explanation. In General Relativity the matter field and the gravitational field are always in interaction; thence it is not possible to separate, for example, the energy of the matter from the energy of the pure gravitational field. At a fundamental level we shall always consider the total conserved quantity, being the matter and gravitational contributions undefined separately, at least without resorting to *ad hoc* prescriptions. This can be clearly seen for all conserved quantities. While off-shell Nöther currents (as well as the superpotentials) receive a separate contribution from the gravitational Lagrangian and a separate contribution from the matter Lagrangian, the same does not happen for the conserved quantities (which are always defined on-shell so that all contributions are mixed up together). Nevertheless, also Nöther currents need some careful attention when they are considered. It is in fact true that they can split off-shell but then the two parts are not separately conserved due to field equations. We shall call this fact the *principle of non-reductionism*: *whenever interactions are present the system is quite different from the mere union of the free parts the system is composed of.*

We also stress that the relation between the Hilbert stress tensor and the restricted canonical tensors can be recast as follows

$$-\hat{H}^{\mu}_{\cdot\,\epsilon} = (T_{\text{mat}})^{\mu}_{\epsilon} + \nabla_{\lambda}\left((T_{\text{mat}})^{\lambda\mu}_{\epsilon} + H_{\epsilon}^{\cdot\mu\lambda}\right) - \mathbb{E}_a \partial^{\mu}_{\epsilon}\lambda^a_b(\mathbb{I})v^b \tag{7.7.36}$$

The matter Nöther current splits as

$$\mathcal{E}(L_{\text{mat}},\xi) = \tilde{\mathcal{E}}(L_{\text{mat}},\xi) + \text{Div}\,\mathcal{U}(L_{\text{mat}},\xi) \tag{7.7.37}$$

where we set

$$\begin{cases} \tilde{\mathcal{E}}(L_{\text{mat}},\xi) = \left[\left(T^{\epsilon}_{\mu} - \nabla_{\nu}T^{[\epsilon\nu]}_{\mu}\right)\xi^{\mu} + T^{(\epsilon\nu)}_{\mu}\nabla_{\nu}\xi^{\mu}\right] \text{d}s_{\epsilon} \\ \mathcal{U}(L_{\text{mat}},\xi) = \frac{1}{2}\left[T^{[\epsilon\nu]}_{\mu}\,\xi^{\mu}\right] \text{d}s_{\epsilon\nu} \end{cases} \tag{7.7.38}$$

Again the reduced current $\tilde{\mathcal{E}}(L_{\text{mat}},\xi)$ does not vanish on-shell due to the Hilbert stress tensor term. However, it vanishes together with the reduced current $\tilde{\mathcal{E}}(L_{\text{gr}},\xi)$ ensuing from the gravitational Lagrangian.

Thence the total superpotential $\mathcal{U}(L,\xi) = \mathcal{U}(L_{\text{gr}},\xi) + \mathcal{U}(L_{\text{mat}},\xi)$ correctly defines the conserved quantities of the whole system.

As a consequence of the covariance identity, the Lagrangian density $\mathcal{L}^*_{\text{mat}}$ satisfies the following identity

$$\mathcal{L}^*_{\text{mat}}\delta^{\mu}_{\epsilon} = -2p^{\mu}_{\cdot\,\epsilon} - p_a\partial^{\mu}_{\epsilon}\lambda^a_b(\mathbb{I})v^b - p^{\lambda}_a\partial^{\mu}_{\epsilon}\lambda^a_b(\mathbb{I})\hat{v}^b_{\lambda} + p^{\mu}_a\hat{v}^a_{\epsilon} \tag{7.7.39}$$

which in view of the definitions above can be recast into the following form

$$-\hat{H}^{\mu}_{\cdot\,\epsilon} = T^{\mu}_{\epsilon} + \nabla_{\lambda}T^{\lambda\mu}_{\epsilon} - \mathbb{E}_a\partial^{\mu}_{\epsilon}\lambda^a_b(\mathbb{I})v^b \tag{7.7.40}$$

(the minus sign depending on the notation used to define the canonical tensor densities). Thus the Hilbert stress tensor, though related to canonical tensor densities, is not a canonical tensor itself. The difference depends on terms vanishing on-shell and on terms which are related to the tensor character of the matter fields under consideration. This will be clearer when looking at the worked examples below.

Klein-Gordon Scalar Field

Let us now specialize to some particularly relevant cases of matter fields. A scalar field is described by a (real or complex) number at each spacetime point. Examples are the temperature in a room, the (classical counterpart of) the π_0 meson or more fundamentally the Higgs field which is considered responsible of the spontaneous symmetry breaking in the Weinberg-Salam electroweak model. The value of the scalar field does not depend on the coordinates. Thence the scalar field is described by a function over spacetime, i.e. either a section of the real 0-form of bundle $A_0(M) = M \times \mathbb{R}$ or the bundle of complex valued 0-forms $A_0(M) \otimes \mathbb{C} = M \times \mathbb{C}$. The bundle is trivial due to the "transformation laws" of 0-forms, i.e. $f' = f$.

Let us choose coordinates (x^μ, φ) on $A_0(M) \otimes \mathbb{C}$. The covariant derivatives of a scalar field coincide with the ordinary partial derivatives, i.e.

$$\nabla_\lambda \varphi = \mathrm{d}_\lambda \varphi \tag{7.7.41}$$

while the Lie derivative reduces to

$$\pounds_\xi \varphi = i_\xi \,\mathrm{d}\, \varphi = \xi(\varphi) \tag{7.7.42}$$

The *(complex) Klein-Gordon Lagrangian* used for complex scalar fields is

$$L_{\mathrm{KG}} = \alpha \left(\nabla_\lambda \varphi^\dagger \, \nabla^\lambda \varphi - m^2 \varphi^\dagger \varphi + V(|\varphi|^2) \right) \sqrt{g}\, \mathrm{d}s \tag{7.7.43}$$

where $\varphi^\dagger$ denotes the complex conjugate field, V is any smooth function, called the *potential* and $\alpha \neq 0$ is a coupling constant. In the real case we set $\varphi^\dagger = \varphi$ and the Lagrangian reduces to the so-called *real Klein-Gordon Lagrangian*

$$L_{\mathrm{KG}} = \alpha \left(\nabla_\lambda \varphi \, \nabla^\lambda \varphi - m^2 \varphi^2 + V(\varphi^2) \right) \sqrt{g}\, \mathrm{d}s \tag{7.7.44}$$

for which we use the same name L_{KG} for the sake of simplicity.

Both Lagrangians are natural; in fact we have for (7.7.43)

$$\begin{aligned}
&p_{\mu\nu} \pounds_\xi g^{\mu\nu} + p\, \pounds_\xi \varphi + p_\dagger\, \pounds_\xi \varphi^\dagger + \hat{p}^\lambda\, \pounds_\xi \hat{\varphi}_\lambda + \hat{p}^\lambda_\dagger\, \pounds_\xi \hat{\varphi}^\dagger_\lambda = \\
&= -2(\alpha\sqrt{g}\, \nabla_{(\mu} \varphi^\dagger \nabla_{\nu)} \varphi - \tfrac{1}{2} \mathcal{L}_{\mathrm{KG}}\, g_{\mu\nu}) \nabla^\mu \xi^\nu + \\
&+ \alpha\sqrt{g}(-m^2 \varphi^\dagger + V'\, \varphi^\dagger) \nabla_\mu \varphi\, \xi^\mu + \alpha\sqrt{g} \nabla_\mu \varphi^\dagger (-m^2 \varphi + V'\, \varphi)\, \xi^\mu + \\
&+ \alpha\sqrt{g} \nabla^\lambda \varphi^\dagger (\xi^\mu \nabla_\mu \hat{\varphi}_\lambda + \nabla_\mu \xi^\lambda \hat{\varphi}_\lambda) + \alpha\sqrt{g}(\xi^\mu \nabla_\mu \hat{\varphi}^\dagger_\lambda + \nabla_\mu \xi^\lambda \hat{\varphi}^\dagger_\lambda) \nabla^\lambda \varphi = \\
&= \alpha\sqrt{g}\, \xi^\mu \Big(-m^2 (\varphi^\dagger \hat{\varphi}_\mu + \hat{\varphi}^\dagger_\mu \varphi) + V'\, (\varphi^\dagger \hat{\varphi}_\mu + \hat{\varphi}^\dagger_\mu \varphi) + \nabla_\mu (\hat{\varphi}^\dagger_\lambda \hat{\varphi}^\lambda) \Big) + \\
&+ \alpha\sqrt{g}\, \nabla_\epsilon \xi^\mu \left(-\hat{\varphi}^{\dagger\epsilon} \hat{\varphi}_\mu - \hat{\varphi}^\dagger_\mu \hat{\varphi}^\epsilon + \hat{\varphi}^{\dagger\epsilon} \hat{\varphi}_\mu + \hat{\varphi}^\dagger_\mu \hat{\varphi}^\epsilon \right) + \mathcal{L}_{\mathrm{KG}}\, \delta^\epsilon_\mu \nabla_\epsilon \xi^\mu = \\
&= \mathcal{L}_{\mathrm{KG}}\, \nabla_\mu \xi^\mu + \nabla_\mu \mathcal{L}_{\mathrm{KG}}\, \xi^\mu = \nabla_\mu \left(\mathcal{L}_{\mathrm{KG}}\, \xi^\mu \right)
\end{aligned} \tag{7.7.45}$$

where V' denotes the derivative of the potential $V(z)$ with respect to its argument z.

Analogous computations can be carried over in the real case (7.7.44). Field equations of the total (complex) Lagrangian $L = L_{1\mathrm{B}} + L_{\mathrm{KG}}$ are thence

$$\begin{cases} \sqrt{g}\, G_{\mu\nu} = \kappa H_{\mu\nu} \\ \bar{G}_{\mu\nu} = 0 \\ \Box\varphi + m^2\varphi = 0 \\ \Box\varphi^\dagger + m^2\varphi^\dagger = 0 \end{cases} \tag{7.7.46}$$

(the fourth one being excluded in the real case $\varphi = \varphi^\dagger$), where we set $\Box = g^{\mu\nu}\nabla_\mu\nabla_\nu$ for the d'Alambert operator. Here we have defined the Hilbert stress tensor of the scalar field as

$$\begin{cases} A_{\mu\nu\alpha} = 0 \\ H_{\mu\nu\alpha} = 0 \\ \hat{H}_{\mu\nu} = -2\alpha\sqrt{g}\,\hat{\varphi}^\dagger{}_{(\mu}\hat{\varphi}_{\nu)} + \mathcal{L}_{\mathrm{KG}}\, g_{\mu\nu} \end{cases} \tag{7.7.47}$$

since the scalar representation $\lambda : (J, \varphi) \mapsto \varphi$ is trivial.

We stress that the Klein-Gordon equations $\Box\varphi + m^2\varphi = 0$ (as well as its complex conjugated) are satisfied by the zero section $\varphi \equiv 0$ and that the Hilbert stress tensor $H_{\mu\nu}$ vanishes for that solution. This is the compatibility condition which is necessary to regard $\bar{G}_{\mu\nu} = 0$ as the field equations of the reference background $(\bar{g}, \bar{\varphi} \equiv 0)$. We also stress that under a fundamental viewpoint the matter background $\bar{\varphi} \equiv 0$ is just one of the infinitely many possible choices. In general, however, one should add to the total Lagrangian a kinetic Klein-Gordon Lagrangian for the background $\bar{\varphi}$ as we did for the Hilbert Lagrangian in Section 7.5.

The total Lagrangian is covariant with respect to any spacetime diffeomorphism. The Nöther current receives a contribution from the Klein-Gordon Lagrangian

$$\mathcal{E}(L_{\mathrm{KG}}, \xi) = T^\epsilon_\mu \xi^\mu\, \mathrm{d}s_\epsilon = \left[\alpha\sqrt{g}(\hat{\varphi}^{\dagger\epsilon}\hat{\varphi}_\mu + \hat{\varphi}^\dagger_\mu\hat{\varphi}^\epsilon) - \mathcal{L}^*_{\mathrm{mat}}\delta^\epsilon_\mu\right] \xi^\mu\, \mathrm{d}s_\epsilon \tag{7.7.48}$$

This does not depend on the derivatives of the field ξ, thence it does not contribute to the superpotential but only to the reduced current. The superpotential of the total Lagrangian equals then the superpotential (7.5.19) of the pure gravitational Lagrangian. Of course this does not mean that the matter field has vanishing conserved quantities; in fact, the metric solution g along which the superpotential is evaluated is a solution of Einstein-Klein-Gordon equations $G_{\mu\nu} = \kappa H_{\mu\nu}$ which contain the matter field through the Hilbert stress tensor. This is typical of all situations in which the matter Lagrangian does not really depend on the connection through the covariant derivatives of the matter field itself. Maxwell-Einstein theory will be another important example.

Let us remark that the complex Klein-Gordon Lagrangian (7.7.43) is invariant also under the phase shift transformation

$$\begin{cases} g' = g & \bar{g}' = \bar{g} \\ \varphi' = e^{i\theta}\varphi \\ \varphi'^{\dagger} = \varphi^{\dagger}e^{-i\theta} \end{cases} \tag{7.7.49}$$

which is the basis of the gauge approach to electromagnetism. That program is however based on a number of physical assumptions about the non-observability of the absolute phase. We can here, on the contrary, regard the transformations (7.7.49) as symmetries of the complex Klein-Gordon Lagrangian. Accordingly, let us define the infinitesimal generator of phase shifts

$$\Xi = \xi\frac{\partial}{\partial\varphi} + \xi^{\dagger}\frac{\partial}{\partial\varphi^{\dagger}} = i\left(\varphi\frac{\partial}{\partial\varphi} - \varphi^{\dagger}\frac{\partial}{\partial\varphi^{\dagger}}\right) \tag{7.7.50}$$

to which the following Nöther current is associated

$$\mathcal{E}_{\text{shift}}(L, \Xi) = i\alpha\left(\varphi^{\dagger}\hat{\varphi}^{\mu} - \hat{\varphi}^{\dagger\mu}\varphi\right)\sqrt{g}\,\mathrm{d}s_{\mu} \tag{7.7.51}$$

We remark that this current depends on the vector field Ξ but not on its derivatives. Thence no superpotential exists for the current $\mathcal{E}_{\text{shift}}(L, \Xi)$. In the next Chapter we shall enlarge the symmetry group to point-dependent phase shifts (and modify appropriately the Lagrangian to allow those transformations to be symmetries). In that context the theory will be gauge natural and it will allow a superpotential.

We stress that in the above scheme nothing forbids to observe the phase of the field φ. This is ultimately related to the fact that there are no compact supported phase shifts. If φ' and φ differ by a phase shift (7.7.49) they are really different everywhere, so that no "hole argument" applies. If phases are observable, two local solutions $\varphi_{(\alpha)}$ and $\varphi_{(\beta)}$ which differ by a phase shift on the intersection $U_{\alpha\beta}$ should be ultimately regarded as describing two different physical situations.

That is why we choose the configuration bundle to be trivial. In fact, two local solutions on two different intersecting patches glue together only if they agree in the overlap of the two patches. Thence transition functions can be chosen to be the identity onto all the intersections $U_{\alpha\beta}$ leading to the trivial bundle.

In the next Chapter we shall be allowed to consider compact supported point-dependent phase shifts and by a "hole argument" the absolute phase will not be observable. Thence two local solutions will be considered to agree on the overlaps if they differ by a point-dependent phase shift. That will lead to physically and mathematically non-trivial situations.

Scalar Density Fields

A scalar density of weight k is a geometric object ϕ which transforms as

$$\phi' = \exp[k \log |\det J|]\, \phi \tag{7.7.52}$$

where ϕ is real or complex valued (see Section 4.4). In general we shall denote by $\mathcal{B}_k$ the bundle $\overset{k}{\mathcal{D}}{}^0_0(M)$ of scalar densities of weight k. In particular, a scalar density of weight $k = -1$ is a section of the bundle $\mathcal{B}_{-1} \simeq A_m(M)$ of m-forms over M, while for $k = 0$ we recover ordinary scalars.

Let us choose $(x^\mu; \phi)$ as fibered coordinates on $\mathcal{B}_k$. The (formal) covariant derivative of a scalar density ϕ is

$$\nabla_\mu \phi = \mathrm{d}_\mu \phi + k \Gamma^\alpha_{\alpha\mu} \phi \tag{7.7.53}$$

(notice the explicit dependence on the connection Γ, which disappears for $k = 0$) while the Lie derivative is

$$\pounds_\xi \phi = \xi^\sigma \nabla_\sigma \phi - k \phi \nabla_\sigma \xi^\sigma \tag{7.7.54}$$

Let us consider as a natural matter Lagrangian the following

$$L_{\mathrm{KGD}} = \mathcal{L}_{\mathrm{KGD}}\, \mathrm{d}s = \alpha \left(\nabla_\lambda \phi^\dagger\, \nabla^\lambda \phi - m^2 \phi^\dagger \phi \right) \sqrt{g}^{\,2k+1}\, \mathrm{d}s \tag{7.7.55}$$

Notice that no (polynomial) potential term (other than the quadratic term already present) can be added if one wants to maintain the naturality of the Lagrangian, unless $k = 0$.

In this case we have

$$\begin{cases} A^{\mu\nu\epsilon} = \alpha\, k\, \nabla^\mu \left(\phi^\dagger \phi\right)\, g^{\nu\epsilon}\, \sqrt{g}^{\,2k+1} \\ H^{\mu\nu\epsilon} = \alpha\, k\, \nabla^\epsilon \left(\phi^\dagger \phi\right)\, g^{\mu\nu}\, \sqrt{g}^{\,2k+1} \equiv A^{\epsilon\mu\nu} \\ \hat{H}_{\mu\nu} = -2\alpha \left(\hat{\phi}^\dagger{}_{(\mu} \hat{\phi}_{\nu)} - k (\mathbb{E}\, \phi + \phi^\dagger \mathbb{E}_\dagger)\, g_{\mu\nu} \right) \sqrt{g}^{\,2k+1} + g_{\mu\nu}\, \mathcal{L}_{\mathrm{KGD}} \end{cases} \tag{7.7.56}$$

where $\mathbb{E} = -\alpha \sqrt{g}^{2k+1} (\Box \phi^\dagger + m^2 \phi^\dagger)$ and $\mathbb{E}_\dagger = -\alpha \sqrt{g}^{2k+1} (\Box \phi + m^2 \phi)$ are field equations for the matter field.

One can easily check that the covariance identity holds so that the Lagrangian is natural. The Nöther current induced by a spacetime vector field is given by

$$\mathcal{E}(L_{\mathrm{KGD}}, \xi) = \left(\hat{p}^{\dagger\mu} \pounds_\xi \phi + \pounds_\xi \phi^\dagger \hat{p}^\mu + \tfrac{1}{2} H^{\alpha\beta\mu} \pounds_\xi g_{\mu\nu} - \xi^\mu \mathcal{L}_{\mathrm{KGD}} \right) \mathrm{d}s_\mu \tag{7.7.57}$$

We remark that it is conserved only together with the Nöther current of the gravitational Lagrangian since field equations are obtained by varying the total Lagrangian $L = L_{1\mathrm{B}} + L_{\mathrm{KGD}}$.

The canonical tensor densities are given by

$$\begin{cases} T^\mu_\sigma = \alpha \left(\hat{\phi}^{\dagger\mu} \hat{\phi}_\sigma + \hat{\phi}^\dagger{}_\sigma \hat{\phi}^\mu \right) \sqrt{g}^{\,2k+1} - \delta^\mu_\sigma \mathcal{L}_{\mathrm{KGD}} \\ T^{\mu\rho}_\sigma \equiv 0 \end{cases} \tag{7.7.58}$$

Thence the superpotential is identically vanishing and consequently what we said above for the scalar field applies to the scalar densities too.

Finally the relation between the canonical tensor densities and the Hilbert stress tensor reads as

$$-\hat{H}^\mu{}_{\cdot\nu} = T^\mu_\nu - k\,\delta^\mu_\nu \left(\mathbb{E}\phi + \phi^\dagger \mathbb{E}_\dagger \right) \tag{7.7.59}$$

We remark that for $k = 0$ these results reproduce the results already obtained for the scalar Klein-Gordon field (with $V \equiv 0$). Once again the Hilbert stress tensor and the canonical tensor coincide at least on-shell, though for scalar densities we see from (7.7.59) that they differ off-shell, since equation (7.7.59) reduces in fact to $-\hat{H}^\mu{}_{\cdot\nu} = T^\mu_\nu$ only for $k = 0$.

We shall see below further examples in which the Hilbert stress tensor and the canonical tensor differ both on-shell and off-shell. In general the difference between the Hilbert stress tensor and the canonical tensor depends on the geometric character of the dynamical field. This is quite a general rule to which few exceptions (the scalar field, the scalar density (though just on-shell) and the electromagnetic field) are encountered.

Matter Vector Fields

Let us now consider vector matter fields. The configuration bundle is $\mathcal{B} = T(M)$ on which we choose local fibered coordinates $(x^\mu; \zeta^\mu)$. The (formal) covariant derivative of a vector field ζ is

$$\nabla_\mu \zeta^\lambda = \hat{\zeta}^\lambda{}_\mu = \mathrm{d}_\mu \zeta^\lambda + \Gamma^\lambda_{\nu\mu} \zeta^\nu \tag{7.7.60}$$

and the Lie derivative is

$$\pounds_\xi \zeta = [\xi, \zeta] \tag{7.7.61}$$

Let us consider the following matter Lagrangian

$$L_{\mathrm{KGV}} = \alpha \left(\nabla_\mu \zeta^\lambda \nabla^\mu \zeta_\lambda + V(|\zeta|^2) \right) \sqrt{g}\, \mathrm{d}s \qquad |\zeta|^2 = \zeta^\lambda\, \zeta_\lambda \tag{7.7.62}$$

which is formally similar to the Klein-Gordon Lagrangian and it will be therefore called the *vector Klein-Gordon Lagrangian*. The potential V is usually chosen as $V = -m^2 \zeta^\lambda\, \zeta_\lambda = -m^2 |\zeta|^2$ and it is interpreted as a mass term for the field ζ. It is a natural Lagrangian thanks to the tensorial character of the objects involved.

The covariance identity reads now as

$$\nabla_\sigma(\xi^\sigma \mathcal{L}_{\text{KGV}}) = p_{\mu\nu}\pounds_\xi g^{\mu\nu} + p_\lambda \pounds_\xi \zeta^\lambda + \hat{p}_\lambda{}^\nu \pounds_\xi \hat{\zeta}^\lambda{}_\nu \tag{7.7.63}$$

where the momenta are defined as follows

$$\begin{cases} p_{\mu\nu} = \alpha\sqrt{g}\,(\nabla_\mu\zeta_\lambda\nabla_\nu\zeta^\lambda - \nabla^\lambda\zeta_\mu\nabla_\lambda\zeta_\nu + V'\zeta_\mu\zeta_\nu) - \frac{1}{2}\mathcal{L}_{\text{KGV}}g_{\mu\nu} \\ p_\lambda = 2\alpha\sqrt{g}\,V'\zeta_\lambda \\ \hat{p}_\lambda{}^\nu = 2\alpha\sqrt{g}\,\nabla^\nu\zeta_\lambda \end{cases} \tag{7.7.64}$$

being V' the derivative of the potential $V(z)$ with respect to its argument z.

As for the case of scalar densities the covariant derivative of ζ depends explicitly on the connections so that it does not commute with the Lie derivative. In fact we have

$$\pounds_\xi \nabla_\nu \zeta^\lambda = \nabla_\nu \pounds_\xi \zeta^\lambda + \pounds_\xi \Gamma^\lambda_{\alpha\nu}\zeta^\alpha \tag{7.7.65}$$

Using Palatini formula (7.7.28), the covariance identity can be recast into the form

$$\nabla_\sigma(\xi^\sigma \mathcal{L}_{\text{KGV}}) = -\frac{1}{2}\hat{H}_{\mu\nu}\pounds_\xi g^{\mu\nu} + \mathbb{E}_\lambda \pounds_\xi \zeta^\lambda + \mathrm{d}_\lambda(\hat{p}_\alpha{}^\lambda \pounds_\xi \hat{\zeta}^\alpha + \frac{1}{2}H^{\alpha\beta\lambda}\pounds_\xi g_{\alpha\beta}) \tag{7.7.66}$$

where we set

$$\begin{cases} A^{\alpha\beta\mu} = 2\alpha\sqrt{g}\,\nabla^\alpha\zeta^\mu\zeta^\beta \\ \mathbb{E}_\lambda = p_\lambda - \nabla_\nu\hat{p}_\lambda{}^\nu = -2\alpha\sqrt{g}\,[\Box\zeta_\lambda - V'\zeta_\lambda] \end{cases} \tag{7.7.67}$$

and the quantities $H^{\alpha\beta\mu}$ and $\hat{H}_{\alpha\beta}$ are given by the general expressions (7.7.30).

The canonical tensors are

$$\begin{cases} T^\lambda_\epsilon = 2\alpha\sqrt{g}\nabla^\lambda\zeta_\alpha\nabla_\epsilon\zeta^\alpha - \mathcal{L}_{\text{KGV}}\delta^\lambda_\epsilon \\ T^{\lambda\nu}_\epsilon = H_\epsilon{}^{\cdot\nu\lambda} - A^{\lambda\nu}{}_\epsilon{}^{\cdot} \end{cases} \tag{7.7.68}$$

and they are related to the Hilbert stress tensor by the following:

$$-\hat{H}^\mu_{\cdot\,\nu} = T^\mu_\nu - \mathbb{E}_\nu\zeta^\mu + \nabla_\epsilon T^{\epsilon\mu}_\nu \tag{7.7.69}$$

We see that they differ both on-shell and off-shell since $T^{\epsilon\mu}_\nu$ does not vanish in general.

Proca Field

The so-called *Proca field* is a section of the cotangent bundle $T^*(M)$ on which we choose local fibered coordinates $(x^\mu; A_\mu)$. The (formal) covariant derivative of a Proca field A is

$$\nabla_\lambda A_\mu = \mathrm{d}_\lambda A_\mu - \Gamma^\alpha_{\mu\lambda} A_\alpha \tag{7.7.70}$$

and the Lie derivative is

$$\begin{aligned} \pounds_\xi A =& \mathrm{d}\, i_\xi A + i_\xi\, \mathrm{d}A = \left(\mathrm{d}_\mu(\xi^\lambda A_\lambda) + \xi^\lambda \mathrm{d}_\lambda A_\mu - \xi^\lambda \mathrm{d}_\mu A_\lambda\right) \mathrm{d}x^\mu = \\ &= \left(\nabla_\mu \xi^\lambda A_\lambda + \xi^\lambda \nabla_\lambda A_\mu\right) \mathrm{d}x^\mu \end{aligned} \tag{7.7.71}$$

Under many respects the Proca field is the dual of a vector field; however, the Proca Lagrangian differs from the Klein-Gordon Lagrangian by a suitable term which is proportional to $\nabla_\alpha A_\beta\, \nabla^\beta A^\alpha$. Let us denote by $F = \frac{1}{2} F_{\mu\nu} \mathrm{d}x^\mu \wedge \mathrm{d}x^\nu = \mathrm{d}A$ the differential of the Proca field, i.e. let us set $F_{\mu\nu} = \nabla_\mu A_\nu - \nabla_\nu A_\mu = \mathrm{d}_\mu A_\nu - \mathrm{d}_\nu A_\mu$; the *Proca Lagrangian* is defined to be

$$L_{\mathrm{Proca}} = \mathcal{L}_{\mathrm{Proca}}\, \mathrm{d}s = \alpha \left(F_{\mu\nu} F^{\mu\nu}_{\cdot\;\cdot} - m^2 A_\mu A^\mu_\cdot\right) \sqrt{g}\, \mathrm{d}s \tag{7.7.72}$$

which is also natural. We remark that potentials more general than the mass term $m^2 A_\mu A^\mu$ can be introduced.

The momenta are

$$\begin{cases} p_{\mu\nu} = \dfrac{\partial \mathcal{L}_{\mathrm{Proca}}}{\partial g^{\mu\nu}} = \alpha\sqrt{g}\, \left(2F_{\mu\sigma} F_\nu^{\cdot\sigma} - m^2 A_\mu A_\nu\right) - \dfrac{1}{2} (\mathcal{L}_{\mathrm{Proca}})\, g_{\mu\nu} \\ P^\mu = \dfrac{\partial \mathcal{L}_{\mathrm{Proca}}}{\partial A_\mu} = -2\alpha\sqrt{g}\, m^2 A^\mu \\ P^{\mu\nu} = \dfrac{\partial \mathcal{L}_{\mathrm{Proca}}}{\partial \nabla_\mu A_\nu} = 4\alpha\sqrt{g}\, F^{\mu\nu} \end{cases} \tag{7.7.73}$$

We used the letters P to avoid any possible confusion between $P^{\mu\nu}$ and $p^{\mu\nu}_{\cdot\;\cdot} \equiv g^{\mu\rho} g^{\nu\sigma} p_{\rho\sigma}$. One can easily check the covariance identity

$$\nabla_\sigma(\xi^\sigma \mathcal{L}_{\mathrm{Proca}}) = p_{\mu\nu}\, \pounds_\xi g^{\mu\nu} + P^\mu\, \pounds_\xi A_\mu + P^{\mu\nu}\, \pounds_\xi \nabla_\mu A_\nu \tag{7.7.74}$$

which can be recast into the standard form

$$\begin{aligned} \nabla_\sigma(\xi^\sigma \mathcal{L}_{\mathrm{Proca}}) =& -\tfrac{1}{2} \hat{H}_{\mu\nu}\, \pounds_\xi g^{\mu\nu} + \nabla_\epsilon \left(\tfrac{1}{2} H^{\alpha\beta\epsilon}\, \pounds_\xi g_{\alpha\beta} + P^{\epsilon\nu}\, \pounds_\xi A_\nu\right) + \\ &+ \mathbb{E}^\mu\, \pounds_\xi A_\mu \end{aligned} \tag{7.7.75}$$

where we set[12]

$$\begin{cases} A^{\alpha\beta\epsilon} = -4\alpha\sqrt{g}\, F^{\alpha\beta} A^\epsilon \\ H^{\alpha\beta\epsilon} = \left(-A^{(\alpha\beta)\epsilon} + A^{(\beta\underline{\epsilon}\alpha)} + A^{\epsilon(\alpha\beta)}\right) \equiv 0 \\ \hat{H}_{\mu\nu} = -2p_{\mu\nu} \\ \mathbb{E}^\mu = -4\alpha\sqrt{g}\, \left(\tfrac{1}{2} m^2 A^\mu - \nabla_\nu F^{\mu\nu}\right) \end{cases} \tag{7.7.76}$$

[12] As before, $\underline{\epsilon}$ means that the index ϵ is excluded from antisymmetrization.

We remark that field equations for a massless Proca field reproduce Maxwell equations. However, as we already discussed above and as we shall reconsider below, field equations are definitely not the ultimate expression of the physical content of a system. Many subtleties about observability of dynamical quantities (and conserved quantities among them) are in fact hidden and completely codified by the geometric character *we assume* for the dynamical fields. As argued in Section 7.6 above, the electromagnetic potential is not a 1-form over spacetime owing to gauge transformations, so that massless Proca fields, though quite similar under some viewpoint, are not a correct description of electromagnetism.

The Nöther current $\mathcal{E}(L_{\mathrm{Proca}}, \xi) = [T^\lambda_\rho \xi^\rho + T^{\lambda\sigma}_\rho \nabla_\sigma \xi^\rho]\, \mathrm{d}s_\lambda$ defines the following canonical tensors

$$\begin{cases} T^\lambda_\rho = 4\alpha\sqrt{g}\, F^{\lambda\mu}\nabla_\rho A_\mu - (\mathcal{L}_{\mathrm{Proca}})\delta^\lambda_\rho \\ T^{\lambda\sigma}_\rho = 4\alpha\sqrt{g} F^{\lambda\sigma} A_\rho \equiv A^{\lambda\sigma\cdot}{}_\rho \end{cases} \tag{7.7.77}$$

which are related to the Hilbert stress tensor by the following identity

$$-\hat{H}^\mu{}_{\cdot\nu} = T^\mu_\nu + \mathbb{E}^\mu A_\nu + \nabla_\lambda T^{\lambda\mu}_\nu \tag{7.7.78}$$

as one can easily check by a direct calculation. Again $\hat{H}$ and T differ both on-shell and off-shell because of $\nabla_\lambda T^{\lambda\mu}_\nu$, which is in general non-vanishing.

8. Metric-Affine Formalism for Gravity

Let us finally consider a different but equivalent formulation of General Relativity. The purely metric Hilbert-Einstein gravitational Lagrangian introduced in Section 7.5 above is by far not the unique Lagrangian which can be introduced for a description of Einstein's theory of gravitation. We shall here recall, as an application of our framework, the classical metric-affine formalism for gravity (i.e. the one which is often improperly called the *Palatini* or *first order approach* to gravity). Many other proposals can be found in literature but in this monograph we shall consider just some of them.

In the metric affine formulation we have two dynamical fields: a metric $g_{\mu\nu}$ and a (symmetric) connection $\Gamma^\alpha_{\beta\mu}$ which *a priori* is not assumed to be the Levi-Civita connection of the metric $g_{\mu\nu}$. We can define the Riemann tensor and the Ricci tensor of Γ by setting as usual

$$R^\alpha{}_{\beta\mu\nu} = \mathrm{d}_\mu\Gamma^\alpha_{\beta\nu} - \mathrm{d}_\nu\Gamma^\alpha_{\beta\mu} + \Gamma^\alpha_{\epsilon\mu}\Gamma^\epsilon_{\beta\nu} - \Gamma^\alpha_{\epsilon\nu}\Gamma^\epsilon_{\beta\mu}\,, \qquad R_{\mu\nu} = R^\alpha{}_{\mu\alpha\nu} \tag{7.8.1}$$

Consider then the so-called *Palatini Lagrangian*

$$L_{\mathrm{Pal}} = \mathcal{L}_{\mathrm{Pal}}\, \mathrm{d}s = \frac{1}{2\kappa} g^{\mu\nu} R_{\mu\nu}(j^1\Gamma)\, \sqrt{g}\, \mathrm{d}s \tag{7.8.2}$$

which is of order zero with respect to the metric and first order in the connection.

From a classical viewpoint this metric-affine formulation is known to be equivalent to the Hilbert one (in dimension $m \neq 2$). In fact, the variation of the Lagrangian is given by

$$\delta L_{\mathrm{Pal}} = p_{\mu\nu}\, \pounds_\xi g^{\mu\nu} + p^{\mu\nu}\, \pounds_\xi R_{\mu\nu} \tag{7.8.3}$$

where we defined the momenta as

$$\begin{cases} p_{\mu\nu} = \dfrac{\sqrt{g}}{2\kappa}\left(R_{\mu\nu} - \dfrac{1}{2}R\, g_{\mu\nu}\right) \equiv \dfrac{\sqrt{g}}{2\kappa} G_{\mu\nu} \\ p^{\mu\nu} = \dfrac{\sqrt{g}}{2\kappa} g^{\mu\nu} \end{cases} \tag{7.8.4}$$

with $G_{\mu\nu}(g,\Gamma) = R_{\mu\nu}(j^1\Gamma) - \frac{1}{2}R(g, j^1\Gamma)\, g_{\mu\nu}$ is called the *Einstein tensor.*[13] The Lie derivative of the Ricci tensor can be treated as follows

$$\pounds_\xi R_{\mu\nu} = \overset{\Gamma}{\nabla}_\alpha \pounds_\xi \Gamma^\alpha_{\mu\nu} - \overset{\Gamma}{\nabla}_\nu \pounds_\xi \Gamma^\alpha_{\mu\alpha} = \overset{\Gamma}{\nabla}_\alpha \pounds_\xi \left(\Gamma^\alpha_{\mu\nu} - \delta^\alpha{}_{(\mu}\Gamma^\epsilon{}_{\nu)\epsilon}\right) = \overset{\Gamma}{\nabla}_\alpha \pounds_\xi u^\alpha_{\mu\nu} \tag{7.8.5}$$

while the variation of the action can be recast as follows

$$\delta L_{\mathrm{Pal}} = p_{\mu\nu}\, \pounds_\xi g^{\mu\nu} - \overset{\Gamma}{\nabla}_\alpha p^{\mu\nu}\, \pounds_\xi u^\alpha_{\mu\nu} + \overset{\Gamma}{\nabla}_\alpha \left(p^{\mu\nu}\, \pounds_\xi u^\alpha_{\mu\nu}\right) \tag{7.8.6}$$

Here the Lie derivative $\pounds_\xi u^\alpha_{\mu\nu}$ is understood to be expressed in terms of the Lie derivative $\pounds_\xi \Gamma^\alpha_{\mu\nu}$ of the connection and $\overset{\Gamma}{\nabla}$ denotes the covariant derivative operator defined by the connection Γ.

Thence we obtain the Euler-Lagrange morphism:

$$\begin{cases} \mathbb{E}_{\mu\nu} = p_{\mu\nu} = \dfrac{\sqrt{g}}{2\kappa} G_{\mu\nu}(g,\Gamma) \\ \mathbb{E}^{\mu\nu}_\alpha = -\overset{\Gamma}{\nabla}_\alpha p^{\mu\nu} + \dfrac{1}{2}\overset{\Gamma}{\nabla}_\epsilon p^{\epsilon\nu}\delta^\mu_\alpha + \dfrac{1}{2}\overset{\Gamma}{\nabla}_\epsilon p^{\mu\epsilon}\delta^\nu_\alpha \end{cases} \tag{7.8.7}$$

By considering the trace of the second equation we obtain as a consequence

$$-\overset{\Gamma}{\nabla}_\epsilon p^{\epsilon\nu} + \frac{m}{2}\overset{\Gamma}{\nabla}_\epsilon p^{\epsilon\nu} + \frac{1}{2}\overset{\Gamma}{\nabla}_\epsilon p^{\nu\epsilon} = \frac{m-1}{2}\overset{\Gamma}{\nabla}_\epsilon p^{\epsilon\nu} = 0 \tag{7.8.8}$$

For $m > 1$, a case to which we restrict in order to speak of spacetime, we obtain field equations

$$\begin{cases} G_{\mu\nu}(g,\Gamma) = 0 \\ \overset{\Gamma}{\nabla}_\alpha g^{\mu\nu} = 0 \end{cases} \tag{7.8.9}$$

---

[13] We set $R(g, j^1\Gamma) = g^{\mu\nu} R_{\mu\nu}(j^1\Gamma)$

If $m > 2$, the second field equations constrains the connection $\Gamma^{\alpha}_{\mu\nu}$, which is *a priori* arbitrary, to coincide *a posteriori* with the Levi-Civita connection of the metric g. By substituting this information into the first field equation the vacuum Einstein equation $G(g) = 0$ is finally obtained for the dynamics of g.[14]

Thence we could say that the metric-affine formulation of gravity as described by (7.8.2) is *dynamically equivalent* to the purely metric formulation of Hilbert and Einstein. We remark that in an eventual quantum field theory of gravity, this dynamical equivalence in principle is not expected to survive beyond Bohr approximations. Of course a quantum gravity theory is not yet at our disposal and, anyway, any possible discussion of this would be far beyond the scope of this monograph.

We also stress that the identification between the dynamical connection $\Gamma^{\alpha}_{\beta\mu}$ and the Levi-Civita connection $\{^{\alpha}_{\beta\mu}\}_g$ of the metric g is typical of the vacuum theory. Let us consider in fact the total Lagrangian $L = L_{\text{Pal}} + L_{\text{mat}}$, with the matter Lagrangian given by $L_{\text{mat}} = \mathcal{L}_{\text{mat}}(g^{\mu\nu}, v^a, \hat{v}^a_\mu)\, \mathrm{d}s$ as in the previous Section. The covariant derivative of the matter field v is defined by the dynamical connection, i.e.:

$$\hat{v}^a_\mu = \nabla_\mu v^a = \mathrm{d}_\mu v^a + \Gamma^a_{b\mu} v^b \,, \qquad\qquad \Gamma^a_{b\mu} = \partial^\alpha_\beta \lambda^a_b(\mathbb{I}) v^b\, \Gamma^\beta_{\alpha\mu} \tag{7.8.10}$$

The variation of the matter Lagrangian is

$$\delta L_{\text{mat}} = -\tfrac{1}{2}\sqrt{g}\, h_{\mu\nu}\delta g^{\mu\nu} + \mathbb{E}_a \delta v^a + \nabla_\lambda(p^\lambda_a \delta v^a) + p^\mu_a \partial^\alpha_\beta \lambda^a_b(\mathbb{I}) v^b\, \delta\Gamma^\beta_{\alpha\mu} \tag{7.8.11}$$

where $\sqrt{g}\, h_{\mu\nu} = -2\frac{\partial \mathcal{L}_{\text{mat}}}{\partial g^{\mu\nu}}$.

Thence field equations of the total Lagrangian are

$$\begin{cases} G_{\mu\nu}(g,\Gamma) = \kappa\, h_{\mu\nu} \\ -\overset{\Gamma}{\nabla}_\alpha p^{\mu\nu} + \dfrac{1}{2}\overset{\Gamma}{\nabla}_\epsilon p^{\epsilon\nu}\delta^\mu_\alpha + \dfrac{1}{2}\overset{\Gamma}{\nabla}_\epsilon p^{\mu\epsilon}\delta^\nu_\alpha = -p^\nu_a \partial^\mu_\alpha \lambda^a_b(\mathbb{I}) v^b \end{cases} \tag{7.8.12}$$

The trace of the second equation is now

$$\overset{\Gamma}{\nabla}_\epsilon p^{\epsilon\nu} = \frac{2}{1-m} p^\nu_a \partial^\alpha_\alpha \lambda^a_b(\mathbb{I}) v^b \tag{7.8.13}$$

By substituting this into the second field equation leads finally to

$$\overset{\Gamma}{\nabla}_\alpha p^{\mu\nu} = \frac{1}{1-m} p^\nu_a \partial^\epsilon_\epsilon \lambda^a_b(\mathbb{I}) v^b \delta^\mu_\alpha + \frac{1}{1-m} p^\mu_a \partial^\epsilon_\epsilon \lambda^a_b(\mathbb{I}) v^b \delta^\nu_\alpha + p^\nu_a \partial^\mu_\alpha \lambda^a_b(\mathbb{I}) v^b \neq 0 \tag{7.8.14}$$

[14] We recall that in the case $m = 2$ field equations are identically satisfied and General Relativity reduces to a topological field theory.

which exactly prevents the connection $\Gamma^{\alpha}_{\beta\mu}$ to be in general the Levi-Civita connection of g. In presence of matter fields we have therefore two natural candidates for the role of dynamical connection. We stress that metric-affine theories are first order theories, so that some consequences of the choice of the dynamical connections are unessential to most issues, e.g. to conserved quantities. Anyway, the standard choice is to choose $\Gamma^{\alpha}_{\beta\mu}$ as the dynamical connection.

Exercises

1- Check directly (7.7.59).
Hint: compute the variational derivative of the matter Lagrangian with respect to the metric.

2- Check directly (7.7.69).
Hint: compute the variational derivative of the matter Lagrangian with respect to the metric.

3- Check directly (7.7.78).
Hint: compute the variational derivative of the matter Lagrangian with respect to the metric.

4- Prove Palatini formula (7.7.28)
Hint: expand both sides in terms of the metric and its first derivatives.

Chapter 8

GAUGE NATURAL THEORIES

Abstract *Gauge Natural field theories* are introduced to generalize natural theories as well as pure gauge theories and to encompass many relevant physical situations. Conserved quantities and superpotentials are considered for an arbitrary gauge natural theory. Some relevant examples in Physics are considered in detail.

The reader should master gauge natural bundles together with their lift properties as well as Lie derivatives (see Chapter 5) and the Lagrangian formalism introduced in Chapter 6. Natural theories in Chapter 7 are very similar to the gauge natural case. The reader should thence also have some ideas about the material contained in Chapter 7.

References

We refer to [KMS93] for the material about gauge naturality. The references about Yang-Mills theories are [GS87], [W72], [B81] and [BD64]. The material about the tetrad formulation of General Relativity can be found in [FFFG98], [FF97], [FF96] and [CDF91]. The Lie derivative for spinors is introduced in [K66]. The Kosmann lift is introduced in [FFFG95]. Spinors are also introduced in [DP86], [BD64], [CDF91], [LM89], [CBWM82].

For topological techniques we refer to [LM89], [CBWM82]. For a general and beautiful setting for Utiyama theorem in the gauge natural framework see [J04].

1. Motivations

Natural Field Theories are field theories which are invariant with respect to spacetime diffeomorphisms. Since in the passive viewpoint spacetime diffeomorphisms are locally identified with changes of coordinates, natural field theories implement the general covariance principle of General Relativity. One could say that Natural Theories are a realization of the first part of General Relativity, i.e. the part which does not talk about gravitation yet. This first part analyzes the mathematical and structural properties that *any* physical theory should have. In the second part of his research about General Relativity, Einstein defined a field theory which met all the criteria of the first part to describe the gravitational field on a geometrical basis.

We remark that the aforementioned structural criteria appear so tightly related to the second part and they are considered so self-evident that all standard empirical tests of General Relativity actually were aimed to test only the second part. Actually this is also due to the fact that the physical content of the symmetry principles, back since Galilei's relativity and up to gauge covariance and supersymmetric theories, is so deeply related to formal mathematical issues that they are considered to be tested by just testing their (more or less) direct consequences.

Up to now we have discussed a number of rather satisfactory applications of this *natural framework* to some fundamental Field Theory, from the theory of General Relativity to the theory of tensor matter fields. In our perspective also *gauge natural theories* rely on a suitable extension of the same principles of natural theories; an extension which has been proven necessary to meet the contemporary phenomenology of fundamental physics. In this broader meaning, also spinor fields, Gauge Theories and supersymmetries are treated in the proper way and under a unifying perspective.

Contemporary physics has definitely put into evidence some characteristic of Nature that most probably will not undergo deep changes in the near future. First of all, we learned that most of what we know about the world can be stated in terms of the structure of symmetry groups. Since we know that there is a *duality* between symmetries and conserved quantities given by Nöther theorem we can say that conserved quantities provide a deep insight into the structure of physical theories.

Secondly, there are certainly some conserved quantities which should be defined in any *generic* theory (mass, momentum and angular momentum), while others are defined just for some specific theory. Within the framework of natural theories this dichotomy is naturally implemented by saying that some symmetries are the lift of diffeomorphisms of space-time. According to the basic prescriptions of General Relativity *any* physical theory ought to allow this kind of symmetries. Then some particular theory can enjoy some additional symmetry and thence admit other conserved quantities. Electric charge, isospin, conformal charges are just some of the fundamental examples. In other words, in natural theories the natural lift identifies a subclass of symmetries, which may be called *horizontal*; mass, momentum and angular momentum are related to horizontal symmetries. In this splitting of symmetries naturality of fields plays an important role.

However, we know that physical fields are hardly ever natural objects, i.e. there is no natural way to associate an automorphism of the configuration bundle to each diffeomorphism of the base manifold. When these more general fields are involved, there is no notion of horizontal symmetry to be related to mass, momentum and angular momentum. Thence we have to require some additional structure which allows us to define at least horizontal infinitesimal symmetries in a canonical way. Infinitesimal symmetries, or equivalently 1-parameter subgroups of symmetries, are the ingredients which Nöther theorem requires in order to define conserved currents.

We believe that these requirements strongly push towards a broad adoption of gauge natural theories as they will be defined in this chapter. We remark that, as we shall show in the sequel, all field theories relevant to fundamental physics (i.e. both to General Relativity and elementary particle physics) are encompassed into this framework. Furthermore, we believe that it is important to identify the necessary structure for all classical field theories, also in view of their possible quantization. As is well known, in fact, standard quantization techniques do not apply to all theories (this is for example the case of General Relativity). We believe that the quantization techniques themselves should be revised in such cases (for example making them independent of the choice of a background metric, which is physically meaningless). On the other hand a deeper analysis of the foundations of classical field theories could also be useful to reveal their intimate and still partially hidden structure.

The situation of nowadays Physics has been compared with the situation of physics of Copernicus before Galilei's and Newton's revolution. We have many theories "in the rough" and we still do not understand the fundamental structure in detail. They give good answers to some problem but uncover many unsolved problems. Furthermore we are not able to question Nature empirically about the correctness of our guesses.

In this situation of confusion all fundamental issues, including also the classical roots of our understanding of the world, should be carefully analyzed in the search for some hints which can make us to find a way out of the *wood*.

2. Gauge Natural Field Theories

Gauge natural field theories are the general framework to discuss Gauge Theories in interaction with gravity and possibly other (natural) matter fields. We shall here introduce gauge natural theories and learn how to cope with the absence of the natural lift which was so important for the definition of conserved quantities in natural theories.

Definition (8.2.1): a *gauge natural field theory* is made of the following items:

- a *structure bundle* $\mathcal{P}$ which is a principal bundle over spacetime M with group G;
- a *configuration bundle* $\mathcal{C}$ which is a gauge natural bundle of order (r, s), $r \leq s$, associated to the structure group G;
- a *Lagrangian* L of order k on $\mathcal{C}$ which is Aut($\mathcal{P}$)-covariant, where Aut($\mathcal{P}$) is understood to act over $\mathcal{C}$ by means of the canonical action defined by (5.1.2).
- two morphisms $\omega : J^k\mathcal{C} \longrightarrow J^1\mathcal{P}/G$ and $\Gamma : J^k\mathcal{C} \longrightarrow J^1L^s(M)/\mathrm{GL}^s(n)$ which associate to any configuration $\rho : M \longrightarrow C$ a principal connection $\omega \circ j^k\rho$ of the structure bundle $\mathcal{P}$ and a principal connection $\Gamma \circ j^k\rho$ of the s-frame bundle $L^s(M)$.

The pair of integers (r, s) is called the *geometric order* of the theory. The structure bundle $\mathcal{P}$ encodes the symmetry structure of the theory. The configuration bundle is assumed to be a gauge natural bundle associated to $\mathcal{P}$ in order to have the canonical action (5.1.2) of the symmetry group Aut($\mathcal{P}$) over $\mathcal{C}$. This action replaces the natural lift of spacetime diffeomorphisms existing in any natural theory. The morphisms ω and Γ are used to define two principal connections, called *dynamical connections*; they together induce a connection on the configuration bundle (see Section 3.6) which in turn allows a definition of infinitesimal horizontal symmetries.

On one hand, gauge natural theories are a direct generalization of *pure gauge theories*, e.g. Yang-Mills theories as introduced in Section 6.7. As we already discussed, in such theories the metric $g_{\mu\nu}$ is regarded as a fixed background and, as a direct consequence, in pure gauge theories just spacetime diffeomorphisms which preserve the background field g may be regarded as symmetries. On the other hand, gauge natural theories generalize natural theories defined in Chapter 7. For any natural theory defined on a natural configuration bundle $\mathcal{C}$, one can, in fact, consider the trivial structure bundle $\mathcal{P} = (M, M, \mathrm{id}; \{e\})$ so that $\mathcal{C}$ is a gauge natural bundle of rank $(0, 1)$ associated to $\mathcal{P}$. Notice that this does not introduce any new information about the theory.

In a less trivial sense, one could say that the symmetry group of a gauge natural theory is enriched with pure gauge (i.e. vertical) transformations. However, we have to stress that a gauge natural theory is not simply a natural theory with some extra additional symmetry given by pure gauge transformations. On the contrary, gauge natural theories are a truly new concept that encompasses the standard frameworks since the group Aut($\mathcal{P}$) of generalized gauge transformations does not simply split into $\text{Aut}_V(\mathcal{P}) \times \text{Diff}(M)$ due to the non-trivial topology of the structure bundle $\mathcal{P}$. We shall spend the rest of this Chapter to prove that gauge natural theories retain all the physically relevant properties of natural and pure gauge theories even in the more general framework of gauge natural theories. In particular, we shall deal with the existence of conserved quantities and superpotentials for a general gauge natural theory.

3. Generalized Bianchi Identities and Superpotentials

Let us first fix some notation. Let $\Xi = \xi^\mu \partial_\mu + \xi^A \rho_A$ be a right invariant vector field on $\mathcal{P}$, i.e. an infinitesimal generator of automorphisms. Using the morphisms $\omega : J^k\mathcal{C} \longrightarrow J^1\mathcal{P}/G$ and $\Gamma : J^k\mathcal{C} \longrightarrow J^1 L^s(M)/\text{GL}^s(n)$ we can define symmetric covariant derivatives:

$$\begin{cases} \nabla_\sigma \xi^\mu = \mathrm{d}_\sigma \xi^\mu + \Gamma^\mu_{\rho\sigma}\xi^\rho \\ \nabla_{\sigma_1\sigma_2}\xi^\mu = \nabla_{(\sigma_1}\nabla_{\sigma_2)}\xi^\mu \\ \dots \end{cases} \qquad \begin{cases} \nabla_\sigma \xi^A = \mathrm{d}_\sigma \xi^A - \omega^A_{B\rho}\xi^B \\ \nabla_{\sigma_1\sigma_2}\xi^A = \nabla_{(\sigma_1}\nabla_{\sigma_2)}\xi^A \\ \dots \end{cases} \tag{8.3.1}$$

According to (5.1.2), the generator Ξ induces a generator $\hat{\Xi}$ of (generalized) gauge transformations on the configuration bundle $\mathcal{C}$. The induced generator $\hat{\Xi}$ is a Lagrangian symmetry (by definition of gauge natural theory) and it induces in turn a weak conservation law by means of Nöther theorem. If we fix a Lagrangian of order k, conserved currents can be thence expanded with respect to symmetric covariant derivatives (analogously to what we proved for natural theories in Section 7.2). The general structure is:

$$\begin{aligned} \mathcal{E}(L,\Xi) &= <\mathbb{F}(L)|j^{k-1}\pounds_\Xi\varphi> - i_\xi \circ L = \\ &= \Big[T^\lambda_\mu \xi^\mu + T^{\lambda\sigma}_\mu \nabla_\sigma \xi^\mu + \ldots + T^{\lambda\sigma_1\ldots\sigma_{s+k-1}}_\mu \nabla_{\sigma_1\ldots\sigma_{s+k-1}}\xi^\mu + \\ &\quad + T^\lambda_A \xi^A + T^{\lambda\sigma}_A \nabla_\sigma \xi^A + \ldots + T^{\lambda\sigma_1\ldots\sigma_{r+k-1}}_A \nabla_{\sigma_1\ldots\sigma_{r+k-1}}\xi^A\Big]\mathrm{d}s_\lambda \end{aligned} \tag{8.3.2}$$

The tensor densities $T^\lambda_\mu, T^{\lambda\sigma}_\mu, \ldots, T^{\lambda\sigma_1\ldots\sigma_{s+k-1}}_\mu$ and $T^\lambda_A, T^{\lambda\sigma}_A, \ldots, T^{\lambda\sigma_1\ldots\sigma_{r+k-1}}_A$ are symmetric with respect to all upper indices but λ. They are (improperly) called *canonical tensors*. The pair $(\alpha, \beta) = (s+k-1, r+k-1)$ is called the *effective order* of the theory.

Analogously, work forms can be expanded as follows:

$$\begin{aligned} \mathcal{W}(L,\Xi) &= - <\mathbb{E}(L)|\pounds_\Xi\varphi> = \\ &= \Big[W_\mu \xi^\mu + W^\sigma_\mu \nabla_\sigma \xi^\mu + \ldots + W^{\sigma_1\ldots\sigma_\alpha}_\mu \nabla_{\sigma_1\ldots\sigma_\alpha}\xi^\mu + \\ &\quad + W_A \xi^A + W^\sigma_A \nabla_\sigma \xi^A + \ldots + W^{\sigma_1\ldots\sigma_\beta}_A \nabla_{\sigma_1\ldots\sigma_\beta}\xi^A\Big]\mathrm{d}s \end{aligned} \tag{8.3.3}$$

where the tensor densities W_μ, W_μ^σ, ..., $W_\mu^{\sigma_1 \ldots \sigma_\alpha}$ and W_A, W_A^σ, ..., $W_A^{\sigma_1 \ldots \sigma_\beta}$ are symmetric with respect to upper indices and they are called *stress tensors*.

We stress that the work form $\mathcal{W}(L, \Xi)$ is actually of order (r, s) and that we have $s \leq \alpha$ and $r \leq \beta$. Consequently, some of the higher stress tensors in (8.3.3) may be identically vanishing. They have been added just for later notational convenience.

Both the Nöther current $\mathcal{E}(L, \Xi)$ and the work form $\mathcal{W}(L, \Xi)$ are linear in the generator Ξ of gauge transformations. We recall that the infinitesimal generator of automorphisms of the structure bundle $\mathcal{P}$ is $\mathcal{G} = T\mathcal{P}/G$ as introduced in Part I (see Section 5.3). Thence, in full analogy with the natural case we can define two variational morphisms

$$\begin{cases} \mathcal{E}(L) : J^{2k-1}\mathcal{C} \longrightarrow J\mathcal{G} \otimes A_{m-1}(M) \\ \mathcal{W}(L) : J^{2k}\mathcal{C} \longrightarrow J\mathcal{G} \otimes A_m(M) \end{cases} \tag{8.3.4}$$

such that

$$\begin{cases} \mathcal{E}(L, \Xi) = < \mathcal{E}(L) \mid j\Xi > \\ \mathcal{W}(L, \Xi) = < \mathcal{W}(L) \mid j\Xi > \end{cases} \tag{8.3.5}$$

As equations (8.3.2) and (8.3.3) show, the use of infinite jet bundles is completely unessential: in fact, both variational morphisms have finite rank, although it is not specified in the notation we choose.

In any case, we can use the splitting Lemmas (6.2.17) and (6.2.30) to obtain the variational morphisms

$$\begin{cases} \mathcal{B}(L) : J\mathcal{C} \longrightarrow \mathcal{G} \otimes A_m(M) \\ \tilde{\mathcal{E}}(L) : J\mathcal{C} \longrightarrow J\mathcal{G} \otimes A_{m-1}(M) \\ \mathcal{U}(L) : J\mathcal{C} \longrightarrow J\mathcal{G} \otimes A_{m-2}(M) \end{cases} \tag{8.3.6}$$

such that

$$\begin{cases} \mathcal{E}(L, \Xi) = < \tilde{\mathcal{E}}(L) \mid j\Xi > + \mathrm{Div} < \mathcal{U}(L) \mid j\Xi > \\ \mathcal{W}(L, \Xi) = < \mathcal{B}(L) \mid j\Xi > + \mathrm{Div} < \tilde{\mathcal{E}}(L) \mid j\Xi > \end{cases} \tag{8.3.7}$$

According to the previous formulae, the Bianchi morphism $\mathcal{B}(L) = [B_\rho \bar{\mathrm{d}}\xi^\rho + B_A \bar{\mathrm{d}}\xi^A] \otimes \mathrm{d}s$ is given by

$$\begin{aligned} B_\rho &= W_\rho - \nabla_{\sigma_1} W_\rho^{\sigma_1} + \ldots + (-1)^\alpha \nabla_{\sigma_1 \ldots \sigma_\alpha} W_\rho^{\sigma_1 \ldots \sigma_\alpha} \\ B_A &= W_A - \nabla_{\sigma_1} W_A^{\sigma_1} + \ldots + (-1)^\alpha \nabla_{\sigma_1 \ldots \sigma_\beta} W_A^{\sigma_1 \ldots \sigma_\beta} \end{aligned} \tag{8.3.8}$$

As a consequence of general considerations it identically vanishes off-shell.

The reduced current $\tilde{\mathcal{E}}(L)$ is thence expanded in the following form

$$\begin{aligned} \tilde{\mathcal{E}}(L, \Xi) = \Big[& \tilde{T}^\lambda_\mu \xi^\mu + \tilde{T}^{\lambda\sigma}_\mu \nabla_\sigma \xi^\mu + \ldots + \tilde{T}^{\lambda\sigma_1 \ldots \sigma_\alpha}_\mu \nabla_{\sigma_1 \ldots \sigma_\alpha} \xi^\mu + \\ & + \tilde{T}^\lambda_A \xi^A + \tilde{T}^{\lambda\sigma}_A \nabla_\sigma \xi^A + \ldots + \tilde{T}^{\lambda\sigma_1 \ldots \sigma_\beta}_A \nabla_{\sigma_1 \ldots \sigma_\beta} \xi^A \Big] \mathrm{d}s_\lambda \end{aligned} \tag{8.3.9}$$

where we set

$$\begin{cases} \tilde{T}^{\sigma}_{\rho} = W^{\sigma}_{\rho} - \nabla_{\sigma_2} W^{\sigma\sigma_2}_{\rho} + \ldots + (-1)^{\alpha-1} \nabla_{\sigma_2\ldots\sigma_\alpha} W^{\sigma\sigma_2\ldots\sigma_\alpha}_{\rho} \\ \tilde{T}^{\sigma\sigma_2}_{\rho} = W^{\sigma\sigma_2}_{\rho} - \nabla_{\sigma_3} W^{\sigma\sigma_2\sigma_3}_{\rho} + \ldots + (-1)^{\alpha-2} \nabla_{\sigma_3\ldots\sigma_\alpha} W^{\sigma\sigma_2\sigma_3\ldots\sigma_\alpha}_{\rho} \\ \ldots \\ \tilde{T}^{\sigma\sigma_2\ldots\sigma_\alpha}_{\rho} = W^{\sigma\sigma_2\ldots\sigma_\alpha}_{\rho} \end{cases} \tag{8.3.10}$$

and

$$\begin{cases} \tilde{T}^{\sigma}_{A} = W^{\sigma}_{A} - \nabla_{\sigma_2} W^{\sigma\sigma_2}_{A} + \ldots + (-1)^{\beta-1} \nabla_{\sigma_2\ldots\sigma_\beta} W^{\sigma\sigma_2\ldots\sigma_\beta}_{A} \\ \tilde{T}^{\sigma\sigma_2}_{A} = W^{\sigma\sigma_2}_{A} - \nabla_{\sigma_3} W^{\sigma\sigma_2\sigma_3}_{A} + \ldots + (-1)^{\beta-2} \nabla_{\sigma_3\ldots\sigma_\beta} W^{\sigma\sigma_2\sigma_3\ldots\sigma_\beta}_{A} \\ \ldots \\ \tilde{T}^{\sigma\sigma_2\ldots\sigma_\beta}_{A} = W^{\sigma\sigma_2\ldots\sigma_\beta}_{A} \end{cases} \tag{8.3.11}$$

We remark that $\tilde{\mathcal{E}}(L)$ is reduced since its coefficients are totally symmetric with respect to upper indices. It also vanishes on-shell since all stress tensors $W^{\sigma\sigma_2\ldots\sigma_n}_{\mu}$ and $W^{\sigma\sigma_2\ldots\sigma_n}_{A}$ do vanish on-shell.

Finally the superpotential $\mathcal{U}(L)$ is the reduced representative of the boundary term of the Nöther current. A general formula can be obtained by suitably combining all the expressions above. It is rather complicated in general. As an example (which covers most cases of physical interest), for a theory of effective order $(2,2)$ the expression of the superpotential turns out to be

$$\begin{aligned} \mathcal{U}(L,\Xi) = \frac{1}{2}\Big[& \Big(T^{[\lambda\sigma]}_{A} - \frac{2}{3}\nabla_\rho T^{[\lambda\sigma]\rho}_{A}\Big)\xi^A + \frac{4}{3} T^{[\lambda\sigma]\rho}_{A} \nabla_\rho \xi^A + \\ & + \Big(T^{[\lambda\sigma]}_{\mu} - \frac{2}{3}\nabla_\rho T^{[\lambda\sigma]\rho}_{\mu}\Big)\xi^\mu + \frac{4}{3} T^{[\lambda\sigma]\rho}_{\mu} \nabla_\rho \xi^\mu \Big] \mathrm{d}s_{\lambda\sigma} \end{aligned} \tag{8.3.12}$$

This includes of course all cases of effective order (α, β) with $\alpha \leq 2$ and $\beta \leq 2$ which are obtained by making the necessary cancellations. We shall hardly ever need more than this in physically reasonable applications.

4. Conserved Quantities

As in general field theories, conserved quantities are defined by integrating the superpotential on a closed spacetime $(m-2)$-region. However, we have to discuss how one can select the symmetry generators Ξ which correspond to the energy, the momentum and the angular momentum (and so on).

We recall that the generators Ξ are vector fields on the structure bundle $\mathcal{P}$, not on spacetime M. In natural theories, symmetry generators $\hat{\xi}$ are the natural lifts of spacetime vector fields so that there is just one symmetry generator $\hat{\xi}$ lying over each spacetime vector field. Thus an observer chooses some coordinates $(x^\mu) = (t, x^i)$ in spacetime and as a consequence of his conventions he measures the energy which is associated to ∂_t, the space momentum which is associated to ∂_i, the angular momentum associated to $(x_\mu \partial_\nu - x_\nu \partial_\mu)$ and so on. In gauge natural theories there are infinitely symmetry generators over the same spacetime vector field ξ and we do not know which one to choose to define, e.g., the energy of the system.

We already remarked that the possibility to use $\Xi = \partial_t$ is globally meaningful only if the structure bundle $\mathcal{P}$ is trivial, while in general this is only locally defined and thence unsatisfactory to define conserved quantities which are non-local (quasi-local) quantities in nature. Since energy, momentum, etc. have to be defined also in a gauge natural theory we have to find some way out.

One possibility consists in regarding some specific example of gauge natural field theory as a natural theory, whenever this is really possible. This is known to be (partly) true for electromagnetism (see Section 7.6) and also a lot of literature about spinors goes in that direction. Electromagnetism works since $U(1)$-connections admit in fact a *natural* interpretation. On the other hand it is hardly clear whether such a prescription exists for $\mathrm{SU}(n)$-connections with $n > 1$. The story about spinors is much more complicated, as we shall try to clarify in Part III.

We believe then that a truly gauge natural answer which is a true extension of the natural prescription should be given also in these particular cases. Perhaps this does not prove that these particular field theories are only "virtually" natural, but at least it proves that the need of a definition of energy-momentum tensors is not a motivation to go towards a purely natural framework.

We notice that in any one of such theories we can define a connection on the structure bundle out of the dynamical fields. In electromagnetism, e.g., the vector potential A_μ is a connection over the structure bundle, while when gravity is involved the Levi-Civita connection of the dynamical metric is a (metric) connection over the frame bundle. Generically speaking, in gauge natural theories the dynamical connections of definition (8.2.1) select in particular a principal connection on the structure bundle. Being the dynamical connection built out of the fields (together with their derivatives up to some finite order) it does not contain any extra unphysical information (as instead happens for the choice of coordinates, frames, etc.) so that it is definitely not necessary to prove the independence on the connection (which in general does not hold true).

We can now give the following prescription: the *energy-momentum tensor is related to symmetry generators Ξ which are horizontal with respect to the dynamical connection.* For example, the energy of the electromagnetic field is related to the vector field $\Xi = \xi^\mu(\partial_\mu - A_\mu \rho) = \omega(\xi)$ (see equation (6.7.18)) with $\xi = \xi^\mu \partial_\mu = \partial_t$. We stress that this is a dynamical prescription, i.e. it is a vector field which depends on the fields and it is not really known until a solution of field equations is provided. Nevertheless, we shall see that this is not really a problem to be overcome since we learned how to represent objects in an off-shell fashion using coordinates on jet prolongations rather than using configurations.

Thence in gauge natural theories we have two main kinds of Nöther currents: the ones related to horizontal symmetries ("horizontal" with respect to the dynamical connection), namely those of the form

$$\mathcal{E}(L, \Xi_{(H)}) =< \mathbb{F} \mid j^{k-1} \pounds_{\omega(\xi)} > -i_\xi L \tag{8.4.1}$$

and the ones related to vertical symmetries, i.e. to pure gauge transformations, namely $\Xi_{(V)} = \xi^A \rho_A$, which are given under the form

$$\mathcal{E}(L, \Xi_{(V)}) =< \mathbb{F} \mid j^{k-1} \pounds_{\Xi_{(V)}} > \tag{8.4.2}$$

The vertical and horizontal parts $\Xi_{(H)}$ and $\Xi_{(V)}$ are defined according to Section 3.5.

Horizontal symmetries are related via Nöther theorem to energy-momentum tensors (mass, angular-momentum, etc.) while pure gauge transformations are related to the so-called *gauge charges*. In electromagnetism we have just one vertical direction to which there corresponds just one gauge charge $Q(L, \rho)$, which is known as the *electric charge*.

We stress that Yang-Mills theories (and Maxwell theory in particular) are not gauge natural theories *tout-court*. In fact, the Yang-Mills Lagrangian (6.7.6) contains a metric background which is used to perform contractions and to define the volume element. As we already did in natural theories, to obtain a genuine gauge natural theory we have to promote the metric field to become a dynamical field so that we have in principle to add a kinetic term for it in the total Lagrangian. In other words, gauge natural theories are not aimed to encompass pure gauge fields (which need a somewhat unphysical background metric field), but rather to describe gauge fields in interaction with gravity.

5. Yang-Mills Theories

Let us fix a Lie group G and a structure bundle $\mathcal{P} = (P, M, \pi; G)$. The configuration bundle $\mathcal{C} = \mathrm{Lor}(M) \times_M \mathcal{C}_\mathcal{P}$ is a gauge natural bundle associated to $\mathcal{P}$. Fibered coordinates on the configuration bundle are $(x^\mu; g_{\mu\nu}, A^A_\mu)$. We consider a total Lagrangian $L = L_{\mathrm{grav}} + L_{\mathrm{mat}}$ which is second order in the metric and first order in the gauge field, i.e. locally $L = \mathcal{L}(x, j^2 g, j^1 A)\, \mathrm{d}s$. The gravitational part is any second-order Lagrangian $L_{\mathrm{grav}}(j^2 g)$; for the sake of simplicity it will be assumed to be the Hilbert Lagrangian L_H introduced above in Section 7.5 or the Palatini Lagrangian L_{Pal} introduced in Section 7.8.

The matter part $L_{\rm mat}$ is assumed to be first order both in the metric and the gauge field, i.e. locally $L_{\rm mat} = \mathcal{L}^*_{\rm mat}(x, j^1 g, j^1 A)\, ds$.

This is a prescription called *minimal coupling principle*:

if a field appears up to order k in the kinetic Lagrangian, it appears at most with order k − 1 in the interaction terms.

This principle is often used in Utiyama-like arguments (see Section 7.7) to obtain a first ansatz to start with. For a general perspective on Utiyama argument for gauge natural operators see [J04]; for specific cases see [FFFG98]. In some sense minimal coupling principle is a definition of what has to be understood as free Lagrangian and what is interaction.

The total Lagrangian (and as a consequence also the matter Lagrangian separately) are gauge natural. By a Utiyama-like argument we have the following proposition:

Proposition (8.5.1): the gauge natural Lagrangian $L_{mat} = \mathcal{L}^*_{\rm mat}\, ds$ depends on its variables only through the metric and the curvature F of the gauge field A, i.e.:

$$L_{mat} = \mathcal{L}_{\rm mat}(g^{\mu\nu}, F^A{}_{\mu\nu})\, ds \tag{8.5.2}$$

Moreover it satisfies the following identity

$$\begin{cases} \mathcal{L}_{\rm mat}\delta^\alpha_\epsilon = -2p^\alpha{}_{\cdot\epsilon} + 2p^{\alpha\nu}_A F^A_{\epsilon\nu} \\ p^{\mu\nu}_A c^A_{\cdot\,BC} F^C_{\mu\nu} = 0 \end{cases} \tag{8.5.3}$$

where $p_{\mu\nu}$ and $p^{\mu\nu}_A$ are the momenta canonically conjugated to the coordinates $g^{\mu\nu}$ and $F^A{}_{\mu\nu}$, respectively.

Proof: let us change coordinates in $\mathcal{C}$ to $(x^\mu, g^{\mu\nu}, \Gamma^\alpha_{\beta\mu}, A^A_\mu, F^A_{\mu\nu})$. The most general matter Lagrangian $L_{\rm mat} = \mathcal{L}_{\rm mat}(x^\mu, g^{\mu\nu}, \Gamma^\alpha_{\beta\mu}, A^A_\mu, F^A_{\mu\nu})\, ds$ is gauge natural if the covariance identity holds for any gauge generator $\Xi = \xi^\mu\partial_\mu + \xi^A\rho_A$, i.e.:

$$d_\sigma(\xi^\sigma \mathcal{L}_{\rm mat}) = p_{\mu\nu}\pounds_\Xi g^{\mu\nu} + p^{\mu\nu}_\alpha \pounds_\Xi\Gamma^\alpha_{\mu\nu} + p^\mu_A \pounds_\Xi A^A_\mu + p^{\mu\nu}_A \pounds_\Xi F^A_{\mu\nu} \tag{8.5.4}$$

Since the metric is a purely natural field we have $\pounds_\Xi g^{\mu\nu} = \pounds_\xi g^{\mu\nu}$ and, consequently, $\pounds_\Xi\Gamma^\alpha_{\mu\nu} = \pounds_\xi\Gamma^\alpha_{\mu\nu}$ (where ξ is the projection of Ξ over M). The Lie derivative of the connection is given by equation (3.8.16) and that of the curvature is defined by equation (3.8.28). We can thence expand the covariance identity as a linear combination of $(\xi^\mu, \xi^A_{(V)})$ together with their (symmetrized) covariant derivatives. We obtain

$$\begin{cases} \nabla_\epsilon\mathcal{L}_{\rm mat} = p^{\mu\nu}_A\nabla_\epsilon F^A_{\mu\nu} + p^\mu_A F^A_{\mu\epsilon} + p^{\beta\mu}_\alpha R^\alpha_{\beta\mu\epsilon} \\ \mathcal{L}_{\rm mat}\delta^\alpha_\epsilon = -2p^\alpha{}_{\cdot\epsilon} + 2p^{\alpha\nu}_A F^A_{\epsilon\nu} \\ 0 = p^{\beta\mu}_\alpha \\ 0 = p^{\mu\nu}_A c^A_{\cdot\,BC} F^C_{\mu\nu} \\ 0 = p^\mu_A \end{cases} \tag{8.5.5}$$

Thence, because of the third equation $p_\alpha^{\beta\mu} = 0$ of (8.5.5) the Lagrangian density $\mathcal{L}_{\rm mat}$ cannot depend directly on the connection $\Gamma^\alpha_{\beta\mu}$. Analogously, because of the fourth equation $p_A^\mu = 0$ the Lagrangian density cannot depend directly on the connection A^A_μ. By substituting the second equation of (8.5.5) into the first, one obtains that $\mathcal{L}_{\rm mat}$ cannot depend on the point x^μ either.

The remaining covariance identities are

$$\begin{cases} \mathcal{L}_{\rm mat}\delta^\alpha_\epsilon = -2p^\alpha_{\cdot\epsilon} + 2p_A^{\alpha\nu} F^A_{\epsilon\nu} \\ p_A^{\mu\nu} c^A_{\cdot BC} F^C_{\mu\nu} = 0 \end{cases} \tag{8.5.6}$$

as claimed. ∎

The Yang-Mills Lagrangian L_{YM} as introduced in Chapter 6 (see equation (6.7.6)) satisfies both these conditions so that it is a gauge natural matter Lagrangian. The variation of the matter Lagrangian L_{YM} with respect to the deformation $X = (\delta F^A_{\mu\nu})\, \partial^{\mu\nu}_A + (\delta g_{\mu\nu})\, \partial^{\mu\nu}$ is in the following form:

$$\begin{aligned} \delta L_{YM} &= p_A^{\mu\nu}\delta F^A_{\mu\nu} + p_{\mu\nu}\delta g^{\mu\nu} = \\ &= 2p_A^{\mu\nu}\nabla_\mu \delta\omega^A_\nu + p_{\mu\nu}\delta g^{\mu\nu} = \\ &= \nabla_\mu\left(2p_A^{\mu\nu}\delta\omega^A_\nu\right) - 2\left(\nabla_\mu p_B^{\mu\nu}\right)\delta\omega^B_\nu + p_{\mu\nu}\delta g^{\mu\nu} = \\ &= \nabla_\mu \ll \mathbb{F}^\mu(L_{YM}) \mid X \gg + \mathbb{E}^\mu_B(L_{YM})\delta\omega^B_\mu - \tfrac{1}{2}\hat{H}_{\mu\nu}(L_{YM})\,\delta g^{\mu\nu} \end{aligned} \tag{8.5.7}$$

where the coefficients $\mathbb{F}^\nu(L_{YM})$ are the components of the Poincaré-Cartan morphism, $\mathbb{E}^\mu_B(L_{YM})$ are the components of the Euler-Lagrange morphism and $\hat{H}_{\mu\nu}(L_{YM})$ is the Hilbert stress tensor (density) of the Yang-Mills Lagrangian L_{YM}. The Hilbert stress tensor is easily seen to be

$$\hat{H}_{\mu\nu}(L_{YM}) \equiv \hat{H}_{\mu\nu} = \sqrt{g}\,\left(F^A{}_{\mu\alpha}F_{A\nu}{}^{\cdot\alpha} - \tfrac{1}{4}F^A{}_{\alpha\beta}F_A{}^{\alpha\beta}g_{\mu\nu}\right) \equiv \sqrt{g}\,H_{\mu\nu} \tag{8.5.8}$$

It agrees also with the usual expression of the *energy-momentum tensor* of the Yang-Mills field.

Einstein-Yang-Mills equations have thence the following local expression:

$$\begin{cases} \sqrt{g}\,G_{\mu\nu} = \kappa H_{\mu\nu} \\ \nabla_\mu F_B^{\mu\nu} = \mathrm{d}_\mu F_B^{\mu\nu} - F_A^{\mu\nu} c^A_{\cdot BC}\omega^C_\mu = 0 \end{cases} \tag{8.5.9}$$

The matter contribution to the Nöther current is

$$\begin{aligned} \mathcal{E}(L_{YM}, \Xi) &= \left(2p_A^{\nu\mu}\pounds_\Xi\omega^A_\mu - \xi^\nu\mathcal{L}_{YM}\right)\,\mathrm{d}s_\nu = \\ &= -\sqrt{g}\,\left[\left(F_A^{\alpha\nu}F^A_{\alpha\mu} - \tfrac{1}{4}F^A_{\alpha\beta}F_A^{\alpha\beta}\delta^\nu_\mu\right)\xi^\mu + F_A^{\nu\mu}\nabla_\mu\xi^A_{(V)}\right]\,\mathrm{d}s_\nu \end{aligned} \tag{8.5.10}$$

By a direct comparison, we have the identity

$$-\hat{H}^\mu_{\cdot\nu} = T^\mu_\nu \tag{8.5.11}$$

The matter Lagrangian does not contribute to the superpotential with any horizontal term containing ξ^μ (nor its derivatives). It contributes however with a vertical term

$$\mathcal{U}(L_{YM}, \Xi) = \frac{1}{2}\sqrt{g}\, F_A^{\nu\mu}\xi^A_{(V)}\, \mathrm{d}s_{\mu\nu} \tag{8.5.12}$$

Finally, Bianchi identities (8.3.8) can be recast as:

$$\nabla_{[\lambda}G_{\mu\nu]} = 0\,, \qquad \nabla_{[\lambda}F^A_{\mu\nu]} = 0 \tag{8.5.13}$$

so that they reproduce, at once, the classical Bianchi identities of General Relativity and of Yang-Mills theory (i.e., those of (M, g) those for the curvature of a principal connection.)

6. Gauge Natural Formulation of the Hilbert Lagrangian

When one introduces the Hilbert Lagrangian in General Relativity the dynamical field is a section of the configuration bundle $\mathrm{Lor}(M)$ which is associated to the frame bundle $L(M)$. Of course, since the Hilbert Lagrangian is natural it is also gauge natural in the trivial sense recalled in general in Section 8.2.

However, being the frame bundle a principal bundle one can argue whether the Hilbert Lagrangian is gauge natural also in a much less trivial sense, i.e. whether it is covariant with respect to *any* principal automorphism of $L(M)$, including those which are not the natural lifts of spacetime diffeomorphisms. For this purpose, let us consider the metric-affine formulation of General Relativity (see Section 7.8) which is described by the Palatini Lagrangian

$$L_{\mathrm{Pal}} = g^{\mu\nu}\, R_{\mu\nu}(j^1\Gamma)\, \sqrt{g}\, \mathrm{d}s \tag{8.6.1}$$

where $R_{\mu\nu} = R^\alpha_{\mu\alpha\nu}$, $R^\alpha_{\beta\mu\nu} = \mathrm{d}_\mu\Gamma^\alpha_{\ \beta\nu} - \mathrm{d}_\nu\Gamma^\alpha_{\ \beta\mu} + \Gamma^\alpha_{\ \gamma\mu}\Gamma^\gamma_{\ \beta\nu} - \Gamma^\alpha_{\ \gamma\nu}\Gamma^\gamma_{\ \beta\mu}$ is the Ricci tensor of the linear connection Γ and the connection is not *a priori* induced by the metric g.

The momenta are thence $p_{\mu\nu} \equiv \frac{\partial\mathcal{L}_{\mathrm{Pal}}}{\partial g^{\mu\nu}} = \sqrt{g}(R_{(\mu\nu)} - \frac{1}{2}Rg_{\mu\nu})$ and $P^{\mu\nu} \equiv \frac{\partial\mathcal{L}_{\mathrm{Pal}}}{R_{\mu\nu}} = g^{\mu\nu}\sqrt{g}$ (see (7.5.3)). Notice that for an arbitrary Γ the Ricci tensor $R_{\mu\nu}$ does not need to be symmetric, so that only $R_{(\mu\nu)}$ enters $p_{\mu\nu}$.

Let $\Xi = \xi^\mu(x)\partial_\mu + \xi^\mu_\nu(x)\rho_\mu^{\ \nu}$ be an infinitesimal generator of automorphisms on $L(M)$. Since $L(M)$ is also natural we can split Ξ into its horizontal (natural) part and its vertical (non-natural) part by setting

$$\Xi = \hat{\xi} \oplus \Xi_{(V)} = (\xi^\mu\partial_\mu + \partial_\nu\xi^\mu\rho_\mu^{\ \nu}) \oplus (\xi^\mu_\nu - \partial_\nu\xi^\mu)\rho_\mu^{\ \nu} \tag{8.6.2}$$

We shall shortly denote by $\hat{\xi}^\mu_\nu = \xi^\mu_\nu - \partial_\nu\xi^\mu$ the components of the vertical part of Ξ.

The Lagrangian (8.6.1) is thence covariant with respect to Ξ if and only if

$$\mathrm{d}_\sigma(\xi^\sigma \mathcal{L}_{\mathrm{Pal}}) = p_{\mu\nu} \pounds_\Xi g^{\mu\nu} + P^{\mu\nu} \pounds_\Xi R_{\mu\nu} \tag{8.6.3}$$

Let us expand the right hand side of this equation obtaining:

$$\begin{aligned} &p_{\mu\nu} \pounds_\Xi g^{\mu\nu} + P^{\mu\nu} \pounds_\Xi R_{\mu\nu} = \\ &= p_{\mu\nu} \pounds_\xi g^{\mu\nu} + P^{\mu\nu} \pounds_\xi R_{\mu\nu} - p_{\mu\nu}(\hat{\xi}^\mu_\lambda g^{\lambda\nu} + \hat{\xi}^\nu_\lambda g^{\lambda\mu}) + P^{\mu\nu}(\hat{\xi}^\lambda_\mu R_{\lambda\nu} + \hat{\xi}^\lambda_\nu R_{\mu\lambda}) = \\ &= \mathrm{d}_\sigma(\xi^\sigma \mathcal{L}_{\mathrm{Pal}}) - 2\sqrt{g}(R_{(\mu\nu)} - \tfrac{1}{2} R g_{\mu\nu}) g^{\nu\lambda} \hat{\xi}^\mu_\lambda + 2\sqrt{g}\, R_{(\lambda\nu)} \hat{\xi}^\lambda_\mu \end{aligned} \tag{8.6.4}$$

In the last equality we used the natural covariance condition for the Hilbert Lagrangian (i.e., $\mathrm{d}_\sigma(\xi^\sigma \mathcal{L}_{\mathrm{Pal}}) = p_{\mu\nu} \pounds_\xi g^{\mu\nu} + P^{\mu\nu} \pounds_\xi R_{\mu\nu}$); indices are lowered and raised by means of the metric g and the explicit expressions of the momenta have been used. Finally we obtain

$$p_{\mu\nu} \pounds_\Xi g^{\mu\nu} + P^{\mu\nu} \pounds_\Xi R_{\mu\nu} = \mathrm{d}_\sigma(\xi^\sigma \mathcal{L}_{\mathrm{Pal}}) + \sqrt{g}\, R g_{\mu\nu} g^{\nu\lambda} \hat{\xi}^\mu_\lambda \tag{8.6.5}$$

Thus we see that the metric-affine Lagrangian is not a gauge natural Lagrangian since it is not covariant with respect to *all* automorphisms of $L(M)$. However, there are some special automorphisms of $L(M)$ which are not natural but still are symmetries, namely the ones for which the condition

$$g^{\lambda(\nu} \hat{\xi}^{\mu)}_\lambda = 0 \tag{8.6.6}$$

holds, so that the extra term of (8.6.5) is automatically killed. For a fixed metric g one can easily show that this condition is in fact equivalent to require Ξ to be tangent to the sub-bundle $\mathrm{SO}(M, g) \hookrightarrow L(M)$ so that Ξ somehow *preserves* the metric structure (M, g). In this sense the metric-affine theory could be appropriately interpreted as a gauge natural theory with $\mathrm{SO}(M, g)$ as a structure bundle, provided the metric g (and thence $\mathrm{SO}(M, g)$ itself) were not an unknown structure to be determined *a posteriori* by field equations.

We shall hereafter present a gauge natural formulation or gravity which is actually related to this last remark.

7. Vielbein Formulation of Gravitational Theories

Let us consider a principal bundle Σ over M having $\mathrm{SO}(\eta)$ as structure group, η being any fixed signature (such bundle Σ always exists since at least the trivial bundle $\Sigma_{\mathrm{triv}} = M \times \mathrm{SO}(\eta)$ is defined). Let us consider, if any, the principal morphisms $i : \Sigma \longrightarrow L(M)$.

When we fix Σ arbitrarily, there it could happen to be no such principal morphism. For example, if M is non-parallelizable[15] and we choose Σ to be

15 A manifold is called *parallelizable* if the frame bundle $L(M)$ is trivial, while it is *non-parallelizable* in other cases. When M is parallelizable we have $TM \simeq M \times \mathbb{R}^m$.

the trivial bundle $\Sigma_{\text{triv}} = M \times \text{SO}(\eta)$, then there is no global principal morphism $i : \Sigma_{\text{triv}} \longrightarrow L(M)$. In fact, if there were such an $i : \Sigma_{\text{triv}} \longrightarrow L(M)$ then a global section σ of Σ_{triv} (which exists since Σ_{triv} is trivial) would produce a global section $i(\sigma)$ of $L(M)$. This would be impossible since M has been assumed to be non-parallelizable. Thence no global principal morphism $i : \Sigma_{\text{triv}} \longrightarrow L(M)$ is allowed in that case.

Definition (8.7.1): A manifold M is called an *η-manifold* if it allows[16] a (global) metric g of signature $\eta = (r, s)$. ∎

If M is an orientable η-manifold then $\Sigma_{\text{orth}} = \text{SO}(M, g)$ is a possible choice for Σ and in that case the canonical embedding $i : \Sigma_{\text{orth}} \longrightarrow L(M)$ is automatically defined. On the other hand, if Σ is given and the morphism $i : \Sigma \longrightarrow L(M)$ exists then the image $i(\Sigma) \subset L(M)$ is a sub-bundle of $L(M)$ and, following the discussion of Section 1.4, there exists a metric g such that $i(\Sigma) \simeq \text{SO}(M, g) \hookrightarrow L(M)$.

Proposition (8.7.2): M is an orientable η-manifold if and only if there exists (at least) one principal bundle Σ such that there exist (global) principal morphisms $i : \Sigma \longrightarrow L(M)$. ∎

From now on, we assume that Σ is chosen so that there exist principal morphisms $i : \Sigma \longrightarrow L(M)$. Any such map $i : \Sigma \longrightarrow L(M)$ will be called a *vielbein on* Σ.

We now wonder whether it is possible to use such a map $i : \Sigma \longrightarrow L(M)$ as a more fundamental description of gravity. First of all we have to regard such a morphism as a (global) section of some suitable bundle over M. For this, let us fix a trivialization $\sigma^{(\alpha)}$ of Σ and a trivialization $V_a^{(\alpha)}$ of $L(M)$ on the same covering $\{U_\alpha\}$ of M. The trivialization on $L(M)$ could be a natural trivialization $\partial_\mu^{(\alpha)}$ induced by a chart on M as well as a more general trivialization not induced by coordinates on M. See Section 1.3 and, in particular, equation (1.3.47).

The morphism $i : \Sigma \longrightarrow L(M)$ is locally expressed by $i(\sigma^{(\alpha)}) = e_a^{(\alpha)} = V_b^{(\alpha)} \, (e_{(\alpha)})_a^b$ where the coefficients $(e_{(\alpha)})_a^b$ form a matrix in $\text{GL}(m)$. By changing trivializations $\sigma^{(\beta)} = \sigma^{(\alpha)} \cdot \gamma$ on Σ with $\gamma \in \text{SO}(\eta)$ and $V_a^{(\beta)} = \bar{J}_a^b \, V_b^{(\alpha)}$ on $L(M)$ with $\bar{J}_a^b \in \text{GL}(m)$, one gets a new local expression for the morphism i, namely $i(\sigma^{(\beta)}) = V_b^{(\beta)} \, (e_{(\beta)})_a^b$. If $V_a^{(\alpha)}$ is the natural trivialization $\partial_\mu^{(\alpha)}$ of $L(M)$ then $\bar{J}_a^b$ is nothing but the inverse Jacobian $\bar{J}_\mu^\nu$ of coordinate change; in general they denote the transition functions of $L(M)$ with respect to the

[16] A metric of signature (r, s) exists on M if and only if the tangent bundle $T(M)$ splits into the direct sum of two sub-bundles $T(M) = E_1 \oplus E_2$ or rank r and s, respectively.

selected trivialization. From now on we suppose for simplicity that $V_a^{(\alpha)}$ is a natural trivialization $\partial_\mu^{(\alpha)}$, leaving to the reader the general case.

The two set of coefficients $(e_{(\alpha)})^\mu_a$ and $(e_{(\beta)})^\mu_a$ are thence related by the following expression

$$(e_{(\alpha)})^\mu_a = \bar{J}^\mu_\nu \, (e_{(\beta)})^\nu_b \, \gamma^b_a \tag{8.7.3}$$

where γ^b_a is the representative for γ induced by the chosen bases.

Now one can regard equations (8.7.3) as the definition of an action ρ of the group $\mathrm{SO}(\eta) \times \mathrm{GL}(m)$ onto the manifold $\mathrm{GL}(m)$ where the coefficients $(e_{(\alpha)})^\mu_a$ live. Shortly, we set $\rho : e \mapsto \bar{J} \, e \, \gamma$ in matrix notation. It is easy to see that the transformations (8.7.3) form a cocycle, so that they can be interpreted as the transition functions of a (unique up to isomorphisms) bundle $\Sigma_\rho = (\Sigma \times_M L(M)) \times_\rho \mathrm{GL}(m)$. The bundle Σ_ρ is associated to the principal bundle $W^{(0,1)}\Sigma = \Sigma \times_M L(M)$ and it is thence by construction a gauge natural bundle of order $(0,1)$ associated to Σ.

Theorem (8.7.4): there is a bijection between principal morphisms $i : \Sigma \longrightarrow L(M)$ and global sections of Σ_ρ.

Proof: if λ is a global section of Σ_ρ then the local expressions of $(e^{(\alpha)})^\mu_a$ are related by the condition

$$(e^{(\beta)})^\mu_a = J^\mu_\nu \, (e^{(\alpha)})^\nu_b \, \bar{\gamma}^b_a \tag{8.7.5}$$

(shortly, $e' = J \cdot e \cdot \bar{\gamma}$), which expresses globality. Here $\bar{\gamma}$ is the inverse of the transition functions γ on Σ. Exactly the same condition ensures that the local morphisms $i^{(\alpha)} : \Sigma \longrightarrow L(M)$ defined by $i^{(\alpha)}(\sigma^{(\alpha)}) = (e^{(\alpha)})^\mu_a \partial^{(\alpha)}_\mu$ glue together to define a global vielbein $i : \Sigma \longrightarrow L(M)$.

Conversely, let $i : \Sigma \longrightarrow L(M)$ be a global vielbein with local expressions $i^{(\alpha)}(\sigma^{(\alpha)}) = (e^{(\alpha)})^\mu_a \, \partial^{(\alpha)}_\mu$. The glueing conditions for such local expressions ensures globality for a section e of Σ_ρ locally defined by

$$e^{(\alpha)} : x \mapsto [\sigma^{(\alpha)}(x), \partial^{(\alpha)}_\mu(x), (e^{(\alpha)})^\mu_a(x)]_\rho \tag{8.7.6}$$

■

From now on, vielbein will be therefore identified with global sections of Σ_ρ. Let us stress, to avoid any misunderstanding that we are not claiming that the gauge natural bundle Σ_ρ always admits global sections. In fact, we already remarked that, depending on the topology of M (and also on the choice of the structure bundle Σ), there could be no principal morphism $i : \Sigma \longrightarrow L(M)$; we stress however that the bundle Σ_ρ is always defined, even if there is no principal morphism $i : \Sigma \longrightarrow L(M)$. Thence theorem (8.7.4) implicitly proves that the bundle Σ_ρ has no global section whenever no morphism $i : \Sigma \longrightarrow L(M)$ exists globally. Of course, in applications to Physics this last case is not of real interest since the system will not admit any global configuration.

However, let us mention that on any manifold M and for any signature η the conditions which ensure the existence of a global metric of signature η also ensure that one can suitably choose Σ so that Σ_ρ allows at least one global section. Further details and a proof of this result will be postponed to Section 9.4 and we assume here this result for granted.

Since M has been already chosen to be an η-manifold let us further assume without loss of generality that Σ is chosen so that at least one global section of Σ_ρ does exist. Any global section of Σ_ρ (as well as any principal morphism $i : \Sigma \longrightarrow L(M)$) is simply called a *vielbein*. Each vielbein induces uniquely a metric on M of signature η which is called the *metric induced by the vielbein*. Locally, the (covariant) induced metric is expressed, in the trivialization induced by a (natural) trivialization on $L(M)$, by the familiar expression:

$$g^{\mu\nu} = e^\mu_a \, \eta^{ab} \, e^\nu_b \tag{8.7.7}$$

where η^{ab} is the canonical diagonal matrix of signature η. We shall raise and lower Greek indices by means of the induced metric $g^{\mu\nu}$; lowercase Latin indices will be raised and lowered by the matrices η^{ab} and η_{ab}. In this way we shall not need a special notation to denote the inverse of the vielbein. With those conventions, in fact, the inverse $\bar{e}^a_\mu$ of the vielbein e^μ_a will coincide with the object obtained from the vielbein e^μ_a by raising the Latin index and lowering the Greek index. In fact, we have

$$e^\mu_a(g_{\mu\nu}e^\nu_b\eta^{bc}) = e^\mu_a(\bar{e}^i_\mu\eta_{ij}\bar{e}^j_\nu)e^\nu_b\eta^{bc} = \eta_{ab}\eta^{bc} = \delta^c_a \tag{8.7.8}$$

so that $g_{\mu\nu}e^\nu_b\eta^{bc}$ and $\bar{e}^c_\mu$ coincide; for simplicity they will be both denoted by e^c_μ.

We remark that giving a vielbein amounts to give a globally covering family $\{e^{(\alpha)}_a\}$ of local frames on M such that the transition functions on the overlaps $U_{\alpha\beta}$ are orthogonal matrices. These correspond only virtually to the *moving frames* sometimes used in General Relativity (also called *tetrads* when $\dim(M) = 4$). In fact, while moving frames are in fact defined patch by patch and are just of local nature unless M is parallelizable, the objects we defined (i.e. the *vielbein*) are global on any η-manifold and the families $\{e^{(\alpha)}_a\}$ are their local representations. In other words, we see that the mathematically correct interpretation of the standard moving frames of General Relativity is not to say that they are global families of local frames but rather global sections of the gauge natural bundle Σ_ρ. There is of course a suitable equivalence, but the latter perspective enhances the understanding of all global aspects of their use.

We can now use the vielbein to provide new formulations of gravity which correspond to both purely metric and metric-affine formulations. Let us thence fix the gauge natural bundle Σ_ρ to be the configuration bundle of the theory. Any vielbein $e : M \longrightarrow \Sigma_\rho$ induces a metric $g_{\mu\nu}$ on M which in turn induces the Levi-Civita linear connection $\Gamma^\alpha_{\beta\mu}$ on $L(M)$. The Levi-Civita connection $\Gamma^\alpha_{\beta\mu}$ can be pulled-back on Σ by means of the fixed vielbein. Locally, the induced connection on Σ is given by

$$\Gamma^a_{b\mu} = e^a_\alpha(\Gamma^\alpha_{\beta\mu} e^\beta_b + \mathrm{d}_\mu e^\alpha_b) \tag{8.7.9}$$

One can define the curvature $R^a_{b\mu\nu}$ of this connection $\Gamma^a_{b\mu}$ as well as its scalar contraction $R(j^2e) = R^a_{b\mu\nu}\, e^\mu_a e^{b\nu}$; they both depend on the second jet prolongation of the vielbein. Let us denote by e the absolute value of the determinant of the vielbein (or, equivalently, the square root of the absolute value of the determinant of the induced metric). We can now choose a second order Lagrangian $L_{\mathrm{pF}} : J^2\Sigma_\rho \longrightarrow A_m(M)$ locally expressed as

$$L_{\mathrm{pF}} = L_{\mathrm{pF}}(j^2e) = R(j^2e)\; e\; \mathrm{d}s \tag{8.7.10}$$

The field theory so defined will be called the *purely frame formulation* of General Relativity.

Classically, it is equivalent to the purely metric formulation. In fact, field equations read as

$$\mathbb{E}^a_\mu = e^{a\nu}(R_{\mu\nu} - \tfrac{1}{2}Rg_{\mu\nu})e \tag{8.7.11}$$

which are clearly equivalent to vacuum Einstein equations for the induced metric (8.7.7). Some classical difference between the purely metric and the purely frame formulation will appear when coupling to particular kind of matter fields will be considered below. In fact, the interactions of matter fields with the vielbein are *in principle* more general than interactions with the induced metric. We shall produce some example when dealing with spinor fields in Chapters 9 and 10.

From the quantum field theory perspective of path integrals (see e.g. [R81], [CDF91]) the two formulations are deeply different also before considering possible couplings to matter. In the metric formulation, in fact, one is lead to sum over all possible metrics (of signature η) on M. In the tetrad formulation one is lead to sum (at least formally) over all vielbeins for a fixed Σ. It may however happen that, in some dimension and for some signature, once a structure bundle Σ is fixed one cannot classically obtain all metrics (of the fixed signature) on M.

For example, let us take $\dim(M) = 3$; let us consider $M = \mathbb{R}^3 - \{0\} \simeq \mathbb{R}^+ \times S^2$ and signature $\eta = (1,2)$. On M let us consider the following two metrics:

- the Minkowski metric $g^{(1)}$ pulled-back along the immersion of $M \hookrightarrow \mathbb{R}^3$. In Cartesian coordinates it is expressed by $g^{(1)} = (\mathrm{d}x^0)^2 - (\mathrm{d}x^1)^2 - (\mathrm{d}x^2)^2$. Notice that there exists a global orthonormal frame on M with respect to $g^{(1)}$ given by $\{\partial_0, \partial_1, \partial_2\}$. Consequently, the metric $g^{(1)}$ is induced by a vielbein defined on the trivial structure bundle $\Sigma_{\mathrm{triv}} = M \times \mathrm{SO}(1,2)$.

- let us consider the global metric $g^{(2)}$ which in polar coordinates on M is locally expressed by $g^{(2)} = \mathrm{d}r^2 - r^2(\mathrm{d}\theta^2 + \sin^2\theta \mathrm{d}\phi^2)$. Notice that there does not exist a global orthonormal frame on M with respect to $g^{(2)}$. In fact, if such a global orthonormal frame existed, then in particular a global spacelike never-vanishing vector field X would exist in M. Because of the form of the metric $g^{(2)}$ this vector field would never be radial being it spacelike. One can thence use an auxiliary Euclidean metric (say $\delta = \mathrm{d}r^2 + r^2(\mathrm{d}\theta^2 + \sin^2\theta \mathrm{d}\phi^2)$) to project X onto a vector field $\hat{X}$ tangent to the unit sphere $r = 1$. By construction the vector field $\hat{X}$ would be a never-vanishing vector field on the sphere S^2 which cannot exist because of the Euler characteristic of S^2.[17] Consequently, the metric $g^{(2)}$ is not induced by any vielbein defined on the trivial structure bundle $\Sigma_{\mathrm{triv}} = M \times \mathrm{SO}(1,2)$.

In other words, this simple but instructive example shows that the space of metrics of a given signature on a manifold M may happen to be disconnected by the choice a structure bundle Σ. A metric that can be obtained as the induced metric of a vielbein defined on Σ is thence called a *Σ-admissible metric.*

There are however cases in which the space of all Σ-admissible metrics (if not empty[18]) coincides with the space of all metrics of the given signature. For example this holds in Euclidean signature $\eta = (m, 0)$ in any dimension m. Suppose in fact that we have two metrics g and $\bar{g}$ and that $\{e_a^{(\alpha)}\}$ is a vielbein for the first metric g. Being the metric g induced by the vielbein, the local frames $e_a^{(\alpha)}$ are orthonormal with respect to g. Then one can use Graham-Schmidt algorithm to build a new family of orthonormal frames $\bar{e}_a^{(\alpha)}$ for $\bar{g}$. Such a family of local orthonormal frames explicitly defines a vielbein to which $\bar{g}$ is associated. By an elementary property of Graham-Schmidt algorithm (see [FF96]), the transition functions of the two local frames are the same, so that they are defined on the same structure bundle Σ. Thence if g is Σ-admissible then any other metric $\bar{g}$ is also Σ-admissible. This is a property typical of Euclidean signature since the Graham-Schmidt algorithm cannot be extended to other signatures.

One can also set up a framework in which the connection $\Gamma^a_{b\mu}$ is not *a priori* induced by the vielbein. This formulation is called *frame-affine formulation* (or, *tetrad-affine formulation*) and it resembles the metric-affine formulation. In this context we choose $\mathcal{C} = \Sigma_\rho \times \mathcal{C}_\Sigma$ as configuration bundle; the dynamical fields are thence a vielbein e^μ_a and any principal connection $\Gamma^a_{b\mu}$ on the structure bundle Σ which is principal. The bundle $\mathcal{C}$ is also a gauge natural bundle associated to the structure bundle Σ. Then one can choose a Lagrangian (which

---

[17] The Euler characteristic is a topological quantity which, on Riemannian manifolds, equals the integral of the scalar curvature suitably normalized.

[18] Recall that this space may be empty as, e.g., if M is non-parallelizable and Σ is trivial.

is 0-order in the frame and 1-order in the connection) by setting:

$$L_{\rm FA} = L_{\rm FA}(e, j^1\Gamma) = e\, R^a_{\ b\mu\nu}\, e^\mu_a e^{b\nu}\, {\rm d}s \tag{8.7.12}$$

where now the curvature $R^a_{\ b\mu\nu}$ depends just on the connection together with its first order derivatives. The vielbein and the connection are now varied independently and the field equations turn out to constrain the connection to be compatible with the vielbein and vacuum field equations for the induced metric to hold. All considerations made for the metric-affine formulation can be now restated for the tetrad-affine formulation. Details are left to the reader as an exercise.

In both the tetrad and tetrad-affine formulation the Lagrangians which have been chosen (i.e. (8.7.10) and (8.7.12), respectively) are covariant with respect to general automorphisms of the structure bundle Σ. Accordingly, both Lagrangians are gauge natural. Thence, using vielbein, we provide a formulation of General Relativity which is truly gauge natural and not simply natural as in the standard formulation. The frame formulation has been extensively used in Physics in the last decades for a number of issues. According to our remark above, the vielbein formulation we introduced here is locally equivalent to the one used in literature from which it differs at the (global) level of the interpretation of families of frames. Our viewpoint generalizes the traditional approach, in which frames are defined just as (local) sections of $L(M)$. The traditional approach is in fact local on general manifolds or it has to be restricted to parallelizable manifolds; this framework introduced here is instead globally well-defined on any η-manifold.

When below we shall be dealing with spinor matter (see Part III), we shall see in detail that also other differences arise between the traditional frame formulation and the vielbein framework presented here. The main difference amounts to a different definition of the covariant derivative of the frame field. In fact, in a geometrical framework the expressions of the covariant derivatives are determined by the configuration bundle. If frames were just local sections of $L(M)$, then their covariant derivative would be (naively and somehow globally incorrectly) defined as

$$\nabla_\mu e^\nu_a = {\rm d}_\mu e^\nu_a + \Gamma^\nu_{\ \lambda\mu} e^\lambda_a \tag{8.7.13}$$

On the contrary, according to the general theory of gauge natural objects, the covariant derivative of vielbein, i.e. global sections of Σ_ρ, is locally given by

$$\nabla_\mu e^\nu_a = {\rm d}_\mu e^\nu_a + \Gamma^\nu_{\ \lambda\mu} e^\lambda_a - \Gamma^c_{\ a\mu} e^\nu_c \tag{8.7.14}$$

The second prescription has been sometimes used in the physical literature on the subject as a heuristic way to indicate that the dynamical fields *feel* the action ρ defined by (8.7.3). Of course, the prescription (8.7.14) is completely unjustified if frames are simply thought as local sections of $L(M)$, while it is instead canonical for the global sections of Σ_ρ.

8. Kosmann Lift

We can use the natural lift to the frame bundle $L(M)$ together with a vielbein $i : \Sigma \longrightarrow L(M)$ to define a canonical non-natural lift of spacetime vector fields up to Σ. Let us denote by g the metric associated with the vielbein i and by $\mathrm{SO}(M, g)$ its orthonormal frame bundle. Let us fix a trivialization σ on Σ and a (possibly non-natural) trivialization V_a on $L(M)$. The trivialization induces fibered coordinates $(x^\mu; e^b_c)$ since any point $e_a \in L(M)$ can be written in the form $e_a = V_b\, e^b_a$. When the trivialization chosen is natural the induced fibered coordinates reduce to natural coordinates $(x^\mu; e^\mu_c)$ on $L(M)$.

Let us further restrict the trivialization $V_a = V^\mu_a \partial_\mu$ so that it is induced by a trivialization of the orthonormal frame bundle $\mathrm{SO}(M, g)$. This amounts to require the frame V_a to be g-orthonormal, something which can always be done. As a consequence of g-orthogonality of V_a, the basis $\rho_b^{\ a}$ of the vertical right invariant vector fields on $L(M)$ is *adapted* to the sub-bundle $\mathrm{SO}(M, g)$. In fact, we can define the vector fields $\sigma_{ab} = \rho^{\ c}_{[a} \eta_{b]c}$ so that they are a basis of vertical right invariant vector fields tangent to $\mathrm{SO}(M, g)$.

We recall that $L(M)$ is a natural bundle and that, depending on the trivialization chosen, we can define two local fiberwise bases of vertical right invariant vector fields, given by

$$\rho_\mu^{\ \nu} = e^\nu_{\ a} \frac{\partial}{\partial e^\mu_a} \qquad \rho_a^{\ b} = e^b_{\ c} \frac{\partial}{\partial e^a_c} \tag{8.8.1}$$

We can thence expand the natural lift $\hat{\xi}$ along these two bases

$$\begin{aligned} \hat{\xi} &= \xi^\mu \partial_\mu + \xi^\mu_{\ \nu} \rho_\mu^{\ \nu} = \\ &= \xi^\mu \partial_\mu + \xi^a_{\ b} \rho_a^{\ b} \end{aligned} \qquad \begin{aligned} \xi^\mu_\nu &= \mathrm{d}_\nu \xi^\mu \\ \xi^a_b &= V^\mu_b (\xi^\nu_{\ \mu} V^a_\nu + \xi^\rho \mathrm{d}_\rho V^b_\nu) \end{aligned} \tag{8.8.2}$$

Thence the natural lift $\hat{\xi}$ can be projected onto $\mathrm{SO}(M, g)$ by means of the canonical operation given by anti-symmetrization; we thence obtain the vector field $\xi_{(K)}$ tangent to the sub-bundle $\mathrm{SO}(M, g)$ locally given by

$$\xi_{(K)} = \xi^\mu \partial_\mu + \xi^{ab} \sigma_{ab} \qquad \xi^{ab} = \xi^{[a}_c \eta^{b]c} \tag{8.8.3}$$

The lift so defined is called the *Kosmann lift (on the bundle* Σ*)* and it is denoted by $K(\xi)$, or briefly by $\xi_{(K)}$. We shall return to this matter later on, in a slightly different context (see Section 9.7). The name of the Kosmann lift is due to the fact that, when dealing with spinor fields, this lift is related to the Lie derivatives of spinors as proposed by Kosmann (in [K66]) with an *ad hoc* prescription.

Proposition (8.8.4): the following identity holds

$$[K(\xi), K(\zeta)] = K([\xi, \zeta]) - \tfrac{1}{2} V^a_\lambda V^c_\mu (\pounds_\zeta g^{\sigma[\lambda} \, \pounds_\xi g^{\mu]\alpha}) g_{\sigma\alpha} \, \sigma_{ac} \tag{8.8.5}$$

Proof: the proof is carried over by direct computation. ∎

Clearly, if ζ or ξ is a Killing vector[19] for g then $[K(\xi), K(\zeta)] = K([\xi, \zeta])$. This property is called *quasi-naturality* (would it hold for any couple of vector fields the Kosmann lift would be natural). Instead, it holds only when at least one vector field is a Killing vectors. This is the deep reason why a wide part of the literature on vielbein is devoted to regard them as natural objects. We will not follow this direction. For us vielbeins are not natural objects; on the contrary they will be regarded in Section 10.1 as gauge natural objects.

We finally remark that the Kosmann lift is canonical since it does not use any additional structure even if it is not natural because of Proposition (8.8.4).

Lie Derivatives of Vielbein

Let us now go back to the vielbein bundle Σ_ρ. Any automorphism $\boldsymbol{\Phi} = (\Phi, \phi)$ of the structure bundle Σ naturally acts over Σ_ρ and thence over its sections, i.e. on vielbein e^μ_a. Thence we can easily define the Lie derivative of vielbein with respect to any infinitesimal generator of automorphisms on Σ which can be locally expressed as

$$\Xi = \xi^\mu(x)\partial_\mu + \xi^{ab}(x)\sigma_{ab} \tag{8.8.6}$$

This induces a vector field $\hat{\Xi}$ over Σ_ρ which is the infinitesimal generator of gauge transformations on vielbeins and which is locally given by

$$\hat{\Xi} = \xi^\mu \partial_\mu + (\partial_\nu \xi^\mu e^\nu_a - e^\mu_b \xi^b{}_{\dot{a}}) \partial^a_\mu \tag{8.8.7}$$

Thence one has the Lie derivative of vielbeins with respect to Ξ which, according to the general theory as exposed in Section 2.6, has the following local expression

$$\pounds_\Xi e^\mu_a = e^\mu_b (\xi_{(V)})^b{}_{\dot{a}} - \nabla_\nu \xi^\mu e^\nu_a \tag{8.8.8}$$

---

19 Actually conformal Killing is enough; see [BG92].

where the connections Γ^{ab}_{μ} and $\Gamma^{\alpha}_{\beta\mu}$ induced by the vielbein e^{μ}_{a} itself have been used to define the vertical part $(\xi_{(V)})^{b}{}_{a}$ and the covariant derivative $\nabla_\nu \xi^\mu$, respectively.

Then one can restrict $\Xi \equiv \xi_{(K)}$ to be the Kosmann lift of a spacetime vector field ξ, i.e.:

$$\Xi \equiv \xi_{(K)} = \xi^\mu \partial_\mu + (e^a_\rho \nabla_\mu \xi^\rho e^\mu_c \eta^{cb} - \Gamma^{ab}_\rho \xi^\rho)\sigma_{ab} \Rightarrow \xi^{ab}_{(V)} = e^{[a}_\mu e^{b]\lambda} \nabla_\lambda \xi^\mu \qquad (8.8.9)$$

In this way, one obtains the Lie derivative of vielbein with respect to a general spacetime vector field, which is locally given by

$$\pounds_\xi e^\mu_a = \pounds_{K(\xi)} e^\mu_a = e^\mu_b \eta_{ac}(e^{[b}_\sigma e^{c]\lambda} \nabla_\lambda \xi^\sigma) - \nabla_\nu \xi^\mu e^\nu_a \equiv \tfrac{1}{2} e_{a\rho} \pounds_\xi g^{\mu\rho} \qquad (8.8.10)$$

An analogous (and less trivial) prescription for spinor fields was first postulated by Kosmann (see Section 9.7 below) by an *ad hoc* argument.

Deformations of Vielbein and Metrics

We conclude this Chapter by pointing out some open problems which bear some importance for both the classical and the quantum aspects of the theory we discussed above. An example will be clarifying. Let us consider again the Lorentzian metrics $g^{(1)}$ and $g^{(2)}$ defined above (see Section 8.7) on the manifold $M = \mathbb{R}^3 - \{0\} \simeq \mathbb{R}^+ \times S^2$. By a degree[20] argument we can show that the two metrics cannot be continuously deformed one into the other. In fact, for each metric g of signature $(1,2)$ on M one can consider the map $\tau : S^2 \longrightarrow S^2$ which associates to each point x in the unit sphere $S^2 \equiv \{r = 1\}$ the timelike eigenvector $\tau(x)$ of the metric g with respect to some auxiliary Euclidean metric δ fixed on M. For the metric $g^{(1)}$ the map $\tau^{(1)}$ is constant, while for the metric $g^{(2)}$ the map $\tau^{(2)}$ is the identity.

Theorem (8.8.11): the metrics $g^{(1)}$ and $g^{(2)}$ cannot be deformed homotopically one into the other.

Proof: let us suppose that such an homotopy F exists and that g_s is the family of metrics joining $g_0 = g^{(1)}$ and $g_1 = g^{(2)}$. One can compute the map τ_s for any such metric g_s and the degree of such a map is constant since the degree is a homotopic invariant. This contradicts the possibility of defining a homotopy joining $g^{(1)}$ and $g^{(2)}$ since $\deg \tau^{(1)} = 0$ and $\deg \tau^{(2)} = 1$. ∎

In this case, on the manifold $M = \mathbb{R}^+ \times S^2$ with signature $(1,2)$ we have provided an example of two metrics which belong to two different classes of admissibility and at the same time cannot be deformed one into the other.

20 The degree of a continuous map counts the parity of the number of the preimages of a generic point.

In other words, this simple example shows that, on a given manifold M we can define at least two interesting equivalence relations in the set of all metrics (of a given signature η). The first one leads to equivalence classes which are *admissibility classes* (i.e. two metrics are equivalent if they are Σ-admissible on some common structure bundle Σ). The second one is the homotopy equivalence relation. In the example we have examined, we have seen the existence of two metrics which are not equivalent in either one of these senses.

The quotient sets of all equivalence classes with respect to these equivalence relations are not yet completely determined for generic M and η, although there is a number of worked results (see [FF96]) in specific cases. A complete classifications is still lacking and in our opinion it deserves to be investigated. In the above example (i.e $M = \mathbb{R}^+ \times S^2$ with Lorentzian signature) there are hints that the two equivalence relations might in fact be equivalent. However, it is not clear to us whether these two equivalence relations define in general or under some specific hypotheses the same equivalence classes, i.e. whether and when they turn out (or not) to be in fact the same equivalence relation.

Exercises

1- Prove that (8.6.6) implies that Ξ given by (8.6.2) is tangent to $\mathrm{SO}(M, g) \longrightarrow L(M)$.
Hint: Consider the gradient of the constraints.

2- Prove that the vector fields $\sigma_{ab} = \rho^{\ c}_{[a}\eta_{b]c}$ defined in Section 8.8 are tangent to $\mathrm{SO}(M, g)$.
Hint: Consider the gradient of the constraints.

3- Write down field equations for the Lagrangian (8.7.10).
Hint: Apply Euler-Lagrange morphism.

4- Prove (8.8.5).
Hint: Expand the objects involved.

PART III

SPINOR FIELDS

Alice thought she had never seen such a curious croquet-ground in all her life; it was all ridges and furrows; the balls were live hedgehogs, the mallets live flamingoes, and the soldiers had to double themselves up and to stand on their hands and feet, to make the arches.

L. Carroll, Alice's Adventures in Wonderland

Introduction

Phenomenological evidence shows that there exists a kind of elementary particles, called *Fermions*, which obey the so-called Fermi-Dirac statistics and which are not geometrically described by natural objects, since they are represented in terms of spinors, which are known to admit no general description as sections of some natural bundle over spacetime. Spinors are globally well defined on Minkowski spacetime by means of an algebraic construction involving the so-called *Clifford algebra* (they exist, in fact, for all $\mathbb{R}^m$ and all signatures $\eta = (r, s)$).

These objects exhibit a peculiar behavior with respect to changes of frame (or, equivalently, with respect to changes of Cartesian coordinates in Minkowski spacetime). When in Minkowski spacetime the spatial frame is rotated through an angle $\theta = 2\pi$, spinors change sign while tensors remain invariant. Spinors are then left unchanged only by rotations which are multiples of 4π. Despite these odd transformation laws one can still try to define the lift to the space of spinors of 1-parameter subgroups of *(linear) isometries* of Minkowski space, i.e. of the Poincaré group.

Following the very basic prescriptions of General Relativity and assuming that spacetime is in fact a curved pseudo-Riemannian manifold (M, g), one can then try to define spinor fields on a curved manifold. In this case one has to look for 1-parameter subgroups of *isometries* of (M, g), i.e. those, if any, generated by Killing vectors. Unfortunately, Killing vectors on a general curved space (M, g) may not exist, depending both on the background metric g chosen and on the topology of M itself. On the other hand, even when spinors do exist globally on a given (M, g) there is no natural way to lift arbitrary transformations of spacetime up to the spinor space, as one can easily see even in the simple case of m-dimensional Minkowski spacetime $\mathbb{R}^m$ with signature $(1, m-1)$. Furthermore, there can also be topological obstructions in solving evolution equation for spinors (usually Dirac equations) owing to the sign ambiguity related to spinor transformations under the Poincaré group. These problems are generally coped with by replacing the non-translational component of the

Poincaré group (i.e., the Lorentz group) by its two-fold covering (which generally is also the universal covering), i.e. the spin group $\mathrm{Spin}(1, m-1)$. To be precise one replaces the connected component of the identity of the Lorentz group with the connected component of the identity of the spin group. However, most results below hold for the whole spin group $\mathrm{Spin}(1, m-1)$ as well as for its connected component to the identity $\mathrm{Spin}_0(1, m-1)$. Thence we shall explicitly indicate the connected components only when they really play an important role, leaving them generally understood. Since our results hold in any dimension and signature we shall denote by $\eta = (r, s)$ the generic signature in dimension $m = r + s$; by an abuse of language we denote by η also the canonical constant diagonal matrix $\eta = \mathrm{diag}(+1, \ldots, +1, -1, \ldots, -1)$ of the given signature. In the traditional approach, global and topological problems are faced by introducing *spin structures*. From a modern viewpoint, spin structures are principal bundle morphisms $\Lambda : \bar{\Sigma} \longrightarrow \mathrm{SO}(M, g)$, where $\bar{\Sigma}$ is a principal bundle having the group $\mathrm{Spin}(\eta)$ as structure group and the bundle $\mathrm{SO}(M, g)$ is the (special) orthonormal frame bundle of the pseudo-Riemannian manifold (M, g) of signature η, which is assumed to be orientable.

These bundles $\bar{\Sigma}$ are shortly called *spin bundles* and they exist if and only if (M, g) meets a number of topological requirements (see below). Spin structures are then used to define spinor fields, which are sections of vector bundles of the type $E_\lambda = \bar{\Sigma} \times_\lambda V$ associated to the spin bundle via a representation λ onto some vector space V. The vector bundle E_λ is then called a *spinor bundle* (not to be confused with the *spin bundle* $\bar{\Sigma}$ which is a principal bundle). More precisely, spin structures are sufficient (as well as just necessary) to overcome the *glueing problems* we mentioned above in defining global Dirac equations on curved spaces (possibly with non-trivial topology).

We shall hereafter recall the standard framework for spinor fields (see Chapter 9). We have to mention that this standard approach is however affected by some drawbacks which are physically relevant and non-trivial. There exists, in fact, because of the relevant constructions, a manifest relation between the metric and an associated spin structure: classically speaking one cannot even define a spin structure without fixing *a priori* a background metric g. Once a metric g has been fixed, moreover, a compatible spin bundle $\bar{\Sigma}$ may exist and be uniquely determined, but it is known that in other cases there can be many possible inequivalent choices for it. This strong dependence and possible non-uniqueness of the spin structure associated to a given metric g is a very serious drawback of the whole construction, at least whenever one really wants to speak coherently and without ambiguities of spinors interacting with gravity. Fixing the metric, in fact, allows one to deal only with field theories on a *given* curved background. However, in all reasonable theories of gravity (interacting with matter), the metric cannot be interpreted as a *fixed* non-dynamical background since the only metric *physically* defined on spacetime is the one directly related to the gravitational field. This metric *has* to be a dynamical variable since the gravitational field is also expected to be effected by spinor fields.

Moreover, since the metric has to be ultimately found as a solution of field equations (conventionally Einstein equations), also the spin structure associated to it cannot be fixed *a priori*. Looking at the problem from a variational perspective, one is therefore led to the conclusion that *both* the metric and the spin structure have to be left undetermined, or, better, one has to consider families of deformations of both in order to allow variations and to coherently define Nöther currents. However, although the problem of deforming metrics is a simple task (since they are geometric objects) there is no classical general construction for the deformations of a spin structure in the classical sense. Because of this we claim that a more general mathematical and physical framework should be adopted.

To this purpose we first establish the following:

Axiom 0: any physically meaningful theory for spin structures ought not to involve non-dynamical background fields.

In other words, we should aim at establishing a theory for spin structures in the category of manifolds, rather than in the category of (pseudo)-Riemannian manifolds.

On the other hand, if the metric g is dynamical and the standard definition for spin structures depends on the metric, the new notion of spin structure itself has to be truly dynamical. As we established in the framework for Variational Calculus any dynamical field has to be a section of some bundle. Thence we state the following:

Axiom 1: there must exist a bundle the sections of which are in one-to-one correspondence with spin structures.

The second problem is strictly related to the fact that the correspondence between metrics and spin structures is in general one-to-many. As a consequence, if the metric is deformed its deformation does not induce in a canonical way a deformation of the corresponding spin structures.

The solution we adopt is very similar to what we already did when defining *vielbein* in Chapter 8: we shall replace spin structures with newly defined fundamental fields (changing first the classical definition of spin structures so that these new fields do not formally depend on the choice of a background metric). These alternative spin structures (which are here called *spin frames*) induce in turn a unique metric, called the *associated metric*. The spin frames are well behaved from a variational viewpoint and admit in a natural way the notion of *deformation*. As a consequence, any deformation of the spin structure, which is canonically defined because of Axiom 1, induces a deformation of the associated metric. In this new framework spin frames are thus the new fundamental variables to describe gravitational field, while the metric geometry of spacetime is an induced structure.

As we shall see, the field theory for spin frames defined in this way is gauge natural, so that we also have a way to produce a direct and canonical treatment for conserved quantities. Furthermore, the spinor field theory obtained can be easily extended to describe the interaction between Bosonic and Fermionic matter and also with a dynamical gravitational field. In other words, we have at our disposal a global Lagrangian theory able to describe all interactions between gravity and spinors. In this way we shall not only describe the influence of gravity onto spinors but we shall also be able to control the opposite, i.e. the gravitational field generated by spinor matter itself.

Moreover, we claim that this new framework is general enough to cover all supersymmetric theories which, in our perspective, reduce in fact to particular cases of gauge natural theories together with their generalized symmetries (in the sense of Section 6.5). The worked examples of Wess-Zumino which end up this monograph, will be enough to support this claim.

As it shall appear, Axiom 0 strongly constrains the geometric theory one can define. If field theories may be seen as "game rules" for the game called *fundamental physics*, then background fields fix the shape of the playground. Within this metaphorical viewpoint, Axiom 0 (or, if one prefers, the basic assumptions of General Relativity itself) claims that the game can be in principle played on any playground, regardless how complicated its shape is. The situation resembles, in a sense, what is known for the fundamental description of gravity in metrical terms. The principles of covariance, of equivalence and of local Lorentz invariance impose that spacetime is a curved Lorentzian manifold (M, g). No further hypotheses are necessary to formulate Einstein theory. However, causality imposes restrictions onto (M, g): it cannot be non-orientable nor it can be compact (otherwise time-like loops would exist). Global solvability of the Cauchy problem for Einstein equations require, e.g., global hyperbolicity, which in turn requires M to have the product structure $\mathbb{R} \times M_3$, and so on. Analogously, once we decide that spin frames constitute the fundamental object, then all possible residual freedom should be left to the theory. Only after, i.e at the level of field equations, motion of particles and conservation laws, one is entitled to discover the existence of further topological restrictions. We believe, in a sense, that the solution given by contemporary Physics is not very far away from the games described by Alice in her travel in Wonderland.

Chapter 9

SPIN STRUCTURES AND SPIN FRAMES

Abstract We shall define *Clifford algebras* and the *spin groups*. Some properties of the Clifford algebras will also be investigated. We shall consider the minimal results which will be needed below. Then we shall introduce *spin structures* in their standard formulation and their relations with the Dirac operator. Thence the *spin frames* will be introduced and used to provide a formulation of gravity in interaction with spinor fields.

References

The main references about spin structures and spinors are [LM89], [GS87], [B81], [BD64]. We refer the reader to [CBWM82], [KN63], [LM89], [DP86], [MS63], [ON83], [GH81], [G70] for topological problems about the existence of global metrics and spinors over spacetime. Causal structure of spacetime is analyzed on [CBWM82], [HE73], [GP78].

A good review on different formulations of General Relativity is [R].

1. Clifford Algebras and Spin Groups

Let (V, η) be a vector space of dimension m endowed with an inner product η of signature $\eta = (r, s)$, $r + s = m$. By an abuse of notation, we shall denote by η the signature, the inner product as well as the canonical diagonal matrix of the fixed signature. Such a matrix is also the matrix associated to the inner product with respect to some orthonormal basis E_i in V, $i = 1, \ldots, m$. Let $\mathcal{T}(V) = \oplus_p T^p_0(M)$ be the (contravariant) tensor algebra of V.

Definition (9.1.1): let $J(\eta)$ be the bilateral ideal generated in $\mathcal{T}(V)$ by the elements of the form $v \otimes v - \eta(v, v)\mathbb{I}$, where $v \in V \hookrightarrow \mathcal{T}(V)$ and $\mathbb{I}$ is the identity. The *Clifford* *algebra* is the quotient $\mathcal{C}(V,\eta) = \mathcal{T}(V)/J(\eta)$. We shall denote by $\diamond$ the product in $\mathcal{C}(V,\eta)$ induced by the tensor product $\otimes$ in $\mathcal{T}(V)$. It will be called *Clifford product.* ■

For example, the Clifford algebra of signature $(0,1)$ is the field $\mathbb{C}$ of complex numbers, while in signature $(0,2)$ one obtains the skew field $\mathbb{H}$ of quaternions.

Since the bilateral ideal $J(\eta)$ is not homogeneous, the $\mathbb{Z}$-grading of $\mathcal{T}(V)$ does not survive in $\mathcal{C}(V,\eta)$. For the same reason, being $J(\eta)$ generated by elements of $\mathcal{T}(V)$ of even degree, the Clifford algebra $\mathcal{C}(V,\eta)$ inherits a $\mathbb{Z}_2$-grading from $\mathcal{T}(V)$. We shall denote by $\mathcal{C}^+(V,\eta)$ the even subalgebra of $\mathcal{C}(V,\eta)$, while $\mathcal{C}^-(V,\eta)$ will be its complementary space which is called the *odd part* of $\mathcal{C}(V,\eta)$.

Let E_i be an orthonormal basis of V with respect to η.

Proposition (9.1.2): the following identity holds

$$E_i \diamond E_j + E_j \diamond E_i = 2\eta(E_i, E_j)\mathbb{I} \tag{9.1.3}$$

Proof: by definition of $J(\eta)$ we have in $J(\eta)$

$$(E_i + E_j) \diamond (E_i + E_j) - \eta(E_i + E_j, E_i + E_j)\mathbb{I} = 0 \tag{9.1.4}$$

The claim is obtained by expanding (9.1.4) modulo $J(\eta)$. ■

As a direct consequence of this proposition the dimension of the Clifford algebra $\mathcal{C}(V,\eta)$ turns out to be finite; more precisely we have $\dim(\mathcal{C}(V,\eta)) = 2^m$. A basis of the Clifford algebra is

$$\mathbb{I}, \quad E_i, \quad \underset{(i<j)}{E_{ij} = E_i \diamond E_j}, \quad \dots, \quad E_{12\dots m} = E_1 \diamond E_2 \diamond \dots \diamond E_m \tag{9.1.5}$$

Using (9.1.3) we can define an involutive operation on $\mathcal{C}(V,\eta)$ called *conjugation.*

Definition (9.1.6): there exists an involutive linear map on $\mathcal{C}(V,\eta)$ which acts on the basis (9.1.5) by

$$\overline{E_{i_1 i_2 \dots i_k}} = E_{i_k \dots i_2 i_1} \tag{9.1.7}$$

Such a map does not depend on the basis chosen. ■

As an example, because of (9.1.3) one has

$$\overline{E_{i_1 i_2}} = E_{i_2 i_1} = -E_{i_1 i_2} + 2\eta_{i_1 i_2}\mathbb{I} \tag{9.1.8}$$

The metric η on V can now be extended to a metric $\hat{\eta}$ in the whole Clifford algebra $\mathcal{C}(V,\eta)$ by setting

$$|E_{i_1 i_2 \dots i_k}|^2 = \hat{\eta}(E_{i_1 i_2 \dots i_k}, E_{i_1 i_2 \dots i_k}) = \overline{E_{i_1 i_2 \dots i_k}} \diamond E_{i_1 i_2 \dots i_k} \tag{9.1.9}$$

and the basis (9.1.5) turns out to be orthonormal with respect to such an extended metric $\hat{\eta}$.

The metric $\hat{\eta}$ has many appealing properties. We here present two Lemmas which will be useful below:

Lemma (9.1.10): just two among all multiples of the identity have unit length.

Proof: let $S = \alpha\mathbb{I} \in \mathcal{C}(V,\eta)$ be a multiple of the identity. We have $\hat{\eta}(S,S) = \alpha^2\hat{\eta}(\mathbb{I},\mathbb{I}) = \alpha^2$. Then $|S|^2$ is positive and $S = \pm\mathbb{I}$. ■

Lemma (9.1.11): the norm induced by $\hat{\eta}$ preserves the Clifford product.

Proof: let us briefly denote by E_A the elements of the Clifford basis (9.1.5). Two elements of the Clifford algebra are thence $v = v^A E_A$ and $u = u^A E_A$. The norm of their product is thence

$$\begin{aligned} |v \diamond u|^2 =& v^A u^B v^C u^D \hat{\eta}(E_A \diamond E_B, E_C \diamond E_D) = \\ =& v^A u^B v^C u^D E_A \diamond E_B \diamond \overline{E}_D \diamond \overline{E}_C = v^A u^B v^C u^D \hat{\eta}_{BD} \hat{\eta}_{AC} = \\ =& |v|^2 |u|^2 \end{aligned} \tag{9.1.12}$$

which proves the claim. ■

As a consequence the product of two unit vectors is again a unit vector, so that the set of all unit vectors is closed with respect to the Clifford product and it will be denoted by $S(V,\eta)$. It is also a subgroup since the inverse of a unit vector $S \in S(V,\eta)$ is given by

$$S^{-1} = \frac{\overline{S}}{|S|^2} \tag{9.1.13}$$

Definition (9.1.14): the group generated in the Clifford algebra $\mathcal{C}(V,\eta)$ by the elements $v \in V \hookrightarrow \mathcal{C}(V,\eta)$ which are also unitary (i.e. $v \in S(V,\eta)$) is denoted by $\mathrm{Pin}(\eta)$. Its intersection with the even Clifford algebra is denoted by $\mathrm{Spin}(\eta) = \mathrm{Pin}(\eta) \bigcap \mathcal{C}^+(V,\eta)$. They are called the *pin group* and the *spin group*, respectively. ■

Because of Lemma (9.1.11) the group $\mathrm{Pin}(\eta)$, and as a consequence also $\mathrm{Spin}(\eta)$, is contained into $S(V,\eta)$.

Proposition (9.1.15): if $S \in \mathrm{Pin}(\eta)$ then there exists an orthonormal transformation $\ell_S \in \mathrm{O}(V, \eta)$ such that for all $v \in V$, $S \diamond v \diamond S^{-1} = \ell_S(v)$.

Proof: the proposition follows easily once it is proved for all $S \equiv w \in V \bigcap \mathrm{Pin}(\eta)$. In that case we have $|w|^2 = \pm 1 \equiv \epsilon$ and $S^{-1} = \epsilon w$.

Let us now decompose the vector $v \in V$ into a part $v_{\parallel}$ parallel and a part $v_{\perp}$ orthogonal to the unit vector $w \in V$. We have

$$v_{\parallel} = \eta(v, w) w \qquad\qquad v_{\perp} = v - v_{\parallel} \tag{9.1.16}$$

Thence it immediately follows

$$S \diamond v \diamond S^{-1} = w \diamond (v_{\parallel} + v_{\perp}) \diamond \epsilon w = v_{\parallel} - v_{\perp} \equiv \ell_w(v) \tag{9.1.17}$$

The linear transformation $\ell_w : V \longrightarrow V$ so defined is orthogonal (it is a reflection along w). Any element of $S \in \mathrm{Pin}(\eta)$ is a product of unit vectors in V and thence ℓ_S is a composition of reflections and thence still an orthonormal transformation. Notice that if $S \in \mathrm{Spin}(\eta)$ then ℓ_S is a proper rotation in $\mathrm{SO}(\eta)$. ∎

Corollary (9.1.18): there exists a group epimorphism $\ell : \mathrm{Spin}(\eta) \longrightarrow \mathrm{SO}(\eta) : S \mapsto \ell_S$ which is a two-fold covering.

Proof: the map ℓ is surjective since any orthogonal transformation Λ is a composition of an even number (say $2k$) of reflections along unit vectors w_i. Then the element $T = w_1 \diamond \ldots \diamond w_{2k} \in \mathrm{Spin}(\eta)$ is a preimage of Λ. Because of the commutation properties of the elements in the basis of $\mathcal{C}^+(V, \eta)$ one can obtain that $\ker \ell = \{\mathbb{I}, -\mathbb{I}\} \simeq \mathbb{Z}_2$. ∎

Corollary (9.1.19): if $S \in \mathcal{C}^+(V, \eta) \bigcap \mathcal{S}(V, \eta)$ and there exists $\Lambda \in \mathrm{SO}(V, \eta)$ such that $S \diamond v \diamond S^{-1} = \Lambda(v)$ then $S \in \mathrm{Spin}(\eta)$

Proof: as in the previous corollary, Λ is the composition of an even number of reflections along the unit vectors w_i and $T = w_1 \diamond \ldots \diamond w_{2k}$ is such that $\ell_T = \ell_S$. As we did in the proof of (9.1.18), using the commutation relations of the elements of $\mathcal{C}(V, \eta)$, we find that $T^{-1} \diamond S = \alpha \mathbb{I}$. As $T \in \mathrm{Spin}(\eta)$ and $S \in \mathcal{S}(V, \eta)$, we necessarily have $\alpha = \pm 1$ and thence $S = \pm T$. Then we finally obtain $S \in \mathrm{Spin}(\eta)$. ∎

The group $\mathrm{Spin}(\eta)$ is always realized as a matrix group, since each element in $\mathrm{Spin}(\eta)$ acts on $\mathcal{C}^+(V, \eta)$ by

$$L_S : \mathcal{C}^+(V, \eta) \longrightarrow \mathcal{C}^+(V, \eta) : T \mapsto S \diamond T \tag{9.1.20}$$

The rank of this representation grows quite rapidly as the dimension of V grows; for dimension $\dim(V) = m$ this standard representation uses $2^{m-1} \times$

2^{m-1} matrices. However, if we find exactly m matrices γ_a in some suitable $GL(k, \mathbb{K})$ (with $\mathbb{K} = \mathbb{R}, \mathbb{C}$ and $k < 2^{m-1}$) such that

$$\gamma_a \gamma_b + \gamma_b \gamma_a = 2\eta_{ab}\mathbb{I} \tag{9.1.21}$$

then they induce a much more manageable representation of the Clifford algebra and, as a consequence, also a representation of the pin and spin groups. Any set of matrices which obey the property (9.1.21) is called a *set of Dirac matrices*.

For example, in dimension 4 and Lorentzian signature $\eta = (1, 3)$ one can define the so-called *Dirac representation*

$$\gamma_0 = \begin{pmatrix} 0 & \mathbb{I} \\ \mathbb{I} & 0 \end{pmatrix} \qquad \gamma_i = \begin{pmatrix} 0 & -\sigma^i \\ \sigma^i & 0 \end{pmatrix} \tag{9.1.22}$$

where σ^i $(i = 1, 2, 3)$ are the so-called *Pauli matrices*

$$\sigma^1 = \begin{pmatrix} 0 & 1 \\ 1 & 0 \end{pmatrix} \qquad \sigma^2 = \begin{pmatrix} 0 & -i \\ i & 0 \end{pmatrix} \qquad \sigma^3 = \begin{pmatrix} 1 & 0 \\ 0 & -1 \end{pmatrix} \tag{9.1.23}$$

Proposition (9.1.24): if $\{\gamma_a\}$ is a set of Dirac matrices and $\Lambda^a_b \in O(\eta)$ then $\{\gamma'_b = \gamma_a \Lambda^a_b\}$ is a set of Dirac matrices, too. ∎

In this way, the group $\mathrm{Spin}(\eta)$ (as well as the group $\mathrm{Pin}(\eta)$) can be identified with a suitable subgroup of $GL(k, \mathbb{K})$; by Corollary (9.1.19), the spin group is identified by means of the following equations

$$S \cdot \gamma_a \cdot S^{-1} = \gamma_b \, \alpha^b_{\ a} \,, \qquad \alpha^b_{\ a} \in SO(\eta) \tag{9.1.25}$$

together with a normalization condition which ensures that $S \in S(v, \eta)$. This normalization is given by the following proposition:

Lemma (9.1.26): if $S \in \mathrm{Spin}_0(\eta)$, the connected component of the identity, then we have $\det(S) = 1$.

Proof: let us first consider a unit vector $v \in V$ and let $\epsilon = |v|^2 = \pm 1$ be its norm. Let us fix a representation and denote by $\tilde{v} \in GL(k, \mathbb{K})$ the matrix representing the element $v \in V$. Then one has $\tilde{v} \cdot \tilde{v} = \epsilon \mathbb{I}$. By taking the determinant of both sides, one has $\det(\tilde{v}) = \pm 1, \pm i$. As each element $S \in \mathrm{Spin}(\eta) \hookrightarrow GL(k, \mathbb{K})$ is a product of unit vectors, one has that $\det(S) = \pm 1, \pm i$ for any element of $\mathrm{Spin}(\eta)$.

Since the function $\det : \mathrm{Spin}(\eta) \longrightarrow \mathbb{K}$ is continuous and it takes its values in a discrete set $\{\pm 1, \pm i\}$, it is locally constant and thence constant on any connected component. In particular, it is constant on the connected component $\mathrm{Spin}_0(\eta)$ of the identity. ∎

Corollary (9.1.27): if $s \in \mathfrak{spin}(\eta)$, the Lie algebra of Spin(η), then $\mathrm{Tr}(s) = 0$.

The immersion of the groups induces the corresponding immersion for the Lie algebras $\mathfrak{spin}(\eta) \hookrightarrow \mathfrak{gl}(k, \mathbb{K})$. An element of the Lie algebra $\dot{S}_0 S_0^{-1} \in \mathfrak{gl}(k, \mathbb{K})$ identified by a curve $S_t : \mathbb{R} \longrightarrow \mathrm{GL}(k, \mathbb{K})$, is also an element of $\mathfrak{spin}(\eta) \subset \mathfrak{gl}(k, \mathbb{K})$ when the following property is satisfied:

Proposition (9.1.28): $\dot{S}_0 S_0^{-1} \in \mathfrak{spin}(\eta)$ if and only if $\dot{S}_0 S_0^{-1} = \frac{1}{8} \dot{\alpha}^i{}_k \bar{\alpha}^k_l \eta^{lj} [\gamma_i, \gamma_j]$ for some $\alpha^i{}_k : \mathbb{R} \longrightarrow \mathrm{SO}(\eta)$.

Proof: by taking the derivative of the equation (9.1.25) we obtain

$$\dot{S}\gamma_a S^{-1} + S\gamma_a \dot{S}^{-1} = \gamma_b \dot{\alpha}^b_a \qquad \Rightarrow \qquad [\dot{S}S^{-1}, \gamma_c] = \gamma_b \dot{\alpha}^b{}_a \bar{\alpha}^a{}_c \tag{9.1.29}$$

By using the commutation relations in the Clifford algebra one obtains

$$\dot{S}S^{-1} = \epsilon \mathbb{I} + \frac{1}{8} \dot{\alpha}^i{}_k \bar{\alpha}^k_l \eta^{lj} [\gamma_i, \gamma_j] \tag{9.1.30}$$

for some real number ϵ.

By taking the trace of this equation and using Corollary (9.1.27), since the trace of a commutator identically vanishes, one has finally that $\epsilon = 0$. ■

For example, the spin group Spin$(1, 3)$ is isomorphic to the group $\mathrm{SL}(2, \mathbb{C})$ of complex 2×2 matrices with determinant equal to one. Other examples will be presented in Section 9.3.

2. Spin Structures

Let (M, g) be an orientable pseudo-Riemannian manifold of signature $\eta = (r, s)$, $r + s = m = \dim(M)$.[1] Let us first give the following:

Definition (9.2.1): a *spin structure* over (M, g) is a pair $(\bar{\Sigma}, \Lambda)$ such that[2]

- a- $\bar{\Sigma}$ is a principal bundle having Spin(η) as structure group. It is called a *spin bundle.*
- b- let us denote by SO(M, g) the orthonormal frame bundle (M, g); then the map $\Lambda : \bar{\Sigma} \longrightarrow \mathrm{SO}(M, g)$ is a principal bundle morphism with respect to the group homomorphism $\ell : \mathrm{Spin}(\eta) \longrightarrow \mathrm{SO}(\eta)$, i.e.

---

[1] Since M is orientable its first Stiefel-Whitney class $w_1(M)$ vanishes.

[2] We shall see that these hypotheses imply further topological restrictions on M.

$$\begin{array}{ccc} \bar\Sigma & \xrightarrow{\Lambda} & \mathrm{SO}(M,g) \\ \downarrow & & \downarrow \\ M & = & M \end{array} \qquad \begin{array}{ccc} \bar\Sigma & \xrightarrow{R_S} & \bar\Sigma \\ \Lambda\downarrow & & \downarrow\Lambda \\ \mathrm{SO}(M,g) & \xrightarrow{R_{\ell(S)}} & \mathrm{SO}(M,g) \end{array} \tag{9.2.2}$$

where R denote both standard right actions and we set $\ell(S) = \ell_S$.

The rest of this Section will be devoted to discuss the existence of spin structures on a fixed (pseudo)-Riemannian manifold (M, g). Let us fix $\{U_\alpha\}$ an open covering (which is not restricted to be an atlas) and a local trivialization $e^{(\alpha)}$ of the orthonormal frame bundle $\mathrm{SO}(M, g)$ (with respect to the open covering U_α); let us denote by $g_{(\alpha\beta)} : U_{\alpha\beta} \longrightarrow \mathrm{SO}(\eta)$ the cocycle of its transition functions which are defined over the intersections $U_{\alpha\beta} = U_\alpha \bigcap U_\beta$.

We want to define on the same covering the transition functions of a principal bundle $\bar\Sigma$ having $\mathrm{Spin}(\eta)$ as a structure group; they will be denoted by $G_{(\alpha\beta)} : U_{\alpha\beta} \longrightarrow \mathrm{Spin}(\eta)$.

Definition (9.2.3): we say that the functions $\{G_{(\alpha\beta)}\}$ are a *lift* of the cocycle $\{g_{(\alpha\beta)}\}$ if and only if $\ell \circ G_{(\alpha\beta)} = g_{(\alpha\beta)}$. ■

Of course, such a family of functions $\{G_{(\alpha\beta)}\}$ need not form necessarily a cocycle in $\mathrm{Spin}(\eta)$ even if they are a lift of a cocycle $\{g_{(\alpha\beta)}\}$. In fact, suppose that we consider a particular lift $\{G_{(\alpha\beta)}\}$ of the cocycle $\{g_{(\alpha\beta)}\}$ which happens to be a cocycle too. We can pick up any single intersection U_{12} and change the sign of the $G_{(12)}$ defined on U_{12}, obtaining a new lift $\{\hat G_{(\alpha\beta)}\}$ of the cocycle $\{g_{(\alpha\beta)}\}$ which is not a cocycle any longer.

Proposition (9.2.4): if $\{G_{(\alpha\beta)}\}$ is a lift of $\{g_{(\alpha\beta)}\}$ and they form a cocycle then (M, g) allows a spin structure.

Proof: because of theorem (1.1.7), the cocycle $\{G_{(\alpha\beta)}\}$ allows one to construct a principal bundle $\bar\Sigma$ with standard fiber $\mathrm{Spin}(\eta)$ together with a trivialization of $\bar\Sigma$ associated to a family of local sections $s^{(\alpha)}$. By construction, such a trivialization allows $\{G_{(\alpha\beta)}\}$ as transition functions.

Then since $\{G_{(\alpha\beta)}\}$ are a lift, the local morphisms

$$\Lambda_{(\alpha)} : \bar\Sigma \longrightarrow \mathrm{SO}(M, g) : s^{(\alpha)} S \mapsto e^{(\alpha)} \ell(S) \tag{9.2.5}$$

glue together to give a global morphism $\Lambda : \bar\Sigma \longrightarrow \mathrm{SO}(M, g)$; the pair $(\bar\Sigma, \Lambda)$ so defined is by construction a spin structure over (M, g). ■

One can easily show that a lift of $\{g_{(\alpha\beta)}\}$ can always be defined. In fact, it is sufficient to remark that the map ℓ is a covering map and thence it has the following standard properties (see [GH81] for a proof):

Lift property of curves (9.2.6): let $\gamma : I = [-1, 1] \longrightarrow \mathrm{SO}(\eta)$ be a curve and let us fix $S \in \ell^{-1}(\gamma(0))$ then there exists one and only one curve $\hat{\gamma} : I \longrightarrow \mathrm{Spin}(\eta)$ such that $\hat{\gamma}(0) = S$ and $\ell \circ \hat{\gamma} = \gamma$. The curve $\hat{\gamma}$ so determined is called the *lift of γ at the point S.* ∎

Homotopy property (9.2.7): let $\gamma_1, \gamma_2 : I \longrightarrow \mathrm{SO}(\eta)$ be two homotopic curves, $S \in \ell^{-1}(\gamma_1(0))$ be a point in the fiber over $\gamma_1(0) = \gamma_2(0)$ and $\hat{\gamma}_1, \hat{\gamma}_2$ be their lifts at S; then the lifts $\hat{\gamma}_1, \hat{\gamma}_2$ are homotopic and in particular $\hat{\gamma}_1(1) = \hat{\gamma}_2(1)$. ∎

By a standard result in topology, it is always possible to choose an open covering $\{U_\alpha\}$ of M so that all the intersections are contractible (see Proposition (1.5.7)). Let us then fix $x_0 \in U_{\alpha\beta}$; as a consequence any other point $x_1 \in U_{\alpha\beta}$ can be joined to x_0 by means of a path $\gamma_{(x_0,x_1)}$ ($U_{\alpha\beta}$ is connected); since $U_{\alpha\beta}$ is simply connected any other path joining x_0 and x_1 is homotopic to $\gamma_{(x_0,x_1)}$.

As we fix $S \in \ell^{-1}(g_{(\alpha\beta)}(x_0))$ in the fiber over $g_{(\alpha\beta)}(x_0)$, by the lift property of curves one can define a lift $\hat{\gamma}_{(\alpha\beta)}$ of the curve $g_{(\alpha\beta)} \circ \gamma_{(x_0,x_1)}$ and, by the homotopy property, $\hat{\gamma}_{(\alpha\beta)}(1)$ does not depend on the representative we choose.

Let us thence define the functions

$$G_{(\alpha\beta)}(x_1) = \hat{\gamma}_{(\alpha\beta)}(1) \tag{9.2.8}$$

which are well defined and are a lift of $\{g_{(\alpha\beta)}\}$ by construction.

In general the maps $\{G_{(\alpha\beta)}\}$ defined by (9.2.8) do not form a cocycle. The only thing we know about them is that the functions

$$\epsilon_{(\alpha\beta\gamma)} = G_{(\alpha\beta)} \cdot G_{(\beta\gamma)} \cdot G_{(\gamma\alpha)} : U_{\alpha\beta\gamma} \longrightarrow \mathrm{Spin}(\eta) \tag{9.2.9}$$

are actually valued into $\ker(\ell) = \mathbb{Z}_2$, since $\ell(\epsilon_{(\alpha\beta\gamma)}) = \mathbb{I}$ and $\{g_{(\alpha\beta)}\}$ is a cocycle. Thence the family $\epsilon = \{\epsilon_{(\alpha\beta\gamma)}\}$ form a Čech 2-cochain with values in $\mathbb{Z}_2$.[3]

Proposition (9.2.10): $\epsilon = \{\epsilon_{(\alpha\beta\gamma)}\}$ is a Čech cocycle.

Proof:

$$(\delta\epsilon)_{(\alpha\beta\gamma\lambda)} = \epsilon_{(\beta\gamma\lambda)} \cdot \epsilon^{-1}_{(\alpha\gamma\lambda)} \cdot \epsilon_{(\alpha\beta\lambda)} \cdot \epsilon^{-1}_{(\alpha\beta\gamma)} = \mathbb{I} \tag{9.2.11}$$

We used the fact that $\epsilon_{(\alpha\beta\gamma)}$ takes values into the center[4] of the group $\mathrm{Spin}(\eta)$. ∎

Proposition (9.2.12): $\{G_{(\alpha\beta)}\}$ is a cocycle if and only if ϵ defined by (9.2.9) is a coboundary, i.e. if and only if ϵ is a representative of the zero class in the homology group $\check{H}^2(M, \mathbb{Z}_2)$.

[3] See [GH81].

[4] The *center* of a group G is the subgroup of the elements which commute with any other element of G.

Proof: if ϵ is a coboundary then there exists a 1-cochain $H_{(\alpha\beta)} : U_{\alpha\beta} \longrightarrow \mathbb{Z}_2$ such that

$$(\delta H)_{(\alpha\beta\gamma)} = \epsilon_{(\alpha\beta\gamma)} \tag{9.2.13}$$

It immediately follows that $G'_{(\alpha\beta)} = G_{(\alpha\beta)} \cdot H_{(\alpha\beta)}$ is a cocycle and it is still a lift of $\{g_{(\alpha\beta)}\}$.
■

The class identified by ϵ in $\check{\mathrm{H}}^2(M, \mathbb{Z}_2)$ is the *second Stiefel-Whitney class*. To be more precise, the obstruction ϵ is exactly the Stiefel-Whitney class if η is strictly Riemannian. In all other signatures the obstruction class ϵ calculated above turns out to be the sum of Stiefel-Whitney class with other classes which however vanish because of the topological conditions imposed onto M by the existence of a global metric of signature η (see [CBWM82]). We can thence state the following result:

Proposition (9.2.14): a (pseudo)-Riemannian manifold (M, g) allows spin structures if and only if its second Stiefel-Whitney class vanishes. ■

If $w_2(M) = 0$ on an orientable manifold M (so that also $w_1(M) = 0$ holds) we say that M is a *spin manifold*. If M is at the same time an η-manifold and a spin manifold we say that M is a *spin η-manifold*. We shall also recall the following important result:

Proposition (9.2.15): in dimension 4 and signature $(1, 3)$, a (pseudo)-Riemannian non-compact manifold (M, g) is a spin manifold if and only if it allows a global orthonormal frame. (see [G70]). ■

This result is important because such hypotheses are satisfied by all *physically* admissible spacetimes. In fact, non-compactness is related to the possibility of having a well-posed Cauchy problem for Einstein equations and with the need of avoiding problems with causality (closed causal geodesics). As we already mentioned, in compact spacetimes one has in fact closed causal geodesics.

3. Example: Spin Structures on Spheres

Let $(W, \tilde{\eta})$ be a vector space endowed with a metric structure $\tilde{\eta}$ of signature $(r+1, s)$, with $m = r + s$, so that $m + 1$ is the dimension of W.

Let us define the sphere $S_+^{(r,s)} = \{v \in W : \tilde{\eta}(v, v) = 1\}$ on which $\tilde{\eta}$ induces a metric η of signature (r, s). All that follows can be repeated for a signature $(r, s+1)$ and by considering the sphere $S_-^{(r,s)} = \{v \in W : \tilde{\eta}(v, v) = -1\}$ on which the metric $\tilde{\eta}$ still induces a metric η of signature (r, s).

Property (9.3.1): the total space of the orthonormal frame bundle $\mathrm{SO}(S^{(r,s)}, \eta)$ is isomorphic to the space $\mathrm{SO}(r+1, s)$.

Proof: let us fix an orthonormal basis (e_α), $\alpha = 0, 1, \ldots, m$, of $(V, \tilde{\eta})$ which will be called the *initial basis*; for each matrix $a \in \mathrm{SO}(r+1, s)$ let E_0, E_1, ..., E_m be the orthonormal basis of $(W, \tilde{\eta})$ obtained by acting by means of a on the initial basis, i.e. $E_\alpha = a \cdot e_\alpha$. The first vector E_0 in such a basis is a vector with positive norm ($\tilde{\eta}(E_0, E_0) = +1$) and it singles out a point $x \in S^{(r,s)}$. The other vectors of the basis $(E_1, \ldots, E_m)$ are tangent to $S^{(r,s)}$ at x and they form an η-orthonormal frame. Thence they define a point of $\mathrm{SO}(S^{(r,s)}, \eta)$. The map $\mathrm{SO}(r+1, s) \longrightarrow \mathrm{SO}(S^{(r,s)}, \eta)$ so defined is bijective. The projection is thence defined by $\pi : \mathrm{SO}(r+1, s) \longrightarrow S^{(r,s)} : a \mapsto E_0$. ∎

Since $\mathrm{SO}(S^{(r,s)}, \eta)$ has a group structure we can algebraically define the group $\mathrm{Spin}(r+1, s)$ which turns out to be a two-fold covering of the orthonormal frame bundle of $(S^{(r,s)}, \eta)$. Thence $\mathrm{Spin}(r+1, s)$ is a natural candidate to be the spin bundle $\bar{\Sigma}$ on which the spin structure is defined. Let us denote by $\tau = \pi \circ \ell : \mathrm{Spin}(r+1, s) \longrightarrow S^{(r,s)}$ the projection over $S^{(r,s)}$ which makes $\mathrm{Spin}(r+1, s)$ a bundle.

Property (9.3.2): the group $\mathrm{Spin}(r+1, s)$ is a principal bundle over $S^{(r,s)}$ having $\mathrm{Spin}(r, s)$ as a structure group.

Proof: let us denote by E_α ($\alpha = 0, 1, \ldots, m$) the elements of an orthonormal basis of $(W, \tilde{\eta})$ and by E_i ($i = 1, \ldots, m$) those which are orthogonal to E_0. Let us then denote by $(W', \tilde{\eta}')$ the vector space spanned by E_i. We have the immersion:

$$(W', \tilde{\eta}') \hookrightarrow (W, \tilde{\eta}) \tag{9.3.3}$$

which induces an immersion of the corresponding Clifford algebras

$$\mathcal{C}(W', \tilde{\eta}') \hookrightarrow \mathcal{C}(W, \tilde{\eta}) \tag{9.3.4}$$

The image of the Clifford algebra $\mathcal{C}(W', \tilde{\eta}')$ into $\mathcal{C}(W, \tilde{\eta})$ is generated by the products of the E_i alone.

By restriction we have the group immersion

$$\mathrm{Spin}(r, s) \hookrightarrow \mathrm{Spin}(r+1, s) \tag{9.3.5}$$

The elements of the image of $\mathrm{Spin}(r, s)$ (generated by the unit vectors orthogonal to E_0) obey the following property

$$S \diamond E_0 \diamond S^{-1} = E_0 \tag{9.3.6}$$

so that $\ell(S)$ fixes E_0. In other words $\ell(S)$ belongs to

$$\mathrm{SO}(r, s) \hookrightarrow \mathrm{SO}(r+1, s) : A \mapsto \begin{pmatrix} 1 & 0 \\ 0 & A \end{pmatrix} \tag{9.3.7}$$

We remark that this last immersion is the one defining the bundle structure of the orthonormal frame bundle $\mathrm{SO}(r+1,s) \longrightarrow S^{(r,s)}$.

We have thence a commutative diagram

$$\begin{array}{ccc} \mathrm{Spin}(r,s) & \longrightarrow & \mathrm{Spin}(r+1,s) \\ \ell \big\downarrow & & \big\downarrow \ell \\ \mathrm{SO}(r,s) & \longrightarrow & \mathrm{SO}(r+1,s) \end{array} \tag{9.3.8}$$

and as a result we have a (right) action of the spin group $\mathrm{Spin}(r,s)$ on the bundle $\mathrm{Spin}(r+1,s) \longrightarrow S^{(r,s)}$. This action is vertical as a consequence of verticality of the action of $\mathrm{SO}(r,s)$ over $\mathrm{SO}(r+1,s)$.

It is now easy to prove that the action is free and transitive on the fibers thence defining a principal bundle. ∎

Corollary (9.3.9): $(\mathrm{Spin}(r+1,s), \ell)$ is a spin structure over $(S^{(r,s)}, \eta)$.

Proof: $\ell : \mathrm{Spin}(r+1,s) \longrightarrow \mathrm{SO}(r+1,s) \simeq \mathrm{SO}(S^{(r,s)}, \eta)$ is a vertical morphism and it preserves the right action because of the commutative diagram (9.3.8). ∎

We shall hereafter consider as an application the spin structures on low dimensional spheres.

Spin Structures on the Sphere S^2

Let us consider the sphere S^2 as the subset of unitary vectors in $\mathbb{R}^3$ endowed with its canonical metric δ of signature $(0,3)$. Let $\{E_0, E_1, E_2\}$ be the canonical basis in $\mathbb{R}^3$. An element in $\mathcal{C}^+(\mathbb{R}^3, \delta)$ is in general in the form

$$S = a\mathbb{I} + bE_{01} + cE_{02} + dE_{12} \tag{9.3.10}$$

Let us denote by

$$\bar{S} = a\mathbb{I} - bE_{01} - cE_{02} - dE_{12} \tag{9.3.11}$$

the conjugate of S so that

$$\alpha = a^2 + b^2 + c^2 + d^2 \tag{9.3.12}$$

is the norm of S. If $S \in S(\mathbb{R}^3, \delta)$ we have that $\alpha = 1$ and the inverse is given by $S^{-1} = \alpha^{-1}\bar{S}$.

Let us compute $\ell(S)$ (see (9.1.15)) and notice that it is an orthogonal matrix. By applying Corollary (9.1.19), we have that the element (9.3.10) in $\mathcal{C}^+(\mathbb{R}^3, \delta)$ is an element of the group $\mathrm{Spin}(0,3)$ if and only if $\alpha = 1$.

In other words, the group Spin(0, 3) is canonically identified with the sphere $S^3 \subset \mathcal{C}^+(\mathbb{R}^3, \delta) \simeq \mathbb{R}^4$. The covering map is

$$\ell : \mathrm{Spin}(0,3) \longrightarrow \mathrm{SO}(0,3) : S \mapsto \ell(S) \tag{9.3.13}$$

where we defined

$$S = a\mathbb{I} + bE_{01} + cE_{02} + dE_{12} \tag{9.3.14}$$

$$\ell(S) = \begin{pmatrix} a^2+d^2-b^2-c^2 & -2(ab+cd) & 2(-ac+bd) \\ 2(ab-cd) & a^2-b^2+c^2-d^2 & -2(ad+bc) \\ 2(ac+bd) & 2(ad-bc) & a^2+b^2-c^2-d^2 \end{pmatrix} \tag{9.3.15}$$

The projection onto $S^2 \subset \mathbb{R}^3$ is thence given by

$$\tau(a\mathbb{I}+bE_{01}+cE_{02}+dE_{12}) \mapsto (a^2+d^2-b^2-c^2, 2(ab-cd), 2(ac+bd)) \tag{9.3.16}$$

By defining two complex numbers by setting $z_1 = a - id$ and $z_2 = b + ic$, that projection can be recast as

$$\tau : S^3 \longrightarrow S^2 : (z_1, z_2) \mapsto (|z_1|^2 - |z_2|^2, 2\bar{z}_1 z_2) \tag{9.3.17}$$

Thence the structure bundle $\bar{\Sigma}$ is the Hopf bundle.

The standard fiber is $U(1) \simeq S^1$ and a trivialization is induced by the following local sections

$$u_N : S^2_N \longrightarrow \bar{\Sigma} : (x,y,z) \mapsto \begin{pmatrix} a = \sqrt{\frac{1+x}{2}}\cos\theta \\ b = 0 \\ c = \sqrt{\frac{1-x}{2}} \\ d = -\sqrt{\frac{1+x}{2}}\sin\theta \end{pmatrix} \qquad x \neq 1 \tag{9.3.18}$$

$$u_S : S^2_S \longrightarrow \bar{\Sigma} : (x,y,z) \mapsto \begin{pmatrix} a = \sqrt{\frac{1+x}{2}} \\ b = \sqrt{\frac{1-x}{2}}\sin\theta \\ c = \sqrt{\frac{1-x}{2}}\cos\theta \\ d = 0 \end{pmatrix} \qquad x \neq -1 \tag{9.3.19}$$

where θ is defined by the relations

$$\begin{cases} \cos\theta = \dfrac{z}{\sqrt{1-x^2}} \\ \sin\theta = \dfrac{y}{\sqrt{1-x^2}} \end{cases} \tag{9.3.20}$$

Spin Structures on S^3

Let the sphere S^3 be the subset of unit vectors in $\mathbb{R}^4$ endowed with the standard metric δ of signature $(0,4)$. Let (E_0, E_1, E_2, E_3) be the canonical basis in $\mathbb{R}^4$.

An element of $\mathcal{C}^+(\mathbb{R}^4,\delta)$ is

$$S = a\mathbb{I} + bE_{01} + cE_{02} + dE_{12} + eE_{03} + fE_{13} + gE_{23} + hE_{0123} \tag{9.3.21}$$

Let us define the conjugation

$$\bar{S} = a\mathbb{I} - bE_{01} - cE_{02} - dE_{12} - eE_{03} - fE_{13} - gE_{23} + hE_{0123} \tag{9.3.22}$$

and the scalars

$$\begin{aligned} \alpha &= a^2 + b^2 + c^2 + d^2 + e^2 + f^2 + g^2 + h^2 \\ \beta &= ah - bg + cf - de \end{aligned} \tag{9.3.23}$$

Since $S \in S(\mathbb{R}^4,\delta)$ one has $\alpha = 1$ and the inverse $S^{-1} = (\alpha^2 - 4\beta^2)^{-1}\bar{S} \diamond (\alpha\mathbb{I} - 2\beta E_{0123})$.

If we compute $S \diamond v \diamond S^{-1}$ $(v \in \mathbb{R}^4)$ we again obtain an element in $\mathbb{R}^4$ if and only if $\beta = 0$. Then the matrix $\ell(S)$ is an orthogonal matrix. By Corollary (9.1.19), the element (9.3.21) in $\mathcal{C}^+(\mathbb{R}^4,\delta)$ belongs the the spin group $\mathrm{Spin}(0,4)$ if and only if $\alpha = 1$ and $\beta = 0$. The group $\mathrm{Spin}(0,4)$ is thence canonically identified to a subset of $\mathcal{C}^+(\mathbb{R}^4,\delta) \simeq \mathbb{R}^8$ by means of the constraints $\alpha = 1$ and $\beta = 0$.

The covering map is

$$\ell : \mathrm{Spin}(0,4) \longrightarrow \mathrm{SO}(0,4) : S \mapsto \ell(S) \tag{9.3.24}$$

where $\ell(S)$ is the matrix which maps the canonical basis of $\mathbb{R}^4$ into the basis

$$\begin{gathered} E'_0 = \begin{pmatrix} a^2 - b^2 - c^2 + d^2 - e^2 + f^2 + g^2 - h^2 \\ 2(ab - cd - ef + gh) \\ 2(ac + bd - eg - fh) \\ 2(ae + bf + cg + dh) \end{pmatrix} \\ E'_1 = \begin{pmatrix} -2(ab + cd + ef + gh) \\ a^2 - b^2 + c^2 - d^2 + e^2 - f^2 + g^2 - h^2 \\ 2(ad - bc + eh - fg) \\ 2(af - be - ch + dg) \end{pmatrix} \\ E'_2 = \begin{pmatrix} 2(-ac + bd - eg + fh) \\ -2(ad + bc + eh + fg) \\ a^2 + b^2 - c^2 - d^2 + e^2 + f^2 - g^2 - h^2 \\ 2(ag + bh - ce - df) \end{pmatrix} \\ E'_3 = \begin{pmatrix} 2(-ae + bf + cg - hd) \\ 2(-af - eb + ch + dg) \\ -2(ag + bh + ce + fd) \\ a^2 + b^2 + c^2 + d^2 - e^2 - f^2 - g^2 - h^2 \end{pmatrix} \end{gathered} \tag{9.3.25}$$

Thence we have the projection onto $S^3 \subset \mathbb{R}^4$

$$\tau(S) = (x, y, z, t) \qquad \begin{cases} x = a^2 - b^2 - c^2 + d^2 - e^2 + f^2 + g^2 - h^2 \\ y = 2(ab - cd - ef + gh) \\ z = 2(ac + bd - eg - fh) \\ t = 2(ae + bf + cg + dh) \end{cases} \tag{9.3.26}$$

Let us define two quaternions[5] $q_1 = a + id + jf + kg$ and $q_2 = h - ie + jc - kb$. The projection can be recast as

$$\tau : \bar{\Sigma} \longrightarrow S^3 : (q_1, q_2) \mapsto (x, y, z, t) \tag{9.3.27}$$

where $2q_1\bar{q}_2 = it - jz + ky$ and $|q_1|^2 - |q_2|^2 = x$.

The standard fiber is $\mathrm{SU}(2) \simeq S^3$ and there exists a global section

$$u : S^3 \longrightarrow \bar{\Sigma} : (x, y, z, t) \mapsto \left(\frac{1+u}{2}, \frac{1-u}{2}\right) \tag{9.3.28}$$

where $u = x + iy - jy + kt$.

Thence the bundle $\bar{\Sigma}$ is the trivial bundle $\bar{\Sigma} = S^3 \times S^3$.

Spin Structures on S^4

Let S^4 be the set of unit vectors in $\mathbb{R}^5$ with the canonical metric δ of signature $(0, 5)$; let $(E_0, E_1, E_2, E_3, E_4)$ be the canonical basis in $\mathbb{R}^5$.

An element of $\mathcal{C}^+(\mathbb{R}^5, \delta)$ is

$$\begin{aligned} S =& a\mathbb{I} + bE_{01} + cE_{02} + dE_{12} + eE_{03} + fE_{13} + gE_{23} + hE_{0123} + \\ &+ AE_{04} + BE_{14} + CE_{24} + DE_{0124} + \\ &+ EE_{34} + FE_{0134} + GE_{0234} + HE_{1234} \end{aligned} \tag{9.3.29}$$

Let us define the quaternions

$$\begin{aligned} q_1 &= a + id + jf + kg \qquad & q_3 &= H - iE + jC - kB \\ q_2 &= h - ie + jc - kb \qquad & q_4 &= A + iD + jF + kG \end{aligned} \tag{9.3.30}$$

The constraints which identify $\mathrm{Spin}(0, 5)$ in $\mathcal{C}^+(\mathbb{R}^5, \delta)$ are the following

$$\begin{aligned} &|q_1|^2 + |q_2|^2 + |q_3|^2 + |q_4|^2 = 1 \\ &\mathrm{Re}(q_1\bar{q}_2 + q_3\bar{q}_4) = 0 \\ &\mathrm{Im}(q_1\bar{q}_4 + q_3\bar{q}_2) = 0 \\ &\mathrm{Re}(q_1\bar{q}_3 - q_4\bar{q}_2) = 0 \end{aligned} \tag{9.3.31}$$

[5] The skew field of quaternions is $\mathbb{R}^4$ equipped with the product induced by the Clifford algebra in the signature $(0, 2)$.

and the projection onto S^4 is

$$\tau : \mathrm{Spin}(0,5) \longrightarrow S^4 : (q_1, q_2, q_3, q_4) \mapsto (x_0, x_1, x_2, x_3, x_4) \tag{9.3.32}$$

where

$$\begin{aligned} q_1\bar{q}_2 + q_3\bar{q}_4 &= i\frac{x_3}{2} - j\frac{x_2}{2} + k\frac{x_1}{2} =: \xi \\ q_1\bar{q}_4 + q_3\bar{q}_2 &= \frac{x_4}{2} =: \zeta \\ |q_1|^2 + |q_3|^2 - |q_2|^2 - |q_4|^2 &= x_0 \end{aligned} \tag{9.3.33}$$

Let us define two quaternion pairs

$$\begin{cases} Q_1 = q_1 + q_3 \\ Q_2 = q_2 + q_4 \end{cases} \qquad \begin{cases} Q_3 = q_4 - q_2 \\ Q_4 = q_1 - q_3 \end{cases} \tag{9.3.34}$$

The projection τ can be obtained in two ways

$$\begin{cases} |Q_1|^2 - |Q_2|^2 = x_0 \\ \quad Q_1\bar{Q}_2 = \zeta + \xi \end{cases} \qquad \begin{cases} |Q_3|^2 - |Q_4|^2 = -x_0 \\ \quad Q_3\bar{Q}_4 = \zeta + \xi \end{cases} \tag{9.3.35}$$

which, apart from a sign, are both projections defining the Hopf bundle $S^7 \longrightarrow S^4$ with the standard fiber S^3.

The spin bundle $\bar{\Sigma} \simeq S^7 \times_{S^4} S^7$ is the fiber product over S^4 of two copies of the Hopf bundle $S^7 \longrightarrow S^4$.

Spin Structures in Different Signatures

As a further exercise, let us finally compute the spin structure over a De-Sitter space

$$S_- := \{v \in \mathbb{R}^{(1,3)} : \delta(v,v) = -1\} \tag{9.3.36}$$

embedded into $\mathbb{R}^{(1,3)} = (\mathbb{R}^4, \delta)$ with signature $(1,3)$. Analogously we shall consider the anti-De-Sitter space

$$S_+ := \{v \in \mathbb{R}^{(2,2)} : \delta(v,v) = +1\} \tag{9.3.37}$$

embedded into $\mathbb{R}^{(2,2)} = (\mathbb{R}^4, \delta)$ with signature $(2,2)$.

In the first case, we have to construct the group $\mathrm{Spin}(1,3)$. Let us denote by

$$S = a\mathbb{I} + bE_{01} + cE_{02} + dE_{12} + eE_{03} + fE_{13} + gE_{23} + hE_{0123} \tag{9.3.38}$$

an element in $\mathcal{C}^+(\mathbb{R}^{(1,3)})$ and define

$$\begin{aligned} \alpha &= a^2 - b^2 - c^2 + d^2 - e^2 + f^2 + g^2 - h^2 \\ \beta &= -2(ah - bg + cf - ed) \end{aligned} \tag{9.3.39}$$

The group $\mathrm{Spin}(1,3)$ is singled out into $\mathcal{C}^+(\mathbb{R}^{(1,3)})$ by the algebraic equations $\alpha = \pm 1$ and $\beta = 0$.

The projection $\tau_- : \mathrm{Spin}(1,3) \longrightarrow S_-$ is

$$\tau_- = (a,b,c,d,e,f,g,h) = (x_1,x_2,x_3,x_4) \tag{9.3.40}$$

where

$$\begin{cases} x_1 = 2(-ae + bf + cg - dh) \\ x_2 = 2(-af + be - ch + dg) \\ x_3 = 2(-ag + bh + ce - df) \\ x_4 = a^2 - b^2 - c^2 + d^2 + e^2 - f^2 - g^2 + h^2 \end{cases} \tag{9.3.41}$$

For the anti-De-Sitter case we have to compute the group $\mathrm{Spin}(2,2)$. Let us denote by

$$S = a\mathbb{I} + bE_{01} + cE_{02} + dE_{12} + eE_{03} + fE_{13} + gE_{23} + hE_{0123} \tag{9.3.42}$$

an element in $\mathcal{C}^+(\mathbb{R}^{(2,2)})$ and define

$$\begin{aligned} \alpha =& a^2 + b^2 - c^2 - d^2 - e^2 - f^2 + g^2 + h^2 \\ \beta =& -2(ah - bg + cf - ed) \end{aligned} \tag{9.3.43}$$

The group $\mathrm{Spin}(2,2)$ is identified into $\mathcal{C}^+(\mathbb{R}^{(2,2)})$ by the algebraic equations $\alpha = \pm 1$ and $\beta = 0$.

The projection reads now as

$$\tau_+ : \mathrm{Spin}(2,2) \longrightarrow S_+ : (a,b,c,d,e,f,g,h) \mapsto (x_1,x_2,x_3,x_4) \tag{9.3.44}$$

where

$$\begin{cases} x_1 = a^2 - b^2 + c^2 - d^2 + e^2 - f^2 + g^2 - h^2 \\ x_2 = 2(-ab + cd + ef - gh) \\ x_3 = 2(-ac + bd + eg - fh) \\ x_4 = 2(-ae + bf - cg + dh) \end{cases} \tag{9.3.45}$$

4. Spin Frames

Spin frames are the lift to the spin group of vielbein defined above in Section 8.7. They are also deeply related to spin structures. Let us fix a spin η-manifold M, i.e. a manifold which we assume to be orientable and to meet the topological conditions which ensure that it is a spin manifold and that exists on it at least one global pseudo-Riemannian metric of a fixed signature

$\eta = (r, s)$.[6] Let us also fix a spin bundle, i.e. a principal bundle $\bar{\Sigma}$ with group $\mathrm{Spin}(\eta)$, which always exists under the above assumptions.

Definition (9.4.1): a *(free) spin frame over* $\bar{\Sigma}$ is a morphism $\Lambda : \bar{\Sigma} \longrightarrow L(M)$, where $L(M)$ denotes the frame bundle over the manifold M such that the following diagrams commute:

$$\begin{array}{ccc} \bar{\Sigma} & \xrightarrow{\Lambda} & L(M) \\ \downarrow & \swarrow & \\ M & & \end{array} \qquad \begin{array}{ccc} \bar{\Sigma} & \xrightarrow{\Lambda} & L(M) \\ {\scriptstyle R_S}\downarrow & & \downarrow{\scriptstyle R_{\hat{\ell}(S)}} \\ \bar{\Sigma} & \xrightarrow{\Lambda} & L(M) \end{array} \qquad \hat{\ell} : \mathrm{Spin}(\eta) \longrightarrow \mathrm{GL}(m) \tag{9.4.2}$$

Thence Λ is a vertical principal morphism with respect to the group homomorphism $\hat{\ell} = i \circ \ell : \mathrm{Spin}(\eta) \longrightarrow \mathrm{GL}(m)$, $m = \dim(M)$. ∎

Clearly spin frames are deeply related to spin structures (see Section 9.2). Before going on we have to clarify in some detail what is new and what is not in this definition of spin frames with respect to spin structures as defined by (9.2.1).

First of all, the codomain of a spin frame is the frame bundle $L(M)$ and not any orthonormal frame bundle $\mathrm{SO}(M, g)$ with respect to some metric, which needs a metric structure (M, g) to be defined first. In this sense, the definition of spin frames is independent of any metric structure which could be fixed *a priori* on M. By this we do not mean that spin frames are unrelated to the existence of a metric structure on M. In fact, once a spin frame is chosen its image is a principal sub-bundle of $L(M)$ having $\mathrm{SO}(\eta)$ as a structure group. Thence for any spin frame $\Lambda : \bar{\Sigma} \longrightarrow L(M)$ there exists one and only one metric g on M for which the frames in the image $\Lambda(\bar{\Sigma}) \subset L(M)$ are g-orthonormal. However, such a metric g is determined *a posteriori* by the spin frame and it is not necessary to know it before defining the spin frame itself. It is as much the same thing that happens for vielbein which are given first and then induce a metric. From a physical viewpoint they are more fundamental variables for a dynamical description of metrics and spinors, since, as we said, it is important to avoid any reference to a fixed metric background (which has no physical meaning). Furthermore, if we succeed in deforming a spin frame we can deform its image in $L(M)$ and thence also the associated metric g. Thence a deformation of a spin frame canonically induces a deformation of the associated metric.

---

[6] See the footnote of Section 8.7, [LM89] and [CBWM82]. The conditions are, as we said above, equivalent to assume $w_1(M) = 0$ and $w_2(M) = 0$.

Secondly, let us remark that the bundle $\bar{\Sigma}$ has been fixed *a priori*: it will be called the *structure bundle*. Of course, in order to make our claims meaningful, we have to prove that fixing $\bar{\Sigma}$ is not equivalent to fixing the metric: if it were the case, in fact, we would have just given nothing but an equivalent formalism for spin structures. For that, we can again use the no-go arguments already presented for vielbein in Section 8.7. If, for instance, the manifold M is non-parallelizable and we choose $\bar{\Sigma}_{\text{triv}} \simeq M \times \text{Spin}(\eta)$ then no global spin frame exists on $\bar{\Sigma}_{\text{triv}}$ (see Section 8.7 for a proof which can be easily adapted to the case under consideration here). Then we have to require the following:

Axiom 2: the spin bundle $\bar{\Sigma}$ has to be chosen so that it allows at least one global spin frame.

If the (orientable) manifold M is an η-manifold, i.e. it admits a metric g of signature η, and at the same time we assume that it is a spin manifold (i.e. it has vanishing second Stiefel-Whitney class) then there exists at least a structure bundle $\bar{\Sigma}$ which satisfies Axiom 2. In fact, if we denote by $i_g : \text{SO}(M, g) \hookrightarrow L(M)$ the canonical embedding and by $(\bar{\Sigma}, \Lambda)$ a spin structure on (M, g), then $i_g \circ \Lambda : \bar{\Sigma} \longrightarrow L(M)$ is a spin frame on $\bar{\Sigma}$. We shall hereafter restrict to that situation.

Under these hypotheses, we can now show that fixing the structure bundle $\bar{\Sigma}$ is much less than fixing a metric. In fact, if we fix $\bar{\Sigma}$ then we can get many different metrics as metrics associated to some spin frame on that given $\bar{\Sigma}$. For instance, we already know that each strictly Euclidean metric is associated to some spin frame on some given spin bundle $\bar{\Sigma}$ of the appropriate signature (see Section 8.7 for a proof, which can be easily adapted to the case of spin frames). Furthermore, the theorem (9.2.15) exactly proves that on a wide class of Lorentzian spacetimes M all metrics are $\bar{\Sigma}$-admissible on trivial structure bundles. In both cases, then, once the signature and the structure bundle $\bar{\Sigma}$ have been fixed, one can still find *all* possible metrics on M as the metrics induced by some spin frame on $\bar{\Sigma}$ (in the appropriate signature).

However, this is by no means the general situation. The example over $M = \mathbb{R}^3 - \{0\} \simeq \mathbb{R}^+ \times S^2$ we considered above in Section 8.7 can in fact be easily adapted to spin frames to prove that, in general, once the structure bundle $\bar{\Sigma}$ has been fixed in non-Euclidean signature just a subset of all metrics are $\bar{\Sigma}$-admissible. Only in particular situations such set can be even empty (e.g. M non-parallelizable and $\bar{\Sigma}$ trivial) or coincide trivially with the set of all metrics (e.g., as above, this holds for Euclidean signature or Lorentzian non-compact four dimensional manifolds). The general classification is still an open problem.

5. The Bundle of Spin Frames

As we did for vielbein we want now to define over any spin η-manifold M a bundle the sections of which correspond to spin frames (see Axiom 1). To this purpose let us define the following action

$$\begin{aligned} \rho : \mathrm{Spin}(\eta) \times \mathrm{GL}(m) \times \mathrm{GL}(m) &\longrightarrow \mathrm{GL}(m) \\ : (S, J, e) &\mapsto J \cdot e \cdot \ell(S^{-1}) \end{aligned} \tag{9.5.1}$$

Definition (9.5.2): the *bundle of spin frames* is the bundle associated to the gauge natural prolongation $W^{(0,1)}\bar{\Sigma} = \bar{\Sigma} \times L(M)$ by means of the action ρ. We shall denote it by

$$\bar{\Sigma}_\rho = W^{(0,1)}\bar{\Sigma} \times_\rho \mathsf{GL}(m) \tag{9.5.3}$$

It is a gauge natural bundle of order $(0,1)$. ∎

We stress the analogy with the vielbein bundle defined in Section 8.7.

Proposition (9.5.4): there is a one-to-one correspondence between global sections of $\bar{\Sigma}_\rho$ and spin frames on $\bar{\Sigma}$.

Proof: the proof is analogous to the proof of theorem (8.7.4). ∎

Since any point in $\bar{\Sigma}_\rho$ have a representative in the form $[s^{(\alpha)}, \partial^{(\alpha)}, e_i^\mu]_\rho$ we have that (x^μ, e_i^μ) are local fibered coordinates.

The associated metric is related to the following bundle epimorphism

$$\begin{array}{ccc} \bar{\Sigma}_\rho & \longrightarrow & \mathrm{Met}(M,\eta) \\ \downarrow & \swarrow & \\ M & & \end{array} \tag{9.5.5}$$

where $\mathrm{Met}(M,\eta)$ denotes the bundle of all metrics of signature η.[7] Let us remark that the above epimorphism is canonical and it is locally given by

$$g_{\mu\nu} = e^a_\mu \, \eta_{ab} \, e^b_\nu \tag{9.5.6}$$

We also stress that $\bar{\Sigma}_\rho$ is not a group bundle since the transition functions do not fix the identity. Axiom 2 then ensures that $\bar{\Sigma}_\rho$ allows at least one global section.

---

[7] See also Section 4.4. For example, $\mathrm{Met}(M, m, 0)$ is the bundle of all strictly Riemannian metrics, while $\mathrm{Met}(M, 1, m-1) = \mathrm{Lor}(M)$ is the bundle of all Lorentzian metrics. The bundle $\mathrm{Met}(M,\eta)$ is a sub-bundle of $S^0_2(M)$, i.e. the bundle of symmetric tensors of rank $(2,0)$. Sections of $\mathrm{Met}(M,\eta)$ are a positive cone of sections $\Gamma(S^0_2(M))$ since λg is a metric if $\lambda > 0$ and g is a metric; this cone might be empty, as we already know.

Infinitesimal Generators of Automorphisms on $\bar{\Sigma}$ and Gauge Transformations

By representing the group $\mathrm{Spin}(\eta)$ as a subgroup of $\mathrm{GL}(k, \mathbb{K})$ identified by the constraints (9.1.25), we can give an expression for infinitesimal generators of automorphisms of $\bar{\Sigma}$.

Let $\boldsymbol{\Phi}_t = (\Phi_t, \phi_t)$ be a 1-parameter subgroup of automorphisms of $\bar{\Sigma}$ locally given by $\Phi_t : [x, S]_\alpha \mapsto [\phi(x), \varphi(x)S]_\alpha$. The infinitesimal generator is

$$\Xi = \xi^\mu \partial_\mu + \dot{\varphi}^\alpha_{0\beta} S^\beta_\gamma \frac{\partial}{\partial S^\alpha_\gamma} = \xi^\mu \partial_\mu + \xi^{ab} \bar{\sigma}_{ab} \tag{9.5.7}$$

where using (9.1.30) we defined

$$\begin{cases} \xi^\mu = \dot{\phi}^\mu_0(x) \\ \xi^{ab} = \partial^\beta_\alpha \ell^{[a}_c(\mathbb{I}) \eta^{b]c} \dot{\varphi}^\alpha_{0\beta} \\ \bar{\sigma}_{ab} = \frac{1}{8}\big([\gamma_a, \gamma_b] S\big)^\alpha_\beta \frac{\partial}{\partial S^\alpha_\beta} \end{cases} \tag{9.5.8}$$

Because of theorem (3.4.1) the fields $\bar{\sigma}_{ab}$ are right invariant and they form a local basis of vertical right invariant vector fields on $\bar{\Sigma}$.

Connections on $\bar{\Sigma}$

Let $\Lambda : \bar{\Sigma} \longrightarrow L(M)$ be a spin frame on $\bar{\Sigma}$ and g its associated metric. Let also (x^μ, u^μ_a) be fibered coordinates on $L(M)$.

The metric g defines the Levi-Civita connection

$$\Gamma = \mathrm{d}x^\mu \otimes (\partial_\mu - \Gamma^\lambda_{\sigma\mu} \rho^\sigma_\lambda) \tag{9.5.9}$$

where $\rho^\sigma_\lambda = u^\sigma_a \partial / \partial u^\lambda_a$ are the right invariant vector fields on $L(M)$ and $\Gamma^\lambda_{\sigma\mu}$ are the Christoffel symbols of the metric g.

The spin frame Λ allows one to define, via *pull-back*, a connection over $\bar{\Sigma}$, called the *spin connection*, which is locally expressed as

$$\omega = \mathrm{d}x^\mu \otimes (\partial_\mu - \Gamma^{ab}_\mu \bar{\sigma}_{ab}) \,, \qquad \Gamma^{ab}_\mu = e^a_\lambda \big(\Gamma^\lambda_{\sigma\mu} e^{\sigma b} + \mathrm{d}_\mu e^{\lambda b}\big) \tag{9.5.10}$$

Theorem (9.5.11): there exist two morphisms

$$\begin{aligned} \Gamma &: J^1 \bar{\Sigma}_\rho \longrightarrow \mathcal{C}_{L(M)} \\ \omega &: J^1 \bar{\Sigma}_\rho \longrightarrow \mathcal{C}_{\bar{\Sigma}} \end{aligned} \tag{9.5.12}$$

which in turn define two connections on $L(M)$ and $\bar{\Sigma}$ out of dynamical fields and their first order derivatives. ∎

Theorem (9.5.13): one has $\nabla_\mu e_a^\nu = \mathrm{d}_\mu e_a^\nu + \Gamma^\nu_{\sigma\mu} e_a^\sigma - \Gamma^{b\,\cdot}{}_{a\mu} e_b^\nu = 0$

Proof: we leave to the reader the easy check that the expression we give for the covariant derivative of a section of $\bar{\Sigma}_\rho$ agrees with the prescription of the general theory of covariant derivatives.

By construction we have then

$$\nabla_\mu e_a^\nu = e_c^\nu e_\sigma^c \left(\mathrm{d}_\mu e_a^\sigma + \Gamma^\sigma_{\rho\mu} e_a^\rho\right) - \Gamma^{c\,\cdot}{}_{a\mu} e_c^\nu = 0 \tag{9.5.14}$$

so that the spin frames are covariantly conserved. ∎

6. Other Formulations of General Relativity

As we already made for purely frame and frame-affine formalisms (see Section 8.7) one can also use spin frames as a fundamental description of gravity. Let us, in fact, choose $\bar{\Sigma}_\rho$ as the configuration bundle with fibered coordinates $(x^\mu; e_a^\mu)$. Then each spin frame $e : M \longrightarrow \bar{\Sigma}_\rho$ can be locally expressed as a map $e(x^\mu) = (x^\mu; e_a^\mu(x))$. It induces a metric which is locally expressed as

$$g_{\mu\nu} = e^a_\mu \, \eta_{ab} \, e^b_\nu \tag{9.6.1}$$

Thence one can define the Levi-Civita connection $\Gamma^\alpha_{\beta\mu}$ which in turn define a principal connection over the structure bundle $\bar{\Sigma}$

$$\Gamma^{ab}_\mu = e^a_\alpha (\Gamma^\alpha_{\beta\mu} e_{b\beta} + \mathrm{d}_\mu e^{b\alpha}) \tag{9.6.2}$$

which is called the *spin connection.*

Thence one can define the curvature $R^{ab}{}_{\mu\nu}$ of such a connection and the Lagrangian

$$L_{sF} = R^{ab}{}_{\mu\nu} \, e_a^\mu e_b^\nu \, e \, \mathrm{d}s \tag{9.6.3}$$

where e denotes (the absolute value of) the spin frame or, equivalently, the square root of the absolute value of the induced metric. It is a second order Lagrangian on the bundle of spin frames and it is covariant with respect to any automorphism of the structure bundle $\bar{\Sigma}$. The theory is thence a gauge natural field theory.

Analogously, one can fix the configuration bundle as the fiber product of the spin frames and the bundle of principal connections on the structure bundle. In this way both the spin frame and the spin connection are (*a priori*) dynamical fields. The Lagrangian (9.6.3) is now to be considered as zero-order in the spin frame and first order in the spin connection. In this way one obtains a formulation very similar to the frame-affine formulation of General Relativity. The Lagrangian is again covariant with respect to any automorphism of the structure bundle and the theory is gauge natural.

These are just some of the possible formulations which can be given for General Relativity. We shall hereafter present a further formulation based on (invertible) tensors of rank $(1, 1)$; the main purpose is to compare such a formulation with spin frame (or frame) formulations. Of course other formalisms exist (twistor, teleparallelism, gauge theories of the Poincaré group just to quote some of them) which will not be treated here.

Invertible $(1, 1)$ tensors

General Relativity can be formulated (in a non-canonical way) by means of invertible spacetime $(1, 1)$ tensors. Let us in fact fix a metric $\bar{g}$ on the spacetime M, with signature η. Thence any other metric g of the same signature η can be obtained by an invertible tensor t^μ_ν as follows

$$g_{\mu\nu} = t^\rho_\mu \, \bar{g}_{\rho\sigma} \, t^\sigma_\nu \tag{9.6.4}$$

In other words, once a reference metric $\bar{g}$ has been fixed, it induces a bundle epimorphism

$$\begin{array}{ccc} T^{\circ 1}{}_1(M) & \xrightarrow{\pi_g} & \mathrm{Met}(M, \eta) \\ \downarrow & & \downarrow \\ M & \equiv\equiv\equiv & M \end{array} \tag{9.6.5}$$

where $T^{\circ 1}{}_1(M)$ denotes the bundle of all invertible tensors of rank $(1, 1)$ in M. Such a morphism allows one to give a new formalism for General Relativity on $T^{\circ 1}{}_1(M)$.

Proposition (9.6.6): there exists a one-to-one correspondence between spin frames and isomorphisms $\bar{\Sigma}_\rho \longrightarrow T^{\circ 1}{}_1(M)$.

Proof: let us fix a global spin frame locally given by $e : M \longrightarrow \bar{\Sigma}_\rho : x \mapsto [s^{(\alpha)}, \partial^{(\alpha)}, a^{(\alpha)}{}^\mu_i]_\rho$. We can thence define the bundle morphism

$$\mu_e : \bar{\Sigma}_\rho \longrightarrow T^{\circ}{}^1_1(M) : \Lambda = [s^{(\alpha)}, \partial^{(\alpha)}, e^{(\alpha)}{}^\mu_i]_\rho \mapsto \mu_e(\Lambda) \tag{9.6.7}$$

where $\mu_e(\Lambda)$ is the $(1, 1)$-tensor defined as follows

$$\mu_e(\Lambda) : \partial^{(\alpha)}_\mu \mapsto \partial^{(\alpha)}_\nu \, e^{(\alpha)}{}^\nu_i \, \bar{a}^{(\alpha)}{}^i_\mu \tag{9.6.8}$$

We remark that globality of the spin frames guarantees the globality of the induced morphism.

■

Thence once a spin frame $e : M \longrightarrow \bar{\Sigma}_\rho$ has been fixed it induces a metric g and a projection $\pi_g : T^{\circ 1}{}_1(M) \longrightarrow \mathrm{Met}(M, \eta)$. Thence we have a non-canonical (i.e. depending on a reference spin frame) isomorphism $\mu_e : \bar{\Sigma}_\rho \longrightarrow T^{\circ 1}{}_1(M)$.

Therefore we have a canonical formalism for General Relativity based on spin frames that, once a reference spin frame $\Lambda : \bar{\Sigma} \longrightarrow L(M)$ has been fixed, factorizes in a non-canonical way through $T^{\circ 1}{}_1(M)$, giving in turn a non-canonical formulation of General Relativity in terms of invertible $(1,1)$-tensors

$$\begin{array}{ccccc} M & \underset{\tau}{\overset{e}{\rightleftarrows}} & \bar{\Sigma}_\rho & \overset{\mu_e}{\dashrightarrow} & T^{\circ}{}^1_1(M) \\ & \nwarrow & \downarrow & \swarrow \pi_g & \\ & & \mathrm{Met}(M,\eta) & & \end{array} \tag{9.6.9}$$

Let us finally remark that $T^{\circ 1}{}_1(M)$ is a group bundle. This is a deep difference with the spin frame bundle $\bar{\Sigma}_\rho$ which is not. The two bundles are isomorphic as bundles even if $T^{\circ 1}{}_1(M)$ carries a non-canonical additional group bundle structure.

7. Kosmann Lift

We can use the natural lift on the frame bundle $L(M)$ and a spin frame $\Lambda : \bar{\Sigma} \longrightarrow L(M)$ to define a canonical non-natural lift of spacetime vector fields to $\bar{\Sigma}$. Let us denote by g the metric induced by the spin frame Λ and by $\mathrm{SO}(M,g)$ the orthonormal frame bundle. Let us fix a trivialization $\bar{\sigma}$ on $\bar{\Sigma}$ and a (non-natural) g-orthonormal trivialization V_a on $L(M)$. We can define the vector fields $\bar{\sigma}_{ab} = \frac{1}{8}([\gamma_a, \gamma_b]S)^\alpha_\beta \frac{\partial}{\partial S^\alpha_\beta}$ so that they form a basis of vertical right invariant vector fields tangent to $\bar{\Sigma}$. In Section 8.8 we defined the Kosmann lift (8.8.3) to the structure bundle Σ of vielbein.

Since we have a local diffeomorphism $\bar{\ell} : \bar{\Sigma} \longrightarrow \Sigma$ locally given by $\bar{\ell} : (x, S) \mapsto (x, \ell(S))$, every vector field over Σ induces a vector field over $\bar{\Sigma}$. In particular we may use this fact to define the so-called *Kosmann lift to* $\bar{\Sigma}$ by means of the following diagram:

$$\begin{array}{ccc} \bar{\Sigma} & & \bar{\xi}_{(K)} = (\bar{\ell}^{-1})_*(\xi_{(K)}) \\ \downarrow \bar{\ell} & & \uparrow \\ \Sigma & & \xi_{(K)} \\ \downarrow & & \uparrow \\ M & & \xi = \xi^\mu \partial_\mu \end{array} \tag{9.7.1}$$

Here the Kosmann lift $\xi_{(K)} = \xi^\mu \partial_\mu + \xi^{ab} \sigma_{ab}$ is defined by $\xi^{ab} = \xi^{[a}{}_c \eta^{b]c}$ in accordance with Section 8.8. The vertical right invariant vector fields $\bar{\sigma}_{ab}$ on $\bar{\Sigma}$ and σ_{ab} on Σ are respectively defined so that $(\bar{\ell})_*(\bar{\sigma}_{ab}) = \sigma_{ab}$ (recall we are assuming that the trivialization on Σ is induced by means of $\bar{\ell}$ by the trivialization fixed on $\bar{\Sigma}$). Thence the Kosmann lift on $\bar{\Sigma}$ is locally given by

$$\bar{\xi}_{(K)} = \xi^\mu \partial_\mu + \xi^{ab}\bar{\sigma}_{ab} \qquad \xi^{ab} = \xi^{[a}{}_c \eta^{b]c} \tag{9.7.2}$$

Proposition (9.7.3): the following identity holds

$$[K(\xi), K(\zeta)] = K([\xi,\zeta]) - \tfrac{1}{2} V^a_\lambda V^c_\mu (\pounds_\zeta g^{\sigma[\lambda} \, \pounds_\xi g^{\mu]\alpha}) g_{\sigma\alpha}\, \sigma_{ac} \tag{9.7.4}$$

Proof: the proof is carried over by direct computation. ∎

Lie Derivative of Spin Frames

Let us now go back to the spin bundle $\bar{\Sigma}_\rho$. Any automorphism $\boldsymbol{\Phi} = (\Phi, \phi)$ of the structure bundle $\bar{\Sigma}$ naturally acts over $\bar{\Sigma}_\rho$ and thence over its sections, i.e. on vielbein e^μ_a. Thence we can easily define the Lie derivative of spin frames with respect to infinitesimal generators of automorphisms on $\bar{\Sigma}$ which can be locally expressed as

$$\Xi = \xi^\mu(x)\partial_\mu + \xi^{ab}(x)\bar{\sigma}_{ab} \tag{9.7.5}$$

It induces a vector field $\hat{\Xi}$ over $\bar{\Sigma}_\rho$ which is the infinitesimal generator of gauge transformations on vielbein and which is locally given by

$$\hat{\Xi} = \xi^\mu \partial_\mu + (\partial_\nu \xi^\mu e^\nu_a - e^\mu_b \xi^b{}_{\cdot a})\partial^a_\mu \tag{9.7.6}$$

Thence one has the Lie derivative of a spin frame with respect to Ξ which, according to the general theory as exposed in Section 2.6, has the following local expression

$$\pounds_\Xi e^\mu_a = e^\mu_b (\xi_{(V)})^b{}_{\cdot a} - \nabla_\nu \xi^\mu e^\nu_a \tag{9.7.7}$$

where the connections Γ^{ab}_μ and $\Gamma^\alpha_{\beta\mu}$ induced via (9.5.9) and (9.5.10) by the vielbein e^μ_a itself have been used to define the vertical part $(\xi_{(V)})^b{}_{\cdot a}$ and the covariant derivative $\nabla_\nu \xi^\mu$, respectively.

Then one can restrict $\Xi \equiv \xi_{(K)}$ to be the Kosmann lift of a spacetime vector field ξ, i.e.:

$$\Xi \equiv \xi_{(K)} = \xi^\mu \partial_\mu + (e^a_\rho \nabla_\mu \xi^\rho e^\mu_c \eta^{cb} - \Gamma^{ab}_\rho \xi^\rho)\sigma_{ab} \tag{9.7.8}$$

Thence one obtains the Lie derivative of vielbein with respect to a general spacetime vector field which is locally given by

$$\pounds_\xi e^\mu_a = \pounds_{K(\xi)} e^\mu_a = e^\mu_b \eta_{ac}(e^{[b}_\sigma e^{c]\lambda} \nabla_\lambda \xi^\sigma) - \nabla_\nu \xi^\mu e^\nu_a \equiv \tfrac{1}{2} e_{a\rho} \pounds_\xi g^{\mu\rho} \tag{9.7.9}$$

Exercises

1- Prove that $\mathcal{C}(0,1) \simeq \mathbb{C}$.
Hint: write down the Clifford product.

2- Prove that $\mathcal{C}(0,2)$ is the skew field of quaternions.
Hint: write down the Clifford product.

3- Prove that $\mathrm{Spin}(1,3) \simeq \mathrm{SL}(2,\mathbb{C})$.
Hint: write down the covering map $\ell : \mathrm{SL}(2,\mathbb{C}) \longrightarrow \mathrm{SO}(1,3)$.

4- Prove that $\bar{\sigma}_{ab}$ defined in (9.5.8) are right invariant vector fields on $\bar{\Sigma}$.
Hint: apply directly the right action.

5- Prove Corollary (9.1.27).
Hint: prove in general that the constraint $\det = 1$ *implies the infinitesimal form* $\mathrm{Tr} = 0$.

6- Prove that $\nabla_\mu e^\nu_a = 0$.
Hint: expand the spin connection Γ^{ab}_μ *in terms of the spin frame and the Levi-Civita connection of the induced metric.*

7- Write down field equations for the Lagrangian (9.6.3).
Hint: use Palatini-like techniques.

Chapter 10

SPINOR THEORIES

Abstract We shall present some example and applications of spinor theories both for commuting and anticommuting spinors. The ordinary Dirac theory, the massless neutrino theory as well as the Wess-Zumino model which deals with Majorana anticommuting spinors.

The reader should be familiar with the material of Chapter 7 (gauge natural theories) and Chapter 9 (spin frames).

References

Standard references about spinor fields are [CDF91], [GS87] and [BD64]. For supersymmetries we refer to [CDF91], [R81], [C90], [M88] and [MS00].

1. Dirac Spinor Field

Let us fix a set of Dirac matrices γ_a which induce a matrix representation of the Clifford algebra by means of $k \times k$ (complex) matrices of $\mathrm{GL}(k, \mathbb{C})$. We shall also constrain our choice by the following requirement

$$\gamma_0 \gamma_a^\dagger \gamma_0 = \gamma_a \tag{10.1.1}$$

which fixes γ_0 to be Hermitian and γ_i to be anti-Hermitian for $i = 1, 2, 3$. That representation induces in turn a representation $\lambda : \mathrm{Spin}(\eta) \times V \longrightarrow V$ of the spin group over $V = \mathbb{C}^k$ given by:

$$\lambda(S, v) = S\, v \tag{10.1.2}$$

Here and in the sequel, when there is no risk of confusion we shall understand the indices due to the matrix character of spinors and more generally of Clifford elements.

We can now define the associated vector bundle $\bar{E}_\lambda = \bar{\Sigma} \times_\lambda V$. Let us denote by $(x^\mu; v^i)$ a set of fibered coordinates on $\bar{E}_\lambda$. The vector bundle $\bar{E}_\lambda$ will be called the *spinor bundle*, not to be confused with the spin bundle which is principal. The sections of the vector bundle $\bar{E}_\lambda$ are called *spinor fields*.

We also define a conjugation operator on spinors given by

$$\bar{v} = v^\dagger \, \gamma^0 \tag{10.1.3}$$

where of course the dagger denotes the transpose complex conjugation. This operation induces an inner product in the space of spinors given by

$$< v \,|\, u >= \bar{v}\, u \tag{10.1.4}$$

The automorphisms of the spin bundle $\bar{\Sigma}$ canonically act on spinor fields. If $\Xi = \xi^\mu(x)\partial_\mu + \xi^{ab}(x)\bar{\sigma}_{ab}$ is a right invariant vector field on $\bar{\Sigma}$ (i.e. an infinitesimal generator of automorphisms) it induces a vector field Ξ_λ on the spinor bundle $\bar{E}_\lambda$ locally given by

$$\Xi_\lambda = \xi^\mu \, \partial_\mu + \tfrac{1}{8}\partial \, [\gamma_a, \gamma_b] v \, \xi^{ab} - \tfrac{1}{8}\xi^{ab} \, \bar{v}[\gamma_a, \gamma_b] \, \bar{\partial} \tag{10.1.5}$$

where we set

$$\partial = \frac{\partial}{\partial v} \,, \qquad \bar{\partial} = \frac{\partial}{\partial \bar{v}} \tag{10.1.6}$$

We can define the Lie derivative of spinor fields along Ξ by

$$\pounds_\Xi v = \xi^\mu \mathrm{d}_\mu v - \tfrac{1}{8}\xi^{ab} \, [\gamma_a, \gamma_b] v = \xi^\mu \nabla_\mu v - \tfrac{1}{8}\xi^{ab}_{(V)} \, [\gamma_a, \gamma_b] v \tag{10.1.7}$$

where both the covariant derivative $\nabla_\mu v = \mathrm{d}_\mu v + \tfrac{1}{8}\Gamma^{ab}_\mu \, [\gamma_a, \gamma_b] v$ and the vertical part $\xi^{ab}_{(V)} = \xi^{ab} + \Gamma^{ab}_\mu \xi^\mu$ are defined with respect to the connection Γ^{ab}_μ induced by the spin frame e^μ_a. In the sequel we shall use also the notation $\nabla_a v = e^\mu_a \nabla_\mu v$ coherently with the general rule according to which spin frames are used to change Latin indices into Greek ones and vice versa.

Let us now define the so-called *Dirac Lagrangian* by setting[8]

$$L_\mathrm{D} = \left[\tfrac{i}{2}\left(\bar{v}\gamma^a \nabla_a v - \nabla_a \bar{v}\gamma^a v\right) - m\, \bar{v}v\right] \sqrt{g}\, \mathrm{ds} = \mathcal{L}_\mathrm{D}(\bar{v}, v, \nabla_a \bar{v}, \nabla_a v, e^\mu_a)\, \mathrm{ds} \tag{10.1.8}$$

where $m \in \mathbb{R}$ is a real constant called the *mass* of the spinor field. This Lagrangian is defined over the bundle

$$J^1 \bar{E}_\lambda \times_M J^1 \bar{\Sigma}_\rho \tag{10.1.9}$$

[8] Here and in the sequel the appropriate position of fields and objects is carefully respected, because of the character of non-commutativity of the algebra.

and it is gauge natural; in fact, the covariance identity reads as

$$p\, \pounds_\Xi v + p^a\, \pounds_\Xi \nabla_a v + \pounds_\Xi \bar{v}\, \bar{p} + \pounds_\Xi \nabla_a \bar{v}\, \bar{p}^a + p^a_\mu\, \pounds_\Xi e^\mu_a = \nabla_\mu(\xi^\mu \mathcal{L}_\mathrm{D}) \tag{10.1.10}$$

where we defined the momenta by

$$\begin{cases} p = -\sqrt{g}\ \left(\frac{i}{2}\nabla_a\bar{v}\gamma^a + m\,\bar{v}\right) \\ p^a = \frac{i}{2}\sqrt{g}\,(\bar{v}\gamma^a) \\ p^a_\mu = -\mathcal{L}_\mathrm{D}\ e^a_\mu \end{cases} \qquad \begin{cases} \bar{p} = \sqrt{g}\ \left(\frac{i}{2}\gamma^a\nabla_a v - mv\right) \\ \bar{p}^a = -\frac{i}{2}\sqrt{g}\,(\gamma^a v) \end{cases} \tag{10.1.11}$$

Notice that the spinor v and its conjugate $\bar{v}$ are as real fields. By expanding the covariance identity (10.1.10) we obtain the following conditions

$$\begin{cases} p\,\nabla_\mu v + p^a\nabla_\mu\nabla_a v + \nabla_\mu\bar{v}\,\bar{p} + \nabla_\mu\nabla_a\bar{v}\,\bar{p}^a = \nabla_\mu\mathcal{L}_\mathrm{D} \\ -\,p^a_\mu e^\lambda_a = \mathcal{L}_\mathrm{D}\delta^\lambda_\mu \\ \frac{1}{8}p\,[\gamma_a,\gamma_b]\ v + \frac{1}{8}p^c\,[\gamma_a,\gamma_b]\,\nabla_c v + p_{[a}{}^{\cdot}\,\nabla_{b]}v + \\ \qquad + \frac{1}{8}\bar{v}\,[\gamma_b,\gamma_a]\,\bar{p} + \frac{1}{8}\nabla_c\bar{v}\,[\gamma_b,\gamma_a]\,\bar{p}^c + \nabla_{[b}\bar{v}\,\bar{p}_{a]}{}^{\cdot} = 0 \end{cases} \tag{10.1.12}$$

The first one holds since the Lagrangian density depends on $(v, \bar{v}, \nabla_a v, \nabla_a\bar{v})$ only, so that one can use Leibniz rule on the right hand side and the equation is identically satisfied. The second one can be easily seen to be identically satisfied. The third one deserves some comment. It has to be expanded and the anti-commutation rules of the Dirac matrices have to be used. By using the identity

$$\gamma_c[\gamma_a,\gamma_b] = [\gamma_a,\gamma_b]\gamma_c - 4\eta_{bc}\gamma_a + 4\eta_{ac}\gamma_b \tag{10.1.13}$$

all terms cancel out and the condition is again identically satisfied. Thus the Dirac Lagrangian L_D is shown to be gauge natural.

Let us now compute the variation of the Lagrangian L_D along a vertical vector field $X = \delta v\,\partial + \bar{\partial}\,\delta\bar{v} + \delta e^\mu_a \partial^a_\mu$.[9] We obtain:

$$< \delta L_\mathrm{D}\,|\,j^1 X >= \left[\mathbb{E}\,\delta v + \delta\bar{v}\,\bar{\mathbb{E}} - \hat{H}_{\mu\nu}e^{a\nu}\delta e^\mu_a\right]\,\mathrm{d}s + \mathrm{Div} < \mathbb{F}(L_\mathrm{D})\,|\,X > \tag{10.1.14}$$

where we defined the *Dirac operators*

$$\mathbb{E} = \mathbb{E}(L_\mathrm{D}) = p - \nabla_a p^a\ , \qquad \bar{\mathbb{E}} = \bar{\mathbb{E}}(L_\mathrm{D}) = \bar{p} - \nabla_a\bar{p}^a \tag{10.1.15}$$

the *Hilbert stress tensor (density)*

$$\begin{aligned} \hat{H}_{\mu\nu} =& \tfrac{i}{2}\left(\nabla_{(\mu}\bar{v}\gamma_{\nu)}v - \bar{v}\gamma_{(\mu}\nabla_{\nu)}v\right)\sqrt{g} + \mathcal{L}_\mathrm{D} g_{\mu\nu} + \\ &+ \tfrac{1}{8}\left(\bar{v}[\gamma_\mu,\gamma_\nu]\bar{\mathbb{E}} - \mathbb{E}[\gamma_\mu,\gamma_\nu]v\right) \equiv \sqrt{g}\,H_{\mu\nu} \end{aligned} \tag{10.1.16}$$

---

[9] See previous footnote.

and the Poincaré-Cartan morphism

$$< \mathbb{F}(L_{\mathrm{D}}) \,|\, X >= \tfrac{i}{2}\sqrt{g}\left[(\bar{v}\gamma^a \delta v) - (\delta\bar{v}\gamma^a v) + \tfrac{1}{2}(\bar{v}\gamma^{abc} v) e_{b\lambda} \delta e^{\lambda}_{c}\right] e^{\mu}_{a}\, \mathrm{d}s_\mu \quad (10.1.17)$$

The Dirac operators provide field equations for v and $\bar{v}$

$$\mathbb{E} = -\sqrt{g}\ (i\nabla_a \bar{v}\,\gamma^a + m\,\bar{v}) = 0\ , \qquad \bar{\mathbb{E}} = \sqrt{g}\ (i\gamma^a\, \nabla_a v - m\, v) = 0 \quad (10.1.18)$$

They are conjugate one of the other thanks to (10.1.1) so that they can be interpreted as a single field equation for the complex field v.

The Hilbert stress tensor (density) is symmetric only on-shell and it shows that spinor fields act as a source for the gravitational field. Notice that $H_{\mu\nu}$ is not proportional to the rest mass m of spinor field, thus meaning that also massless spinor fields generate a gravitational field. When one considers a total Lagrangian, e.g. $L = L_H + L_{\mathrm{D}}$, describing gravity and a spinor field in interaction, then together with the Dirac equation (10.1.18) a field equation for the spin frame is defined, namely

$$R_{\mu\nu} - \tfrac{1}{2} R g_{\mu\nu} = \kappa H_{(\mu\nu)} \quad (10.1.19)$$

which ultimately justifies the interpretation in terms of gravitational sources.

The Poincaré-Cartan morphism is necessary to define conserved quantities via Nöther theorem. The covariance condition (10.1.10) can be in fact recast as the following weak conservation law

$$\mathrm{Div}\, \mathcal{E} = \mathcal{W} \quad (10.1.20)$$

where we defined the quantities

$$\begin{cases} \mathcal{E} =< \mathbb{F}(L_{\mathrm{D}}) \mid \pounds_\Xi > - i_\xi L_{\mathrm{D}} = \\ \quad = \left[\tfrac{i}{2}\sqrt{g}\left((\bar{v}\gamma^a \pounds_\Xi v) - (\pounds_\Xi \bar{v}\gamma^a v) + \tfrac{1}{2}(\bar{v}\gamma^{abc} v) e_{b\mu} \pounds_\Xi e^{\mu}_{c}\right) e^{\lambda}_{a} - \xi^\lambda \mathcal{L}_{\mathrm{D}}\right] \mathrm{d}s_\lambda \\ \mathcal{W} = - < \mathbb{E}(L_{\mathrm{D}}) \mid \pounds_\Xi >= \\ \quad = \left(\hat{H}_{\mu\nu} e^{a\nu} \pounds_\Xi e^{\mu}_{a} - \mathbb{E}\, \pounds_\Xi v - \pounds_\Xi \bar{v}\, \bar{\mathbb{E}}\right)\, \mathrm{d}s \end{cases} \quad (10.1.21)$$

By integrating covariantly by parts the work form $\mathcal{W}$ one obtains

$$\mathcal{W} = (B_\mu\, \xi^\mu + B_{ab} \xi^{ab}_{(V)})\, \mathrm{d}s + \mathrm{Div}\, \tilde{\mathcal{E}} \quad (10.1.22)$$

The two quantities B_μ and B_{ab} identically vanish

$$\begin{cases} B_\mu = (\nabla_\lambda \hat{H}_{\mu}{}^{\lambda} - \mathbb{E}\, \nabla_\mu v - \nabla_\mu \bar{v}\, \bar{\mathbb{E}}) \equiv 0 \\ B_{ab} = (\hat{H}_{\mu\nu} e^{\mu}_{[a} e^{\nu}_{b]} + \tfrac{1}{8}\mathbb{E}[\gamma_a, \gamma_b] v - \tfrac{1}{8}\bar{v}[\gamma_a, \gamma_b]\bar{\mathbb{E}}) \equiv 0 \end{cases} \quad (10.1.23)$$

These are the Bianchi identities, which imply that the Hilbert stress tensor is symmetric on-shell and conserved on-shell. The reduced current is given by

$$\tilde{\mathcal{E}} = -\hat{H}_\mu{}^\lambda \xi^\mu \, \mathrm{d}s_\lambda \tag{10.1.24}$$

The Nöther current $\mathcal{E}$ can be covariantly integrated by parts obtaining

$$\mathcal{E} = \tilde{\mathcal{E}} + \mathrm{Div}\, \mathcal{U} \tag{10.1.25}$$

where the superpotential reads as

$$\mathcal{U} = \tfrac{i}{8}\sqrt{g}(\bar{v}\gamma_\mu{}^{\lambda\nu} v)\xi^\mu \, \mathrm{d}s_{\lambda\nu} \tag{10.1.26}$$

2. Neutrinos

Let us now consider the Clifford element $\gamma = (i)^\epsilon \gamma_0 \gamma_1 \dots \gamma_{m-1}$, where ϵ is an integer. The following property holds true

$$\gamma \cdot \gamma = (-1)^{[\frac{m}{2}]+s+\epsilon} \mathbb{I} \tag{10.2.1}$$

where $[\cdot]$ denotes the integer part, m is the dimension of the base manifold M and (r, s) is the fixed signature. One can thence choose $\epsilon = -[\frac{m}{2}] - s$ so that we have, in any dimension and signature, $\gamma \cdot \gamma = \mathbb{I}$. Furthermore, the matrix γ commutes with all γ_a in odd dimensions $m = 2\mu + 1$, while it anticommutes in even dimensions $m = 2\mu$; in fact we have:

$$\gamma \cdot \gamma_a = (-1)^{m-1} \gamma_a \cdot \gamma \tag{10.2.2}$$

Using γ we can define two Clifford operators

$$\pi_L = \tfrac{1}{2}(\mathbb{I} + \gamma) \ , \qquad \pi_R = \tfrac{1}{2}(\mathbb{I} - \gamma) \tag{10.2.3}$$

which are called *chiral operators*. They are a complete set of projectors, i.e. the following properties hold true

$$\begin{gathered} \pi_L + \pi_R = \mathbb{I} \\ \pi_L \circ \pi_L = \pi_L \ , \qquad \pi_R \circ \pi_R = \pi_R \\ \pi_R \circ \pi_L = 0 \ , \qquad \pi_L \circ \pi_R = 0 \end{gathered} \tag{10.2.4}$$

A spinor v is a *left spinor* if

$$v = \pi_L v \qquad \Leftrightarrow \qquad \pi_R v = 0 \tag{10.2.5}$$

Analogously it is called a *right spinor* when

$$v = \pi_R v \qquad \Leftrightarrow \qquad \pi_L v = 0 \tag{10.2.6}$$

We shall denote by v_L all left spinors and by v_R all right spinors.

The standard fiber V of the spinor bundle $\bar{E}_\lambda = \bar{\Sigma} \times_\lambda V$ splits into two subspaces $V = V_L \oplus V_R$ defined as the subspaces of left spinors (right spinors, respectively). In other words, we have an exact sequence of vector spaces which splits (as all sequences of vector spaces do), i.e. we have

$$0 \longrightarrow V_L \underset{i}{\longrightarrow} V \underset{\pi_R}{\longrightarrow} V_R \longrightarrow 0 \qquad (\pi_L : V \to V_L) \tag{10.2.7}$$

The splitting condition reads exactly as the properties $\pi_L v = v$ for all $v \in V_L$, i.e. equation (10.2.5).

Lemma (10.2.8): let w be a generator of the group Pin(η), i.e. a unit vector in the Clifford algebra; the following property holds

$$[w, \gamma] = \begin{cases} 0 & (m = 2\mu + 1) \\ 2w \cdot \gamma & (m = 2\mu) \end{cases} \tag{10.2.9}$$

Proof: we simply have

$$[w, \gamma] = w^a (\gamma_a \cdot \gamma - \gamma \cdot \gamma_a) = w^a (1 + (-1)^m) \gamma_a \cdot \gamma \tag{10.2.10}$$

■

Corollary (10.2.11): both V_L and V_R are invariant under the action of the group Spin(η).

Proof: let us consider $v \in V_L$, i.e. $v_L = \gamma\, v_L$, let w be a generator of the Pin(η) group as above and let us define $v' = w\, v_L$. In even dimension $m = 2\mu$ we have

$$\pi_R v' = \tfrac{1}{2}(\mathbb{I} - \gamma) v' = \tfrac{1}{2}(w - \gamma \cdot w) v_L = \tfrac{1}{2}[w, \gamma] v_L = w \cdot \gamma\, v_L = w\, v_L = v' \tag{10.2.12}$$

so that v' is a right spinor. Thence in even dimension we have $w(V_L) = V_R$. Analogously one can prove that, still in even dimension, we have $w(V_R) = V_L$.

In odd dimension $m = 2\mu + 1$ we have instead

$$\pi_R v' = \tfrac{1}{2}(\mathbb{I} - \gamma) v' = \tfrac{1}{2}(w - \gamma \cdot w) v_L = \tfrac{1}{2}[w, \gamma] v_L = 0 \tag{10.2.13}$$

so that v' is a left spinor. Thence in odd dimension $w(V_L) = V_L$. Analogously, we have $w(V_R) = V_R$.

In either case, when one considers $S \in \mathrm{Spin}(\eta)$, i.e. $S = w_1 \cdot \ldots \cdot w_{2l}$ a product of an even number of generators of the $\mathrm{Pin}(\eta)$ group, we have $S(V_L) = V_L$ and $S(V_R) = V_R$. ■

Being the two subspaces V_L and V_R preserved by the action of the spin group one can restrict to left spinors (right spinors, respectively) by constructing the *bundle of left spinors* $\bar{E}^{(L)}_\lambda = \bar{\Sigma} \times_\lambda V_L$ (the *bundle of right spinors* $\bar{E}^{(R)}_\lambda = \bar{\Sigma} \times_\lambda V_R$, respectively), which is globally well defined since the transition functions of $\bar{E}_\lambda$, by construction, take their values in $\mathrm{Spin}(\eta)$. In this way, from a kinematical point of view, we can define *chiral spinors*.

On the other hand, we have to check whether the dynamics preserves chirality. Let us thence consider Dirac equations

$$\begin{cases} i\gamma^a \nabla_a v = mv \\ i\nabla_a \bar{v}\, \gamma^a = -m\bar{v} \end{cases} \tag{10.2.14}$$

Notice that the covariant derivative of spinors preserves chirality, i.e.

$$\pi_L \nabla_a v_L = \nabla_a v_L \ , \qquad \pi_R \nabla_a v_R = \nabla_a v_R \tag{10.2.15}$$

Thence applying the projector π_L to the Dirac equations one obtains in even dimension $m = 2\mu$

$$\begin{cases} i\gamma^a \nabla_a v_R = mv_L \\ i\nabla_a \bar{v}_R\, \gamma^a = -m\bar{v}_L \end{cases} \tag{10.2.16}$$

while in odd dimension $m = 2\mu + 1$ one has

$$\begin{cases} i\gamma^a \nabla_a v_L = mv_L \\ i\nabla_a \bar{v}_L\, \gamma^a = -m\bar{v}_L \end{cases} \tag{10.2.17}$$

Thence in even dimension chirality is preserved by dynamics only if $m = 0$, while in odd dimension chirality is always preserved. As a consequence one can evolve the massless left (resp., right) spinor in even dimension, as well as left (resp., right) spinor (possibly with mass) in odd dimension. These fields are a model for (massless) chiral neutrinos. For example, in dimension 4 and Lorentzian signature $\eta = (1, 3)$ the equations for a massless left spinors are

$$\begin{cases} \gamma^a \nabla_a v_L = 0 \\ \nabla_a \bar{v}_L \gamma^a = 0 \end{cases} \tag{10.2.18}$$

and they can be derived form the so-called *Weyl Lagrangian* on $J^1 \bar{E}^{(L)}_\lambda$

$$L_W = \tfrac{i}{2} \left[\bar{v}_L \gamma^a \nabla_a v_L - \nabla_a \bar{v}_L \gamma^a v_L \right] \sqrt{g}\, \mathrm{ds} \tag{10.2.19}$$

We stress that this Lagrangian is obtained by pull-back of the Dirac Lagrangian (10.1.8) along the map induced by the fiberwise embedding $i_L : V_L \longrightarrow V$.

3. (Reduced) Electroweak Model

In this Section we shall collect most of the results we obtained above when discussing simple examples and applications. We are aimed to showing that also relatively complex systems can be successfully studied by suitably glueing the partial results we already know for the single component fields. Our purpose is here to provide a classical starting point for the description of the electroweak model (hereafter simplified by reducing the number of lepton families to just one) coupled to a gravitational field.

Let us then fix a spacetime M, i.e. a manifold of dimension $m = 4$ which is assumed to be a spin manifold and to allow global metrics of Lorentzian signature $\eta = (1, 3)$, and let us also fix a spin bundle $\bar{\Sigma}$ over it. Let us first define spin frames as in Section 9.5 and use the first order Lagrangian in its purely frame formulation

$$L_{\text{grav}} = \frac{1}{2\kappa}\left[g^{\alpha\beta}R_{\alpha\beta}\sqrt{g} - \nabla_\mu(g^{\alpha\beta}w^\mu_{\alpha\beta}\sqrt{g}) - \bar{g}^{\alpha\beta}\bar{R}_{\alpha\beta}\sqrt{\bar{g}}\right]\mathrm{d}s \qquad (10.3.1)$$

This Lagrangian is written on the second jet prolongation of the bundle $\mathcal{C}_{\text{grav}} = \bar{\Sigma}_\rho \times_M \bar{\Sigma}_\rho$ which is a gauge natural bundle of order $(0, 1)$ associated to the spin bundle $\bar{\Sigma}$. We shall use fiber coordinates $(x^\mu; e^\mu_a, \bar{e}^\mu_a)$.

Since we shall also need a $U(1)\times\mathrm{SU}(2)$ gauge field, let us also fix a principal bundle $\mathcal{P}$ having the group $G = U(1)\times\mathrm{SU}(2)$ as a gauge group. Let us consider the bundle $\mathcal{C}_{\text{ew}} = \mathcal{C}_{\mathcal{P}}$ of all the principal connections over $\mathcal{P}$, which, according to (5.2.31), is a gauge natural bundle of order $(1, 1)$ associated to $\mathcal{P}$. Since the group G is the direct product of two semisimple groups we can choose fiber coordinates $(x^\mu; \theta_\mu, \omega^A_\mu)$ (with spacetime indices $\mu = 0, 1, 2, 3$ running over M and Lie algebra indices $A = 1, 2, 3$ running over the Lie algebra $\mathfrak{su}(2)$; no algebra index is associated to $\mathfrak{u}(1) \simeq i\mathbb{R}$). The so-called *field strengths* $\mathfrak{F}_{\mu\nu}$ and $\mathfrak{F}^A_{\mu\nu}$ (see Section 6.7) are given in terms of the gauge connection fields by

$$\begin{aligned}\mathfrak{F}_{\mu\nu} &= \mathrm{d}_\mu\theta_\nu - \mathrm{d}_\nu\theta_\mu \\ \mathfrak{F}^A_{\mu\nu} &= \mathrm{d}_\mu\omega^A_\nu - \mathrm{d}_\nu\omega^A_\mu + c^A{}_{BC}\,\omega^B_\mu\omega^C_\nu\end{aligned} \qquad (10.3.2)$$

where $c^A{}_{BC}$ denote the structure constants of the algebra $\mathfrak{su}(2)$ with respect to any chosen set of generators T_A. Here Greek indices have been moved up and down by means of the induced metric $g_{\mu\nu}$ over M, while the capital Latin indices are moved by means of the Cartan-Killing metric k_{AB} over SU(2). We shall hereafter use the standard k-orthonormal basis of the Lie algebra $\mathfrak{su}(2)$ given by $T_A = \frac{i}{2}\sigma_A$, where σ_A are the Pauli matrices (9.1.23); thus one has

$$T_1 = \frac{i}{2}\begin{pmatrix}0 & 1\\ 1 & 0\end{pmatrix} \qquad T_2 = \frac{i}{2}\begin{pmatrix}0 & -i\\ i & 0\end{pmatrix} \qquad T_3 = \frac{i}{2}\begin{pmatrix}1 & 0\\ 0 & -1\end{pmatrix} \qquad (10.3.3)$$

Accordingly, we have $k_{AB} = \delta_{AB}$ and the structure constant $c^A{}_{BC} = -\epsilon^A{}_{BC}$ are the completely antisymmetric Ricci symbols.

The Lagrangian to be chosen is the sum of a Maxwell Lagrangian for $U(1)$ and a Yang-Mills Lagrangian for SU(2), i.e:

$$L_{\text{ew}} = -\tfrac{1}{4}\left(\mathfrak{F}_{\mu\nu}\mathfrak{F}^{\mu\nu} + \mathfrak{F}^A_{\mu\nu}\mathfrak{F}^{\mu\nu}_A\right)\sqrt{g}\,\mathrm{d}s \tag{10.3.4}$$

Matter fields will be affected both by G-gauge transformations over $\mathcal{P}$ and spin transformations over $\bar{\Sigma}$. Let us thence introduce the so-called *structure bundle* which is simply given by the principal bundle $\bar{\Sigma} \times_M \mathcal{P}$. Spin frames e^μ_a are insensitive to the G sector, namely the bundle $\mathcal{C}_{\text{gr}}$ is trivially associated to $\mathcal{P}$. On the other side, the bundle $\mathcal{C}_{\text{ew}}$ can be formally regarded as an associated bundle to $\bar{\Sigma} \times_M \mathcal{P}$ even though it is in fact trivially associated to $\bar{\Sigma}$, gauge fields being insensitive to spin transformations. On the contrary the bundles of matter fields will be associated to the whole structure bundle $\bar{\Sigma} \times_M \mathcal{P}$ since matter fields usually are affected both by gauge and spin transformations.

Let us first consider a Dirac spinor field ϵ to describe the electron and a massless left spinor ν to describe the (electronic) neutrino.

To obtain a full description of the standard model one should here in principle add two further copies of the lepton family $(\mu, \nu^{(\mu)})$ and $(\tau, \nu^{(\tau)})$. Furthermore, one can add a SU(3) sector to describe strong interactions, the corresponding gauge field connection, as well as the three families of quarks[10] (u, d), (s, c), (b, t) to describe the so-called *standard model*. Here we restrict to this simpler model since no additional theoretical difficulties are encountered by extending our simplified framework to the standard model, which is thence obtained by just making the necessary additions.

Let us thence fix $V = \mathbb{C}^2 \times \mathbb{C}^4$ as the standard fiber; coordinates over V are divided into two groups (ν, ϵ). The first group of coordinates will eventually describe the electron and the second will describe the (electron) neutrino. We shall also use another set of coordinates over V given by splitting the Dirac spinor $\epsilon = \lambda \oplus \rho$ into its chiral components. In this way we shall parametrize the lepton sector by means of the left hand doublet $\chi = {}^t(\nu, \lambda)$ and the right hand singlet ρ.

The groups Spin(η), $U(1)$ and $SU(2)$ all have to act on (ν, λ, ρ). Accordingly, we proceed to fix a representation

$$\phi : \text{Spin}(\eta) \times U(1) \times \text{SU}(2) \times V \longrightarrow V \tag{10.3.5}$$

according to which the spin group acts by matrix multiplication both on ϵ and ν, separately. It can be regarded as choosing a set of Dirac matrices $\gamma^{(\epsilon)}_a$

[10] Quarks are Fermions constituting the particles which are sensitive to strong interaction. As far as electroweak interactions are concerned quarks are very similar to electron and neutrinos.

related to the action of the spin group on ϵ, a different set of Dirac matrices $\gamma_a^{(\nu)}$ related to the action of the spin group on ν and then to define the reducible representation on V induced by the block-diagonal Dirac matrices

$$\gamma_a = \begin{pmatrix} \gamma_a^{(\nu)} & 0 \\ 0 & \gamma_a^{(\epsilon)} \end{pmatrix} \tag{10.3.6}$$

Equivalently, we can choose two sets of 2×2 Dirac matrices $\gamma_a^{(L)}$ and $\gamma_a^{(R)}$ defining the left hand and right hand representations, respectively. The corresponding Dirac matrices to be used with the chiral coordinates (ν, λ, ρ) are thence

$$\gamma_a = \begin{pmatrix} \gamma_a^{(L)} & 0 & 0 \\ 0 & \gamma_a^{(L)} & 0 \\ 0 & 0 & \gamma_a^{(R)} \end{pmatrix} \tag{10.3.7}$$

Since it will be clear from the context which specific representation one is using for the Dirac matrices, we shall omit to indicate the superscripts (L) and (R).

The $U(1)$ group acts by phase shifting according to the following matrix representation

$$e^{i\theta} \mapsto \begin{pmatrix} e^{-\frac{i}{2}q_1\theta} & 0 & 0 \\ 0 & e^{-\frac{i}{2}q_1\theta} & 0 \\ 0 & 0 & e^{-iq_1\theta} \end{pmatrix} \tag{10.3.8}$$

with chiral coordinates (ν, λ, ρ); here q_1 is a real coupling constant.

Finally the SU(2) group acts on the chiral doublet $\chi = {}^t(\nu, \lambda)$ by block-matrix multiplication, while it leaves the right singlet unchanged. Namely if $U \in SU(2)$ we have

$$U = e^{-\frac{i}{2}q_2\xi^A\sigma_A} = \begin{pmatrix} \alpha & \beta \\ -\beta^* & \alpha^* \end{pmatrix} \tag{10.3.9}$$

with the condition $|\alpha|^2 + |\beta|^2 = 1$, $*$ denoting complex conjugation. The action on the chiral coordinates (λ, ν, ρ) is defined by means of the matrix representation

$$\begin{pmatrix} \alpha & \beta \\ -\beta^* & \alpha^* \end{pmatrix} \mapsto \begin{pmatrix} \alpha & \beta & 0 \\ -\beta^* & \alpha^* & 0 \\ 0 & 0 & 1 \end{pmatrix} \tag{10.3.10}$$

so that the action on the lepton sector is given by

$$\begin{pmatrix} \nu' \\ \lambda' \\ \rho' \end{pmatrix} = \begin{pmatrix} \alpha\nu + \beta\lambda \\ -\beta^*\nu + \alpha^*\lambda \\ \rho \end{pmatrix} \tag{10.3.11}$$

Using the representation $\phi : \mathrm{Spin}(\eta) \times U(1) \times \mathrm{SU}(2) \times V \longrightarrow V$ so defined one can build the bundle $\mathcal{C}_{\mathrm{lep}}$ associated to the structure bundle $\bar{\Sigma} \times_M \mathcal{P}$. Because of the representation chosen, the covariant derivative of the left hand spinor doublet is

$$\nabla_\mu \chi = \mathrm{D}_\mu \chi - \tfrac{i}{2} q_1 \theta_\mu \chi - \tfrac{i}{2} q_2 \omega^A_\mu \sigma_A \chi \tag{10.3.12}$$

while for the right hand singlet one has

$$\nabla_\mu \rho = \mathrm{D}_\mu \rho - i q_1 \theta_\mu \rho \tag{10.3.13}$$

where we set for short

$$\mathrm{D}_\mu = \mathrm{d}_\mu + \tfrac{1}{8} \Gamma^{ab}_\mu [\gamma_a, \gamma_b] \tag{10.3.14}$$

By using the explicit expression of the algebra generators T_A and of the Dirac matrices we obtain the covariant derivatives

$$\begin{aligned} \nabla_\mu \nu &= \mathrm{D}_\mu \nu - \alpha_1 Z_\mu \nu - \alpha_2 A_\mu \nu - \tfrac{i}{2} q_2 \sqrt{2} W^-_\mu \lambda \\ \nabla_\mu \lambda &= \mathrm{D}_\mu \lambda - \alpha_3 Z_\mu \lambda - \alpha_4 A_\mu \lambda - \tfrac{i}{2} q_2 \sqrt{2} W^+_\mu \nu \\ \nabla_\mu \rho &= \mathrm{D}_\mu \rho - i q_1 \sin(\theta_w) Z_\mu \rho - i q_1 \cos(\theta_w) A_\mu \rho \end{aligned} \tag{10.3.15}$$

where we set $W^+_\mu = \frac{1}{\sqrt{2}}(\omega^1_\mu + i\omega^2_\mu)$ and $W^-_\mu = \frac{1}{\sqrt{2}}(\omega^1_\mu - i\omega^2_\mu)$. For later purposes we have also introduced the *Weinberg angle* θ_w together with the new fields

$$\begin{cases} Z_\mu = \cos(\theta_w)\omega^3_\mu + \sin(\theta_w)\theta_\mu \\ A_\mu = -\sin(\theta_w)\omega^3_\mu + \cos(\theta_w)\theta_\mu \end{cases} \Leftrightarrow \begin{cases} \omega^3_\mu = \cos(\theta_w) Z_\mu - \sin(\theta_w) A_\mu \\ \theta_\mu = \sin(\theta_w) Z_\mu + \cos(\theta_w) A_\mu \end{cases} \tag{10.3.16}$$

and the corresponding new coupling constants

$$\begin{cases} \alpha_1 = \frac{i}{2}\left(q_1 \sin(\theta_w) + q_2 \cos(\theta_w)\right) \\ \alpha_2 = \frac{i}{2}\left(q_1 \cos(\theta_w) - q_2 \sin(\theta_w)\right) \\ \alpha_3 = \frac{i}{2}\left(q_1 \sin(\theta_w) - q_2 \cos(\theta_w)\right) \\ \alpha_4 = \frac{i}{2}\left(q_1 \cos(\theta_w) + q_2 \sin(\theta_w)\right) \end{cases} \tag{10.3.17}$$

Notice that one cannot separately define the covariant derivative of the spinor fields λ and ν: only the covariant derivative of the left hand doublet has a geometrical meaning. This is the very deep meaning of unification theories; here the fields λ and ν are two aspects of the same *unique* lepton field. Deciding which one is to be called the electron and which one is to be called the neutrino is, under most viewpoints, mainly a matter of convention. We stress in fact that a continuous family of gauge transformations is defined (namely the $SU(2)$ group transformations) exchanging the λ and ν sectors.

As a Lagrangian for the lepton sector let us thence choose

$$L_{\text{lep}} = \tfrac{i}{2}\left[\bar\chi \gamma^a \nabla_a \chi - \nabla_a \bar\chi \gamma^a \chi + \bar\rho \gamma^a \nabla_a \rho - \nabla_a \bar\rho \gamma^a \rho\right] \sqrt{g}\, \mathrm{d}s \tag{10.3.18}$$

Here we understand why it was necessary to introduce the chiral splitting of the spinor ϵ. As a matter of Nature, the weak interaction is only coupled to the left hand component of leptons. The new vector bosons $(A_\mu, Z_\mu, W_\mu^+, W_\mu^-)$ will be at the very end identified with the phenomenological bosons (photon, Z_0, $W^\pm$) seen in high-energy colliders. Since, in particular, A_μ has eventually to be identified with the electromagnetic field we shall constrain the parameters of the theory in order to reproduce the correct known coupling between the lepton sector and the electromagnetic field.

By inserting the explicit expression of covariant derivatives given by equations (10.3.13) and (10.3.15) into the lepton Lagrangian (10.3.18) and defining the following lepton currents

$$\begin{aligned} J_\mu &= -(\bar\epsilon\gamma_\mu\epsilon)\ , & J^3_\mu &= (\bar\chi\sigma^3\gamma_\mu\chi) = (\bar\nu\gamma_\mu\nu) - (\bar\lambda\gamma_\mu\lambda) \\ J^+_\mu &= 2(\bar\nu\gamma_\mu\lambda)\ , & J^-_\mu &= 2(\bar\lambda\gamma_\mu\nu) \end{aligned} \tag{10.3.19}$$

one obtains (modulo a pure divergence term which is here omitted since it is inessential for our present purposes)[11]:

$$\begin{aligned} L_{\text{lep}} = \Big[& i(\bar\epsilon\gamma^a D_a\epsilon) + i(\bar\nu\gamma^a D_a\nu) + \tfrac{1}{2\sqrt{2}} q_2 (W^+_\mu J^-_\mu + W^-_\mu J^+_\mu) + \\ & + \tfrac{1}{2}\left(q_1\cos(\theta_w) - q_2\sin(\theta_w)\right) J^3_\mu A^\mu - q_1\cos(\theta_w) J_\mu A^\mu \\ & + \tfrac{1}{2}\left(q_1\sin(\theta_w) + q_2\cos(\theta_w)\right) J^3_\mu Z^\mu - q_1\sin(\theta_w) J_\mu Z^\mu \Big] \sqrt{g}\,\mathrm{ds} \end{aligned} \tag{10.3.20}$$

The electromagnetic field is known to be coupled to spinors through the current J_μ and not through the *axial current* J^3_μ. Furthermore the coupling with electromagnetic field has to be assumed in the form $-eJ_\mu A^\mu$ in order to reproduce the usual electrodynamics. Thence we assume the coupling constants to satisfy the following conditions

$$q_1\cos(\theta_w) = q_2\sin(\theta_w) = e \tag{10.3.21}$$

so that the lepton Lagrangian is of the form

$$\begin{aligned} L_{\text{lep}} = \Big[& i(\bar\epsilon\gamma^a D_a\epsilon) + i(\bar\nu\gamma^a D_a\nu) + \tfrac{1}{2\sqrt{2}} g (W^+_\mu J^-_\mu + W^-_\mu J^+_\mu) + \\ & - eJ_\mu A^\mu + \tfrac{1}{2}({q_2}^2 + {q_1}^2)^{\frac{1}{2}} J^0_\mu Z^\mu \Big] \sqrt{g}\,\mathrm{ds} \end{aligned} \tag{10.3.22}$$

where we set $J^0_\mu = J^3_\mu - 2\sin^2(\theta_w) J_\mu$ for the *neutral current*.

Our model depends now on the parameters (q_1, q_2, θ_w, e) among which just two are independent because of the relations (10.3.21). Everything depends on just two free parameters: the coupling constants, as the masses of the vector

[11] This divergence term, of course, plays a role when conservation laws are considered.

bosons and of the leptons which will appear in a low energy regimes will have to obey a number of relations which have to be satisfied in order to validate the model. These relations among the parameters in the effective low energy limit is the most important test a unification theory has to undergo.

We remark that all the spinor fields in the lepton Lagrangian (10.3.18) and the gauge fields in the Yang-Mills Lagrangian are originally massless and they gain a finite rest mass as a consequence of the so-called *spontaneous symmetry breaking*, owing to the presence of a further scalar field called the *Higgs boson*. In the beginning it is a doublet of scalar complex fields $\Phi = {}^t(\phi, \varphi)$ on which the spin group acts trivially (i.e. by means of the identity representation), while the $U(1)$ group and the SU(2) group act by phase shifting and matrix multiplication, respectively, i.e.

$$e^{i\theta} \in U(1) \ , \qquad \begin{pmatrix} \phi' \\ \varphi' \end{pmatrix} = \begin{pmatrix} e^{\frac{i}{2}q_1\theta}\phi \\ e^{\frac{i}{2}q_1\theta}\varphi \end{pmatrix} \tag{10.3.23}$$

and

$$\begin{pmatrix} \alpha & \beta \\ -\beta^* & \alpha^* \end{pmatrix} \in \mathrm{SU}(2) \ , \qquad \begin{pmatrix} \phi' \\ \varphi' \end{pmatrix} = \begin{pmatrix} \alpha\phi + \beta\varphi \\ -\beta^*\phi + \alpha^*\varphi \end{pmatrix} \tag{10.3.24}$$

The doublet Φ is thence a section of the associated bundle $\mathcal{C}_{\text{Higgs}} = \mathcal{P} \times \mathbb{C}^2$ with fiber coordinates $(x^\mu; \phi, \varphi)$. The covariant derivative is

$$\nabla_\mu \Phi = \mathrm{d}_\mu \Phi + \tfrac{i}{2} q_1 \theta_\mu \Phi - \tfrac{i}{2} q_2 \omega^A_\mu \sigma_A \Phi \tag{10.3.25}$$

which, by using the explicit expression (9.1.23) of the generators σ_A, can be recast as

$$\begin{aligned} \nabla_\mu \phi &= \mathrm{d}_\mu \phi + \alpha_3 Z_\mu \phi + \alpha_4 A_\mu \phi - \tfrac{i}{2} q_2 \sqrt{2} W^-_\mu \varphi \\ \nabla_\mu \varphi &= \mathrm{d}_\mu \varphi + \alpha_1 Z_\mu \varphi + \alpha_2 A_\mu \varphi - \tfrac{i}{2} q_2 \sqrt{2} W^+_\mu \phi \end{aligned} \tag{10.3.26}$$

Again these two covariant derivatives cannot be decoupled.

Let us choose the Higgs Lagrangian to be

$$L_{\text{Higgs}} = \left[\nabla_\mu \Phi^\dagger \nabla^\mu \Phi - \mu^2 \Phi^\dagger \Phi - \alpha (\Phi^\dagger \Phi)^2\right] \sqrt{g}\, \mathrm{ds} \tag{10.3.27}$$

Finally a coupling between leptons and Higgs fields is added in the form

$$L_{\text{lep}-\text{Higgs}} = -G(\bar{\chi}\Phi\rho + \bar{\rho}\Phi^\dagger\chi) \sqrt{g}\, \mathrm{ds} \tag{10.3.28}$$

We can now consider the (reduced) Weinberg-Salam model on the configuration bundle $\mathcal{C} = \mathcal{C}_{\text{gr}} \times_M \mathcal{C}_{\text{ew}} \times_M \mathcal{C}_{\text{lep}} \times_M \mathcal{C}_{\text{Higgs}}$, which is a gauge natural bundle of order $(1,1)$ associated to $\mathcal{P} \times_M \bar{\Sigma}$. The total Lagrangian

$$L = L_{\text{gr}} + L_{\text{ew}} + L_{\text{lep}} + L_{\text{Higgs}} + L_{\text{lep}-\text{Higgs}} \tag{10.3.29}$$

is gauge natural and the dynamical connections are obviously defined. Thence we are considering a gauge natural theory.

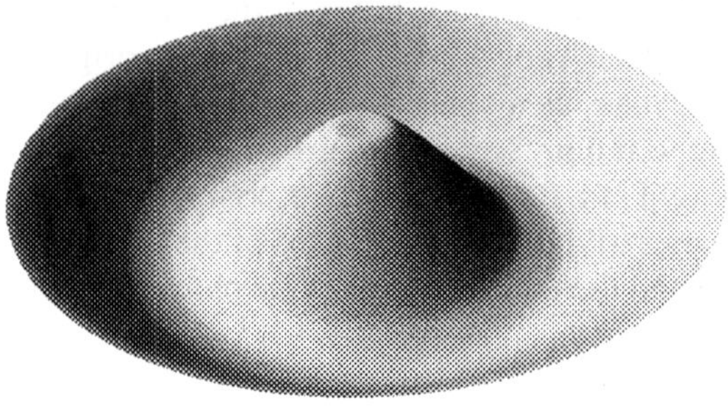

Fig. 27 – The potential for the Higgs sector.

The so-called spontaneous symmetry breaking occur because of the particular form of the Higgs potential shown in *Fig.* 27. In fact, quantum field theory is perturbative in its nature and it needs a stable equilibrium point (the so-called *vacuum state*) around which one can consider small oscillations. The oscillation modes are then identified with the phenomenological particle fields one can observe in a collider. For the Higgs field the extremum $\Phi = 0$ is unstable, so that one should move to a minimum, e.g. $\Phi_0 = {}^t(0, \varphi_0)$ where we set $\varphi_0 = \mu/\sqrt{2\alpha}$. All the other choices are equivalent to this one modulo a linear transformation in the space where Φ lives. Small oscillations around the point Φ_0 are thence given by

$$\begin{pmatrix} \xi \\ \zeta \end{pmatrix} = \begin{pmatrix} \phi \\ \varphi - \varphi_0 \end{pmatrix} \tag{10.3.30}$$

There exists a SU(2)-gauge transformation after which one has

$$\Phi - \Phi_0 = \begin{pmatrix} 0 \\ \eta \end{pmatrix} \tag{10.3.31}$$

with η a real scalar field. This proves that the complex field ξ and the imaginary part of ζ are pure gauge degrees of freedom and they can be gauged away. The Higgs Lagrangian can be then written with respect to the new fields (up to quadratic terms) as

$$\begin{aligned} L_{\text{Higgs}} \simeq \Big[&\tfrac{1}{2}(\nabla_\mu \eta \nabla^\mu \eta + m^2 \eta^2) + m_W^2 (W^+)_\mu (W^-)^\mu + \tfrac{1}{2} m_Z^2 Z_\mu Z^\mu + \\ &+ q_\pm \eta (W^+)_\mu (W^-)^\mu + \tfrac{1}{2} q_0 \eta Z_\mu Z^\mu \Big] \sqrt{g}\, \mathrm{ds} \end{aligned} \tag{10.3.32}$$

having set, for notational convenience, $(W^-)^\mu = g^{\mu\nu}\,(W^-)_\nu$ and $Z^\mu = g^{\mu\nu}\,Z_\nu$. By means of the same transformation we obtain

$$L_{\text{lep-Higgs}} = -m_\epsilon \bar{\epsilon}\epsilon - G'\eta\bar{\epsilon}\epsilon \tag{10.3.33}$$

Thence, as a consequence of the symmetry breaking, only a scalar Higgs boson survives out of the Higgs doublet and it gains a mass. As a side effect, the electron ϵ and the vector bosons Z, $W^\pm$ also gain a mass and the Higgs boson self-interacts by means of weak interactions. The masses of vector bosons Z, $W^\pm$ can be empirically determined and they put some further constraints on the original coupling constants q_1 and q_2.

4. Anticommuting Majorana Spinor Fields

In the canonical quantization procedure one imposes the canonical commutators of fields and momenta eventually obtaining a quantum field theory which describes Bosons which obey to the Bose-Einstein statistics. When quantizing spinors one imposes instead anticommutators in order to obtain fields which obey the Fermi-Dirac statistics. On the other hand, in a path integral approach to quantum field theory there is nothing to be imposed *a priori* and thence spinor fields have to be *anticommuting* from the very beginning in order to implement the Pauli principle. For this reason let us first introduce a framework for classical anticommuting spinors which may be considered as the starting point for a quantum field theory of spinor fields.

Let us hereafter consider

$$V = [\Lambda_-(\mathbb{R}^n) \otimes \mathbb{C}]^k \tag{10.4.1}$$

where k is any positive integer and $\Lambda_-(\mathbb{R}^n)$ is the odd part of the exterior algebra of $\mathbb{R}^n$. Since $\Lambda_-(\mathbb{R}^n) \otimes \mathbb{C}$ is a (complex) vector space, then V is a (complex) vector space too. Let us denote by e^A $(A = 1, \dots, 2^{n-1})$ a basis of $\Lambda_-(\mathbb{R}^n) \otimes \mathbb{C}$, so that a point $\omega \in \Lambda_-(\mathbb{R}^n) \otimes \mathbb{C}$ is in the form $\omega = \omega_A e^A$. Then a point $\psi \in V$ may be alternatively regarded as a k-tuple $\psi = (\psi^a)$ (with $a = 1, \dots, k$) of elements in $\Lambda_-(\mathbb{R}^n) \otimes \mathbb{C}$ as well as a list of (complex) numbers $\psi = (\psi^1_A, \psi^2_A, \dots, \psi^k_A)$. The ψ^a are called *anticommuting components* of ψ, since we have that $\psi^a \phi^b = -\phi^b \psi^a$ (the wedge product is understood).

Since V is a complex vector space, once a set of Dirac matrices γ_a has been fixed, we can build the associated bundle $\bar{E}_\lambda = \bar{\Sigma} \times_\lambda V$ where λ is the obvious representation of $\mathrm{Spin}(\eta)$ defined on V by γ_a. Such a representation is given by block multiplication of anticommuting spinors, i.e.

$$\lambda(S, \psi) = S^a_b\, \psi^b \tag{10.4.2}$$

We remark that the vector structure of $\Lambda_-(\mathbb{R}^n) \otimes \mathbb{C}$ is here used explicitly.

We are hereafter interested in Majorana spinors, which can be shown to exist only in specific dimensions (see [CDF91]Vol I pg. 529). For this reason, from now on we shall restrict to dimension $m = 4$ and Lorentzian signature, a case in which Majorana spinors do exist. In the relevant Clifford algebra $\mathcal{C}(\mathbb{R}^4, (1,3))$ we use the *charge conjugation operator* defined (up to a constant factor) by the condition

$$C\gamma_a C^{-1} = -{}^t\gamma_a \tag{10.4.3}$$

Let us choose the representation induced by Dirac matrices (9.1.22); the charge conjugation operator is then defined by

$$C = \begin{pmatrix} -i\sigma^2 & 0 \\ 0 & i\sigma^2 \end{pmatrix} \tag{10.4.4}$$

which satisfies equation (10.4.3), as one can easily check.

By means of the charge conjugation operator above we can now define the charge conjugation on spinors by setting

$$(\psi)^c = C\,{}^t\bar{\psi} \tag{10.4.5}$$

We remark that the name of *charge conjugation* is here somewhat improper, since we are dealing with neutral spinor fields (in fact no $U(1)$-gauge invariance has been assumed).

A spinor ψ is called a *Majorana spinor* if it is *self charge-conjugate*, i.e. if and only if $\psi = (\psi)^c$. By an explicit calculation we obtain that a Majorana spinor is of the general form

$$\psi_{\mathrm{Maj}} = \begin{pmatrix} \alpha \\ i\sigma^2\,{}^t\alpha^\dagger \end{pmatrix} \tag{10.4.6}$$

where α is any two-component spinor.

In other words, the vector space V decomposes into the two subspaces of self-charge-conjugated spinors V_+ and of anti-self-charge-conjugated spinors V_-, i.e. we have the following exact sequence

$$0 \longrightarrow V_- \longrightarrow V \longrightarrow V_+ \longrightarrow 0 \tag{10.4.7}$$

The action of the spin group $\mathrm{Spin}(1,3)$ is compatible with the Majorana condition, in the sense that Majorana spinors are preserved by the action. In fact, if ψ is a Majorana spinor and $S \in \mathrm{Spin}(\eta)$ is a element of the spin group, then $\psi' = S\psi$ is again a Majorana spinor:

$$(\psi')^c = (S\psi)^c = C\,{}^t(\bar{\psi}S^{-1}) = C\,{}^tS^{-1}C^{-1}(\psi)^c = S(\psi)^c \tag{10.4.8}$$

Thence the exact sequence (10.4.7) of vector spaces induces an exact sequence of vector bundles

$$0 \longrightarrow \Sigma \times_\lambda V_- \longrightarrow \Sigma \times_\lambda V \longrightarrow \Sigma \times_\lambda V_+ \longrightarrow 0 \qquad (10.4.9)$$

This means that we can define the *bundle of Majorana spinors* $\bar{E}_\lambda^{\mathrm{Maj}} = \Sigma \times_\lambda V_+$.

We shall now prove the so-called *Majorana flip identities*, which will be often used below.

Proposition (10.4.10): let us consider two Majorana spinors ψ and φ; then the following identities hold true

$$\begin{array}{ll} \bar{\psi}\varphi = \bar{\varphi}\psi \;, & \bar{\psi}\gamma\varphi = \bar{\varphi}\gamma\psi \\ \bar{\psi}\gamma^a\varphi = -\bar{\varphi}\gamma^a\psi \;, & \bar{\psi}\gamma^a\gamma^b\varphi = \bar{\varphi}\gamma^b\gamma^a\psi \\ \bar{\psi}\gamma^{ab}\varphi = -\bar{\varphi}\gamma^{ab}\psi \;, & \bar{\psi}\gamma^a\gamma^b\gamma\varphi = \bar{\varphi}\gamma\gamma^b\gamma^a\psi \\ \bar{\psi}\gamma^a\gamma\varphi = \bar{\varphi}\gamma^a\gamma\psi & \end{array} \qquad (10.4.11)$$

Proof: we shall prove the first two of them only; the others can be proved in a similar way. The first one is

$$\bar{\psi}\varphi = \bar{\psi}C\,{}^t\bar{\varphi} \qquad (10.4.12)$$

Now, since each bilinear form in spinors is an element of $\Lambda_-(\mathbb{R}^n) \otimes \mathbb{C}$ and the spinors are anticommuting, then the transpose operation just acts by a change of a sign, so that

$$\bar{\psi}\varphi = -\bar{\varphi}\,{}^tC\,{}^t\bar{\psi} = \bar{\varphi}C\,{}^t\bar{\psi} = \bar{\varphi}(\psi)^c = \bar{\varphi}\psi \qquad (10.4.13)$$

which proves the first one. The second is

$$\bar{\psi}\gamma^a\varphi = \bar{\psi}\gamma^a C\,{}^t\bar{\varphi} = -\bar{\varphi}\,{}^tC\,{}^t\gamma^a C^{-1} C\,{}^t\bar{\psi} = \bar{\varphi}C^{-1}\,{}^t\gamma^a C(\psi)^c = -\bar{\varphi}\gamma^a\psi \qquad (10.4.14)$$

which proves the second identity. ■

As a corollary, one can use the Majorana flip identities and the Clifford commutation rules to prove that

$$\varphi\bar{\psi} - \psi\bar{\varphi} = -\tfrac{1}{2}(\bar{\psi}\gamma_a\varphi)\gamma^a + (\bar{\psi}\gamma_{ab}\varphi)\gamma^{ab} \qquad (10.4.15)$$

Before reverting to a field theory for the Majorana spinors we have to prove that also the dynamics preserves the subspace of Majorana spinors. Let us then start by proving first that the covariant derivative of a Majorana spinor in the form (10.4.6) is again a Majorana spinor; in fact

$$\nabla_\mu\psi = \nabla_\mu\begin{pmatrix}\alpha\\ \beta\end{pmatrix} = \begin{pmatrix}\mathrm{d}_\mu\alpha + \Gamma_\mu\alpha\\ \mathrm{d}_\mu\beta + \bar{\Gamma}_\mu\beta\end{pmatrix} \;, \qquad (10.4.16)$$

where we set

$$\begin{cases} \Gamma_\mu = \frac{1}{2}\Gamma^{0i}_\mu \sigma^i - \frac{i}{4}\epsilon_{ijk}\Gamma^{ij}_\mu \sigma^k \\ \bar{\Gamma}_\mu = -\frac{1}{2}\Gamma^{0i}_\mu \sigma^i - \frac{i}{4}\epsilon_{ijk}\Gamma^{ij}_\mu \sigma^k \end{cases} \tag{10.4.17}$$

When ψ is a Majorana spinor we have $\beta = i\sigma^2\, {}^t\alpha^\dagger$. Furthermore, since the following identity holds true

$$\bar{\Gamma}_\mu = \sigma^2\, {}^t\Gamma^\dagger_\mu\, \sigma^2 \tag{10.4.18}$$

then the covariant derivative of a Majorana spinor is still a Majorana spinor, as we claimed.

As a consequence of this result the Dirac equation splits into two equations

$$\begin{cases} -i\nabla_0\alpha + i\sigma^i\nabla_i\alpha + m\beta = 0 \\ -i\nabla_0\beta - i\sigma^i\nabla_i\beta + m\alpha = 0 \end{cases} \tag{10.4.19}$$

When (the complex conjugate of) the second equation is multiplied on the left by $i\sigma^2$ the first equation is recovered. Thence we can regard the entries of α as coordinates on V_{Maj} and the equations

$$-i\nabla_0\alpha + i\sigma^i\nabla_i\alpha + im\sigma^{2t}\alpha^\dagger = 0 \tag{10.4.20}$$

are called *Majorana field equations.*

The Dirac Lagrangian can be thence pulled-back along the immersion induced by $i : V_{\mathrm{Maj}} \longrightarrow V$ obtaining the *Majorana Lagrangian*

$$\begin{aligned} L_{\mathrm{Maj}} =& i[\alpha^\dagger(\nabla_0\alpha) - (\nabla_0\alpha^\dagger)\alpha + (\nabla_i\alpha^\dagger)\sigma^i\alpha - \alpha^\dagger\sigma^i(\nabla_i\alpha) + \\ &+ m({}^t\alpha\sigma^2\alpha - \alpha^\dagger\sigma^{2t}\alpha^\dagger)]\,\sqrt{g}\,\mathrm{d}s \end{aligned} \tag{10.4.21}$$

One can verify that this Lagrangian correctly produces (10.4.20) as field equations. We stress, however, that the Majorana Lagrangian is the pull-back of the Dirac Lagrangian, i.e. it is obtained by inserting Majorana spinors (10.4.6) into the usual Dirac Lagrangian (10.1.8) (taking now into account that spinors are anticommuting).

Even if there is no phenomenological evidence of elementary particles described by Majorana spinors, this framework is theoretically important. Furthermore, Majorana spinors enter many supersymmetric models as we shall see in next Section.

5. Wess-Zumino Model

Wess-Zumino model is the simplest supersymmetric model and it may be considered as a starting point for *supergravity theories*, i.e. gauge theories of supersymmetries. Let us hereafter consider this simple model as a further example of generalized symmetries in field theory. Let us assume as fields the following: a Majorana (anticommuting) field ψ and four scalar densities commuting fields (A, B, C, D) on a spacetime manifold M. Let the weights of the densities be $(\alpha, \beta, \gamma, \delta)$, respectively, and let us also fix a spin frame e_a^μ on M. As a consequence of the spin frame fixing the manifold M is endowed with an induced metric structure g and a compatible spin structure. Let us thence consider the Lagrangian

$$\begin{aligned} L_{\mathrm{WZ}} =& \tfrac{1}{2}\left[(\nabla_\mu A)(\nabla^\mu A)\, e^{2\alpha+1} + D^2\, e^{2\delta+1} + 2mAD\, e^{\alpha+\delta+1}\right]\, \mathrm{d}s+ \\ &+ \tfrac{1}{2}\left[(\nabla_\mu B)(\nabla^\mu B)\, e^{2\beta+1} + C^2\, e^{2\gamma+1} - 2mBC\, e^{\beta+\gamma+1}\right]\, \mathrm{d}s \\ &- \bar\psi\,(i\gamma^a\nabla_a\psi + m\psi)\, e\, \mathrm{d}s \end{aligned} \tag{10.5.1}$$

Here the covariant derivatives are considered with respect to the Levi-Civita connection induced by the metric, which is in turn induced by the spin frame.

This Lagrangian is invariant (modulo divergence terms) under the infinitesimal transformations

$$\begin{cases} \delta A = \frac{a}{2}(\bar\epsilon\psi)\, e^{-\alpha} \quad , \quad \delta C = -\frac{a}{2}(\bar\epsilon\gamma\gamma^a\nabla_a\psi)\, e^{-\gamma} \\ \delta B = -i\frac{a}{2}(\bar\epsilon\gamma\psi)\, e^{-\beta} \quad , \quad \delta D = i\frac{a}{2}(\bar\epsilon\gamma^a\nabla_a\psi)\, e^{-\delta} \\ \delta\psi = \frac{a}{2}\left[i(\gamma^a\epsilon)\,\nabla_a A\, e^\alpha + (\gamma^a\gamma\epsilon)\,\nabla_a B\, e^\beta + i(\gamma\epsilon)\, C\, e^\gamma + (\epsilon)\, D\, e^\delta\right] \\ \delta\bar\psi = \frac{a}{2}\left[-i\, e^\alpha\nabla_a A\,(\bar\epsilon\gamma^a) - e^\beta\nabla_a B\,(\bar\epsilon\gamma\gamma^a) + i\, e^\gamma C\,(\bar\epsilon\gamma) + e^\delta\, D\,(\bar\epsilon)\right] \end{cases} \tag{10.5.2}$$

Notice that the relative coupling constants appear to be fixed by the requirement that the infinitesimal transformations (10.5.2) are symmetries of the Lagrangian L_{WZ}. Here the transformation parameter ϵ is a Majorana (anticommuting) spinor which is assumed to be covariantly conserved, i.e. $\nabla_\mu\epsilon = 0$. We stress that the condition for ϵ to be covariantly conserved is a very strong one. It corresponds more or less to a point-independent transformation which gauge theories are based on. The model we present here is a first step towards a field theory which is covariant with respect to point-dependent transformations. Such a theory is called *supergravity* and one has to re-define the Lagrangian and to introduce auxiliary fields to obtain covariance with respect to this wider class of transformations. Here we are not interested in these generalizations, for which we refer the interested reader to [CDF91]. Under these conditions the transformation (10.5.2) is global only when trivial structure bundles $\Sigma = M \times \mathrm{Spin}(\eta)$ are considered.

One can now define the infinitesimal (vertical) generalized symmetry

$$\Xi = (\delta A)\frac{\partial}{\partial A} + (\delta B)\frac{\partial}{\partial B} + (\delta C)\frac{\partial}{\partial C} + (\delta D)\frac{\partial}{\partial D} + (\delta \psi)\frac{\partial}{\partial \psi} + (\delta \bar{\psi})\frac{\partial}{\partial \bar{\psi}} \quad (10.5.3)$$

which is called the *supersymmetry generator* and which can be shown to leave the Lagrangian invariant modulo the following divergence term:

$$\delta L_{\mathrm{WZ}} = \mathrm{Div}(\theta) \quad (10.5.4)$$

where we set

$$\begin{aligned} \theta =& \tfrac{a}{4}\Big[\Big(2imA(\bar{\epsilon}\gamma^\mu\psi) + 2\nabla^\mu A(\bar{\epsilon}\psi) - \nabla_\nu A(\bar{\epsilon}\gamma^\nu\gamma^\mu\psi)\Big)\, e^{\alpha+1} + \\ &+ \Big(2mB(\bar{\epsilon}\gamma\gamma^\mu\psi) - 2i\nabla^\mu B(\bar{\epsilon}\gamma\psi) + i\nabla_\nu B(\bar{\epsilon}\gamma\gamma^\nu\gamma^\mu\psi)\Big)\, e^{\beta+1} + \\ &+ iD(\bar{\epsilon}\gamma^\mu\psi)\, e^{\delta+1} + C(\bar{\epsilon}\gamma^\mu\gamma\psi)\, e^{\gamma+1}\Big]\mathrm{d}s_\mu \end{aligned} \quad (10.5.5)$$

Notice that the supersymmetry generators are not closed with respect to the commutator. One can in fact check that, given two supersymmetries generated by ϵ_1 and ϵ_2, their commutator is the Lie derivative with respect to the vector field defined on the structure bundle Σ by:

$$\hat{\xi} = \xi^\mu(\partial_\mu - \Gamma^{ab}_\mu \sigma_{ab}) \oplus (e^a_\mu \nabla_\nu \xi^\mu e^{b\nu})\sigma_{ab}\,, \qquad \xi^\mu = i\tfrac{a^2}{2}(\epsilon_2\gamma^\mu\epsilon_1) \quad (10.5.6)$$

We recall that the vector field (10.5.6) is the so-called *Kosmann lift* of the spacetime vector field $\xi = \xi^\mu\partial_\mu$; see Section 9.7.

Once we take into account both the vector field (10.5.6) and the supersymmetry generators they form an algebra with the following commutation rules

$$[\delta_1, \delta_2] = \pounds_{\hat{\xi}}\,, \qquad [\delta_1, \pounds_{\hat{\xi}}] = 0\,, \qquad [\pounds_{\hat{\zeta}}, \pounds_{\hat{\xi}}] = 0 \quad (10.5.7)$$

Of course these vector fields do not span an ordinary Lie algebra, since some of the parameters are actually anticommuting, while the ordinary Lie algebra parameters are scalars. In fact, one should regard them as generators of a graded Lie algebra (also called a *superalgebra*).

6. Rarita-Schwinger Model

In the previous Section we regarded Wess-Zumino supersymmetries as generalized symmetries of a suitable gauge natural theory. Except for a better geometrical understanding of the frames (which was developed to deal with spinors) and a canonical off-shell treatment of conserved currents, we just restated there some standard results within the new gauge natural framework. In that case the gauge natural formalism does not produce any new result, even if it better clarifies problems which are present in the classical formulation.

It should be investigated whether supersymmetry transformations can be understood directly as gauge natural transformations. The first problem to overcome is the condition $\nabla_\mu \epsilon = 0$ imposed to the transformation parameter. This condition implies that ϵ is *(covariantly) constant.* In other words Wess-Zumino supersymmetries are analogous to *"global gauge transformations"* (e.g. $\phi' = \exp(i\theta)\phi$ with θ constant, i.e. independent of the spacetime point x). Accordingly, one should aim at regarding Wess-Zumino supersymmetries as the action of automorphisms of some structure (super)-bundle; then ϵ necessarily needs to be point dependent since the condition $\nabla_\mu \epsilon = 0$ would either turn out to be non-kinematical (since it would depend on dynamical fields), or it would need a background fixing (which we would wish to avoid), or finally it would depend on the trivialization of the structure (super)-bundle (and as such it would not be global in general).

However, the condition $\nabla_\mu \epsilon = 0$ is avoided in other well known models, as e.g. the Rarita-Schwinger model. Let us briefly sketch this last example in order to enucleate which further problems arise despite all standard results can be recovered. In the Rarita-Schwinger model one deals with a spin frame e^μ_a, a principal torsionless connection ω^{ab}_μ over the structure bundle $\bar\Sigma$ and a 3/2-spin anticommuting spinor field ψ_μ. The bundle $\bar\Sigma_\rho$ of spin frames is defined as usual. The principal connection ω^{ab}_μ is a section of $\mathcal{C}_{\bar\Sigma}$. The torsionless condition is

$$T = e_{c\lambda}(\nabla_\mu e^c_\nu - \tfrac{i}{2}\bar\psi_\mu \gamma^c \psi_\nu)\, \mathrm{d}x^\lambda \wedge \mathrm{d}x^\mu \wedge \mathrm{d}x^\nu = 0 \qquad (10.6.1)$$

where as usual the Clifford products are understood and covariant derivatives are with respect to the principal connection induced on $\bar\Sigma \times_M L(M)$ by ω^{ab}_μ and the Levi-Civita connection of the metric induced by the spin frame. According to torsionless condition, the connection ω can be expressed as a function of the spin frame (and its first derivatives) and the spinor field.

Let us define $V = [\Lambda_-(W) \otimes \mathbb{C}]^k \otimes (\mathbb{R}^m)^*$ and the action

$$\lambda : \mathrm{GL}(m) \times \mathrm{Spin}(\eta) \times V \longrightarrow V : (J, S, \psi_\mu) \mapsto S \cdot \psi_\nu \bar J^\nu_\mu \qquad (10.6.2)$$

where, as usual, $\bar J^\nu_\mu$ denotes the inverse of J^μ_ν. Thence we can define the bundle $\bar E_\lambda = (L(M) \times_M \bar\Sigma) \times_\lambda V$ for anticommuting 3/2-spinors. The set of torsionless connections is compatible with the action (5.2.31) so that one can define the total configuration bundle:

$$\mathcal{C} = \bar{\Sigma}_\rho \times_M \mathcal{C}_{\bar{\Sigma}} \times_M \bar{E}_\lambda \tag{10.6.3}$$

and local fibered "coordinates" on it are $(x^\mu; e^\mu_a, \omega^{ab}_\mu, \psi_\mu)$. Of course these are not coordinates since ψ_μ are anticommuting (and constrained by the Majorana conditions) and ω^{ab}_μ is constrained by the torsionless condition.

Let us consider the Lagrangian

$$L_{\rm RS} = (e\, R^{ab}_{\mu\nu} e^\mu_a\, e^\nu_b + \alpha \epsilon^{\mu\nu\rho\sigma} \bar{\psi}_\mu \gamma_5 \gamma_\sigma \nabla_\nu \psi_\rho)\, \mathrm{d}s \tag{10.6.4}$$

where $R^{ab}_{\mu\nu}$ is the curvature tensor of the connection ω^{ab}_μ, e is, as usual, the (absolute value of the) determinant of the spin frame and α is a suitable coupling constant to be determined so that the supersymmetry conditions below hold. We remark that the field equations for the connection ω turn out to be identically satisfied as a consequence of the torsionless constraint. It is a gauge natural Lagrangian since all automorphisms of $\bar{\Sigma}$ are symmetries.

Let us now consider the generalized vector field

$$\Xi = (\delta e^\mu_a)\frac{\partial}{\partial e^\mu_a} + (\delta \omega^{ab}_\mu)\frac{\partial}{\partial \omega^{ab}_\mu} + (\delta \psi_\mu)\frac{\partial}{\partial \psi_\mu} \tag{10.6.5}$$

defined by

$$\begin{cases} \delta e^a_\mu = i\bar{\epsilon}\gamma^a \psi_\mu \\ \delta \psi_\mu = \nabla_\mu \epsilon \end{cases} \tag{10.6.6}$$

where ϵ is a fermionic Majorana parameter as in Wess-Zumino model. This time no restriction on the parameter ϵ is imposed. The action of supersymmetries on the connection ω is determined by the expression of ω as a function of the spin frame and the spinor field. In many fundamental expressions (e.g. in the expression of Nöther currents) the expression of infinitesimal supersymmetries on the connection appears multiplied by the connection field equations (which are identically satisfied). Accordingly those terms automatically vanishes off-shell because of the torsionless constraint.

The Lagrangian $L_{\rm RS}$ is invariant with respect to the generalized vector field Ξ modulo pure divergence terms *and* terms vanishing on-shell (see [CDF91] for details). Hence we say that the supersymmetries (10.6.6) are *generalized on-shell symmetries.* We stress however that on-shell symmetries are not yet satisfactorily characterized in literature. In fact, simply extending definition (6.5.31) by adding a term vanishing on-shell cannot be the ultimate solution. In fact, if one defines a *symmetry on-shell* as a vector field Ξ for which:

$$\pounds_{j\Xi}\, (\pi^\infty_{2k-1})^*\, \Theta = \mathrm{d}\alpha + \omega + f(\mathbb{E}) \tag{10.6.7}$$

holds for some $f : V^*(\mathcal{C}) \longrightarrow A_m(M)$ vanishing on the zero section of $V^*(\mathcal{C})$, then *any* vector field Ξ would be a symmetry (as it can be easily proved by using the first variation formula, see (6.4.15)).

Technically speaking this is not a *mistake* (in fact, Nöther theorem would produce in these cases a trivial conservation law, namely $\mathcal{E} \equiv 0$ off-shell). However, it would be beautiful to have a characterization of these trivial on-shell symmetries able to distinguish them from the non-trivial ones, like e.g. the non-trivial supersymmetries (10.6.6). This characterization is presently still lacking and it will deserve further investigations.

More seriously, the same Lagrangian and the same vector field could produce two different splittings of the form (10.6.7) depending on how integrations by parts are carried over. Hence the same symmetry could produce more than one conservation law.

Even once this problem were solved and a precise satisfactory definition of *on-shell symmetry* were provided, then one should ask whether the vector fields (10.6.6) could be interpreted as infinitesimal generators of automorphisms in some suitable generalized sense. In fact, by construction one would need at least to extend principal bundles to allow a supergroup as a standard fiber.

The standard approach seems to be to extend spacetime manifolds to allow supermanifolds and to regard supersymmetries as natural transformations on them. It is not clear to us whether one can more simply provide, at least in some relevant case, a gauge natural framework for supersymmetric theories allowing supermanifolds as fibers but keeping ordinary manifolds as base spacetimes.

Exercises

1- Prove the other Majorana flip identities.
Hint: Get inspiration from the two identities we proved.

2- Compute the commutators (10.5.7).
Hint: expand the objects involved.

Final word

We would like here to collect and summarize our viewpoint about classical Field Theory.

Gauge natural formalism is the general framework necessary to take into account the observer freedom.

The only thing one really requires when describing a physical system is that, despite any observer is free to choose its own conventions[1], in the end we are able to compare the results of different observers. Once we are able to compare different observations this defines an equivalence relation on possible observation results. Each of the corresponding equivalence classes contains all the possible results of an observation one for each possible observer. The classes are thence independent of the observers and thence they depend on the physical situation only. Thence there exists *one physical system* but each observer has its own *description* of the same physical situation. Such observer's viewpoint is made of a number of conventions, which are private to the observer and in some sense they *define* the physical observer itself. The conventions are necessary to reduce the physical situations to a number of values which describe the physical world.

The change of observer's conventions has to be regarded as a map giving the observation of one observer as a function of the observation of another observer. It is clear that, at least in a wide class of actual physical systems, the possible values of observations may be endowed with a manifold structure F and the change of observer's convention can be regarded as transition functions. Each observer lives in a patch and a set of observers thence define a trivialization of the configuration bundle. This is all one needs to justify

[1] We are not going to treat the observer as a person; in a physical sense a photomultiplier or any other set of apparata may be considered as an observer. We do not want to enter the problem of quantum observers since we are discussing classical physics.

the geometrical field framework: to be able to describe locally the physical situation and to know the transformation laws of the resulting description.

The changes of convention form a group which utimately justifies the approach via principal bundles. Finally, the changes of convention represent the change of observers so that all the physical quantities do not depend on the particular convention adopted. On one hand the conventions are identified with trivializations, on the other hand, actively speaking, they are automorphisms of the structure bundle. Such transformations have to be symmetries of the system so that all the physically relevant quantities are gauge invariant. Roughly speaking this leads to gauge natural field theories.

Because of physical evidence, some part of the observer's convention are the choice of time and space coordinates, leading directly to general covariance and homogeneous formalism. Under this viewpoint General Relativity simply encodes the possibility of describing the real physical world independently on any observer's convention. Moreover, the observer needs other conventions (e.g. the choice of a representative for each class of gauge equivalent vector potentials in Maxwell theory) to describe more complex physical systems. The gauge natural framework is the natural extension of the covariant formalism to encompass this further freedom.

Manifest (gauge) covariance has to be retained as long as possible.

At least for aestethical reasons, manifest covariance has to be preserved. However, there are also more important reasons:

- When we are computing, e.g., the energy of a complex system we have various contributions from different sub-systems. It is of primary importance to be able to say whether the contribution of a particular sub-system is gauge covariant or not. In the first case, the notion of partial energy of the subsystem is independent of the observer and well-posed; in the second case such a notion is simply *undefined.*
- If covariance is mantained, although manifest covariance is lost, each time we define something we have to spend some (or a lot of) effort to prove that such a quantity is actually well-defined. On the contrary, when manifest covariance is retained from the beginning one can distinguish among quantities which are well-defined and quantities which are not.

The Lagrangian contains all the information about the system.

This is the most obvious (and perhaps unnoticed) feature of our attitute towards field theory. We have proved that all physical issues, from field theories to conserved quantities, from covariance to Hamiltonian framework, can be successfully analyzed in terms of the Lagrangian.

Still we did not take a position on whether Lagrangians differing by pure divergence terms have to be considered as physically equivalent of not. The reason is that we still do not know the answer. Divergences do not influence the equations of motion but they actually influence conserved quantities; this is due to our strategy in the subject. There are other approaches in which one can get rid of these ambiguities by introducing other ambiguities (boundary conditions, constraints on deformations, or other) which eventually can be fixed by particular choices (asymptotical flatness, on-shell deformations). Sometimes these restrictions lead to a less general framework which cannot be applied to all situations. Sometimes one can show this framework to be equivalent to the framework we presented here but this is not always the case. We refer to [FR] for a detailed discussion on the subject.

The final word on this point will be given in terms of Quantum Field Theory. In fact, two classically equivalent Lagrangians have the same classical field equations but the corresponding actions differ off-shell. It is Quantum Field Theory which has to predict whether quantum effects can distinguish between the two (classically equivalent) Lagrangians.

All contributions have to be kept and considered.

Because of the situation described in the previous paragraph we always considered two Lagrangians which differ by a pure divergence term as two different Lagrangians, *tout court*. Equivalence between the two (if it holds in any given sense) has to be proved *a posteriori* and not assumed *a priori*.

For these reasons all terms emerging from calculations, e.g. in conserved quantities, have to be kept and considered to the very end and eventually discarded when it is possible. This applies to pure divergences, terms which vanish on-shell, and terms which vanish for topological reasons or for restrictions on deformations.

We still hardly know what can really happen in our Universe.

The gauge natural framework claims that we actually never experienced non – triviality of the topology of the structure bundle because we live within a *small* spatial region where fields are weak and almost stationary. For example, there are solutions of Einstein equations in which this *quasi classical heaven* collapses and one would be exposed to all kind of paradoxes, up to time travels in the past which contradict causality. Of course, the last word on whether these situations can actually occur in our world is out of observations. These observations are beyond our present scope so the answer has to be postponed. The final fact is that nowadays we are almost completely ignorant about the situations which really may occur in our Universe. On a mathematical ground, one can decide to restrict *a posteriori* solutions of Einstein equations in order to preserve e.g. causality; from a physical point of view these restrictions may sound reasonable but presently we do not really know if they really reflect how our world goes on.

Presently, these situations are banned *a posteriori* as non-physical for two main reasons: first, causality violations have never been experienced; second, the most commonly accepted tool for canonical quantization is Dirac-Bergman theory of constrained systems which strongly relies on ADM splittings.

We have to remark, however, that Dirac-Bergman framework has not proved to bring to a satisfactory solution for quantum gravity, yet. Furthermore, in the last decades a new approach to quantization based on path integrals has appeared and grown in importance. Both approaches are still under investigation and despite they are equivalent on standard quantum systems nobody knows whether they are always equivalent or not.

References

[ADM62] – R. Arnowitt, S. Deser, C.W. Misner, in: *Gravitation: An Introduction to Current Research*, L. Witten ed. Weley, New York, 1962, 227

[AKO93] – I.M. Anderson, N. Kamran, P.J. Olver, *Internal, External, and Generalized Symmetries*, Advances In Mathematics, **100**, 1993, 53

[B97] – C. Bar, Acta Phys. Pol.B **29**(4), 1998, 891

[BG92] – J.-P. Bourguignon, P. Gauduchon, Comm. Math. Phys., **144**(3), 1992, 581

[CB79] – Y. Choquet-Bruhat, in: Proceedings of the Second Marcel Grossmann Meeting on General Relativity, Part A, B (Trieste, 1979), North-Holland, Amsterdam-New York, 1982, 167

[CBCF78] – Y. Choquet-Bruhat, D. Christodoulou, M. Francaviglia, *Problème de Cauchy sur une variété. (French)*, C. R. Acad. Sci. Paris Sér. A-B **287**(5), (1978), A373

[CFFML97] – K. Chu, C. Farel, G. Fee, R. McLenaghan, Fields Inst. Comm. **15**, 1997, 195

[CFT79] – D. Christodoulou, M. Francaviglia, W.M. Tulczyjew, Gen. Relativity Gravitation **10**(7), 1979, 567

[DP86] – L. Dąbrowski, R. Percacci, Commun. Math. Phys. **106**, 1986, 691; L. Dąbrowski, R. Percacci, J. Math. Phys. **29**, 1987, 580

[E81] – D.J. Eck, Mem. Amer. Math. Soc., **33**, 1981, 247

[F84] – M. Ferraris, Geometrical Methods in Physics, UJEP Brno, Krupka Ed. 1984, 61; M. Ferraris, M. Francaviglia, Quaderni di Matematica, Università di Torino, Quaderno n. **86**, Torino, 1984

[F98] – L. Fatibene, *Ph.D. Thesis*, University of Torino, Italy, 1998

[FF88] – M. Ferraris, M. Francaviglia, in: *8th Italian Conference on General Relativity and Gravitational Physics*, Cavalese (Trento), August 30 – September 3, World Scientific, Singapore, 1988

[FF91] – M. Ferraris, M. Francaviglia, in: *Mechanics, Analysis and Geometry: 200 Years after Lagrange*, Editor: M. Francaviglia; Elsevier Science Publishers B.V., (1991), Amsterdam

[FF96] – L. Fatibene, M. Francaviglia, in: Seminari di Geometria 1996-1997, ed. S. Coen, 1998, 69

[FF97] – L. Fatibene, M. Francaviglia, in: *Procs. Gauge Theories of Gravitation*, Jadwisin Sept. 1997, Acta Phys. Pol. B**29** (4), 1998, 915

[FF01] – L. Fatibene, M. Francaviglia, in: Seminari di Geometria 2001-2002, ed. S. Coen, 2001, 123

[FFF97] – L. Fatibene, M. Ferraris, M. Francaviglia, J. Math. Phys. **38** (8), 1997, 3953

[FFFG95] – L. Fatibene, M. Ferraris, M. Francaviglia, M. Godina, in: Proceedings of *"6th International Conference on Differential Geometry and its Applications, August 28–September 1, 1995"*, (Brno, Czech Republic), Editor: I. Kolář, MU University, Brno, Czech Republic 1996; gr-qc/9608003

[FFFG98] – L. Fatibene, M. Ferraris, M. Francaviglia, M. Godina, Gen. Rel. Grav. **30**(9), 1998, 1371; gr-qc/9609042

[FFML02] – L. Fatibene, M. Ferraris, M. Francaviglia, R.G. McLenaghan J. Math. Phys. **43**(6), 2002, 3147

[FFFR99] – L. Fatibene, M. Ferraris, M. Francaviglia, M. Raiteri, Annals of Phys., **275**, (1999) 27; hep-th/9810039

[FFFR00] – L. Fatibene, M. Ferraris, M. Francaviglia, M. Raiteri, Annals of Phys. **284**, 2000, 197; gr-qc/9906114

[FFFR01] – L. Fatibene, M. Ferraris, M. Francaviglia, M. Raiteri, J. Math. Phys. **42**(3), 2001, 1173; gr–qc/00030019

[FFRxx] – M. Francaviglia, M. Ferraris, C. Reina, Ann. Inst. Henri Poincarè, Vol XXXVIII(4), 1983, 371

[FFS92] – M. Ferraris, M. Francaviglia, I. Sinicco, Il Nuovo Cimento, **107B** (11), 1992, 1303

[G70] – R. Geroch, J. Math. Phys. **9** (11), 1968, 1739

[GFF88] – L. Gatto, M. Ferraris, M. Francaviglia, Rend. Sem. Mat. Univers. Politecn. Torino, vol. 46, 3, 1988, 309

[GIMMPSY] – M.J. Gotay, J. Isemberg, J.E. Marsden, R. Montgomery, J. Śniatycki, P.B. Yasskin, *Momentum Map and Classical Relativistic Fields*, (preprint); see also physics/9801019

[GP78] – W. Greub, H.R. Petry, in: *Lecture Notes in Mathematics* **676**, Springer - Verlag, NY, 1978, 271

[GS78] – H. Goldschmidt, D. Spencer, J. Differential Geom. **13**(4) (1978), 455

[HHP98] – S.W. Hawking, C.J. Hunter, D.N. Page, Phys.Rev. D**59**, 1999, 044033; hep-th/9809035

[K59] – A. Komar, Phys. Rev., **113**, 934, (1959)

[K66] – Y. Kosmann, Ann. di Matematica Pura et Appl. **91**, 1972, 317;
Y. Kosmann, Comptes Rendus Acad. Sc. Paris, série A, **262**, 1966, 289;
Y. Kosmann, Comptes Rendus Acad. Sc. Paris, série A, **262**, 1966, 394;
Y. Kosmann, Comptes Rendus Acad. Sc. Paris, série A, **264**, 1967, 355

[K84] – I. Kolář, J. Geom. Phys. **1**(2), 1984, 127

[K85] – J. Katz, Class. Quantum Grav., **2**, 1985, 423

[J04] – J. Janyška, *Reduction theorems for general linear connections*, Diff. Geom. and Its Appl., 2004 (to appear);
see also http://www.math.muni.cz/ janyska/publikace.html

[M63] – C.W. Misner, J. Math. Phys., **4**, (1963) 924

[P99] – M. Palese, *Ph.D. Thesis*, University of Torino, Italy, 1999

[PT77] – R.S. Palais, C.L. Terng, Topology **16** (1977), 271

[R] – A. Borowiec, M. Francaviglia, in: *Proc. Int. Sem. Math. Cosmol., Potsdam, 1998*, M. Rainer and H.-J. Schmidt (eds.), WSPC Singapore (gr-qc/980611);
K. Hayashi, T. Shirafuji, Phys. Rev. D **19**(12), 1979;
F.W. Hehl, in: *Proc. 8th M. Grossmann Meeting*, T. Piran (ed.) World Scientific, Singapore 1998 (gr-qc/9712096);
R.B. Mann, *Summary of Session A6: Alternative Theories of Gravity* gr-qc/9803051;
P. Peldan, Class.Quant.Grav. **11** (1994) 1087 (gr-qc/9305011)

[R97] – C. Rovelli, *Half way through the woods*, in: "The Cosmos of Science", J. Earman and J.D. Norton eds., University of Pittsburgh Press: Universitaets Verlag Konstanz, 1997

[R00] – M. Raiteri, *Ph.D. thesis*, Università di Torino, 2000

[RT74] – T. Regge, C. Teitelboim, Annals of Physics **88**, (1974) 286

[T51] – A.H. Taub, Ann. Math. **53**, 1951, 472;
E.T. Newman, L. Tamburino, T. Unti, J. Math. Phys. **4**, 1963, 915.

[T72] – A. Trautman, in: *Papers in honour of J. L. Synge*, Clarenden Press, Oxford, 1972 UK, 85

[V96] – R. Vitolo, *Ph.D. Thesis*, University of Firenze, Italy, 1996

Bibliography

[A79] – M.F. Atiyah, *Geometry on Yang-Mills fields*, Scuola Normale Superiore Pisa, Pisa, 1979.

[AMD69] – M.F. Atiyah, MacDonald, *Introduction to Commutative Algebra*, Addison-Wesley Publishing Co., 1969

[B81] – D. Bleecker, *Gauge theory and variational principles*, Addison-Wesley Publishing Company, Inc., Massachussetts, 1981

[BD64] – J.D. Bjorken, S.D. Drell, *Relativistic quantum mechanics*, McGraw-Hill, New York, 1964.

[BM94] – J. Baez, J.P. Munian, *Gauge Fields, Knots and Gravity*, World Scientific, Singapore, 1994

[CDF91] – L. Castellani, R. D'Auria, P. Fré, *Supergravity and Superstrings. A Geometrical Perspective*, World Scientific, Singapore, 1991

[CB68] – Y. Choquet-Bruhat, *Géométrie Différentielle et Systémes Extérieurs*, Dunod Paris, 1968

[CBWM82] – Y. Choquet-Bruhat, C. DeWitt-Morette, *Analysis, Manifolds and Physics*, North-Holland, Amsterdam, 1982

[C90] – R. Cianci *Introduction to Supermanifolds*, Bibliopolis, Napoli, 1990

[F88] – M. Francaviglia, *Elements of Differential and Riemannian Geometry*, Bibliopolis, Napoli, 1988 Italy

[F91] – M. Francaviglia, *Relativistic Theories*, Quaderni del CNR, 1991 Italy

[FLS66] – R.P. Feynman, R. Leinghton, M. Sands, *The Feynman Lectures on Physics*, Addison-Wesley Publishing Co., Tokyo, 1966

[FR] – M. Francaviglia, M. Raiteri, *Lagrangian and Hamiltonian Formulation of Relativistic Field Theories*, (in preparation)

[FP99] – M. Francaviglia, M. Palese, *Geometric Foundations of Calculus of Variations: Variational Sequences, Symmetries and Jacobi Morphisms*, (in preparation)

[GH81] – M.J. Greenberg, J.R. Harper, *Algebraic topology : a first course*, Reading, Mass., 1981.

[GP74] – V. Guillemin, A. Pollack *Differential Topology*, Prentice – Hall, Inc., Eaglewood Cliffs, New Jersey, 1974

[GS87] – M. Göckeler, T. Schücker, *Differential Geometry, Gauge Theories and Gravity*, Cambridge University Press, New York, 1987 USA

[HE73] – S.W. Hawking, G.F.R. Ellis, *The large scale structure of space-time*, Cambridge University Press, Cambridge, UK, 1973

[HS79] – H. Herrlich, G.E. Strecker, *Category theory : an introduction*, Heldermann, Berlin, 1979

[K82] – I. Kolář, Colloquia Mathematica Societatis János Bolyai, 3.1 Differential Geometry, Budapest 1979, North – Holland, 317-324 (1982)

[KMS93] – I. Kolář, P.W. Michor, J. Slovák, *Natural Operations in Differential Geometry*, Springer – Verlag, New York, 1993 USA

[KN63] – S. Kobayashi, K. Nomizu, *Foundations of differential geometry*, John Wiley & Sons, Inc., New York, 1963 USA

[KSHMC80] – D. Kramer, H. Stephani, E. Herlt, M. MacCallum, *Exact solutions of Einstein's Field equations*, Cambridge University Press, Cambridge UK, 1980

[LM89] – B.H. Lawson, M.-L. Michelsohn, *Spin Geometry*, Princeton University Press, New Jersey, 1989

[MS63] – J. Milnor, J.D. Stasheff, *Characteristic Classes*, Ann. of Math. Studies **76**, Princeton University Press, Princeton 1963

[M88] – Y.I.Manin, *Gauge Field Theory and Complex Geometry*, Springer – Verlag, Berlin Heidelberg, Germany, 1988

[ML71] – S. Mac Lane, *Categories for Working Mathematician*, Springer – Verlag, New York 1971

[MM92] – K.B. Marathe, G. Martucci, *The mathematical foundations of gauge theories*, North – Holland, Amsterdam, 1992 Holland

[MS00] – L. Mangiarotti, G. Sardanashvily, *Connections in Classical and Quantum Field Theory*, World Scientific, Singapore, 2000

[N90] – M. Nakahara, *Geometry, Topology and Physics*, Institute of Publishing, Bristol, 1990 UK

[ON83] – B. O'Neill, *Semi-Riemannian Geometry*, Academic Press, USA, 1983

[R81] – P. Ramond, *Field Theory: A Modern Primer*, Addison/Wesley, 1981

[S51] – N. Steenrod, *Topology of Fibre Bundles*, Princeton University Press, Princeton, 1951 USA

[S66] – E.H. Spanier, *Algebraic Topology*, McGraw – Hill, Berkeley, 1966 USA

[S89] – D.J. Saunders, *The Geometry of Jet Bundles*, Cambridge University Press, Cambridge, 1989 UK

[S93] – G. Sterman, *An Introduction to Quantum Field Theory*, Cambridge University Press, Cambridge, 1993

[S95] – G. Sardanashvily, *Generalized Hamiltonian Formalism for Field Theory*, World Scientific, Singapore, 1995, Singapore

[T75] – B.R. Tennison, *Sheaf Theory*, London Mathematical Society Lectu re Notes Series, **20** Cambridge University Press, Cambridge 1975

[W72] – S. Weinberg, *Gravitation and Cosmology: Principles and Applications of the General Theory of Relativity*, Wiley- VCH, New York 1972

[W80] – R.O. Wells, *Differential Analysis on Complex Manifolds*, Graduate Texts in Mathematics, Springer – Verlag, New York 1980

Analitic Index

-A-

Action of a Lie group: 24 27 83 139 212 216 283 317 330 333–334 340 345.
(see also *Right, Left, Adjoint*)
Action functional: **(152)** 146–147 154–155 169 171–172 176 198 266 340.
(see also *Lagrangian*)
Adjoint action:**(83)** 86 104 112 139.
Adjoint matrix: 139.
ADM conserved quantities: 236.
ADM splitting:**(229)** 230–232 241.
Affine bundle:**(24)** 10 23–26 34 37 56 91 94 102 123 134.
Affine parameter: 229 232.
Affine transformation: 23–26 31 56 75 92 94 118 134.
Affine space: 23–24 26 123 134.
Anomalous factor: 236 243 245.
Associated bundle:**(43)** 44–45 50–51 56 83 87 98–99 101 104 110 115–116 118 124 126–127 132 138–139 215–216 224 233 249–250 272 277 280 283 286 296 317 325 332–334 337 339.
Asymptotically flat: 235–236 241–242 244–246.

-B-

Background field: 173 206 211 214–215 217 233–234 236 238 240 242–245 253 255–256 259 271–272 277 295–298 315. (see also *Reference*)
Base manifold: 12–13 16 18 21–22 24 28 33–34 40–41 56 66 70 94 101 103 107 109 113 115 126 138–141 145–146 148 152 161 167 170 181 183 271 329.
Bianchi identities: **(225)** 110 209–210 224–226 248 273–275 279.
– in General Relativity: 110 225 280 328.
Black hole: 236 241–242 244.
Boson: 251–257 298 335–336 338–339.
BRST transformations: 213–217.
BTZ solution:**(244)** 234 244–245.
Bundle:**(12)** 13.
– of automorphisms:**(83)** 84 104.
– of configurations:
(see *Configuration bundle*)
– of connections:**(93)**.
– of curvatures:**(109)**.
– of infin. gen. of autom. (IGA):**(135)** 87 105.
– of jets: (see *Jet*)
– of metrics:**(123)** 234 317.
– of principal connections: **(94)** 93–95 124 126 134–135 139 205 212 249 319.
– of spin frames:**(317)** 317–317 319 321.
– of tensor densities:**(122)**.
– of tensors:**(43)** 74 78 102 113 115 122 317 320.
– with structure group:**(34)** 10 24 34–35 37–38 44–45 86 99 113–116 125–127 130 134 249.
Bundle morphism: **(13)** 10 19 22–26 28 39 41–42 57–58 66 75 80 102 114–116 118 126–127 130 132–133 149 152–153 155 187 220 283 305 314 320.

-C-

Canonical right action:**(27)** 26–29 32 36 45 73 77 80–81 86 89–90 93–94 99 126 128–131 133 142 248 287 304 309 317 323.
Cartesian coordinates: 25–26 34 190 198 202 212 230 241 285 295.
Cartesian product: 10 147.
Category:**(13)** 19 30 33–34 41–42 58 63–66 105 113–115 125–127 130 132 297.
Cauchy data: 229–231.
Cauchy problem: 222 230 298.
Cauchy surface:**(229)** 230–231.
Causality: 187 230 298 307.
(see also *Determinism*)
Čech cohomology: 306.
Charge conjugation:**(339)** 340.
Charged particle: 212–213.
Clifford algebra:**(299)** 295–302 304 308 312 325 330 339 341.
Cohomology: 148 168 227–228 231.
Configuration: 3 5 45 91 139 141–142 152 154 187 221 239 272 283.
Configuration bundle:**(3)** 4 9 34 76 139 147 152–153 173 180 182 205–206 213 216 220–221 223–224 227 233–234 240 248 251 260 262 271–273 277 280 284 286–287 319 337.
Configuration space:**(72)** 78 152 187 190 197.
Connection:**(88)** 89 91 93.
Principal –: (see *Principal*)
Fibered –: (see *Fibered*)
Spin –: (see *Spin*)
Levi-Civita –: (see *Levi-Civita*)
Conserved quantity:**(226)** 146–148 181 192–196 201 203–204 210–211 217 226–229 232–234 236 238–245 247 255–257 259 264 267 270 272 275–277 298 328.
Contact form:**(60)** 60 62–63 67 78 149 173–174 176–179 185–186.
Extended –: (see *Extended*)
Pure –: (see *Pure*)
Contact part of a form: 62.
Coordinate curve:**(17)**.
Coordinates: 17 30 32 34 42 50.
Fibered –: (see *Fibered*)

Natural –: (see *Natural*)
Cartesian –: (see *Cartesian*)
Cosmological constant:**(234)** 241 244.
Cotangent bundle:**(41)** 34–43 49 113 115 264.
Cotangent functor:**(42)** 115.
Covariance: 182 185 190 193 197–203 215 229 231 233 238 240 252–253 259 280–281 287 298 319.
– identity:**(182)** 183 185 187 192–193 201–203 206–208 212 235 252–253 255 257 261–264 278–279 326.
Gauge –: (see *Gauge*)
General –: (see *General*)
Covariant conservation: **(226)** 225 231–232 319 343.
Covariant derivative: 95 102 109 162–163 170 201 215 235 240 252–253 255 258–259 266–267 318 343.
– of a curvature: 109.
– of frames: 287.
– of sections:**(107)** 107–108 110–111 318.
– of spinors: 326 331 334–337 341.
– of vector fields: 111–112 262–263 289 322.
Covering (of open set): 4 10 12–14 16 21 24 38–39 282 305–306.
Covering space: 295 302 308.
Covering map: 305 309 311 323.
Curvature:**(98)** 95 102 109–112 141 182 205 209 214 234 250 278 280 285 287 319.
(see also *Field strength*)
Transformation laws: (see *Transformation*)
Curve:**(16)** 17 72 105–106 152 186–188 192 197–198 201–202 233 303 305–306.

-D-

Deformation: 69 147 153–154 169 172 207–208 215 236–237 240 242 279 290 297 315.
Determinism: 222 228. (see also *Causality*)
Differential equation:**(75)** 74 105 145 172 229.
Differential operator: 20 54 59 75–77 208 213 215 224.
– symbol: (see *Symbol*)
Dirac:**(327)** 145 295–296 330–331 342.
– matrices:**(302)** 303 325 327 333–335 339–340.
– monopole: 141–142 217 250.
– spinors: 214 325–329 333.
Direct family:**(64)** 65.
Direct limit:**(64)** 64–65.
Direct topology: 65.
Divergence operator: 156 237.
Pure –: (see *Pure*)
Domain:**(14)** 69 154. (see also *Region*)
Double fibration:**(48)** 58.
Dual bundle: 41.
Dual vertical bundle: 49.
Dynamic connection:**(220)** 221 225 233–234 238 240 254 267 272 276–277 337.

-E-

Electric field: 210 227 231.
Electric charge:**(227)** 211 226–229 270 277.
Elliptic operator:**(76)** 77.
Euler-Lagrange morphism: **(170)** 75 172 179 266 279 291.
Euler-Lagrange equations: **(172)** 154 169–173 176 178 180 190 194–195 197–198 204 207 226 237.
Extended contact forms:**(60)** 61–62.
Exterior algebra: 43 96 213–214 339.
(see also *Forms*)
Exterior covariant differential: **(97)** 208.
(see also *Horizontal*)
Exterior forms: (see *Forms*)

-F-

Fermion: 295 298 333.
Fiber:**(12)** 3–5 10–14 18–19 21–27 29–32 34–35 37 41–42 44 46 48 50 56 73 76 80 90–91 94 105 135 152 233 305–306 310 312–313 329 333.
Fiber bundle:(see *Bundle*)
Fibered connection:**(157)** 158 160–161 165 167 170–171 177 179 184 225.
Fibered coordinates:**(22)** 22–23 47–48 53–54 61 67–68 73 75–78 84–85 87–88 91 93 99–100 103–105 108 118 133–135 137–139 142 152 156 170 224 261–262 264 277 288 317–319 325.
Fibered product: 39–39 139.
Field equations: 75 141 145–148 152 154–155 172–173 183 185 207–209 215 221–223 225 229–231 234 236–239 241 244 246 248 253–254 256 258–259 261 264 266–267 276 281 285 287 291 297–298 323 327–328 342.
(see also *Euler-Lagrange*)
Field strength:**(205)** 217 225 247 250.
(see also *Curvature*)
Formal covariant derivative:**(108)** 62 261 264.
Formal derivative:**(54)**.
Formal differential:**(59)** 62 156.
Formal Lie derivative:**(70)** 67–71 224.
Forms: 38 59–62 66 97 152–153 174 177–178 183.
Frame: 37 44 100–101 104 210 286–287 295.
(see also *Spin frame*)
– bundle:**(32)** 32–33 44–45 49 51 92 99–101 110 276 280–281 288 295 308 314–315 321.

– – invariant vector: 100 104 110 124 288 318.
– – trivialization: (see *Trivialization*)
– – transition functions: (see *Transition*)
– functor:**(33)**.
Free particle: 190–194 197–198 203 212.
Function: 16–17 19 24 29 43 48 53–55 59 65 72 75–76 90 96–97 100 134 147 154 170 182 186 196 200 215 230 240–241 248 258 303.
Functor:**(19)** 30 33 42 57–58 73 114–115 117 119 124–127 130–132 138 142.
Cotangent functor: (see *Cotangent*)
Frame functor: (see *Frame*)
Jet functor: (see *Jet*)
Natural functor: (see *Natural*)
Gauge natural functor: (see *Gauge natural*)

-G-

Gauge covariance: 205 214–215 220 270.
Gauge fixing:**(96)** 201 213–214 236.
Gauge natural :
– bundle:**(132)** 86 95 114 118 125–135 137–139 148 249–250 272 277 283–284 286 317 332 337.
– – trivialization: (see *Trivialization*)
– – transition functions: (see *Transition*)
– field theory:**(272)** 138 148–149 168 184 206 211 213 220 260 270–273 275–277 281 287 298 319 337.
– functor:**(126)** 127 132.
– Lagrangian:**(272)** 278–279 281 287 326–327 337.
– lift: 138.
– morphism:**(127)**.
– object: 127 287 289.
– – Lie derivative: 84–85 88 93 95 110 138–138.
– – order: 132–133.
– prolongation of a group: **(128)** 128–130 132–135.
– prolongation of a bundle: 129 132 317.
Gauge potential: 182. (see also *Vector pot.*)
Gauge transformation: **(36)** 34 45 138–142 148–149 208–209 213 222 248 264 272–274 277 289 318–318 322 333 335 338.
Gauss theorem: 211 227.
General covariance: 220–222 225 229 270.
General Relativity: 4 6 113 145–146 148–149 196 201 203 208 211 220 222–223 225 228–229 232 234–241 251 256 270–271 280 284 295 298 319.
– frame affine formulation:**(286)** 287 319.
– metric affine formulation:**(148)** 265–266 280 284 286–287.
– pure frame formulation: 285 287 332.
– pure metric formulation: 246 266 284–285.
– covariant first order formulation: 173 234 236 240 245.
Generalized symmetry:**(186)** 67 70 173 190–193 195–196 212–213 216 298 343.
Generalized vector field: **(67)** 67–70 108 154 185–186 195 216 224.
Geodesics:**(190)** 194 198 233.
Geometric bundle: 34 36–37 45.
Geometric object:**(115)** 3 115 261 297. (see also *Natural object*)
– Lie derivative: 73–74 78 120–121 124 225 248 252 255 258 261–264 266.
– order: 224.
Ghost field: 213–215.
Grahm-Schmidt: 38 286.

-H-

Hamilton principle:**(154)** 152 171.
Harmonic oscillator: 75 180.
Heat equation: 76.
Higgs: 258 336–338.
Higher order frames: 103.
– bundle: 73–73 103 113 115–116 118 133 224 249 272 288.
– functor:**(73)**.
Hilbert Lagrangian:**(234)** 235 237 239 259 265 277 280–281.
Hilbert stress tensor: **(208)** 254–255 257 259 261–263 265 279 327–328.
Hole argument: 222 228 230 248 260.
Homotopy: 40 69 106 239 290 306.
Homology: 148 168 227–228 231.
– group: 306.
Hopf bundle: 20–22 31–31 77 112 141–142 310 313.
Horizontal differential:**(97)** 109. (see also *Exterior covariant differential*)
Horizontal form:**(59)** 97 109 153 156 183 197.
Horizontal lift:**(89)** 91 93 105–106.
Horizontal part: 62 89 95 135 174–175 178 195 208–209 277 280.
Horizontal vector:**(89)** 88 98 276.
Hyperbolic: 230 236.

-I-

IGA:(see *Infinitasimal gen. of autom.*)
Induced connection: 95 98–99 104 109–110 137 284. (see also *Associated connection*)
Infinite jet bundle: 61 63 66–67 216 274.
Infinitesimal generator of autom.: 85–88 95 126 135–138 208 273–274 280 289 318 322 326.

Infinitesimal symmetry: 183 190 225 256 271.
Initial conditions: 105 141 222 229–230.
Inverse family:**(63)** 64 66.
Inverse limit: 66–67.
Inverse topology: 64.

-J-

Jacobian: 30 32 42 50–51 100 113–114 122 131 158 177–178 181 198 250 282.
Jet bundle: 53–56 61–63 66–67 72 74–75 93 124 142 149 172–173 216 332.
– natural coordinates: 54 61 76 93 102 129.
– trivialization: (see *Trivialization*)
– transition functions: (see *Transition*)
Jet functor:**(57)** 58 116.
Jet prolongation:(see *Prolongation*)
Jet prolongation of a bundle:(see *Jet bundle*)

-K-

Kepler problem: 195–196.
Kerr solution: 234–236 243–244.
Klein-Gordon: 258–260 262.
– real scalar field: 258.
– complex scalar field: 258.
– vector field: 262–263.
– Lagrangian: 258–260 262 264.
Komar potential: 235–236 240 245 247.
Kosmann lift: 288–290 321–322 344.

-L-

Lagrangian:**(152)** 75 154–155 170 172–176 178–179 182–183 185–195 197–203 205–208 212–215 217 220 225 230 233–234.
(see also *Action functional*)
– density: 153 169 187–188 190 252–253 257 278 327.
Hilbert –: (see *Hilbert*)
Palatini –: (see *Palatini*)
Purely metric –: (see *General relativity*)
Purely frame –: (see *General relativity*)
Metric affine–: (see *General relativity*)
Frame affine–: (see *General relativity*)
Proca –: (see *Proca*)
Maxwell –: (see *Maxwell*)
Yang-Mills –: (see *Yang-Mills*)
Matter –: (see *Matter*)
Klein-Gordon –: (see *Klein-Gordon*)
Particle –: (see *Mechanics, Free particle*)
Dirac –: (see *Dirac*)
Weyl –: (see *Weyl*)
Majorana –: (see *Majorana*)
Wess-Zumino –: (see *Wess-Zumino*)
BRST –: (see *BRST*)
Laplace vector: (see *Runge-Lenz*)
Laplacian operator: 77 230.
Left action: 27 35 38 43–44 90 115 123–124 127 132 134. (see also *Adjoint*)
Left invariant vector field: 81 112.
Levi-Civita connection: 102 112 201 221 225 235–238 240 252–254 265–267 276 284 318–319 323 343.
Lie algebra: 20 43 46 81 86 95–97 109 117 128 138 205 247 344.
– structure constants:**(81)**.
– generators: 81 87 139 142 205.
– $\mathfrak{gl}(n)$: 96 99 303.
– $\mathfrak{so}(r, s)$: 192.
– $\mathfrak{su}(n)$: 332.
– $\mathfrak{u}(1)$: 142 209 332.
– $\mathfrak{spin}(r, s)$: 303–304.
Lie derivative: 85 109–110 124 182 185 188 190 195 216 224 344.
– of forms: 78 176 248 264.
– of connections: 93 95 109 212 255 266 278.
– of sections:**(70)** 67 70–71 84–85 88 120–121 124 138–138.
– of spin frames: 322–322.
– of spinors: 288 326.
– of tensors: 74 78 266.
– of vector fields: 73–74.
– of Vielbein: 289–290.
Lie group: 23 26 35 80–81 128 205 216 247 249 277.
– action: (see *Action*)
– adjoint action: (see *Adjoint*)
– homomorphism: 26 28 80 249.
– left translation: 23 26 29 31–32 34–35 37 80 86 131.
– right translation: 29 130.
Linear group: 18 23 25 30 32–33 37–38 44–45 76 96 99 102 109 122–124 129–131 133 135–136 215 220–221 249–251 282–283 302–303 315 317–318 325.
Linear differential operator: 20 75.
Local diffeomorphism:**(114)** 22 39–115 118 126 222 321.
Local coordinates:(see *Coordinates*)

-M-

Magnetic field: 210 231.
Magnetic monopole:(see *Dirac monopole*)
Majorana identities:**(340)** 341 347.
Majorana Lagrangian:**(342)**.
Majorana spinor: 339–343.
Manifold: 3–4 10–13 16–18 27 29–30 32 34 38–39 42–43 46 50–51 54 56 63–64 66 72–73 77 105 114–115 123 125 127 132 145–146 148

152 174 201 209 213–214 222–223 229 248–250 281 283 286–287 290 295 298 307 314–316 332 343.
η-manifold:**(282)** 284 287 307 314 316–317.
Matter Lagrangian: 251–253 255–256 259 261–262 267–268 278–279.
Maurer-Cartan identity:**(97)**.
Maxwell: 3 209–210 247 259 277.
– equations:**(209)** 210 217 230–231 247–248 264.
– field:**(139)** 141 209 247.
– Lagrangian: 206 332.
Mechanics: 3 6 72 75 146 152 160 173 177 186–187 189 197 204 233.
Metric: 102 123 139 182 190–191 198 201 205–209 212 221 223 234 236–238 240–242 244–247 249 251–255 259 265–268 271–272 276–278 280–288 290 295–297 300 307 309 311–312 314–321 323 332 343.
(see also *Riemannian metric*)
Metric connection: 112. (see also *Levi-Civita*)
Metric theory:(see *General Relativity*)
Momenta: 170–171 175 194 199–200 234 252 262 264 266 280–281 326 339.
Multiple fibration: 48 56.
(see also *Double Fibration*)

-N-

Natural bundle:**(115)** 18 33 101 113–116 118–126 148–149 220 224 233 249–251 288.
– coordinates: 30 34 47 54 61 71–72 76–77 86 93 100 102 119–122 129 156 182 197 288.
– covariance: 281.
– trivialization: (see *Trivialization*)
– transition function: (see *Transition*)
– morphism: 115 118.
(see also *Natural functor, Natural lift*)
Natural field theory:**(220)** 138 149 168 184 211 220–222 224–226 233–236 238 240 250 254 270 272–273 275–276.
Natural functor:**(114)** 115–116 120 126.
Natural Lagrangian:**(220)** 252 261–262 280.
Natural lift: 117–119 121 149 224 250 270 272 275 288.
– of frames: 280 288 321.
– of higher order frames: 118–119.
– of vector fields: 124 188.
Natural object: 102 115 121–123 138 247 250 252 271 278 289 295. (see also *Geometric object*)
Neutrinos: 329–331 333 335.
Nöther current: 183–184 186 188 191–192 202–204 208–209 224–226 232 241 255–257 259–261 265 274–275 279 328.
Nöther theorem:**(186)** 192 203–204 216 224–225 271 273 277 328.

-O-

Orthogonal group: 38 51 192 249 281 284 302–303 307 309 311 315.
Orthogonal frame: 38 45 249 285 288 304–305 307–308 321.

-P-

Palatini: 255 263 265 268 277 280 323.
Parallel transport: 105–106 112.
Parallelizable:**(281)** 284 286–287 315–316.
Partition of unity:**(24)** 38 123 222.
Pauli matrices:**(303)** 332.
Pin group:**(301)** 301–303 330.
Poincaré-Cartan form: **(174)** 147 173–180 184 186 189–190 195 200 216 220.
Poincaré-Cartan morphism:**(170)** 177 179 183 220 279 327–328.
Principal automorphism:**(28)** 26 28–29 33 36 73 80 82 84–85 118 125–127 130–136 139 142 280–284 295 304 315.
Principal bundle:**(26)** 20 23 26–29 31–32 34–35 37 43–45 73 80–81 89–90 92 95–96 98–100 104 109–110 112 125–134 137–138 142 149 205–206 209 214 248–250 272 280–283 295–296 304–305 308–309 314 332–333.
Principal connection: **(89)** 92–95 98–100 104 109–112 124–126 134–135 137 139 141–142 149 209 212 214 247–248 250 272 276 286 319 332.
Principal symbol:**(76)** 77.
Proca field: 264–265.
Proca Lagrangian: 264.
Projectable vector field: **(46)** 58 66 68–70 103 117 120 178 182 195 201.
Prolongation of morphisms:**(57)** 57–60 66 73 77.
Prolongation of sections:**(58)** 66 75 152 165.
Prolongation of vector fields: 66 77 131–132.
Pull-back bundle:**(39)** 39–40 67 105.
Pull-back forms: 59 61–62 65–68 96 172 183 200 217 284 318 331 342.
Pure contact forms:**(60)** 61.
(see also *Contact forms*)
Pure divergence: 163 172 184–186 195 215 224 245 247 336 343–344.

-Q-

Quasi-linear operator:**(75)**.
Quaternions:**(312)** 300–313.

-R-

Reduced current: **(204)** 209 224–225 240–241 249 257 259 274–275 328.
Reference background field: **(237)** 234–238 240 242–246 251 253–254 259.
(see also *Background*)
Region: 147 152–154 165 172 180 222 226 228 232 243–244 275. (see also *Domain*)
Regular conserved quantity: 226.
Riemannian manifold: 190 201 217 233 246 285 295 297 304–305 307.
Riemannian metric: 38 45 102 112 198 234 249–250 314.
Riemann tensor: 110 182 237 242 265.
Right action: (see *Canonical right action*)
Right invariant vector field: 81 84–85 87 92 100 112 131–132 136 142 212 273 318 321 323 326.
Runge-Lenz vector: 196.

-S-

Schwarzschild solution: 234 241–245.
Section of a bundle:**(14)** 3–15 19 22 24–25 28 43 53 58 66 69–70 75 78 83 86–88 93 108–110 123 134 152–154 156 172–173 180 185 197 208 215 222 248 250 261 281–284 296 312 317 322 325 337.
Semisimple:**(139)** 205 247 332.
Solution:**(154)** 75–76 152 180 217 230 234 241 243–244 246 297.
Sphere: 4 20 50 141–142 217 227 242 245 285 290 307 309 311.
Spin connection: 318–319 323.
Spin frame: 297–298 314–323 326 328 332–333 343.
Spin group: 295 299 301–306 308–309 311–314 318 325 330 333 336 339–340.
Spin manifold:**(307)** 295 314 332.
Spin structure:**(304)** 295–305 307–309 313 315–316 343.
Spinors: 149 183 213 220 270 276 285 287–288 290 295–296 298 315 325–326 328–329 335–336 339–340.
Anticommuting –: 339 341–343.
Chiral –:**(330)** 329–331 333.
Dirac –:(see *Dirac*)
Stiefel-Witney classes: 304 307 316.
Stokes theorem: 172.
Strong conservation laws:**(225)**.
Strong automorphisms:**(13)** 57–58.
Structure bundle:**(272)** 35–37 44–45 51 86–87 118 125–127 132–135 139–141 148 206 214 216 250 274–277 281 283 285–287 289–290 310 315–316 319 321–322 333–334 343–344.
Structure constants: (see *Lie algebra*)
Structure group: 10 23–24 35–38 45 51 95 99 109 126–127 130 272 281 295 304–305 308 315. (see also *Bundle with structure group*)
Superpotential: **(204)** 205 209 220 225–226 232 234–236 240 242 245 254 256–257 259–261 272 275 279 328.
Supersymmetry: 203 213 220 270 298 342–344.
Swap map: 73 77.
Symbol:**(76)**. (see also *Principal symbol*)
Symmetrized covariant derivative: **(157)** 170–171 253 273 278.
Symmetry:**(180)** 45 141 147–148 173 176–181 183 185 187–195 198 201 203 206–210 212 220 222 224–225 231 236 240 243 245 247 252 255 258 260 270–273 275–277 338 343.
– group: 34 201 203 252 260 270 272.

-T-

Tangent bundle: **(17)** 4 16–19 30 32–33 37–38 42–46 50 73 77 104 106 112–115 118–119 249 282.
– trivialization: (see *Trivialization*)
– transition function: (see *Transition*)
Tangent map:**(19)** 16 30 33 42 81 98 114 126.
Tangent vector: 4 19 37 46 50 88 105–108.
TAUB-Bolt solution:**(246)** 234 246–247.
Tensor density:**(122)** 122 208 230 251 255–257 261–263 273 327–328 343.
Tensor field:**(43)** 74 78 101–102 121–123 182 190 220 223 238 317 319–320.
Tensor product:**(41)** 43 299.
Tetrad: 238. (see also *Vielbein*)
Time conservation: 232.
Torsion:**(101)** 102 110 236–237.
Transition functions:**(13)** 5–17 21 23–26 29–32 34–36 38 41–42 44–45 47–48 50 57 65 77 80 85 87 92 94 96 100–101 113 117 122 124 134–135 137 142 145 222 249 260 282–284 286 305 317 330. (see also *– transition functions for specific examples*)
– of frame bundle: 32–33.
– of gauge natural bundles: 86–87.
– of jet bundles: 54–57.
– of natural bundles: 50 86 201.
– of tangent bundles:**(18)** 30 37–38 42 44–45.
Transformation action: 17 19 25–26 43 86 121 139 220 272.
Transformation laws: 69 85 92–93 100 111 134–137 158 160 168 182 258 295.
– of curvature: 109 247.
– of connections: 92–94 100 124.
Transpose connection:**(101)** 102.
Trivial bundle:**(10)** 3–11 16 23 30 40 43 54 56–57 72 75 139–140 187 260 281 312.

Trivialization:(12) 4 12–17 21–31 34–37 39 41 43 46–48 50 53–54 57 72 75 80–82 85–88 92–93 96 99 111 113 142 157–158 187 249 282 288 305 310.
- of frame bundles:**(32)** 32–33 37–38 44–45 92 100–101 118 288 321.
- of gauge natural bundles: 126 130 132–133 135 137–138 249 282 288 305 321.
- of jet bundles: 54–55.
- of natural bundles: 17 30 32–33 50 92 100–101 107 113 118 122–124 133 156 249 282 284 288.
- of principal connections: 134.
- of tangent bundle: 30 37–38.

-U-

Utyhama theorem: 252–253 278.

-V-

Variational morphism: **(155)** 155–170 173 177 184–186 225 274.
- boundary term: 159 161 163 165–167 224.
- splitting: 158–167 170 220 224 274.
- reduced: **(160)** 161–167 169–170 177 184 209 224.
- volume term: 159 161–163 165–167 224.

Vector bundle:(25) 23 25–26 30–31 34 37 41–43 46–49 51 58–59 75 91 96 102 133 135 155–157 160 164–165 237 251 253 296 325 340.(see also *Tangent*)

Vector field:(19) 20 37 43 46–47 66 68 70–71 73 77 84–85 87–88 92 94–95 100 103–105 107–110 117 119–121 124 131–132 136–138 142 153 174 178 185 188 194 201–202 212 224–227 229–232 235–236 240 242–243 254 260–262 275–276 285 288–291 321–322 326 344.
- invariant:(see *Left (Right) invariant*)
- generalized:(see *Generalized*)
- projectable:(see *Projectable*)
- vertical:(see *Vertical*)

Vector field commutator: 73 82 111–112 120 138 216–217 344.

Vector potential:(95) 96 100 276.

Vertical automorphism: 47 83–84 87–88 110 126 138 153.

Vertical bundle:(46) 56 71 121.

Vertical morphism:(13) 102 309.

Vertical vector: 46–49 69–70 88 104 107 119 132 153 170 174 176 180 187 207 212 327.

Vielbein: **(282)** 283–290 297 314–315 317 321–322.

-W-

Wave equation: 76.

Weak conservation laws:(183) 185–186 204 224 235 273 328.

Weinberg-Salam: 258 337.

Wess-Zumino: 298 343–344.

Weyl Lagrangian: 331.

Work form:(183) 186 224 235 273–274 328.

-Y-

Yang-Mills: 139 206 209–211 213 225 236 238 247 269 272 277 280.
- equations: 208 279.
- field: 139 205 208 213–214 228 279.
- Lagrangian: 206 213 277 279 332 336.

MIX
Papier aus verantwortungsvollen Quellen
Paper from responsible sources
FSC® C105338

If you have any concerns about our products,
you can contact us on
ProductSafety@springernature.com

In case Publisher is established outside the EU,
the EU authorized representative is:
Springer Nature Customer Service Center GmbH
Europaplatz 3, 69115 Heidelberg, Germany

Printed by Libri Plureos GmbH
in Hamburg, Germany